Communications in Computer and Information

Zhenhua Li   Xiang Li
Yong Liu   Zhihua Cai (Eds.)

Computational Intelligence and Intelligent Systems

6th International Symposium, ISICA 2012
Wuhan, China, October 27-28, 2012
Proceedings

 Springer

Volume Editors

Zhenhua Li
China University of Geosciences
School of Computer Science
Wuhan, China
E-mail: zhli@cug.edu.cn

Xiang Li
China University of Geosciences
School of Computer Science
Wuhan, China
E-mail: lixiang@cug.edu.cn

Yong Liu
The University of Aizu
School of Computer Science and Engineering
Aizu-Wakamatsu, Japan
E-mail: yliu@u-aizu.ac.jp

Zhihua Cai
China University of Geosciences
School of Computer Science
Wuhan, China
E-mail: zhcai@cug.edu.cn

ISSN 1865-0929   e-ISSN 1865-0937
ISBN 978-3-642-34288-2   e-ISBN 978-3-642-34289-9
DOI 10.1007/978-3-642-34289-9
Springer Heidelberg Dordrecht London New York

Library of Congress Control Number: 2012949326

CR Subject Classification (1998): F.1.1, F.2.1, F.4.1, G.1.6, G.2.1, H.2.8, I.2.4, I.2.6, I.2.11

Typesetting: Camera-ready by author, data conversion by Scientific Publishing Services, Chennai, India

Printed on acid-free paper

Springer is part of Springer Science+Business Media (www.springer.com)

Preface

This volume comprises the proceedings of the 6th International Symposium on Intelligence Computation and Applications (ISICA 2012) held in Wuhan, China, October 27–28, 2012. ISICA 2012 successfully attracted over 277 submissions. Through rigorous reviews, 72 high-quality papers are included in this volume, CCIS 316. The ISICA conferences are one of the first series of international conferences on computational intelligence that combine elements of learning, adaptation, evolution and fuzzy logic to create programs as alternative solutions to artificial intelligence. The last six ISICA proceedings including two volumes of CCIS have been accepted in both the Index to Scientific and Technical Proceedings (ISTP) and Engineering Information (EI).

Following the success of the past five ISICA events, ISICA 2012 persisted in exploring the new fields of computational intelligence by exploiting the intelligence of ancient Chinese mathematics. The ancient mathematician Hui Liu calculated the ratio of a circle's area to the square of its radius π to 3.1416 more than a thousand years ago. The methodology used in calculating π actually shares a similar principle with computational intelligence concerning the evolution of solutions, and inspires the new concepts in computational intelligence. One of ISICA's missions is to pursue the truth that the complex system inherits the simple mechanism of evolution, while the simple models could produce the evolution of complex morphologies.

CCIS 316 featured the most up-to-date research in analysis and theory of evolutionary algorithms, neural network architectures and learning, fuzzy logic and control, predictive modeling for robust classification, swarm intelligence, evolutionary system design, evolutionary image analysis and signal processing, and computational intelligence in engineer design. ISICA 2012 provided a venue to foster technical exchanges, renew everlasting friendships, and establish new connections.

On behalf of the Organizing Committee, we would like to thank warmly the sponsor, China University of Geosciences, who helped in one way or another to achieve our goals for the conference. We wish to express our appreciation to Springer for publishing the proceedings of ISICA 2012. We also wish to acknowledge the dedication and commitment of the CCIS editorial staff. We would like to thank the authors for submitting their work, as well as the Program Committee members and reviewers for their enthusiasm, time, and expertise. The invaluable help of active members from the Organizing Committee, including Hengjian Tong, Yinwen Gong, Changhe Li, Xiaoyue Wang, Wenbin Fan, and

Lingling Wang in setting up and maintaining the online submission systems, assigning the papers to the reviewers, and preparing the camera-ready version of the proceedings is highly appreciated. We would like to thank them personally for helping to make ISICA 2012 a success.

October 2012

Zhenhua Li
Xiang Li
Yong Liu
Zhihua Cai

Organization

ISICA 2012 was organized by the School of Computer Science, China University of Geosciences, sponsored by China University of Geosciences, and supported by Springer.

General Chair

Fang Hao   China University of Geosciences, China

Program Chairs

Zhihua Cai   China University of Geosciences, China
Yong Liu   University of Aizu, Japan

Publication Chairs

Xiang Li   China University of Geosciences, China
Zhenhua Li   China University of Geosciences, China

Local Arrangements Chair

Hui Li   China University of Geosciences, China

Program Committee

Andrea Cavallaro	University of London, UK
Tughrul Arslan	The University of Edinburgh, UK
Javad Sohafi Bonab	Islamic Azad University, Iran
Tan Kay Chen	National University of Singapore, Singapore
Carlos A. Coello	LANIA, Mexico
Guangming Dai	China University of Geosciences, China
Kalyanmoy Deb	Indian Institute of Technology, India
Yaochu Jin	Honda Research Institute Europe, Germany
Pavel Kromer	Technical University of Ostrava, Czech Republic
Yuanxiang Li	Wuhan University, China
Zhenhua Li	China University of Geosciences, China
Steffen Limmer	Friedrich Alexander University Erlangen Nürnberg, Germany

Shiow-Jyu Lin   National Taiwan Normal University, Taiwan
Charles X. Ling   The University of Western Ontario, Canada
Bob Mckay   Seoul National University, Korea
Ryszard Tadeusiewicz   AGH University of Science and Technology, Krakow,
                            Poland
Hamid R. Tizhoosh   The University of Waterloo, Canada
Dong-Min Woo   Myongji University, Korea
Zhijian Wu   Wuhan University, China
Shengxiang Yang   University of Leicester, UK
Xin Yao   University of Birmingham, UK
Gary G. Yen   Oklahoma State University, USA
Sanyou Zeng   China University of Geosciences, China
Harry Zhang   University of New Brunswick, Canada
Qingfu Zhang   University of Essex, UK
Xiufen Zou   Wuhan University, China

Local Co-chairs

Guangming Dai   China University of Geosciences, China
Sifa Zhang   China University of Geosciences, China
Yi Zeng   China University of Geosciences, China

Local Committee

Shuanghai Hu   China University of Geosciences, China
Li Zhang   China University of Geosciences, China
Xiaolan Guo   China University of Geosciences, China
Huili Zhang   China University of Geosciences, China

Secretaries

Xiaoyue Wang   China University of Geosciences, China
Wenbin Fan   China University of Geosciences, China
Lingling Wang   China University of Geosciences, China

Sponsoring Institutions

China University of Geosciences, Wuhan, China

Table of Contents

Section IV: Data Mining

Section V: Evolutionary Multi-objective and Dynamic Optimization

Section VI: Intelligent Computation

Section VII: Intelligent Learning Systems

Section VIII: Neural Networks

Section IX: Real-World Applications

Dynamic Weapon Target Assignment Method Based on Artificial Fish Swarm Algorithm

Chengfei Wang, Zhaohui Zhang, Runping Xu, and Ming Li

Naval Academy of Armament
Beijing 100036, P.R. China
wang.chengf@163.com

Abstract. Aiming at the problem of weapon target assignment when multiple weapon units on diverse battle platforms head off a group of targets, the mathematic model of this problem is established. A weapon target assignment method based on Artificial Fish Swarm Algorithm is proposed for solving the dynamic weapon target assignment problem, and a typical instance is designed for simulation experiment. The simulation results demonstrate that this method is superior in the condition of time restraint comparing with the Genetic Algorithm, and validate the effectiveness of employing the Artificial Fish Swarm Algorithm in resolving dynamic weapon target assignment problem.

Keywords: Optimize, Weapon Target Assignment, Artificial Fish Swarm Algorithm.

1 Introduction

The problem of weapon target assignment (WTA) is an important theoretical problem in military operational research, and one of the most pivotal issues in the domain of combat command and decision. It demonstrates specifically which firepower installations should be used to fire at which targets, and what shooting plans should be adopted. The WTA problem is to coordinate various fire units' combat behaviors, so it is also named as fire distribution or target assignment. The procedure of WTA is a process of multifactor analysis and decision-making. Therefore, before the ending of target assignment, the optimization of assignment is dynamic, and the assignment draft will be adjust properly according to the changes of current battle situation, targets' threat extent, and the preparation and execution conditions of each fire units.

Initially, the means to resolve WTA problem are based upon linear programming, dynamic programming, games theory, and graph theory. These are commonly named as the traditional algorithm. At the middle of 1980s, WTA problem was proved to be a NP-complete problem[1][2], and this means there are not some deterministic algorithms can be used to obtain the optimal solution in the polynomial time, or say, only through exhaustive methods can the optimal solution of WTA be obtained. Moreover, exhaustive methods are inapplicable to the WTA problems bigger than a middle scale, then approximation

Z. Li et al. (Eds.): ISICA 2012, CCIS 316, pp. 1–7, 2012.

algorithms should be employed for obtaining the suboptimal solutions. With the development and application of generalized heuristic search techniques, several methods for solving WTA problem are put forward one after the other, such as Simulated Annealing Algorithm, Genetic Algorithm, Tabu Search, Artificial Immune Algorithm, Ant Colony Algorithm, Particle Swam Optimization, and heuristic hybrid algorithm.

The emergence of target group is a stochastic process, and the states of weapon systems change in real time. Therefore the problem of dynamic weapon target assignment (DWTA) is a research focus currently. In this paper, a weapon target assignment method based on Artificial Fish Swarm Algorithm (AFSA) is proposed, and it can meet the real-time demands of DWTA problem.

2   Problem Formulation

The basic mission of WTA problem is to assign N weapon units to attack M targets, so as to maximize the whole anticipant attack efficiency, or minimize the threat to defenders. Several factors need to consider in WTA problem include the characters and status of weapon units in our various battle platforms, the characters and status of targets, and the assigning rules such as to maximize the damage extent to targets, to maximize our survival probability, and to minimize the consumptions of our ammunition. Furthermore, WTA problem embodies some mathematical characters in itself like that it is NP-complete and discrete, and its objective function is nonlinear, and it is a random problem and belongs to the problem of dynamic programming.

Then observe the mathematic model of WTA problem. Consider the WTA problem that N weapon units on our distributed platforms are involved in attacking M targets. Within given time, each platform receives the threat value w_j of target j and single shot damage probability $P_{i,j}$ of weapon unit i to target j by the C^3I system. Thereinto w_j relates to the aggression intentions of targets and the worth of units under fire potentially, and $P_{i,j}$ is determined by weapon units' technical performance and targets' position and kinetic parameters[4]. For the whole battle operation, w_j and $P_{i,j}$ vary according to time. But for WTA problem in the restricted time interval, both can be assumed to be constant.

Define the assignment matrix as $\boldsymbol{Y} = [y_{i,j}]_{N \times M}$, the decision-making variable $y_{i,j}$ is a boolean variable, viz., $y_{i,j} \in \{0, 1\}$, where 1 denotes that the i-th weapon unit attacks the j-th target, and 0 denotes not attack. Thus the whole threat value of this batch of target is:

$$f(\boldsymbol{Y}) = \sum_{j=1}^{M} \left[w_j \cdot \prod_{i=1}^{N} (1 - P_{i,j})^{y_{i,j}} \right] \tag{1}$$

The objective of each assignment scheme is to minimize the whole threat value of arriving targets in the restricted time interval, thereby maximizing the survival probability of our units under fire. Hence, the mathematic model of WTA problem can be described as:

$$min \sum_{j=1}^{M} \left[w_j \cdot \prod_{i=1}^{N} (1 - P_{i,j})^{y_{i,j}} \right] \tag{2}$$

$$s.t. \sum_{i=1}^{N} y_{i,j} = 1, \quad j = 1, ..., M \tag{3}$$

In this model, the formula (2) is the objective function, and the formula (3) denotes the restriction condition of the decision-making variable.

3 Resolving Based on Artificial Fish Swarm Algorithm

The AFSA is employed to resolve the DWTA problem. The AFSA[5] is a swarm intelligence optimization method based on animal behavior. It imitates the fish swarm behavior for searching the global optimum in solution space.

3.1 Designing for AFSA

Artificial Fish (AF) is a fictitious entity of true fish. Define the current state of an AF to be $X = (x_1, x_2, ..., x_N)$, where $x_i = j$ donates that the i-th weapon unit is assigned to attack the j-th target, and $j \in [1, M], i = 1, ..., N$. There existe a mapping relation between the state vector X and the assignment matrix Y of WTA problem: if $x_i = j$, then $y_{i,j} = 1$ and $y_{i,k} = 0, k \neq j$. According to this mapping relation, a state X can be used to work out all decision-making variable $y_{i,j}$, thus gain targets' whole threat value W under this state X through the formula (1). And regard W as the objective function value, then the the objective function is a minimization problem by the state vector X.

Prey Behavior. Let X_p be the AF current state and select a state X_q randomly in its visual distance. If $W_q < W_p$, the AF goes forward a step in this direction. Otherwise, select a state X_q randomly again and judge whether it satisfies the forward condition. If the AF cannot yet satisfy after try several times, it moves a step randomly.

Swarm Behavior. Let X_p be the AF current state, n_f be the number of all companions in its visual distance and X_c be the center position. If $W_c < W_q$ and $n_f/n_T < \delta$ (n_T is the size of total fish swarm, δ is a predeterminate crowding factor), which means that there are more foods in the companion center and not crowded, the AF goes forward a step to the companion center. Otherwise, the prey behavior is executed.

Follow Behavior. Let X_p be the AF current state, and it explores the individual X_{min} which has the least threat value in its visual distance. If $W_{min} < W_p$ and the number of all companions in its visual distance n_f satisfies $n_f/n_T < \delta$, the AF goes forward a step to the companion X_{min}. Otherwise, the prey behavior is executed.

Three behaviors mentioned above are the basic behaviors of AF. The AFs visual distance, moving step and crowding factor need to be set when initialization of algorithm, or change self-adaptively in the execution process of algorithm.

The AF will judge the current state after each moving. Firstly compare the threat value of state after swarming with ones of following, and select the behavior with the less threat value to execute. If neither of these two behaviors threat values improve, the prey behavior is executed. The AFs go forward to the optimal direction by means of this behavior selection. Finally they congregate around several local extremums, and there are more individuals of AFs around the superior extremum region.

3.2 Procedure of DWTA Method

The problem of static weapon target assignment leaves the time factor out of account, whereas owing to the characteristic of real time, the DWTA problem corresponds with practical application. The concept of temporal restriction of

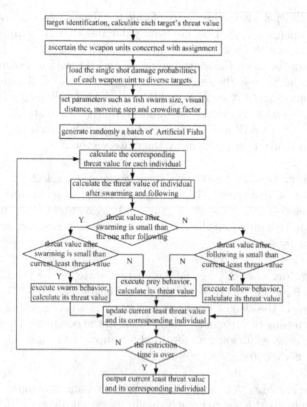

Fig. 1. The flow chart of DWTA method based on AFSA

assignment is introduced in this paper to satisfy the demands of real time. All the process of searching optimum is carried out in the restriction time. When time exceeds the limit of restriction time, assignment stops, and output the current optimal assignment scheme.

The flow of DWTA method based on AFSA is described such as Fig. 1. Firstly, the preprocess is operated, including: target identification, calculating each targets threat value; ascertaining the weapon units on battle platforms concerned with assignment; loading the single shot damage probabilities of each weapon uint to diverse targets. Then execute the artificial fish swarm optimization iteratively in restriction time, and update the current least threat value and chalk up the corresponding individual. When the restriction time is over, output the current least threat value and its corresponding AF individual. The state of this optimal individual can form a weapon target assignment scheme, which is regarded as the solution of DWTA problem.

4   Computational Experiments

For examining the validity of the DWTA method based on AFSA in this paper, an instance of WTA problem is designed for simulation and calculation. The algorithm of this method is programed by Matlab. And the instance and its parameters are described as follows: there are 8 usable weapon launching units on our distributed platforms; 12 enemy targets can be separated in 4 teams and enforce saturation attack, and they are in the order of $\{1, 2, 3\}, \{4, 5, 6, 7\}, \{8, 9, 10\}$ and $\{11, 12\}$ according to the arriving orientation; the threat value of targets gained through C^3I system are shown as formula (4):

$$w_j|_{j=1,2,...,12} = \{0.09, 0.13, 0.17, 0.21, 0.19, 0.26, 0.23, 0.11, 0.13, 0.15, 0.09, 0.10\} \tag{4}$$

The successive single shot damage probability of 8 weapon units to 12 targets can be expressed as the damage probability matrix $P_{8\times12}$,

$$P_{8\times12} = \begin{bmatrix} 0.67 & 0.72 & 0.85 & 0.86 & 0.85 & 0.64 & 0.53 & 0.65 & 0.33 & 0.41 & 0 & 0 \\ 0.89 & 0.77 & 0.81 & 0.71 & 0.76 & 0.68 & 0.66 & 0.59 & 0.52 & 0.42 & 0 & 0 \\ 0.82 & 0.78 & 0.65 & 0.77 & 0.89 & 0.61 & 0.65 & 0.79 & 0.53 & 0.35 & 0.21 & 0 \\ 0.90 & 0.86 & 0.86 & 0.69 & 0.72 & 0.85 & 0.83 & 0.89 & 0.65 & 0.71 & 0.52 & 0.58 \\ 0.61 & 0.67 & 0.71 & 0.81 & 0.82 & 0.67 & 0.66 & 0.74 & 0.81 & 0.53 & 0.66 & 0.47 \\ 0.49 & 0.62 & 0.52 & 0.78 & 0.77 & 0.71 & 0.52 & 0.78 & 0.71 & 0.73 & 0.86 & 0.62 \\ 0 & 0.35 & 0.56 & 0.88 & 0.71 & 0.90 & 0.88 & 0.67 & 0.74 & 0.81 & 0.69 & 0.71 \\ 0 & 0 & 0.21 & 0.45 & 0.42 & 0.57 & 0.61 & 0.70 & 0.73 & 0.85 & 0.81 & 0.92 \end{bmatrix} \tag{5}$$

where '0' denotes that this target is out of the shooting region of corresponding weapon.

The assignment method proposed in this paper is used to resolve the instance, and compared its performance with which of the method based on Genetic Algorithm. Each weapon is restricted to intercept two targets at most as a result of the shooting diversion time and the shot load. This instances optimal weapon target assignment scheme calculated by simulation is shown as Table 1. And the statistical results of least threat value got respectively through operations of two algorithms are shown in Fig. 2.

Table 1. Optimal weapon target assignment scheme

Serial number of weapons	1	2	3	4	5	6	7	8
Serial number of targets	3, 4	1	5	2, 8	9	11	6, 7	10, 12

The corresponding least threat value of optimal assignment scheme is 0.2374, and its assignment matrix is presented as follows:

$$P_{8\times12} = \begin{bmatrix} 0&0&1&1&0&0&0&0&0&0&0&0 \\ 1&0&0&0&0&0&0&0&0&0&0&0 \\ 0&0&0&0&1&0&0&0&0&0&0&0 \\ 0&1&0&0&0&0&0&1&0&0&0&0 \\ 0&0&0&0&0&0&0&0&1&0&0&0 \\ 0&0&0&0&0&0&0&0&0&0&1&0 \\ 0&0&0&0&0&1&1&0&0&0&0&0 \\ 0&0&0&0&0&0&0&0&0&1&0&1 \end{bmatrix} \tag{6}$$

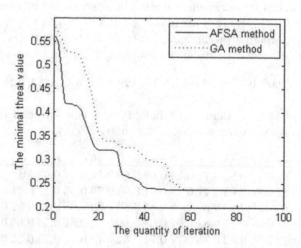

Fig. 2. The statistical results of least threat value of two algorithms

Contrast the statistical results of two methods by simulation, and we can see that both of them can converge on the optimal objective function value, but the convergence speed of method based on AFSA is faster than ones based on GA. The AFSA method obtains the least threat value at the 39-th iteration, and the GA method needs averagely 57 iterations. Therefore, the DWTA method based on AFSA proposed in this paper has advantages in computational efficiency.

5   Conclusion

The WTA problem that is to assign several multiple weapon units distributed on our diverse platforms to head off a group of targets has important significance in theoretical research and martial application. In this paper, the characteristic of real time is considered for the DWTA problem, and a DWTA method based on AFSA is proposed in the condition of time restraint. Moreover, a simulation is implemented for analysing the reliability of this method and validating its effectiveness; thereby a viable means is supplied to resolve the DWTA problem.

References

1. Lloyd, S.P., Witsenhausen, H.S.: Weapon Allocation Is NP-Complete. In: IEEE Summer Simulation Conference, pp. 1054–1058. IEEE Press, Nevada (1986)
2. Hosein, A.P., Athans, M.: Some Analytical Results for the Dynamic Weapon-Target Allocation Problem. MIT Laboratory for Information and Decision System, LIDS-P-1944, pp. 1–28 (1990)
3. Zhang, J.Y., Yao, P.Y., Wang, X.: Multiple Platforms Coordinated Target Assignment Method Based on Time Restraint. Systems Engineering and Electronics 33(6), 1287–1292 (2011)
4. Beaumont, P., Chaib-draa, B.: Multiagent Coordination Techniques for Complex Environment: the Case of a Fleet of Combat Ships. IEEE Transactions on Systems, Man and Cybernetics 37(3), 373–385 (2007)
5. Li, X.L., Shao, Z.J., Qian, J.X.: An Optimizing Method Based on Autonomous Animats: Fish-swarm Algorithm. Systems Engineering Theory & Practice 22(11), 32–38 (2002)

An Agent-Based Model for Simulating Human-Like Crowd in Dense Places[*]

Muzhou Xiong[1], Yunliang Chen[1], Hao Wang[2], and Min Hu[3]

[1] School of Computer Science
China University of Geosciences
Wuhan 430074, P.R. China
[2] School of Material Science and Chemistry
China University of Geosciences,
Wuhan 430074, P.R. China
[3] Chinese PLA Defense Information Academy
mzxiong@gmail.com

Abstract. Crowd simulation has been becoming an efficient tool to study the crowd behaviour and its movement. Compared with macroscopic model, microscopic model is able to generate a fine grain simulation result. This paper proposes a agent-based model for crowd simulation for dense area. The simulation model considers how agent selects a goal as its moving destination, how avoids collision with neighboured agents, how leads or follow a group of agents, and how avoids high dense area to reduce its traveling time. Simulation results show the proposed agent-based model is able to simulate agent navigating and move around dense simulation environment. The simulation performance is also efficient.

Keywords: Crowd Simulation, Agent-based Simulation, Microscopic Model.

1 Introduction

Crowd behavior and its movement is an ubiquitous and fascinating social phenomenon. Every pedestrian in a crowd may have her/his own desire and execute her/his decision independently. Crowd with high density may even lead to severe disasters [1]. It has become a mesmerizing question, for psychologists, sociologists, computer scientists, and even for architecture designers and urban planners, on how the crowd behavior and movement can be predicted and depicted. Simulation methods and techniques like [2] then plays an important role for effective analyzing the movement pattern of crowd in diverse scenarios. It has

[*] This study was supported in part by National Natural Science Foundation of China (grant No.61103145), and the Fundamental Research Funds for the Central Universities, (China University of Geosciences (Wuhan), No.CUG100314 and No.CUG120409).

Z. Li et al. (Eds.): ISICA 2012, CCIS 316, pp. 8–19, 2012.

been widely applied to transportation systems [3], sporting and general spectator occasions [4], and fire escapes [5].

The core problem of crowd simulation is to build a efficient and accurate simulating model for simulating the movement of the whole crowd and each agent in crowd. The existing behavior models for crowd simulation can be roughly divided into two categories: *microscopic* and *macroscopic* approach. The two categories of approaches focus on different level of crowd and result in different detail-of-level simulation results. Microscopic approaches emphasize on the issues of individual aspect, including pedestrian's psychological and social behaviors (e.g., intention and emotion), communication among pedestrians, and individual decision making process. It attains a fine-grained simulation result with high resolution, since it describes how every pedestrian behaves in every step. Cellular Automata [6,7], agent-based model [8,9], and particle system [10,11] belong to this category. On the contrary, the macroscopic approach attains a coarse-grained simulation result with lower resolution. It takes the crowd as a whole and emphasizes on how external or environmental factors may affect the behavior of crowd. A typical example of this category is flow-based model [4,12]. Individual issues of each pedestrian in a crowd are not considered in this approach. Different considerations of the two approaches cause different features. Accuracy is one of the key features of microscopic model. However, it is achieved at the cost of high computation complexity. With the increase of the scale of crowd simulation, the microscopic model cannot make an effective and smooth progress for the simulation. As for macroscopic model, it cannot bring about a precise simulation result for every pedestrian, but a tendency of how the whole crowd behaves because of its lack of concerns about social and psychological elements, and the process of decision making for each pedestrian.

This paper aims to simulate how pedestrians navigate and move around dense environment. In order to simulate such scenarios in detailed level, the paper proposes an agent-based model for simulating the behavior and movement of agents. During the simulation process, each agent needs to select a goal as its destination, and then choose a behavior, i.e., leader or follower, to achieve its selected destination. After this, it also needs to make a response for the oncoming collisions with neighbored agents. Since agent moves around and dense area, it needs to avoid dense area to reduce its traveling time.

The rest of the paper is organized as follows. A brief overview of prior work is given in Section 2. Followed the overview of the proposed agent-based model in Section 3, the detailed design issues are depicted in Section 4. Simulation results and performance evaluation goes in Section 5. The paper is concluded in Section 6.

2 Related Work

There are four main-stream types of models for crowd simulation in terms of how the model imitates the behavior of human in crowd: *flow-based model, particle system, agent-based model,* and *cellular automata model,* and each of them belongs to either microscopic or macroscopic approach. Flow-based model, a macroscopic approach, considers crowd as a whole part and adopts some physical-law-like equations (e.g. fluid equations) to demonstrate the movement of crowd. As a result, flow rate and crowd density become two key parameters to characterize the feature of crowd movement. Some further research [13,14,15] about the relationship between the two parameters shows that the mutual interaction between them may further reveal the regulation of crowd movement. Naturally, flow-based model cannot provide a fine-grained result with detailed individual properties. But the simulation still benefits from it since its computational complexity is not affected by the scale of crowd.

Similar to particle system, agent-based model simulates each individual in the crowd individually. Its key issue exists in how to imitate and demonstrate the decision-making process of individual in a crowd. Existing framework of decision-making process includes Bayesian networks [16], fuzzy logic [17], neural networks [18], BDI [19], and decision networks [20]. However, these methods focus on the mathematical and computational aspects of decision-making process rather than the imitation of the real process that human does. Moreover, the psychological and social factors could cause significant effects on the process of human's decision making. For example, an individual in a crowd may behave quite differently as when she/he is alone. Hence some cognitive and psychological models [8,9,21,22] are proposed to avoid the limitation of those frameworks mentioned above. Since agent-based model simulates crowd in the way real human takes, it obtains a result more realistic than that of particle system. However the scalability limitation of crowd size still remains in this approach.

The Cellular Automata model for crowd simulation is characterized by the adoption of Cellular Automata [23] with a discrete spatial representation and discrete time-steps, to represent the simulated environment and the entities it comprises. Typically, each individual in a crowd can occupy only one cell, and she/he must obey the predefined transit rules determining which neighbored cell she/he selects as the destination for the next simulation step. The rules are usually expressed as the probability of the agent moving to each of the neighbored cells. Moreover, with well-designed format, some regulations from microscopic approaches, such as social force, can be combined into Cellular Automata approach [24,25]. Such combination expands its simulation capability, and it can be applied into more complicated scenarios. However only the microscopic features can be reflected in the model, not including the macroscopic ones. Again, crowd size is still limited by the high computational complexity.

3 Overview of Agent-Based Model

An agent-based model is adopted as the microscopic model, in which every simulated pedestrian is considered as an agent using the same behavior model introduced below. The model simulates how a pedestrian behaves in dense crowd. The basic assumption is that a pedestrian does not have much of his own idea about his movement, but imitates the behavior of pedestrians nearby [26]. The behavior imitation includes how the agent selects an exit and how it gets to its selected exit.

The simulation process of the agent-based model is depicted in Figure 1, which is executed in each simulation step. There are two movements for an agent in every simulation step. Firstly an agent will select an exit as its goal to leave. Before making the first movement in the step, the agent still needs to decide whether there exists a lower density area nearby. If there does, it will execute *Density-based Navigation*, in which it makes the first movement. If not, *Leading* or *Following* is executed as the first movement, and its position and velocity are updated accordingly. After that, the agent executes *Neighbor Avoidance*, within which the second movement is performed.

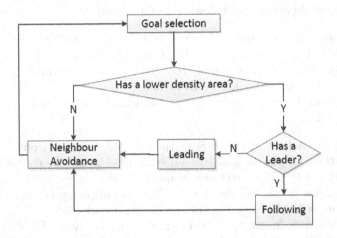

Fig. 1. Agent-Based Behavior Model

4 Model Design

Detailed model design is discussed in this section, including goal selection, following behaviour, leading behaviour, neighbour avoidance and navigation in dense areas.

4.1 Goal Selection

Each agent needs a goal as its destination for movement. Regarding agent's purpose of movement, the environment typically holds several positions acting as a goal. Agent needs to select one of them as its goal. For example, if there exists several exit for agent to leave the simulation environment, agent should choose one of them as its goal to evacuate the environment.

It can be observed that crowd movement always holds a moving pattern, which means that most agents in a crowd move towards similar direction. Regarding this, the agent needs to obtain the goals of agents nearby and follow one of them which is held by most of agents nearby. This can effectively reduce the probability of collision with other agents and then make the traveling time shorter.

In order to simulate this behaviour, the agent-based model imitates the agents nearby and selects the goal that most of the agents around it selected.

4.2   Leader Selection and Following

If agent find a goal following most of its nearby agents, (i.e., most of agents holding similar goal for movement), it agent will select a leader to follow. This behaviour can be explained as it follows the crowd movement.

A leader should satisfy the following requirements:

1. the leader has the same goal as the agent, which means the leader should not move towards the opponent direction to the agent's desire;
2. the angle between agent's preferred direction and the direction from the agent to the leader should be small enough;
3. the distance between the two agents should be the smallest one with the same angle.

Here, the preferred direction refers to the selected direction from the current position of the agent to the goal (like an exit). In this case, we choose the value of the angle as $\pm 10°$. If more than one agent satisfies the requirements, the one with the smallest angle will be selected. When multiple agents have the same angle, the one with the closest distance will then be selected.

Once a leader is found, the agent executes the behavior of *Following*. The execution of following behaviour is performed by the method of *steerForSeek* from *OpenSteer*[1].

4.3   Leading

If the agent does not find a leader satisfying the above requirements, it will move forward along the preferred direction without imitating others.

It still needs to find a goal first. Once the goal is determined, the needs to navigate around the dense simulation environment. This means agent needs to plan its path for its current location to the destination. In order to reduce its traveling time, the path is first shaped as a line, with the direction from current location to destination. However, agent may encounter high density areas along the path, it will avoid these areas to maintain as a relative high moving speed. Since the environment is dense, it is highly probable to collision with other agent around. Hence a collision avoidance motion should also be responded once encountering such scenario.

---

[1] http://opensteer.sourceforge.net/

The leading behaviour can then be depicted as how agent moving directly to its selected goal by the preferred velocity, and how agent avoid high dense area and collision. The selection of preferred velocity includes preferred speed and direction. Preferred speed depends on the density of agent's current location and will be discussed in the next section. The preferred direction, it can be obtained by calculating the direction vector from current location to the selected goal. Neighbour avoidance behavior is executed by the method of *steerToAvoidNeighbors* from *OpenSteer*. As for high density avoidance, they will be discussed in the next two subsections.

4.4 Density Avoidance

If the density in the agent sensed-area (a circle area with a predefined radius and the center at the position of the agent) is larger than a threshold (e.g., $0.5/m^2$), the density-based navigation described below will be performed.

The process of density-based navigation is designed as follows. If the density of the *area 2* is much larger than that of the *area 1* as shown in Figure 2, the direction of velocity will be adjusted to the direction of v_1 with the same value of speed. Otherwise, the direction of v_2 is selected.

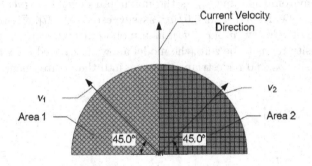

Fig. 2. Density-Based Navigation

Through this direction adjustment, agent changes its preferred direction temporally, until it find an sparse area along the original path. During its movement along the temporal direction, agent also attempts to seek new goal along the new direction. If there exists such a new goal, the old goal will be substituted by the new one.

5   Case Study

5.1   Scenario Description

The environment is set to $50 \times 20m$. Agents come into the environment from the left edge of the environment. At the beginning of the simulation, there exist only one exit located at the right side of the environment. This means there is only one candidate goal for all agents in the simulation to leave the environment. When simulation time comes to $4s$, another exist located at the top edge opens. The new exit provides another choice for agent's goal selection. In addition, there is a round obstacle located at the centre of the environment.

5.2   Agent Input

The dynamic input determines how many pedestrians coming into the environment during the simulation. The flow-based model does not require the dynamic input as it only works under stable conditions where the flow rate does not change. It is essential for agent-based model to determine the number of agents moving into the environment. Here we adopt Greenshields' model [13], in which it assumes that the magnitude of velocity (i.e., speed) is a linearly decreasing function of the density, as given by:

$$|\boldsymbol{v}(\rho(x,y,t))| = v_f(1 - \frac{\rho(x,y,t)}{\rho_m}) \tag{1}$$

where v_f is the free speed (i.e., the maximum speed a pedestrian is able to move in the environment) and ρ_m is the maximum density. Typical values for v_f and ρ_m may be $1.4m/s$ and $5.0/m^2$ as suggested by [27]. Figure 3 shows the magnitude of velocity $|\boldsymbol{v}(\rho(x,y,t))|$ as a monotonically decreasing function. When the density tends to be zero, the model allows free speed v_f, while for the maximum density ρ_m, no pedestrian can move into the environment.

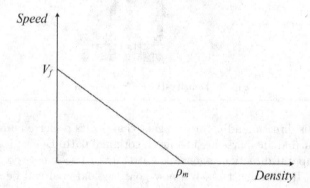

Fig. 3. Relationship between speed and ρ

The number of agents that enter into the environment (from the entrance) is calculated by the supplement number of agent for keep the fixed density around the environment entrance.

5.3   Simulation Results

At the beginning of the simulation, all agents entering from the left side of environment desire to leave the environment. The only goal can be selected is the right side exit. Hence, all agents' preferred direction is towards the right side, illustrated in Figure 4 where each agent is represented as a solid triangle. The preferred speed is determined by Formula 1. Due to the presence of the round obstacle at the centre of the environment, agents also need to avoid the obstacle.

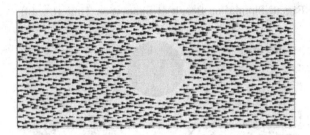

Fig. 4. Simulation Result at Simulation Time($t = 0$)

As simulation time comes to $4s$, another exit locate at the top edge of the environment with the length $10m$ opens. Then agents hold two candidate goals for selection. Agent near the original exit still choose it for goal selection; on the contrary, agent near the new exit will immediately choose the new one as its desired destination. Other agents needs to imitate agents around for selection. This means agent near an exit will make an impact on other agent's decision for selecting a goal. In the mean time, these agents will also play a role of leader to lead other agent to leave the environment through the selected exit.

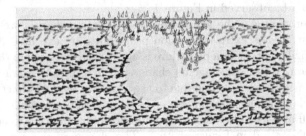

Fig. 5. Simulation Result at Simulation Time ($t = 4$)

The simulation result is reported in Figures 5 to 7, where solid triangles represent agents holding the original exit as their goals and hollow triangles are agents with the new exit for their destination.

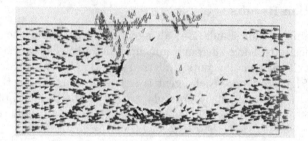

Fig. 6. Simulation Result at Simulation Time ($t = 10$)

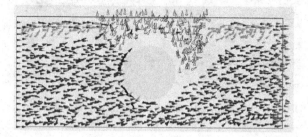

Fig. 7. Simulation Result at Simulation Time ($t = 15$)

5.4   Performance Evaluation

The performance evaluation is also conducted for the proposed agent-based model based upon the scenario applied in this section. The simulation cost over simulation time is shown in Figure 8. At the beginning of the simulation, the simulation cost is relative low due to all agents holding the same goal and moving in order. This means rare collisions are detected during agent's movement. It also implies that agents should be distributed uniformly with an steady dynamic input, which can be observed in Figure 4.

At simulation time $4s$, the simulation cost increases sharply when the new exit opens. This means extra simulation procedures are executed during each simulation step. The first extra procedure is agent need to compare the two exits for goal selection. When some agents changes their directions towards the new exist, more collision may be arisen due to the orderless moving direction. Hence, agent also needs to avoid neighboured collisions. In the meanwhile, the change of direction will also leads the density from uniform distribution to non-uniform distribution (illustrated in Figures 5 to 7. The change of density distribution causes the agents changing its moving direction for high density avoidance. The above mentioned extra model procedures leads the increase of simulation cost.

After the sharp increase of simulation cost, it drops back to a relative stable cost for each simulation time step for around $7s$. Agent is able to impact on other agents' goal selection and moving trajectories since agent first imitates other agent's behaviour. After a period of time for the availability of the new

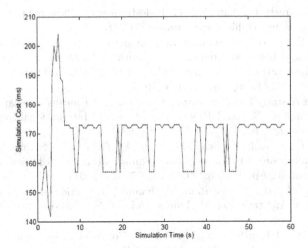

Fig. 8. Simulation Performance Evaluation

exit, agents gradually selects similar goals for similar locations due to the imitation behaviour. Hence, the probability of collision again drops. This means the above mentioned extra model procedure will not be triggered to execute. Simulation cost consequently decreases. Since there is no other new change for the simulation, the simulation cost keeps stable for the remain simulation time.

6   Conclusion

In this paper, an agent-based model is proposed for simulating human-like crowd's movement in dense places. Aiming at the behaviour and movement of each pedestrian in crowd, the model is able to produce a low detail-of-level simulation results. In each simulation time step, agent needs to select a position as its goal and subsequently determine its preferred moving direction. The preferred speed depends on the density of agent's current location. The preferred velocity (composed of preferred direction and speed) then becomes the agent's motion in each simulation time step, upon which agent's position updating can be calculated. Agent may encounter dense area during its movement. In this case, the behaviour of density area avoidance is conducted. As for the on-coming collision with neighboured agent, the behaviour of neighbour avoidance is necessary to be executed. The simulation results from case study indicates that the model is able to simulate crowd behaviour and movement in dense area with efficient execution cost.

References

1. Dynamics, C.: Crowd disaster (June 2008),
   http://www.crowddynamics.com/main/crowddisasters.html
2. Thalmann, D., Musse, S.R.: Crowd Simulation. Springer (2007)

3. Daly, P., McGrath, F., Annesley, T.: Pedestrian speed/flow relationships for underground stations. Traffic Engineering and Control 32(2), 75–78
4. Bradley, G.E.: A proposed mathematical model for computer prediction of crowd movements and their associated risks. In: Smith, R.A., Dickie, J.F. (eds.) Proceedings of the International Conference on Engineering for Crowd Safety, pp. 303–311. Elsevier Publishing Company, London (1993)
5. Tanaka, T.: A study for performance based design of means of escape in fire. In: Cox, G., Langford, B. (eds.) Proceedings of the 3rd International Symposium on Fire Safety Science, pp. 729–738. Elsevier (1991)
6. Burstedde, C., Klauck, K., Schadschneider, A., Zittartz, J.: Simulation of pedestrian dynamics using a two-dimensional cellular automaton. Physica A: Statistical Mechanics and its Applications 295(3-4), 507–525 (2001)
7. Muramatsu, M., Irie, T., Nagatani, T.: Jamming transition in pedestrian counter flow. Physica A: Statistical Mechanics and its Applications 267(3-4), 487–498 (1999)
8. Nguyen, Q.H., McKenzie, F.D., Petty, M.D.: Crowd behavior cognitive model architecture design. In: Proceedings of the 2005 Behavior Representation in Modeling and Simulation (BRIMS) Conference, Universal City, CA, pp. 55–64 (2005)
9. Pelechano, N., O'Brien, K., Silverman, B., Badler, N.: Crowd simulation incorporating agent psychological models, roles and communication. In: Proceedings of the First International Workshop on Crowd Simulation, EPFL, Lausanne-Switzerland (November 2005)
10. Brogan, D., Hodgins, J.: Group behaviors for systems with significant dynamics. Autonomous Robots 4(1), 137–153 (1997)
11. Helbing, D., Farkas, I., Vicsek, T.: Simulating dynamical features of escape panic. Letters to Nature 407, 487–490 (2000)
12. Chenney, S.: Flow tiles. In: Proceedings of the 2004 ACM SIGGRAPH/ Eurographics Symposium on Computer Animation, San Diego, California, USA, July 26-27, pp. 233–242 (2004)
13. Greenshields, B.D.: A study in highway capacity. Highway Research Board Proceedings 14, 448–478 (1934)
14. Smith, R.A.: Volume flow rate of densely packed crowds. In: Smith, R.A., Dickie, J.F. (eds.) Proceedings of the International Conference on Engineering for Crowd Safety, pp. 313–319. Elsevier Publishing Company, London (1993)
15. Smith, R.A.: Density, velocity and flow relationships for closely packed crowds. Safety Science 18(4), 321–327 (1995)
16. Pearl, J.: Probabilistic Reasoning in Intelligent Systems: Networks of Plausible Inference. Morgan Kaufman, San Mateo (1998)
17. Zadeh, L.A.: Fuzzy logic. IEEE Computer 22(4), 83–93 (1998)
18. Hassoun, M.H.: Fundamentals of Artificial Neural Networks. MIT Press (1995)
19. Rao, A.S., Georgeff, M.P.: An abstract architecture for rational agents. In: Proceedings of Knowledge Representation and Reasoning, pp. 439–449 (1992)
20. Yu, Q., Terzopoulos, D.: A decision network framework for the behavioral animation of virtual humans. In: Proceedings of Eurographics/ACM SIGGRAPH Symposium on Computer Animation 2007, San Diego, USA, pp. 119–128 (August 2007)
21. Rymill, S.J., Dodgson, N.A.: A psychologically-based simulation of human behaviour. In: Proceedings of Theory and Practice of Computer Graphics 2005, pp. 35–42 (2005)

22. Fridman, N., Kaminka, G.A.: Towards a cognitive model of crowd behavior based on social comparison theory. In: Proceedings of the Twenty-Second AAAI Conference on Artificial Intelligence, Vancouver, British Columbia, Canada, pp. 731–737 (July 2007)
23. Wolfram, S.: Theory and applications of Cellular Automata. World Press (1986)
24. Kirchner, A., Schadschneider, A.: Simulation of evacuation processes using a bionics-inspired cellular automaton model for pedestrian dynamics. Physica A: Statistical Mechanics and its Applications 312(1-2), 260–276 (2002)
25. Henein, C.M., White, T.: Macroscopic effects of microscopic forces between agents in crowd models. Physica A: Statistical Mechanics and its Applications 373, 694–712 (2007)
26. Festinger, L.: A theory of social comparison processes. Human Relations, 117–140 (1954)
27. Pushkarev, B., Zupan, J.M.: Capacity of walkways. Transportation Research Record (538), 1–15 (1975)

A Novel Heuristic Filter Based on Ant Colony Optimization for Non-linear Systems State Estimation

Hadi Nobahari and Alireza Sharifi

Sharif University of Technology
P.O. Box 11365-11155, Tehran, Iran
nobahari@sharif.edu
alirezasharifi@ae.sharif.ir

Abstract. A new heuristic filter, called Continuous Ant Colony Filter, is proposed for non-linear systems state estimation. The new filter formulates the states estimation problem as a stochastic dynamic optimization problem and utilizes a colony of ants to find and track the best estimation. The ants search the state space dynamically in a similar scheme to the optimization algorithm, known as Continuous Ant Colony System. The performance of the new filter is evaluated for a nonlinear benchmark and the results are compared with those of Extended Kalman Filter and Particle Filter, showing improvements in terms of estimation accuracy.

Keywords: Non-linear Systems State Estimation, Heuristic Filter, Ant Colony Optimization, Particle Filter.

1 Introduction

In many engineering applications, one needs to estimate the states of a dynamic system. A state estimation problem is defined as follows: given the mathematical model of a dynamic system, it is desired to estimate the time-varying states using a noisy measurement. Estimation problems are often categorized as prediction, filtering and smoothing, depending on intended objectives and the available observations[1]. Here, the domain of focus is filtering, which is usually referred as the extraction of true signal from the observations. Filters are usually minimizing a given objective function, while they are working. Such filters are called optimal filters[2].

Optimal filters are categorized to recursive and batch filters[1][3]. A batch filter, e.g. least square filter, uses the complete history of measurements to estimate unknown states. A Recursive filters, in comparison, has the ability to receive and process measurements sequentially. Recursive filters consist of two essentially stages: prediction and update[3]. Prediction uses the estimated states of the previous time step to produce an initial estimate of the current step. This stage is also known as the priori state estimation because it does not use the observations, obtained in the current time step. In update stage, the priori state

Z. Li et al. (Eds.): ISICA 2012, CCIS 316, pp. 20–29, 2012.

estimation is combined with the current observation to refine the state estimation. This improved estimation is also termed as the posterior state estimation. The dynamic states can be estimated using the posterior Probability Density Function (PDF), obtained based on the received measurement. If either the system or measurement model is nonlinear, the posterior PDF will not be Gaussian, even if the measurement and the process noises are assumed to be Gaussian.

Several recursive filters can be found within the literature, the most well-known of which are Kalman Filter (KF)[4], Extended Kalman Filter (EKF)[5], Unscented Kalman Filter (UKF)[6], Particle Filter (PF) [7] and etc.

Recursive filters can also be categorized to linear and nonlinear filters[1][3]. In a linear filter, such as KF, both system and measurement models are linear. KF assumes the posterior PDF to be Gaussian and can be characterized by a mean and a covariance. In opposite, a nonlinear filter, such as EKF, UKF and PF, is used to estimate the states of a nonlinear dynamic system when either the system or the measurement model is nonlinear.

Analytical approximation and states sampling are two common approaches in nonlinear filtering. In the first approach, the nonlinear functions of the mathematical model are linearized and then a linear filter such as KF is used as well. EKF is an example of filters, work based on analytical approximation. Unlike to EKF, UKF is a sample based filter. It does not approximate the nonlinear mathematical model. Instead, it approximates the posterior PDF by a set of deterministically chosen samples. UKF is also referred to as a linear regression Kalman filter, because it is based on statistical linearization rather than analytical ones[3].

The authors categorize sample based filters to mathematical and heuristic approaches. UKF can be taken a mathematical sample based filter to account, since it uses a deterministic sampling process, the general estimation mathematics and the mathematical operators such as unscented transform. In comparison, there are several sample based filters that utilize heuristic algorithms to sample the particles and to improve the position of them. These filters can be called heuristic filters.

PF is an example of heuristic filters. It works based on point mass (or particle) representation of the probability densities[8]. Unlike to UKF, PF represents the required posterior PDF by a set of random samples instead of deterministic ones. Also, it uses a re-sampling procedure to reduce the degeneracy of particle set. The standard re-sampling procedure copies the important particles and discards insignificant ones based on their fitness. This strategy suffers from the gradual loss of diversity among the particles, known as sample impoverishment. Different re-sampling strategies have been proposed in the literature, such as Binary Search[9], Systematic Re-sampling[10] and Residual Re-sampling[8].

PF has several variants with different sampling and re-sampling procedures. All sampling procedures, utilized in PFs, can be derived from the Sequential Importance Sampling (SIS) algorithm[11] by the appropriate choice of importance sampling density[3]. The combination of SIS and Systematic Re-sampling is called Generic PF (GPF)[3]. Sampling Importance Resampling (SIR) filter[9],

Boostrap Particle Filter (BPF)[12], Auxiliary Sampling Importance Resampling (ASIR) filter[13], Unscented Particle Filters (UPF)[14], Extended Particle Filters (EPF) [10], Multiple-model Particle Filter (MMPF)[12], Regularized Particle Filter (RPF)[15] and Markov Chain Monte Carlo (MCMC)[16] are example variants of PF.

Recently, some heuristic optimization algorithms have been augmented with PFs. Genetic Algoritm (GA) and PF have been combined to increase diversity of samples after re-sampling[17][18]. Simulated Annealing (SA) has been introduced into PF to improve its performance[19]. A local search method has been inserted into particle filter to reduce the sample size and improve the efficiency[20]. Particle Swarm Optimization (PSO) has been introduced into PF to solve the particle impoverishment and sample size dependency problems[21]. Ant Colony Optimization (ACO) has been utilized to improve the re-sampling process[22][23]. ACO for Real domains (ACOR) has been incorporated into PF to optimaize the sampling process[24].

The state estimation problem can be formulated as a stochastic dynamic optimization problem. Therefore, different ideas of heuristic optimization can be extended and modified to solve this problem. Here, the authors have proposed a new heuristic filter for non-linear systems state estimation, based on their previously developed metaheuristic, known as Continuous Ant Colony System (CACS)[25]. The proposed filter is called Continuous Ant Colony Filter (CACF). CACF can be categorized as a heuristic sample based filter. It utilizes a colony of moving ants, the average positions of which is returned as the current estimation. In this filter the estimation of the current states is formulated as a stochastic dynamic optimization problem and an optimization algorithm, based on CACS is utilized to iteratively find and track the best estimation.

This paper is organized as follows: A state estimation problem is formulated in section 2. Section 3 is devoted to a detailed description of the new estimation algorithm. The experimental results are provided in section 4. The final Conclusion is made in section 5.

2   Problem Formulation

The problem is to estimate the states of a nonlinear dynamic system. Discrete-time state space approach is utilized to model the evolution of system and the noisy measurements. The states are assumed to be evolved according to the following stochastic model:

$$x_k = f_k(x_{k-1}, v_{k-1}) \tag{1}$$

where f_k is a known, possibly nonlinear function of the state vector x_{k-1}, v_{k-1} is referred to as the process noise, and k is the time counter. The objective of a nonlinear filter is to recursively estimate x_k from the available measurements, z_k. The measurements are related to the states via the measurement equation, stated as follows:

$$z_k = h_k(x_k, w_k) \tag{2}$$

where h_k is a known, possibly nonlinear function and w_k is the measurement noise. The noise sequences, v_k and w_k, are assumed to be white, with known probability density functions and mutually independent. A graphical illustration of the evolution and the measurement models can be depicted as in Fig. 1. The initial state, x_0, is assumed to have a known PDF $p(x_0)$ and to be independent of the noise sequence.

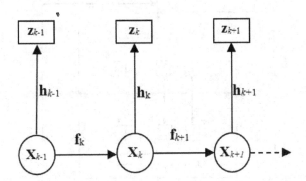

Fig. 1. Process and measurement models of a dynamic system

3 Continuous Ant Colony Filter

In this section, the new heuristic filter is introduced as a tool for nonlinear systems state estimation. First, the general structure of the filter is presented. Then each constructive module is discussed in detail.

3.1 General Setting Out of the Algorithm

Fig. 2 shows the general iterative structure of CACF. A high level description of the sequential steps is shown in this figure. The parameters of CACF and the initial position of ants are set during the initialization, as discussed in the section 3.2.

CACF has two loops: a main outer loop, iterating every time a new measurement is entered, and an inner loop, iterates to find the best estimation of the current states, corresponding to the entered measurement. The inner loop propagates the initial distribution of ants, at first. Then, the output, estimated by each ant, is made. The estimated outputs are compared with the real measurement and each ant is assigned a cost, based on the quality of its estimation. Ants use their experience to update the state space pheromone distribution. As in CACS[25], a Gaussian function is utilized to model the pheromone distribution over the continuous state space. Ants use this pheromone distribution to move from their current position toward the minimum cost destinations. The destinations are chosen using a normal PDF. The inner loop is terminated after a predefined number of iterations. Finally, the current state estimation is made using a mean operator. In the following subsections, these steps are discussed in detail.

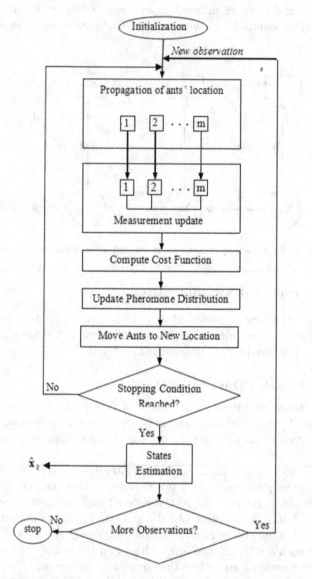

Fig. 2. Continuous ant colony filter (CACF) algorithm

3.2   Initialization

The new algorithm has some control parameters that must be set before the execution of the algorithm. The inner loop is terminated after q iterations. Moreover, the initial position of ants is initialized using a uniform random generator.

3.3   Propagation and Measurement Update

At the beginning of the i-th iteration of the inner loop, the position of ant j at time $k - 1$, defined as, $x_k^{i,j}$ is propagated as follows:

$$\hat{x}_k^{i,j} = f_k(x_{k-1}^{i,j}, v_{k-1}^{i,j}) \tag{3}$$

Then the current output, estimated in iteration i by ant j at time k, noted by $\hat{z}_k^{i,j}$, is calculated as follows:

$$\hat{z}_k^{i,j} = h_k(\hat{x}_k^{i,j}) \tag{4}$$

3.4   Compute Cost Function

Each ant is assigned a cost, based on the quality of its current position. The cost function is defined as the square error between the estimated output, $\hat{z}_k^{i,j}$, and the real measurement, z_k. Therefore, the cost, assigned in iteration i to ant j at time k, is calculated as follows:

$$f_k^{i,j} = (\hat{z}_k^{i,j} - z_k)^2 \tag{5}$$

In this way, the cost function is calculated in different points of the state space and some knowledge about the problem is acquired, used later to update the pheromone distribution.

3.5   Update Pheromone Distribution

CACF utilizes the same pheromone model and pheromone updating rule, as in CACS[25]. During any iteration, phermone distribution will be updated using the acquired knowledge from the evaluated points by ants. Pheromone updating rule of CACF can be stated as follows: during any iteration, the cost is calculated for the new points, explored by the ants. Then, the best point, found up to the ith iteration, at time $k - 1$, is assigned to $x_{k-1,min}^i$.

Also, the standard deviation of the pheromone distribution (σ_{k-1}^i) is updated based on the cost of the evaluated points and the aggregation of those points around $x_{k-1,min}^i$. To satisfy simultaneously the fitness and aggregation criteria, the concept of weighted variance, proposed in [25], is defined as follows:

$$(\sigma_{k-1}^i)^2 = \frac{\sum_{j=1}^m \frac{1}{f_{k-1}^{i,j} - f_{k-1,min}^i}(x_{k-1}^{i,j} - x_{k-1,min}^i)^2}{\sum_{i=1}^m \frac{1}{f_{k-1}^{i,j} - f_{k-1,min}^i}} \tag{6}$$

Here, m is the number of ants. This strategy means that the center of region, discovered during the subsequent iteration, is the last best point and the narrowness of its width depends on the aggregation of the other competitors around the best point. It should be noted that after termination of the inner loop, the standard deviation of the pheromone distribution is increased by an Expansion Factor (EF) to increase the exploration of the filter when the new measurement is entered.

3.6 Movement of the Ants

During any iteration, ants move from their current position to their destination using the current pheromone distribution. Pheromone distribution is modeled using a normal PDF, the center of which is the best point ($x_{k-1,min}^i$), found from the first iteration and its variance depends on the aggregation of other ants around the best one. Normal PDF permits all points of the continuous state space to be chosen, either close to or far from the best point. As stated in section 3.2, in the first iteration, the position of ants is initialized using a uniform random generator, whereas for all subsequent iterations, ants chose their destination using the updated pheromone distribution, based on equation (6).

3.7 Stopping Condition

CACF has two loops, each with its own specific stopping condition. The inner loop stops when the maximum number of iterations (q) is reached. The outer loop stops when the measurements are finished.

3.8 States Estimation

After termination of the inner loop, the states are estimated based on the average position of top ants:

$$\hat{x}_k = \frac{1}{m_t} \sum_{j=1}^{m_t} \hat{x}_k^{q,j} \tag{7}$$

where m_t denote number of top ants.

4 Results and Discussion

In this section, the performance of the new filter is investigated for a benchmark, taken from the literature. This study is intended to provide a comparison of the proposed state estimation method with more established approaches. Table 1 shows the tuned parameters of CACF.

A nonlinear single variable economic model[14], defined by (8) and (9), is employed to test the performance of CACF and compare it with that of EKF and Generic PF.

$$x(t+1) = 1 + sin(4 \times 10^{-2}\pi t) + 0.5x(t) + v(t) \tag{8}$$

$$z(t) = \begin{cases} \frac{x(t)^2}{5} + w(t) & t \le 30 \\ -2 + \frac{x(t)}{2} + w(t) & t > 30 \end{cases} \tag{9}$$

where $v(t)$ and $w(t)$ stand for zero-mean white noise and Gamma distribution, respectively[22]. The variance of $w(t)$ is 1×10^{-5} and the parameters k and θ of Gamma distribution are equal to 7 and 2, respectively[26].

Considering the measurement and process noises, the state sequence x_t is estimated using CACF and the results are compared with those of EKF and Generic PF, as reported in[21][22]. To make the results comparable with those of[22], the simulations are done from t = 1 to 60 and the average performance, obtained for 30 different runs are compared. Fig. 3 represents a sample output of CACF and shows that this filter can track the true signal accurately.

Table 2 shows the mean of the Root Mean Square Error (RMSE) obtained three filters. It can be observed that CACF produces comparable and even better results than EKF and Generic PF.

Table 1. Parameters of CACF

Parameter	Value
Numbers of Ants(m)	200
Maximum Number of Iterations(q)	10
Expansion Factor(EF)	2
Number of Top Ants	80

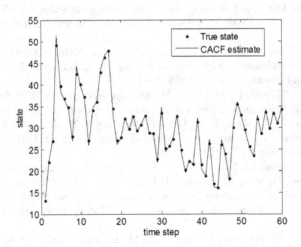

Fig. 3. The estimation history of CACF for the nonlinear single variable economic model

Table 2. Comparison of CACF with EKF and Generic PF

Filter	Mean RMS Error	RMSE Percentage (EKF=100%)
EKF[22]	0.9809	100
Generic PF[22]	0.7792	79
CACF	0.6513	66.398

5   Conclusion

In this paper a new heuristic sample based filter, was proposed for non-linear systems state estimation. The proposed filter, called CACF, models the estimation problem as a stochastic dynamic optimization problem and an optimization scheme, based on CACS, was utilized to solve this problem. CACF was tested over a benchmark to compare its results with those of EKF, as a mathematical nonlinear approach and PF, as a heuristic approach. The overall results show that CACF can properly compete with these well-known filters. One of the most important features of CACF is its simplicity.

References

1. Siouris, J.: An Engineering Approach to Optimal Control and Estimation Theory. Air Force Institute of Technology, New York (1995)
2. Brayson, J., Ho, Y.: Applied Optimal Control. Blaisdell Publishing Company, Waltham (1969)
3. Ristic, B., Arulampalam, S., Gordon, N.: Beyond the Kalman Filter. Artech House, London (2004)
4. Kalman, R.E.: A New Approach to Linear Filtering and Prediction Problems. Transaction of the ASME Journal Basic Engineering 82(Series D), 35–45 (1960)
5. Jazwinski, A.H.: Stochastic Processes and Filtering Theory. Academic Press, New York (1970)
6. Julier, S.J., Uhlmann, J.K.: A New Extension of the Kalman Filter to Nonlinear Systems. In: AeroSense 11th International Symposium Aerospace Defense Sensing, Simulation and Controls, pp. 182–193 (1997)
7. Carpenter, J., Clifford, P., Fernhead, P.: An improved particle filter for non-linear problems. IEE Proceedings, Radar Sonar and Navigation 146, 2–7 (1997)
8. Arulampalam, M.S., Maskell, S., Gordon, N., Clapp, T.: A Tutorial on Particle Filters for Online Nonlinear/Non-Gaussian Bayesian Tracking. IEEE Transactions on Signal Processing 50(2), 174–188 (2002)
9. Gordon, N.J., Salmond, D.J., Smith, A.F.M.: Novel Approach to Nonlinear/Non-Gaussian Bayesian State Estimation. IEE Proceedings F 140(2), 107–113 (1993)
10. Doucet, A., Freitas, D., Gordon, N.J.: Sequential Monte Carlo Methods in Practice. Springer Series in Statistics for Engineering and Information Science, New York (2001)
11. Liu, J.S., Chen, R.: Sequential Monte Carlo Methods for Dynamic Systems. J. Am. Statist. Ass. 93, 1032–1044 (1998)
12. McGinnity, S., Irwin, G.W.: Multiple model bootstrap filter for maneuvering target tracking. IEEE Transactions on Aerospace and Electronic Systems 36(3), 1006–1012 (2000)

13. Pitt, M., Shephard, N.: Auxiliary Particle Filters. J. Am. Statist. Ass. 94(446), 590–599 (1999)
14. Merwe, R., Freitas, N., Doucet, A., Wan, E.: The Unscented Particle Filter. Cambridge University Engineering Department, Cambridge (2000)
15. Musso, C., Oudjane, N., LeGland, F.: Improving Regularised Particle Filter. Sequential Monte Carlo Methods in Practice. Springer, New York (2001)
16. Gilks, W.R., Berzuini, C.: Following a Moving Target, Monte Carlo Inference for Dynamic Bayesian Models. Journal of the Royal Statistical Society, 127–146 (2001)
17. Higuchi, T.: Monte Carlo Filter Using the Genetic Algorithm Operators. Journal of Statistics Computation and Simulation 59(1), 1–23 (1997)
18. Park, S., Hwang, J.P., Kim, E., Kang, H.: A New Evolutionary Particle Filter for The Prevention of Sample Impoverishment. IEEE Transection on Evolutionary Computation 13(4), 801–809 (2009)
19. Clapp, T.C.: Statistical Methods for the Processing of Communication Data. University of Cambridge, Cambridge (2000)
20. Troma, P.: Szepesvari. C.: LS-N-IPS: An Improvement of Particle Filters By Means of Local Search. In: Proc. Non-Linear Control Systems (NOLCOS 2001), St. Petersburg, Russia (2001)
21. Tong, G., Fang, Z., Xu, X.: A Particle Swarm Optimized Particle Filter for Nonlinear System State Estimation. In: IEEE Congress on Evolutionary Computation, pp. 438–442 (2006)
22. Zhong, J., Fung, Y., Dai, M.: A Biologically Inspired Improvement Strategy for Particle Filter: Ant Colony Optimization Assisted Particle Filter. International Journal of Control, Automation and Systems 8(3), 519–526 (2010)
23. Hao, Z., Zhang, X., Yu, P., Li, H.: Video Object Tracing Based on Particle Filter with Ant Colony Optimization, pp. 232–236. IEEE (2010)
24. Yu, Y., Zheng, X.: Particle Filter with Ant Colony Optimization for Frequency Offset Estimation in OFDM Systems with Unknown Noise Distribution. Signal Processing 91, 1339–1342 (2011)
25. Pourtakdoust, S.H., Nobahari, H.: An Extension of Ant Colony System to Continuous Optimization Problems. In: Dorigo, M., Birattari, M., Blum, C., Gambardella, L.M., Mondada, F., Stützle, T. (eds.) ANTS 2004. LNCS, vol. 3172, pp. 294–301. Springer, Heidelberg (2004)
26. Lukacs, E.: A Characterization of the Gamma Distribution. The Annals of Mathematical Statistics 26(2), 319–324 (1955)

New Proofs for Several Combinatorial Identities

Chuanan Wei[1] and Ling Wang[2,*]

[1] Department of Information Technology
Hainan Medical College
Haikou 571199, P.R. China
[2] College of Sciences
Hebei Polytechnic University
Tangshan 063009, P.R. China
mcombinatorics@yahoo.cn

Abstract. It is well known that inversion techniques have an important role in the development of combinatorial identities. In 1973, Gould and Hsu[8] gave a pair of surprising inverse series relations. Combing Gould-Hsu inversions just mentioned with Vandermonde's formula, four known terminating $_4F_3$−series identities are recovered. In addition, new proofs for two extensions of q-Chu-Vandermonde identity due to Fang[6] and two formulae on Stirling numbers of the second kind due to Chu and Wei[4] are also offered by means of combinatorial techniques.

Keywords: Gould-Hsu Inversions, Hypergeometric Series, Basic Hyper-geometric Series, Stirling Numbers of The Second Kind.

1 Gould-Hsu Inversions and $_4F_3$−series Identities

For a complex variable x and a natural number n, define the shifted-factorial by:

$$(x)_0 = 1 \ and \ (x)_n = x(x+1) \cdots (x+n-1) \ when \ n = 1, 2, \ldots$$

Following Bailey[3], the hypergeometric series can be defined by:

$$_{1+r}F_s \left[\begin{array}{c} a_0, a_1, \ldots, a_r \\ b_1, \ldots, b_s \end{array} \middle| z \right] = \sum_{k=0}^{\infty} \frac{(a_0)_k (a_1)_k \ldots (a_r)_k}{k!(b_1)_k \ldots (b_s)_k} z^k$$

Where $\{a_i\}_{i\geq 0}$ and $\{b_j\}_{j\geq 0}$ are complex parameters such that no zero factors appear in the denominators of the summand on the right hand side. There are many interesting hypergeometric series identities in the literature. Thereinto, the famous Vandermonde's formula[3] can be expressed as:

$$_2F_1 \left[\begin{array}{c} -n, b \\ c \end{array} \middle| 1 \right] = \frac{(c-b)_n}{(c)_n} \tag{1}$$

---

* Corresponding author.

Z. Li et al. (Eds.): ISICA 2012, CCIS 316, pp. 30–39, 2012.

For a complex variables x and two complex sequences $\{a_k, b_k\}_{k\geq 0}$, define a polynomial sequence by

$$\phi(x; 0) \equiv 1 \ and \ \phi(x; m) = \prod_{i=0}^{m-1} (a_i + xb_i), when \ m = 1, 2, \ldots$$

The Gould-Hsu inversions[8] can be stated as follows:

$$f(n) = \sum_{k\geq 0} (-1)^k \binom{n}{k} \phi(k; n) g(k) \tag{2}$$

$$g(n) = \sum_{k\geq 0} (-1)^k \binom{n}{k} \frac{a_k + kb_k}{\phi(n; k+1)} f(k) \tag{3}$$

Subsequently, we shall employ this pair of inverse series relations to prove several known terminating $_4F_3$-series identities.

1.1 The First $_4F_3$−series Identity

Theorem 1[2].

$$_4F_3 \left[\begin{matrix} -\frac{n}{2}, \frac{1-n}{2}, a, & -a \\ \frac{1}{2}, & c, 1-c-n \end{matrix} \middle| 1 \right] = \frac{(c+a)_n + (c-a)_n}{2(c)_n}$$

Proof For $\varepsilon = 0, 1$, it is not difficult to show that:

$$_2F_1 \left[\begin{matrix} -\varepsilon - 2n, c+a \\ c-n \end{matrix} \middle| 1 \right] + _2F_1 \left[\begin{matrix} -\varepsilon - 2n, c-a \\ c-n \end{matrix} \middle| 1 \right] = \frac{2(a)_n(-a)_n}{(c)_n(1-c)_n}(1-\varepsilon) \tag{4}$$

by means of Vandermonde's formula(4). Reformulate the last equation as:

$$\sum_{k=0}^{\varepsilon+2n} (-1)^k \binom{\varepsilon+2n}{k} (1-c-k)_n \frac{(c+a)_k + (c-a)_k}{2(c)_k} = \frac{(a)_n(-a)_n}{(c)_n(1-c)_n}(1-\varepsilon)$$

which matches with (2) perfectly and (3) creates the dual relation:

$$\sum_{k\geq 0} \binom{n}{2k} \frac{1}{(1-c-n)_k} \frac{(a)_k(-a)_k}{(c)_k} = \frac{(c+a)_n + (c-a)_n}{2(c)_n}$$

It is equivalent to Theorem 1 and so we complete the proof.

1.2 The Second $_4F_3$-series Identity

Theorem 2[2].

$$_4F_3 \left[\begin{matrix} -\frac{n}{2}, \frac{1-n}{2}, 1+a, & -a \\ \frac{3}{2}, & c, & 1-c-n \end{matrix} \middle| 1 \right] = \frac{(c+a)_{n+1} - (c-a-1)_{n+1}}{(1+n)(1+2a)(c)_n}$$

Proof. In terms of Vandermonde's formula (4), we have

$$_2F_1\left[\begin{matrix}-\varepsilon-2n, & c+a\\ & c-\varepsilon-n\end{matrix}\Big|1\right]-_2F_1\left[\begin{matrix}-\varepsilon-2n, & c-a-1\\ & c-\varepsilon-n\end{matrix}\Big|1\right]=\frac{(1+a)_n(a)_n}{(c)_n(2-c)_n}\frac{(1+2a)\varepsilon}{(1-c)}$$

Where $\varepsilon=0,1$. It can be manipulated as:

$$\sum_{k=0}^{\varepsilon+2n}(-1)^k\binom{\varepsilon+2n}{k}(1-c-k)_{n+\varepsilon}\frac{(c+a)_k-(c-a-1)_k}{(1+2a)(c)_k}=\frac{(1+a)_n(-a)_n}{(c)_n}\varepsilon$$

which fits into (2) ideally and (3) gives the dual relation:

$$-\sum_{k\geq0}\binom{n}{1+2k}\frac{1}{(1-c-n)_{k+1}}\frac{(1+a)_k(-a)_k}{(c)_k}=\frac{(c+a)_n-(c-a-1)_n}{(1+2a)(c)_n}$$

Performing the replacement $n\to n+1$ for the last equation we achieve Theorem 2 to finish the proof.

1.3   The Third $_4F_3$−series Identity

Theorem 3[2].

$$_4F_3\left[\begin{matrix}-\frac{n}{2}, & \frac{1-n}{2}, & a, & c-a\\ -n, & \frac{1+c}{2}, & \frac{2\pm c}{2}\end{matrix}\Big|1\right]=\frac{(a)_{n+1}-(c-a)_{n+1}}{(2a-c)(1+c)_n}$$

Proof. For $\varepsilon=0,1$, it is not dicult to show that

$$\frac{(a)_{n+1}}{(1+c)_n}(_2F_1)\left[\begin{matrix}-\varepsilon-n, & 1+a+n\\ & 1+a+n\end{matrix}\Big|1\right]$$

$$-\frac{(c-a)_{n+1}}{(1+c)_n}(_2F_1)\left[\begin{matrix}-\varepsilon-n, & 1+c-a+n\\ & 1+c+n\end{matrix}\Big|1\right]$$

$$=\frac{(a)_n(c-a)_n}{(1+c)_{2n}}(1-\varepsilon)(2a-c)$$

by means of Vandermonde's formula (4). Rewrite the last equation as

$$\sum_{k=0}^{\varepsilon+n}(-1)^{k+n}\binom{\varepsilon+2n}{k+n}(-k-n)_n\frac{(a)_{k+n+1}-(c-a)_{k+n+1}}{(2a-c)(1+c)_{k+n}}$$

$$=\frac{(2n!)(a)_n(c-a)_n}{n!(1+c)_{2n}}(1-\varepsilon)$$

Shifting the summation index $k\to k-n$, we get the identity

$$\sum_{k=0}^{\varepsilon+2n}(-1)^k\binom{\varepsilon+2n}{k}(-k)_n\frac{(a)_{k+1}-(c-a)_{k+1}}{(2a-c)(1+c)_k}$$

$$=\frac{(2n!)(a)_n(c-a)_n}{n!(1+c)_{2n}}(1-\varepsilon)$$

which matches with (2) perfectly and (3) offers the dual relation:

$$\sum_{k \geq 0} \binom{n}{2k} \frac{1}{(-n)_k} \frac{(2k!)(a)_k(c-a)_k}{k!(1+c)_{2k}} = \frac{(a)_{n+1} - (c-a)_{n+1}}{(2a-c)(1+c)_n}$$

It is equivalent to Theorem 3 and so we complete the proof.

1.4 The Fourth $_4F_3$-series Identity

Theorem 4[9].

$$_4F_3\left[\begin{matrix} -\frac{n}{2}, & \frac{1-n}{2}, & a, & c-a \\ & 1-n, & \frac{c}{2}, & \frac{1+c}{2} \end{matrix}\middle| 1\right] = \frac{(c-a)_n + (a)_n}{(c)_n}$$

Proof. In terms of Vandermonde's formula (4), we have

$$\frac{(c-a)_n}{(c)_n}(_2F_1)\left[\begin{matrix} -\varepsilon - n, & c-a+n \\ & c+n \end{matrix}\middle| 1\right] + \frac{(a)_n}{(c)_n}(_2F_1)\left[\begin{matrix} -\varepsilon - n, & a+n \\ & c+n \end{matrix}\middle| 1\right]$$

$$= \frac{(c-a)_n(a)_n}{(c)_{2n}}(2-\varepsilon)$$

Where $\varepsilon = 0, 1$. It can be restated as

$$\sum_{k=0}^{\varepsilon+n}(-1)^{n+k}\binom{\varepsilon+2n}{k+n}(-k-n)_n\frac{(c-a)_{k+n} + (a)_{k+n}}{(c)_{k+n}}$$

$$= \frac{(\varepsilon+2n)!(c-a)_n(a)_n}{(\varepsilon+n)!(c)_{2n}}(2-\varepsilon)$$

Shifting the summation index $k \to k-n$ for the last equation, we obtain the identity:

$$\sum_{k=0}^{\varepsilon+2n}(-1)^k\binom{\varepsilon+2n}{k}(-k)_n\frac{(c-a)_k + (a)_k}{(c)_k}$$

$$= \frac{(\varepsilon+2n)!(c-a)_n(a)_n}{(\varepsilon+n)!(c)_{2n}}(2-\varepsilon)$$

Which fits into (2) ideally and (3) produces the dual relation:

$$\sum_{k \geq 0} \binom{n}{2k} \frac{1}{(-n)_k} \frac{2(2k)!(c-a)_k(a)_k}{k!(c)_{2k}}$$

$$- \sum_{k \geq 0} \binom{n}{1+2k} \frac{k-(1+2k)}{(-n)_{k+1}} \frac{(1+2k)!(c-a)_k(a)_k}{(1+k)!(c)_{2k}}$$

$$= \frac{(c-a)_n + (a)_n}{(c)_n}$$

It is equivalent to Theorem 4 and therefore we finish the proof.

2   Two Extensions of q-Chu-Vandermonde Identity

For two complex numbers x and q, define the q-shifted factorial by

$$(x;q)_0 = 1 \text{ and } (x;q)_n = \prod_{k=0}^{n-1}(1 - xq^k) \text{ when } n = 1, 2, ...$$

The multi-parameter form of it reads as

$$(\alpha, \beta, ..., \gamma; q)_n = (\alpha; q)_n (\beta; q)_n ... (\gamma; q)_n$$

Following Slater[10], define the basic hypergeometric series by

$$_{1+r}\Phi_1 \left[\begin{matrix} a_0, a_1, ..., a_r \\ b_1, ..., b_s \end{matrix} \middle| q, z \right] = \sum_{k=0}^{\infty} \frac{(a_0, a_1, ..., a_r; q)_k}{(q, b_1, ..., b_s)_k} z^k$$

Then q-Chu-Vandermonde identity[10] can be stated as

$$_2\Phi_1 \left[\begin{matrix} q^{-n}, b \\ c \end{matrix} \middle| q; \frac{q^n c}{b} \right] = \frac{(c/b; q)_n}{(c; q)_n}$$

2.1   The First Extension of q-Chu-Vandermonde Identity

Theorem 5[6].

$$\sum_{k=0}^{n} \begin{bmatrix} n \\ k \end{bmatrix} (-1)^k q^{\frac{k^2-k}{2}} \frac{(b, a_1, a_3, ..., a_{2t+1}; q)_k}{(c, a_2, a_4, ..., a_{2t+2}; q)_k} \left(\frac{ca_2 a_4 ... a_{2t+2}}{ba_1 a_3 ... a_{2t+1}} \right)^k$$

$$= \frac{(c/b; q)_n}{(c; q)_n} \sum_{0 \leq n_{t+1} \leq n_t \leq ... \leq n_1 \leq n_0} \prod_{i=0}^{t} \frac{(q^{-n_i}, a_{2i-1}, a_{2i+2}/a_{2i+1}; q)_{n_i+1}}{(q, a_{2i+2}, q^{1-n_i} a_{2i-1}/a_{2i}; q)_{n_i+1}} q^{n_i+1}$$

where $t = -1, 0, 1, 2, ..., a_{-1} = b, a_0 = c$ and $n_0 = n$.

Proof. Let two complex sequences $\{\alpha_r, \beta_r\}_{r \geq 0}$ satisfy the relation:

$$\beta_n = \sum_{r=0}^{n} \frac{\alpha_r}{(q; q)_{n-r}(aq; q)_{n+r}}$$

Then we have

$$\beta'_n = \sum_{r=0}^{n} \frac{\alpha'_r}{(q; q)_{n-r}(aq; q)_{n+r}}$$

Where $\{\alpha'_r\}_{r \geq 0}$ and $\{\beta'_r\}_{r \geq 0}$ are respectively defined by

$$\alpha'_r = \frac{(P_1; q)_r (P_2; q)_r}{(aq/P_1; q)_r (aq/P_2; q)_r} \left(\frac{aq}{P_1 P_2} \right)^r \alpha_r$$

$$\beta'_r = \sum_{j=0}^{n} \frac{(P_1; q)_j (P_2; q)_j (aq/P_1 P_2; q)_{n-j}}{(q; q)_{n-j}(aq/P_1; q)_n (aq/P_2; q)_n} \left(\frac{aq}{P_1 P_2} \right)^j \beta_j$$

A pair of sequences (α_n, β_n) satisfying (5) is called Bailey pair[2]. The power of the Bailey lemma lies in the following fact. A Bailey pair (α_n, β_n) subject to (5) yields another Bailey pair (α'_n, β'_n) subject to (6). Thus the iterating process leads to the following Bailey chain:

$$(\alpha_n, \beta_n) \to (\alpha'_n, \beta'_n) \to (\alpha''_n, \beta''_n) \to \cdots$$

Combing Bailey chain with the q-Dougall-Dixon formula[10]:

$$_6\phi_5\left[\begin{array}{cccccc} a, & q\sqrt{a}, & -q\sqrt{a}, & b, & c, & q^{-n} \\ & \sqrt{a}, & -\sqrt{a}, & qa/b, & qa/c, & aq^{n+1} \end{array} \middle| q; \frac{q^{n+1}a}{bc}\right] = \frac{(qa, qa/bc; q)_m}{(qa/b, qa/c; q)_n}$$

we can derive, without diculty, the following sum:

$$_{2t+8}\phi_{2t+7}\left[\begin{array}{ccccccc} a, & q\sqrt{a}, & -q\sqrt{a}, & b, & c, & \{a_s\}_{s=1}^{2t+2}, & q^{-n} \\ & \sqrt{a}, & -\sqrt{a}, & qa/b, & qa/c, & \{qa/a_s\}_{s=1}^{2t+2}, & qa^{n+1} \end{array} \middle| q; \frac{a^{t+2}q^{n+t+2}}{bc\prod_{s=1}^{2t+2} a_s}\right]$$

$$= \left[\begin{array}{cc} qa, & qa/bc \\ qa/b, & qa/c \end{array} \middle| q\right]_n \times$$

$$\sum_{0 \le n_{t+1} \le n_t \le \ldots \le n_1 \le n_0} \prod_{i=0}^{t} \frac{(q^{-n_i}, a_{2i-1}, a_{2i}, qa/a_{2i+1}a_{2i+2}; q)_{n_{i+1}}}{(q, qa/a_{2i+1}, qa/a_{2i+2}, q^{-n_i}a_{2i-1}a_{2i}/a; q)_{n_{i+1}}} q^{n_{i+1}}$$

which can also be deduced form a known result due to Andrews[1]. Performing the replacements $c \to qa/c$ and $a_{2j} \to qa/a_{2j}$ with $j = 1, 2, \ldots, t+1$ for the last equation and then letting $a \to 0$, we attain Theorem 5 to complete the proof.

2.2 The Second Extension of q-Chu-Vandermonde Identity

Theorem 6[6].

$$\sum_{k=0}^{n} \begin{bmatrix} n \\ k \end{bmatrix} \frac{(b; q)_k}{(c, q)_k} \left(\frac{-c}{b}\right)^k q^{\frac{k^2-k}{2}} \sum_{k_1=0}^{k} \frac{(q^{-k}, a_1; q)_{k_1}}{(q, b; q)_{k_1}} (qa_2b)^{k_1} \times$$

$$\sum_{0 \le k_{t+1} \le t_k \le \ldots \le k_1} \prod_{i=1}^{t} \frac{(q^{k_i}, a_{2i+1}; q)_{k_{i+1}}}{(q, q^{1-k_i}/a_{2i-1}; q)_{k_{i+1}}} \left(\frac{qa_{2i+2}}{a_{2i-1}}\right)^{k_{i+1}}$$

$$= \frac{(c/b; q)_n}{(c; q)_n} \sum_{n_1=0}^{n} \frac{(q^{-n_1}, a_1; q)_{n_1}}{(q, q^{1-n}b/c; q)_{n_1}} (qa_2b)^{n_1} \times$$

$$\sum_{0 \le n_{t+1} \le n_t \le \ldots \le n_1} \prod_{i=1}^{t} \frac{(q^{-n_i}, a_{2i+1}; q)_{n_{i+1}}}{(q, q^{1-n_i}/a_{2i-1}; q)_{n_{i+1}}} \left(\frac{qa_{2i+2}}{a_{2i-1}}\right)^{n_{i+1}}$$

Where $t = 1, 2, \ldots$.

Proof. Let $U(n)$ stand for the following double sum:

$$U(n) = \sum_{k=0}^{n} \begin{bmatrix} n \\ k \end{bmatrix} \frac{(b; q)_k}{(c, q)_k} \left(\frac{-c}{b}\right)^k q^{\frac{k^2-k}{2}} \sum_{k_1=0}^{k} \frac{(q^{-k}, a_1; q)_{k_1}}{(q, b; q)_{k_1}} (qa_2b)^{k_1} \Omega(k_1)$$

Where

$$\Omega(k_1) = \sum_{0 \le k_{t+1} \le k_t \le \dots \le k_1} \prod_{i=1}^{t} \frac{(q^{-k_i}, a_{2i+1}; q)_{k_{i+1}}}{(q, q^{1-k_i}/a_{2i-1}; q)_{k_{i+1}}} \left(\frac{qa_{2i+2}}{a_{2i-1}}\right)^{k_{i+1}}$$

Interchanging the summation order for $U(n)$ and then shifting the summation index $k \to k + k_1$, we have

$$U(n) = \sum_{k_1=0}^{n} \frac{(q^{-n}, a_1; q)_{k_1}}{(q, c; q)_{k_1}} (-a_2 c)^{k_1} q^{nk_1 - \frac{k_1^2 - k_1}{2}} \Omega(k_1) \times$$

$$_2\Phi_1 \left[\begin{matrix} q^{-n+k_1}, bq^{k_1} \\ cq^{k_1} \end{matrix} \middle| q; \frac{q^{n-k_1}c}{b} \right]$$

Employing (4) to the last equation, we gain the identity:

$$U(n) = \frac{(c/b; q)_n}{(c; q)_n} \sum_{k_1=0}^{n} \frac{(q^{-n}, a_1; q)_{k_1}}{(q, q^{1-n}b/c; q)_{k_1}} (qa_2 b)^{k_1} \Omega(k_1)$$

This finishes the proof of Theorem 6.

3   Two Formulae on Stirling Numbers of the Second Kind

For a natural number m, let M be the nite set given by $M := \{x_1, x_2, \dots, x_m\}$. Then the number of partitions of M into n nonempty parts is the classical Striling number of the second kind $S(m, n)$. Its explicit formula can be found in[5]:

$$S(m, n) = \frac{1}{n!} \sum_{k=0}^{n} (-1)^{n-k} \binom{n}{k} k^m$$

3.1   The First Formula on Stirling Numbers of the Second Kind

Theorem 7[4]. For two elements x_i and x_j from M, define their distance by $|i - j|$, the distance between their subscripts i and j. Let $S_l(m, n)$ stand for the number of partitions of M into n nonempty parts with $|i - 1| \ge l$. Then we have the explicit formula:

$$S_l(m, n) = \sum_{k=0}^{n-l} \frac{(-1)^{n-k-l}}{(n-l)!} \binom{n-l}{k} k^{m-l} \text{ where } n \ge max\{1, l\}$$

Proof. Consider the number $S_l(1 + m, n)$ of partitioning $M \cup \{x_{m+1}\}$ into n nonempty parts with the distance between any two members in the same part begin greater than l. The last element x_{m+1} either stays alone or joins some part. For the former case, we have $S_l(m, n - 1)$ partitions of M into $n - 1$ nonempty parts. While for the latter case, x_{m+1} can be inserted into the parts of the partitions of M enumerated by $S_l(m, n)$ except for those containing one of $x_m, x_{m-1}, ..., x_{m-l+1}$ as an element, which counts $n - l$ times $S_l(m, n)$. Thus we achieve the following recurrence relation:

$$S_l(m + 1, n) = S_l(m, n - 1) + (n - l)S_l(m, n) \text{ where } m \geq n \geq max\{1, l\}$$

Based on the recurrence relation just established, we can manipulate the generating function as follows:

$$S_n^l(x) = \sum_{m \geq n} S_l(m, n)x^m + x \sum_{m \geq n} S_l(m + 1, n)x^m$$

$$= x^n + x \sum_{m \geq n} \{S_l(m, n - 1) + (n - l)S_l(m, n)\}x^m$$

$$= x \sum_{m \geq n-1} S_l(m, n - 1)x^m + (n - l)x \sum_{m \geq n} S_l(m, n)x^m$$

$$= xS_{n-1}^l(x) + (n - l)xS_n^l(x)$$

This leads us to the functional equation:

$$S_n^l(x) = \frac{x}{1 - (n - l)x}S_{n-1}^l(x)$$

By iterating this relation for $n - l$ times, we deduce

$$S_n^l(x) = \frac{x^{n-l}}{\prod_{k=1}^{n-l}(1 - kx)}S_l^l(x)$$

Noting that for $m \geq n = l$, we have

$$S_l(l, l) = 1 \text{ and } S_l(m, l) = 0 \text{ when } m \neq l$$

This can be justified as follows. For any partition of M, the first l elements $x_1, x_2, ..., x_l$ must stay in l different parts among which no one can contain x_{l+1} as an element. In general for $0 < n \leq l$, we have

$$S_l(n, n) = 1 \text{ and} S_l(m, n) = 0 \text{ when } m \neq n$$

Therefore, we derive the following generating function:

$$S_n^l(x) := \sum_{m=n}^{\infty} x^m = \frac{x^n}{\prod_{k=1}^{n-l}(1 - kx)} \text{ where } n \geq max\{1, l\}$$

Comparing the coecients of x^m on both sides of the last equation, we get hold of the following explicit formula:

$$S_l(m, n) = \sum_{k=0}^{n-l} \frac{(-1)^{n-k-l}}{(n - l)!} \binom{n - l}{k} k^{m-l}$$

This completes the proof of Theorem 7.

3.2   The Second Formula on Stirling Numbers of the Second Kind

Theorem 8[4]. For two elements x_i and x_j from M, define their distance by $|i - j|$, the the distance between their subscripts i and j. Let $T_l(m, n)$ stand for the number of partitions of M into n nonempty parts with $|i - j| \neq l$. Then we have the explicit formula:

$$T_l(m, n) = \sum_{k=1}^{n} \frac{(-1)^{n-k}}{n!} \binom{n}{k} k^l (k-1)^{m-l} \; where \; m \geq l > 0$$

Proof. Consider the number $T_l(1 + m, n)$ of partitioning $M \cup \{x_{m+1}\}$ into n nonempty parts such that the distance of any two members in the same part differs from f. The last element x_{m+1} either stays alone or joins some other part. The former case corresponds to $T_l(m, n - 1)$ partitions of M into $n - 1$ parts. While in the latter case, x_{m+1} can join every part of the partitions of M enumerated by $T_l(m, n)$ except for the one containing x_{1+m-l}. So we gain the following recurrence relation:

$$T_l(1 + m, n) = T_l(m, n - 1) + (n - 1)T_l(m, n)$$

Based on the recurrence relation just founded, we can manipulate the generating function as follows:

$$T_n^l(x) = \sum_{m \geq n} T_l(m, n)x^m = x^n + x \sum_{m \geq n} T_l(m + 1, n)x^m$$

$$= x^n + x \sum_{m \geq n} \{T_l(m, n - 1) + (n - 1)T_l(m, n)\}x^m$$

$$= xT_{n-1}^l(x) + (n - 1)xT_n^l(x)$$

We have therefore the functional equation:

$$T_n^l(x) = \frac{xT_{n-1}^l(x)}{1 - x(n - 1)}$$

Iterating this last relation for $n - 1$ times, we obtain the generating function:

$$T_n^l(x) = \frac{x^{n-1}T_1^l(x)}{\prod_{k=1}^{n-1}(1 - kx)} = \frac{x^n\{1 - x^l\}}{(1 - x)\prod_{k=1}^{n-1}(1 - kx)}$$

where we have employed the following easily derived expression:

$$T_1^l(x) = \frac{1 - x^l}{1 - x}x$$

Comparing the coefficients of x^m on both sides of the last equation, we catch hold of the following explicit formula:

$$T_l(m, n) = \sum_{k=1}^{n} \frac{(-1)^{n-k}}{n!} \binom{n}{k} k^l (k-1)^{m-l}$$

This finishes the proof of Theorem 8.

Acknowledgment. The work is supported by the Natural Science Foundation of China (N0. 11126213) and the Foundation of Hainan Medical College (No. HYP201116).

References

1. Andrews, G.E.: Problems and prospects for basic hypergeometric functions. Theory and application for basic hypergeometric functions. In: Askey, R.A. (ed.) Math. Res. Center, Univ. Wisconsin, Publ. No. 35, pp. 191–224. Academic Press, New York (1975)
2. Andrews, G.E., Burge, W.H.: Determinant identities. Pacific J. Math. 158, 1–14 (1993)
3. Bailey, W.N.: Generalized Hypergeometric Series. Cambridge University Press, Cambridge (1935)
4. Chu, W., Wei, C.: Set partitions with rescrictions. Discrete Math. 308, 3163–3168 (2008)
5. Comtet, L.: Advanced Combinatorics. Dordrecht, Holland, The Netherlands (1974)
6. Fang, J.: Extensions of q-Chu-Vandermonde's identity. J. Math. Anal. Appl. 339, 845–852 (2008)
7. Gould, H.W.: Some generalizations of Vandermonde's convolution. Amer. Math. Month. 63, 84–91 (1956)
8. Gould, H.W., Hsu, L.C.: Some new inverse series relations. Duke Math. J. 40, 885–891 (1973)
9. Krattenthaler, C.: Determinant identities and a generalization of the number of totally symmetric self-complementary plane partitions. Electronic J. Combinatorics 4, #R27 (1997)
10. Slater, L.J.: Generalized Hypergeometric Functions. Cambridge University Press, Cambridge (1966)
11. Wei, C., Gong, D.: New proofs for Whipple's transformation and Watson's q-Whipple transformation. Ars Combin. 102, 257–262 (2011)
12. Wei, C., Gong, D., Li, J.: Duplicate Form of Gould-Hsu Inversions and Binomial Identities. In: Liu, B., Chai, C. (eds.) ICICA 2011. LNCS, vol. 7030, pp. 145–152. Springer, Heidelberg (2011)

Construction of Standard College Tuition Model and Optimization

Cuirong Chen, Sa Zhao, and Chengyu Hu*

Higher Education Research Institute
China University of Geoscience
Wuhan 430074, China
huchengyucug@gmail.com

Abstract. The reasonable sharing of the costs of education and tuition pricing standard are at present major problems for higher education. This paper has considered some universities area per capita, per capita costs of training and education quality, under conditions of constraint, established a mathematical model with the minimum difference between personal and social satisfaction and the greatest overall satisfaction, by using particle swarm algorithm to optimize the model, has given a reasonable charge tuition standard and national finance investment standard recommendations.

1 Introduction

Higher education is related to the training of high-quality talents, enhancement of national innovation ability, the building of a harmonious society in the overall situation. So the party, government and all aspects of society pay great attention to it. Training quality is a core indicator of higher education, different disciplines, professional in the setting of different training objectives, and its quality need to have the funds as safeguard.

According to the Chinese education cost sharing theory and principles, government per capita allocations and tuition are closely related, when government per capita allocations grow, tuition will decrease accordingly. However, equity of education has decided the government funds and tuition should not be too high or too low. On one hand, higher education in any country is not compulsory, if higher education funded entirely by the government burden, cost of higher education is all paid by taxpayers, there is only a partial beneficiaries. In that case, it loses the fairness of higher education. Personal satisfaction degree is higher, while social satisfaction degree is lower; On the other hand, if the level of tuition fees is too high, because there is a huge difference of dweller income level in the market economy, presenting a state of non-equilibrium, exorbitant tuition exceed most residents' ability to pay which will cause the unfairness of education opportunities, violate the fair education principles and objectives, and also trigger a new round of unfair distribution of income. The government can

---

* Corresponding author.

Z. Li et al. (Eds.): ISICA 2012, CCIS 316, pp. 40–49, 2012.

put much more money into more people for their welfare, social (all taxpayers) satisfaction degree is larger. Unable to pay higher tuition fees, the individual (family) satisfaction degree is smaller. Therefore, establishing higher education tuition standard and national finance investment standard model which is optimized and developing a tuition standard that enables both the individuals and society can be simultaneously satisfied is very necessary.

Guo Zhen-ping considered the students per capita costs of training, personal and social sensitivity coefficient, in personal and social conditions, introduced the student individual and social satisfaction, established a total maximum satisfaction as the goal of university tuition standard optimization model[1]. Zhang Wei-dong and others in the analysis of the impact of university tuition standard on the foundation of main factor, use the national per capita allocations, training fees, family income and other indicators as parameters, established the higher school double objective optimization model[2]. YeJun, apply linear programming method for the quantitative analysis, according to the different regions, different professional higher education tuition standard problem modeling and analysis[3]. Xu Xin-sheng see the government financial allocation and family pay tuition relationship as "competition", "cooperation", established cross programming model between the two ones[4].TianWei has made the formulation of China's higher education tuition standard bi-directional optimization model, using the related data to the model to make empirical analysis.[5]

The above study is inadequate, the model is relatively simple.It adopts existing toolbox to solve the model which is lack of flexibility, its specific decision-making reference is also insufficient. In this paper, according to the National Bureau of statistics of China statistical bulletin, Finance Department of the Ministry of education statistics, taking into account the different regions of the urban and rural residents per capita annual income, the training cost and running quality comprehensive evaluation index, the paper has established a higher education tuition and the country's standard model to the funding costs, with the target of personal and social satisfaction, using the standard particle swarm algorithm to get every college specific tuition and national investment standards.

2 Model Establishment

2.1 Model Analysis

Higher education, which is not compulsory education, its money Q is funded by government financial allocation F, the school itself, social donations and tuition income T etc. In our country, government funding F and tuition income T are the main sources of funds. The school itself and social donations in the small proportion of university funding sources, in a short period of time, this part of the funds is constant, Set the value for D. Assuming that university funds just can be met, there is:

$$Q = F + T + D \tag{1}$$

According to the above analysis, in a shorter period of time, and D and Q as a fixed value, so there is an inverse trend between government financial funding F and tuition income T.

2.2   Model Establishment

Based on the above analysis, here comes two concepts of the family satisfaction and government (social) satisfaction. Family satisfaction S_f refers to the family satisfaction for their tuition; government(social) satisfaction S_g refers to the government satisfaction degree for the students' tuition.

Therefore, we establish a target function $|max(S_f + S_g) + min(S_f - S_g)|$ to solve what tuition and national investment on education standards that make family and government satisfaction the biggest, while making their satisfaction consistency. In this way, from national macroscopic point of view, it is conducive to the assurance of the citizen life quality and social stability, but also beneficial to the normal operations of the government, so this tuition standard is the best.

Family Satisfaction Sub Model. Family satisfaction S_f depends on what the tuition, family income level as well as the children's college education quality. Training quality refers to universities which are available for students to cultivate and develop strength after the completion of the results.

For S_f to quantify, introducing two factors. Bearing capacity factor α and factor β in return. characterization of a family of tuition capacity. An average family income is higher, less on tuition, fees to bear ability more. Therefore α can be defined as the ratio of average annual household income is I and t.

$$\alpha = \frac{I}{t} \tag{2}$$

β characterization of a students pay tuition for units training quality. College of education quality is high, the tuition is lower, and higher the value of is β, Therefore β can be defined as the ratio of training quality and tuition.

$$\beta = \frac{Q}{t} \tag{3}$$

Training quality refers to universities which are available for students to cultivate and develop strength after the completion of the results. The university quality comprehensive evaluation index E can reflect the quality of running a university, the larger values of E, the higher the quality of running a school. According to the 2010 China University ranking data, scientific research, personnel training, the reputation of the school and other indices are normalized, and then averaged to determine the comprehensive evaluation index E.

C is higher education per capita cultivation expenses, K is government financial per capita allocation. In accordance with the relevant state policies and higher education cost sharing theory, government financial funds are generally accounted for 75% of the cost of higher education.

$$C = \frac{K}{75\%} \tag{4}$$

The comprehensive evaluation index of university quality is higher, the cultivation ability is stronger, then cultivation quality will be better; Per capita higher cultivation costs means that the training process configuration to each person's resources are more and better, so does the final training quality. According to the University of E value gap, the training quality is defined as:

$$Q = CE^{\frac{1}{3}} \tag{5}$$

According to the analysis, bearing capacity factor is bigger, family satisfaction is higher; return factor is bigger, the higher the family satisfaction is. For people of different economic conditions and different ideas of education, economic capacity and returns to education in their minds on the different degree, then can get the following model:

$$S_f = \alpha^u \beta^{1-u} \tag{6}$$

Among them, u and 1-u represent bearing capacity factor , return factor that in the minds of imbalance degree. In our country, urban and rural income level and concepts of education are quite different, so the higher education tuition standard has large different impact on rural families and city families. Generally speaking, rural residents per capita annual household income is lower, the tuition is relatively sensitive, bearing capacity factor has great influence on their satisfaction degree. Urban residents per capita annual family income is higher, the expenditure on education of sensitivity is weak in the education quality, and is greatly influenced by education quality. So we can get the family satisfaction model:

$$S_{fr} = \alpha^u \beta^{1-u}$$
$$S_{fc} = \alpha^{1-u} \beta^u \tag{7}$$

For rural households, in considering the tuition fees, economic capacity is more important than education repayment; For urban households, education repayment is more important than economy bearing capacity. After several times of simulations and calculations, when $\mu = 0.6$, the effect of the model is the most ideal.

According to the reference[6]we consider approximately that the rural students and urban students in the ratio of 1: 4, so the family average satisfaction is:

$$S_f = 0.2S_{fr} + 0.8S_{fc} \tag{8}$$

Eventually the family satisfaction model is:

$$S_f = 0.2(\frac{I_r}{t})^{0.6} (\frac{\frac{K}{75\%}E^{\frac{1}{3}}}{t})^{0.4} + 0.8(\frac{I_c}{t})^{0.4} (\frac{\frac{K}{75\%}E^{\frac{1}{3}}}{t})^{0.6} \tag{9}$$

Government Satisfaction Model. Government satisfaction S_g depends on government per capita allocation and students'tuitions. Per capita allocation less, students pay more tuition in education, the government's financial burden is lighter, and government satisfaction is higher. We use the tuition and national per capita education investment ratio as measurable indicators of S_g

$$S_g = \delta \frac{t}{K} \qquad (10)$$

δ is a constant coefficient, through the simulation and calculation: when $\delta = 16$, the effect of the model is the most ideal, so we take $\delta = 16$. Integrated above several steps of the analysis process and results, the optimization model is established as follows:

$$|max(S_f + S_g) + min(S_f - S_g|)| \qquad (11)$$

The constraint conditions are:

$$t \le 3 \times \frac{I_r + I_c}{2} \times 50\%t + K + j \ge \Omega \qquad (12)$$

Constraint 1 refers to the general family member number is 3, then the total family income = member number ×per capita income; we believe that the tuition costs are not more than 50% of the total income of the family.

Constraint 2 refers to the fees t, per capita education funds K, other per capita incomes (social donors, school-run workshop etc.) j are not less than individual cultivation cost W.

3   Based on Improved Particle Swarm Optimization Algorithm

3.1   Particle Swarm Optimization

Particle swarm optimization(PSO)is a kind of evolutionary computation technique based on swarm intelligence, by Dr Eberhart and Dr Kennedy who invented a new evolutionary algorithm for global optimization, beginning from the flock of birds, fish, wild goose and other animal dynamic aggregation behavior of thinking[7].Because its algorithm is simple, easy to implement, and the concept is very clear, therefore, it causes the concern of the majority of scholars and has achieved remarkable achievements in function optimization, neural network training, intelligent control, scheduling and many other fields. [8]

In the particle swarm algorithm, each particle can be viewed as a potential solution of optimization problems, the flight of particle swarm optimization is the process of optimizing process. Particle swarm optimization mathematical model can be expressed by the following two formulae: particle velocity updating model and location update model. Eventually all the particles will gather in the approximate optimal solution through continuous iteration.

$$V_{t+1} = W \times V_t + c_1 \times r_1 \times (P_i - X_i) + c_2 \times r_2 \times (P_g - X_i) X_{i+1} = X_i + V_{t+1} \quad (13)$$

The particle X_i represents a solution to a point in space. P_i is currently the best position of particle i, P_g is the best particle. W is the inertia weight, c_1, c_2 are learning factors and the standard uniform distribution of the random variable. Particle velocity updating is divided into three parts:

1. $W \times V_i$: inertial portion, maintain the previous iteration speed;
2. $c_2 \times r_2 \times (P_g - X_i)$:social interactions, imitating other surrounding individual successful behaviors;
3. $c_1 \times r_1 \times (P_i - X_i)$: cognitive components, following the previous successful experiences.

The core of particle swarm optimization lies in three points: random, memory, learning. The three reasons make particle swarm algorithm in optimization show excellent performance, so the algorithm has got great development.

3.2   PSO Based on the Penalty Factor

Fitness Function and Constraint Function. Considering the objective function need to solve the maximum value and the minimum value, so it is necessary to convert, In this paper, transform the formula $max(S_f + S_g)$ of into $min(t) + min(k)$, In which t, k are respectively the tuition fees which family should pay and government funding. Obviously, these two parameters is smaller, the value of $max(S_f + S_g)$ is bigger. In order to be consistent with $min\|S_f - S_g\|$ number grade, t, k need to be normalized, use t, k directly on solution-vector coding. Therefore, the fitness function can be defined as:

$$f = min\|S_f - S_g\| + 0.001 \times t + 0.0001 \times k \tag{14}$$

The constraint conditions by using the penalty function method, penalty function is:

$$\varphi = f + p_1 + p_2 \tag{15}$$

$$p_1 = \begin{cases} 0, & t \leq 3 * 50\% * 0.5(I_r + I_c) \\ 30 * 50\% * (t - 0.5(I_r + I_c)), & t \geq 3 * 50\% * 0.5(I_r + I_c) \end{cases} \tag{16}$$

$$p_2 = \begin{cases} 0, & t + K + j \geq \Omega \\ 10(\Omega - (t + K + j)), & t + K + j < \Omega \end{cases} \tag{17}$$

Experimental Settings and Results of Simulation. In the simulation experiment, proposed by Clerc with convergence factor of the improved particle swarm algorithm [9], population 10, iterative termination conditions for the operation of 300 generation algorithm, and the algorithm runs 10 times.

In the algorithm, involving different areas of the towns' annual per capita income and rural residents' per capita income, due to the huge different incomes of areas of urban residents and rural residents, the article only take Shanghai, Beijing and other high income area 985 universities and 211 colleges as an example, discuss on the country's appropriate input and tuition standard. According to the National Bureau of statistics of China 2010 statistical data and school-running quality comprehensive evaluation index[10,11], we can calculate the national financial investment in education and tuition fee standard.

Table 1. The Annual Income and Cost per Resident and Quality of Universities

	University	Location	income (Urban)	income (Resident)	Cost	Quality
985	BUAA	Beijing	30673.68	11668.59	35000	0.356
985	BIT	Beijing	30673.68	11668.59	35000	0.296
985	RUC	Beijing	30673.68	11668.59	30000	0.530
985	BNU	Beijing	30673.68	11668.59	30000	0.431
985	SJTU	Shanghai	32402.97	12482.94	45000	0.531
211	BCUT	Beijing	30673.68	11668.59	30000	0.12
211	ZUC	Beijing	30673.68	11668.59	25000	0.113
211	BJU	Beijing	30673.68	11668.59	30000	0.164
211	ECNU	Shanghai	32402.97	12482.94	30000	0.318
211	ECUST	Shanghai	32402.97	12482.94	30000	0.222

Table 2 is the results of particle swarm algorithm optimization. From table 2 we can see, tuitions of different universities should be different, in nature, engineering colleges in personnel training costs are high, as a result, the tuition should also be high, investment of the government also should be higher too, but for the same region of all the engineering colleges, according to the quality of running a school, tuition pricing should be different, apparently, for the famous universities, because of the good quality in education, so the tuition fee shall be higher than non famous colleges, and government spending on education financial input should be slightly lower. But the reality is not the case, government tuition is basically the same in the same area, but the financial investment is biased in favor of the famous universities.

Table 2. Simulation Result

	University	Location	standard (tuition)	actual (tuition)	should Investment	Actually Investment
985	BUAA	Beijing	5556.03	5500	29444	36399
985	BIT	Beijing	5281.56	5500	29718.4	37034
985	RUC	Beijing	5457.33	5500	24542.7	31372
985	BNU	Beijing	5161.93	5500	24838.1	30579
985	SJTU	Shanghai	6980.51	5000	33019.5	43741
211	BCUT	Beijing	3628.5	5500	26371.5	18667
211	ZUC	Beijing	3082	5500	21918	16700
211	BJU	Beijing	3959.26	5500	26040.7	15116
211	ECNU	Shanghai	4804.11	5000	25195.9	23960
211	ECUST	Shanghai	4353.63	5000	25646.4	14805

Figure 1 is the particle swarm optimization algorithm for Beijing University of Aeronautics and Astronautics tuition and financial investment in the process of optimization the best fitness value of curve and the particle mean distance curve.

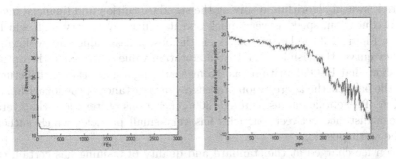

Fig. 1. Simulation Results For BUAA,the left figure is the curve of the best particle fitness value, and the right figure is the curve of the mean fitness value

Figure 2 is the particle swarm algorithm for Renmin University of China tuition and financial investment in the process of optimization the best fitness value of curve and the particle mean distance curve.

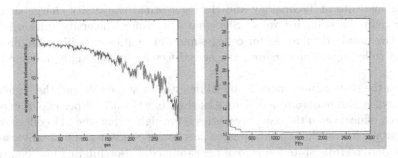

Fig. 2. Simulation Results For RUC,the left figure is the curve of the best particle fitness value,and the right figure is the curve of the mean fitness value

Figure 3 is the particle swarm algorithm for East China Normal University tuition and financial investment in the process of optimization the best fitness value of curve and the particle mean distance curve.

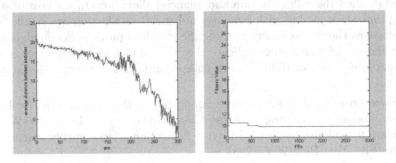

Fig. 3. Simulation Results For ECNU,the left figure is the curve of the best particle fitness value,and the right figure is the curve of the mean fitness value

Particle swarm algorithm in solving the standard tuition and financial investment in education, speed is very quick and its efficiency is very high which is shown in Figure 1-3 the left picture A , is the best fitness value and evolution of frequency curve, the visible after 1000 adaptation values is to reach the optimal value, show that PSO algorithm has strong searching ability; the right picture B means the particle the aggregation process average distance of the logarithm and evolution of algebraic curves, through 300 generations of iteration, the average Euclidean distance between particles has very small particles which gathered near the merit position.

Tuition fee charged at the standard and quality of training has certain connection, cultivate high quality, fees charged at the standard is relative taller, this also shows that the model is reasonable. From the results of Table 2, according to the model, every college fees are not the same, but the actual charges are much the same, it also indicates that the current university fees are unreasonable. As for the part of 985 colleges and universities in Beijing and Shanghai, tuition and the standard value in Renmin University of China and Beijing University of Aeronautics and Astronautics are not much difference, while Beijing Institute of Technology and Beijing Normal University charge slightly higher than the standard value, Shanghai Jiao Tong University tuition is far below the standard value. As for colleges and universities in Beijing and Shanghai 211 colleges and universities, the actual fees are higher than the standard value.

From the state of university financial investment standards and the standard of tuition, it can maintain in 4: 1 relationship, but the actual per capita expenditure on education in the 985 universities is far higher than the 211 colleges and universities, it also shows that, quite a part finance incomes of 985 universities are not derived from state funds, real per capita education funding bias big, thus make the quality of running a school higher.

4   Summary

Considering the different districts of urban and rural residents' annual average income, the training cost and running quality comprehensive evaluation index, we have established the college tuition and financial allocations to the cost of standard optimization model, with personal and social satisfaction are small, total satisfaction as the fitness function, using the standard particle swarm algorithm solving the model, the results indicate that this model has certain rationality, providing some basis for different area college tuition and financial allocations.

Acknowledgment. The Project was Supported by the Special Fund for Basic Scientific Research of Central Colleges, China University of GeosciencesWuhan (CUG090119) and also by Hubei Province Education science outline research of Eleventh Five-Year Plan(2009B298).

References

1. Guo, Z.-P., Yang, W.-X., Fan, Y.-J.: Higher Education Tuition Pricing Model Based on the Maximum Satisfaction. Journal of Wuhan University of Technology 37, 178–181 (2010)
2. Zhang, W.-D.: The Construction of Mathematics Model for the Higher Education Tuition Standard. Journal of Xi'an Polytechnic University 25(3), 384–388 (2011)
3. Ye, J., Zhang, J.-F., Li, Y.-M.: The Model for the Higher Education Tuition Fee Standard. Science and Technology Information (10), 189 (2009)
4. Xu, X.-S., Li, J.-D., Zhao, K.: Cross Programming Model For Higher Education Tuition. Journal of Binzhou University (6), 39–43 (2009)
5. Tian, W., Li, L.: The Formulation of Bi-directional Optimization Model for China's Higher Education Tuition Standard. Journal of Honghe College (5), 24–30 (2009)
6. Wen, D.-M., Yan, G.-J., Bao, W.: Survey Report of the Capital Higher Education Quality. In: Beijing Institute of Higher Education 2007 Academic Annual Conference (2007)
7. Eberhart, R.C., Kennedy, J.: A New Optimizer Using Particle Swarm Theory. In: Proceedings of the Sixth International Symposium on Micro Machine and Human Science, pp. 39–43 (1995)
8. AlRashidi, M.R., El-Hawary, K.M.: A Survey of Particle Swarm Optimization Applications in Electric Power Systems. IEEE Transactions on Evolutionary Computation 13(4), 913–918 (2009)
9. Clerc, M., Kennedy, J.: The Particle Swarm: Explosion, Stability, and Convergence in a Multi-dimensional Complex Space. IEEE Transactions on Evolutionary Computation 6(1), 58–73 (2002)
10. National Bureau of statistics of China. China Statistical Yearbook 2010. China Statistics Press, Beijing (2010)
11. The National Bureau of Statistics of Social Science Statistics Division. China Education Statistical Yearbook 2009. People Education Press, Beijing (2010)

Research and Realization of N–Queens Problem Based on the Logic Language Prolog

Baolei Gu

School of Computer Science
China University of Geosciences
Lumo Road 388, Wuhan 430074, P.R. China

Abstract. To solve the 8-queen problem in Chess is always the classical problem in mathematics and computer fields. Based on brief introduction of logic programming language Prolog, the author uses Prolog program to solve the 8-queen up to N-queen problem and improve running speed and simplify the algorithm complexity by the agency of features of Prolog logic language and through optimizing recursive algorithm.

Keywords: N Queens, Prolog, Recursion, Logic Programming.

1 Introduction

In the international chess pattern, eight queens are placed on the chessboard of 8×8 such that any two queens are unable to attack each other (not in the same row or column or diagonal), N-Queen problem is the expansion and extension of 8-Queen problem.

Prolog logic programming language has been widely used in artificial intelligence research for rapidly developing program, mainly used to build expert systems, natural language understanding and intelligent knowledge base and so on.

2 Main Characteristics of Prolog

2.1 Statements without Specific Operating Sequence and Control Flow

Program is computer instructions which executes according to certain order. But Prolog program is carried out not determined by programmer; as a descriptive language, it describes a problem using a specific method, and then find out the answer to this question by computer automatically. There are not control flow statements if and for, etc. in Prolog. In general, programmer only concerns whether or not the program description is complete. Prolog also provides some control flow methods, but there is big difference between these methods and those of other language.

Z. Li et al. (Eds.): ISICA 2012, CCIS 316, pp. 50–56, 2012.

2.2   Data Management

Prolog program and data are highly conformable. Program and data are diffi-
cultly to be distinguished in the Prolog program. Generally speaking, all code
in Prolog has the same form, data is program and program is data.

The Prolog program is the relational database which built on the basis of query
relationship; it can processes data easily. FACT is the simplest PREDICATE in
this language. It is similar to the records in the relationship database.

The query in this language is achieved based on pattern matching. The query
template is called goal; query is successful if a FACT matches a goal, and query
is failed if there is no matched FACT.

The following FACTS are taken for example:

```
QueenQ ( [0,1] ). % placing a queen on row 0 and 1st column
QueenQ ( [1,2] ). % placing a queen on 1st row and 2nd column
QueenQ ( [3,1] ). % on 3rd row without placing the queens
```

Enquiry can be made according to the FACTS above:

```
? - queenQ ( X ). % situation about placing all queens
? - queenQ (2, X ). % situation about placing queens in 2nd row
```

The Prolog program is the predicate database. The interpreter firstly transfers
all the clauses to the memory when the program runs. So these clauses written in
the program are fixed. But Prolog also provides the function of dynamic control
of the clause in memory. This means that program can also change itself while
it operates by using some internal predicate.

The function of predicate assert(X) is that transfers the first clause deemed
clause X as predicate of this clause into the dynamic database. It and the control
of I/O internal predicate are the same. For example: If there have been the
following FACTS in memory:

```
queenQ([0,1]).
queenQ([1,2]).
queenQ([3,-1]).
```

After running asserta $(queenQ([3, 4]))$, the clauses of $queenQ/1$ in memory
are as follows:

```
queenQ([0,1]).
queenQ([1,2]).
queenQ([3,-1]).
queenQ([3,4]).
```

The opposite from the function of asserta(X) is retract(X) predicate, and its
function is to delete clause X from dynamic database .

2.3   List Recursion and Arithmetic

List Prolog introduced List data structure for expressing a set of data. The
list is assembling of a set of data elements. The elements can be any data type,

including structure and list. A list without element is called null list, expressed as use "[ ]". For example:

```
[0,1] % placing a queen on row 0 and 1st column
[1,2] % placing a queen on the 1st row and 2nd column
[3,1] % No queen being placed on the 3rd row
```

General form of List: $[X|Y]$, this list can match with any list, after successful match, the first value of element of the list bound by X, namely the head , and rest lists are bound by Y , namely the end table (tail):

```
?- [1|[2,3,4]] = [1,2,3,4].
```

The matching above is successful, because the lists of both sides of equal-sign are equivalent.

```
?- [X|Y] = [].
```

This is a failed match which is commonly used for boundary detection in the course of recursive process. As long as the list is not null, it can match with $[X|Y]$. When the list is null, it can not be matched, expressing boundary condition has been reached.

Predicate append/3 links two lists into one list: parameter 1 and 2 connect values of list and provide the value to parameter 3.

Boundary condition: empty list linking with a list is the list itself.

```
append([],X,X).
append([H|T1],X,[H|T2]) :- append(T1,X,T2).
```

Connection list of recursive predicate append realization: list $[H|T1]$ connects with list X from which the head of the new list is H, end table (tail) is a list of list T1 linking with list X.

According to the result above: Prolog has powerful recursive function, recursion is a very concise description, and it can solve many problems efficiently; in Prolog, strong points of recursion has been fully embodied.

Arithmetic and Logic Operations. Support arithmetic calculations, but there is big difference between method of calculation mathematical expression and pattern matching; use internal predicate 'is' to calculate mathematical expression. Its grammatical form: X is < math expression >. In addition to predicate 'is', it also provides some operational characters which compare the size: $>, <, >=, =<$ etc.

2.4   Cutting

Algorithm provides the function of matching and recursion, but not all matching results and recursion are necessary. Manually controlling matching and recursion process is required. Prolog provides predicate cut (expressed as symbol !) to effectively eliminate some unnecessary matching and recursion. When cut appears, it has changed the procedures of program implementation process, directly sends the control to the predicate at upper stage instead of the predicate on the left.

3   Key Technology for Realization 8-Queen

Because queen can attack other queens in the same row or same column, the situation of two queens in the same row or same column is not allowable. Therefore use a list to express all the Queen's position, the position of element in the list represents a column in which a queen is, and the value of the element represents a row in which a queen is. The list will not have elements with the same value.

After using the list in which the elements are not the same, all the queens are safe in horizontal and vertical direction. We need to do next is to determine if the queen is safe in a diagonal direction; this is a typical method of "choose and verify". The following predicate safe/1 determines if its parameters is in a safe arrangement:

```
safty([Q|Qt]):- safty(Qt),not(attackcheck(Q,Qt)).
safty([]).
```

The codes above are designed with recursion method. Firstly determine if the tail list will attack each other, and then judge queens at both head and tail of the list will attack each other. Codes of attack/2 are shown below:

```
attackcheck(X,Xt):-  attackcheck(X,1,Xt).
attackcheck(X,N,[Y|Yt]):- X is Y+N;X is Y-N.
attackcheck(X,N,[Y|Yt]):- Nt is N+1,attackcheck(X,Nt,Yt).
```

The statement above describes two predicates, attackcheck/2 and attackcheck/3; process of its judgment implementation is: if the sum or difference between a value (line) and its location (column) of an element in the list equal to the value (line) of considered queen, then queens attacking each other exist. For example, if detect list [2,3,4,1], after firstly dividing it into 2 and [3,4,1], 2= 4-2, where the 2 is the location of elements 4. So the queens at first column and fourth column can attack each other. When the list is divided into 3 and [4,1], 3 = 2+1, where the 2 is the location of elements 1, so the queens at second column and the third column can attack each other.

The situation above is that eight Queens has been already set, then judge whether they are safety or not. Therefore the following statements must be used to achieve the Queen's placement and removal:

```
queen(N,Qt):-
    getrange(1,N,Nt),getList(Qt,Nt),safty(Qt).
    getrange(M,N,[M|Nt]):- M<N,Mt is M+1,getrange(Mt,N,Nt).
    getrange(N,N,[N]).
    getList([],[]).
    getList([A|X],Y) :- plusele(A,Y,Yt),getList(X,Yt).
    plusele(A,[A|X],X).
    plusele(A,[B|X],[B|Y]) :- plusele(A,X,Y).
```

Where, getrange/3 generates a list of number from 1 to N.

4 Optimization of 8-Queen Realization Technology with Prolog

In the codes above, Prolog solves the N queens problem, but the arrangement of eight queens problem have 8! =40320 kinds, plus judging for each placement, then executing times required is more than 200000. If the queen number N is changed into 9, 10 or more, then the calculating times required will be more.

4.1 One Kind of Optimized Algorithm

The method selection and check above can solve the eight queens problem, but its efficiency is low. The reason is that each verifying will be done after all queens are in place. If the first queen and second queen can attack each other, there is no need to consider the location of the third. Therefore, method for improving the efficiency is: put the verifying part and selecting part together, once selecting complete, immediately check is done, this can reduce many checks of alignment.

The following is part of the program codes:

```
queen(N,Qt):- getrange(1,N,Nt),queen(Nt,[],Qt).
                % main programs, use getrange/3 to generate
                  a list first, then use the Queen/3 to place
                  the queen.
queen(UnplacedQt,SafeQt,Qt):-
                % UnplaceQt: queen not to be placed, SafeQt:
                  list with queens to be placed
plusele(Q,UnplacedQt,UnplacedQt1),
                % Select a queen from UnplaceQs list with
                  plusele/3 first;
not(attackcheck(Q,SafeQt)),
                % Check if it can attack with other queen to be
                  placed each other
queen(UnplacedQt1,[Q|SafeQt],Qt).
                % If they cannot to attack each other, then put
                  this queen into the SafeQt list, and begin to
                  select the next queen;
queen([],Qt,Qt).
                % When all queens are placed, send second
                   parameter to the third parameter.
... ...          % The predication attackcheck, plusele and
                  getrange are the same as above.
```

The statement above is beginning from the rightmost queen. After a line (plusele/3 is used to select the row position) is selected, then select the next queen; using the recursion method, recursion end condition is that all queens are in place; when the row in which queen to be placed in a column cannot be found, algorithm will automatically be recursive backtracking and rearrange the queen on the right.

4.2   Another Kind of Optimization Algorithm

Efficiency of optimization algorithm in item 4.1 is higher, generally sharp rise of calculation time with the increasing of queen number N does not appear; if the positions with Queen placed are stored in the program, becoming the Prolog's fact, i.e. a part of the program, when need calculating, read the relative information directly without computing again, thus the calculation efficiency can be efficiently improved to a degree. During the operation of Prolog program, change of the program itself is achieved through two inner predicates assert and retract, namely, the relevant FACT. This is basically conformable to the algorithm stated in item 4.1. The main design idea is as Fig.1:

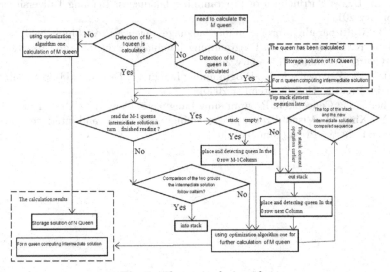

Fig. 1. The main design idea

Five queens are computed on the basis of having calculated 4 queens and saved the intermediate solution of computing 4 Queen:

Read [0,0], [0,0] [1,2]; after comparing, push [0,0] into stack; read [0,0] [1,3]; after comparing, using "optimized algorithm" stated in 4.1 to go on calculating; read [0,0][1,3][2,1]; after comparing, push [1,3] into stack; read [0,1], after comparing, go on computing [0,0][1,3][2,1] with "optimized algorithm" stated in 4.1; comparing top stack[1,3]and [0,1], get the top stack [1,3] out of the stack; continue to calculate [0,0][1,3] with "optimized algorithm" stated in 4.1 under the condition that queen can be placed after detecting [2,4]. Finally calculate with "optimized algorithm" stated in 4.1 beginning with [0,4]. At the same time, save the intermediate solution and 5 queens solution with predicate asserta.

5   Conclusion and Prospect

Use Prolog language solving eight queens problem, at the same time that improve efficiency and reduce the frequency of test, optimization Prolog recursion is fully used so that the program becomes more laconic. With the method of dynamically mending program and save intermediate solution of related calculation, operation speed of the program is increased and recursion test times are reduced. But with the increase of the number of the queen, the intermediate data required to be saved increase too, thus larger buffer memory will be also required. Unnecessary intermediate solutions in main research next will be reduced, therefore space for the program running is lessened.

References

1. Mai, Z., Lv, W.: Principles of programming languages. Beihang University press (February 2011)
2. Gu, B.: Multi-paradigm programming language. Science Press (June 2006)
3. Louden, K.C., Lambert, K.: Programming Languages Principles and Practice, 3rd edn. Course Technology (January 2011)
4. Pratt, T.W., Zelkowitz, M.V.: Programming Languages Design and Implementation, 4th edn. Prentice-Hall (September 2000)
5. Mitchell, J.C.: Concepts in Programming Languages. Cambridge Univ. Press (2003)
6. Lei, Y., Xing, Q.: Visual Prolog programming environment and interface. National Defence Industry Press (January 2004)

Enterprise Information Management Based on Cloud Platform

Mengyu Hua and Junkai Yang

International School
Beijing University of Posts and Telecommunications
West Tucheng Road. 10, 100876 Beijing, P.R. China
{dreamrain91,yangjunkai91}@gmail.com

Abstract. As a new application of Internet, cloud computing has became a dominant way in procuring services and information. Based on
the cloud computing technology, the paper introduced the concept of
cloud data management system, analysis the basic principles of BigTable,
Hbase and Sector / Sphere which are the main Internet cloud data management system at present in depth. Then propose the idea that cloud
computing can apply to the enterprise which would bring a great deal
of information in its production and management, and the in-depth application of this information by the enterprise could provide necessary
management means, and thus raise its economic benefit. By using cloud
computing technology, the design and implementation of cloud platform
for comprehensive information management of the enterprise are studied,
the physical and technical frameworks, the function design and coding
rule of the could platform analyzed, the design thinking and implementation of the cloud platform also given.

Keywords: Cloud Computing, Data Management, Framework, Comprehensive Information.

1 Introduction

In the latest 10 years, business community and academic groups tries to take
advantage of network computing and storage resources to achieve the relative
wide range of collaboration and the sharing of resources. Thus, the using of these
to achieve high efficiency and low cost computing objective have been proposed,
such as grid computing, on-demand computing, performance computing (Utility
computing), Internet Computing, software as a service, platform as a Service
and other similar cloud computing concepts and models.

Group companies are generally far more than its subsidiaries or accounting
entity, there may be investment companies, commercial companies,
production companies, research companies as its subsidiaries. The covering areas
of these subsidiaries are able to cover the real estate development, production
and sales of drugs, alcohol production and sales, agricultural crop production
and management of goods of the accounting entity. However, the business of

Z. Li et al. (Eds.): ISICA 2012, CCIS 316, pp. 57–65, 2012.

them is different, mainly for the independent business accounting, the accounting entity is mostly distributed in different regions, the Group in the human, financial and material aspects of the need to belong accounting entity effective management. Group also needs to clear their own operation and management of a comprehensive and timely way. This requires the group is able to not only find the problems in time but also analyze the management of the company from a global perspective.

In this way, the Group - owned subsidiaries or accounting units for the production and marketing should be able to manage and analysis the physical measurement, capital measured using the currency in all aspects, that is, their subsidiaries should not only accounting, but also to production, operation and management of various aspects of timely supervision, all this is in order to identify problems and timely improvements to achieve cost savings, greater efficiency purposes.

Compared to group companies, a single accounting enterprise survives in the market in early times are based on the accounting treatment system (a computerized accounting system) to form the accounting treatment on the basis, and then expand into the enterprise resource planning (ERP) system. The ERP system is made up of the accounting treatment, including supply chain management, production management and other functional modules [1], each functional module is organic associated and made up as an entity. At the same time, they have been adapted to group enterprises, centralized deployment of ERP systems. However, for the subsidiary business is different, the Group is involved in the kind of information that is more like performance evaluation, process monitoring, the bulk of centralized procurement of raw materials, office automation, accounting treatment, the portal network management, marketing management, decision analysis, budget management, management accounting, also banking, taxation, finance and other departments, information exchange, the effect of this centralized deployment of the ERP system is not ideal. Reference [2] gives a general business use of cloud computing technology to implement an information management strategy. The paper researches on a group of companies, taking into account the requirements of an integrated information management, as well as to avoid the subsidiary repeat investment, take advantage of cloud computing technology, construction, a subsidiary of the Group headquarters applications integrated information management cloud platform design and implementation.

2   Cloud Computing

Cloud computing is a new series of concept of shared information technology infrastructuremethods and composition in recent years. Users can obtain what they need at any time through the network, it is on demand, can extend and pay according to the use of resources, including hardware, software, network bandwidth, services, etc. Cloud computing generally consists of three levels of service: Infrastructure as a Service (IAAS), platform as a service (PAAS) and software as a service (SAAS). [3] Software as a service mainly refers to provide software and service model through the network, application software and functional modules unified deployment in server based on cloud platform. Users are able to order application software and services according to their own work from the cloud platform. What is more, users no longer need to build their own software system, also the need for software maintenance, all this will be managed with full authority by a professional team. In the meanwhile, cloud platform also provides off-line operation of the software and local data storage, allowing users to use customized software and services at any time[4].

The cloud platform infrastructure as a service(IAAS), including servers, storage devices, network virtualization software, database software, operating systems, cloud computing software, etc.

3   Cloud Platform Framework

3.1   Cloud Platform Technology Framework

According to cloud computing system framework [5], cloud security framework system, the information system security risk analysis [6], the cloud platform security analysis [7], the enterprise information management cloud platform based on cloud computing framework system, cloud security framework system, information system security and cloud platform security design, these will be made up of application (Software as a Service), platform (platform as a service), infrastructure services (infrastructure services), as well as cloud security, cloud management, etc. It is shown in Fig. 1.

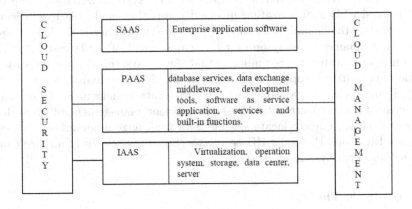

Fig. 1. Cloud platform technology framework

IAAS: Comprehensive virtualization of server, storage and network resources, then aggregate these resources and accurately distribute to application on demand according to the priority, which is the support of enterprises and subsidiaries to use cloud computing technology.

PAAS: platform services can be seen as two aspects, on the one hand, to provide the necessary infrastructure support for the user, on the other hand, to put the software system used by headquartered on the platform, such as information sharing, decision analysis, comprehensive inquiry and comprehensive analysis, etc.

SAAS: Application services is the service provided to the subsidiary or accounting entity, it mainly related to the subsidiary or the accounting entity in terms of content, such as account processing , inventory management, order management, etc.

3.2   Physical Structure of Cloud Platform

The physical structure of the cloud platform is centralized layout manner, using virtualization technology to realize cloud computing functionality. In the aspect of storage, using storage area network frameworks (SAN), using high-performance server or minicomputer as a data server, mid-range servers as application servers, the number of SAN, database servers and application servers will determined in terms of the amount of enterprise information, using virtual storage virtualization technology, database servers, application servers, use the vCenter and vCloud server to manage the whole virtualization platform.

4   The Basic Principle of the Cloud Data Management System(CDMS)

Although GFS, HDFS, S3 and other distributed file systems well solve the problem of massive data organization in cloud computing, and able to efficiently read and write the "cloud" huge amounts of data, it still needs the help of specialized data management system for the management of structured data. The relationship between them is similar to that the file system which is responsible for file organization in operating system, or the database management system (DBMS)[8] which is responsible for structured data management. Cloud data management must effectively address the efficient management of large data sets in the cloud; quickly locate specific data from huge amounts of data, etc. Google's BigTable, Hadoop's HBase, Sector/Sphere is relatively mature cloud data management system at present.

4.1   BigTable Theory

BigTable improves the data read efficiency by optimizing the R/W operation of the data and using column storage. The basic elements of BigTable include the row, column families and Timestamps. Among them, the line keywords can be arbitrary strings, in a row keyword, each read and write operations are atomic operations (regardless of how many different columns are), so that when the concurrent operation occur in the same line, it is easier for users to understand and control the behavior of the system. The column family is organized by a group of the same type of column keyword, which is the basic unit of access control. We should first create the column family, and then store the data in the column keyword, when a column family is created, any column keyword in this family can be used.

In the sample, one web page is shown in Figure 2 [9] , the line name is a reversed URL (namely com.cnn.www), column family "contents" is used to store web content, column family "anchor" is used to store anchor link text which is referenced to the page. Here, the Sports Illustrator " (CNN's sports programs) and "MY-look" take use of the home page, the line contains the name "anchor: cnnsi.com" and "anchor: my.look.ca" column.

The content of the BigTable is divided by row, composed by multiple lines of a Tablet, then we are able to save it into a tablet server node. In the physical layer, data storage format is SSTable, each SSTable contains a series of data block in size 64 KB (enabled to be configured). Figure 3 shows the framework of BigTable.

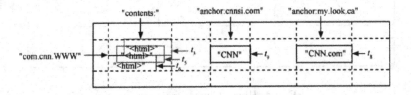

Fig. 2. Web page stored sample for BigTable

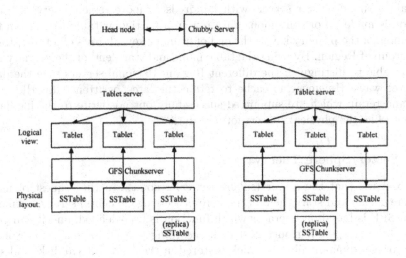

Fig. 3. Framework for BigTable

4.2   HBase Theory

HBase[10] is a subproject of Hadoop[11], which is one of the mature cloud data management as well as open source solutions. HBase's data model is very similar to the Bigtable. Users are able to store data row in a labeled table, one data line have a sort of primary key or sortable key and any number of columns.

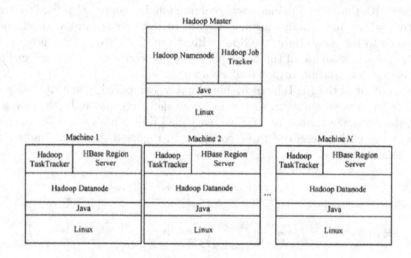

Fig. 4. Hbase Cluster framework

Fig. 4 shows a simple master-slave server framework, each Hbase clusters contain a single master server, with hundreds or more regional servers. Each Region is made up of continuous data row in a particular table rows, from the beginning of the primary key to the end of primary key, all rows of a table stored in a group of Region. By using the table name and start/end of the primary key we are able to distinguish the different Region. Regional servers save the data in three ways: Hmemcache: cache to retain the latest written data; Hlog: log files and retain which are submitted successfully but not write to the file data; Hstores: file, the physical storage form of data.

4.3   Sector/Sphere Theory

Fig. 5 shows that the server (sphere server) response to the request of users (sphere client) and start the SPE (sphere processing elements) services. Among them, SPE is based on operator which function is user-defined, and it can generate the corresponding output stream according to the input stream. Sphere operator is a dynamic library which is stored in the server's local disk, and sector server is responsible for the management it. Data segment in sphere can be one or a set of data records and can also be a file.

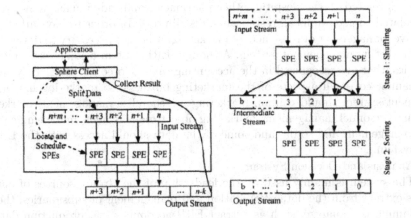

Fig. 5. Sphere Data management processes 5 Platform

5 Platform Functionality

Group integrated information management cloud platform is designed to meet the business needs of Group and affiliated subsidiaries, in can implemented through software design, and these software designed based on the software, principle, service idea of cloud computing, software will developed according to users own need. All the features of the entire platform shall be a series of independent, network services which is up and down independent, and follow a common service specification, data exchange norms and operation norms.

Mainly functions of cloud platform:

1. ERP System
   ERP system is the foundation of enterprise information; it should generally be included accounting arrangement, supply chain management, human resources management and so on, which is one of the most frequently used business information systems by Group's subsidiaries or accounting entities. It is evolved from early accounting arrangement, some of the existing products are basically covering all aspects of enterprise resource information, but these products are mostly the standard edition, that is to say, it is designed for the general public enterprises and enterprise should be cooperate it with its own cloud platform. It is necessary to carry out the secondary development of ERP, this will according to the needs of enterprises and completed by the ERP vendors.

2. Marketing Management System
   Group companies, both production companies, also commercial companies, the product of production companies and commercial companies there may diverse, such as a group of enterprises under the jurisdiction of a number of pharmaceutical companies, pharmacies, liquor manufacturing and sales, the drug or the sale of alcoholic involved in a very wide range, such as drug

sales related to the hospital, other pharmacies and individuals. Wine sales related to geographical, restaurants, distributors. In order to have an effective management of these acts of marketing, it is very necessary to develop a marketing management system. Although, ERP system also has sales modules, that are recognized in the accounting sales. Moreover, an ERP system cannot cover all aspects of the marketing business, so marketing management system should contain marketing merchandise management, marketing personnel management, marketing object management, marketing costs management and so on, and some of the data should access with the ERP system.

3. Analysis and Decision System

The system is mainly used by the leadership of the Group, sources of data is extract from the data of operation and management of subsidiaries, then summarize, analyze, such as extract ERP accounts data, production data, human resources management data, then form the decision-making information, and can be illustrated using intuitive graphic. The function of this system should include centralized purchasing of bulk raw materials, financial analysis, marketing analysis, integrated query, time, quality, cost comparison analysis, there should also be some prediction function.

Other functions will not list in this paper.

6   Encoding Rules and the Network Set Up

1. Encoding Rules

Group companies related to a wide variety of materials and other resources, such as production, sales, and procurement. In order to unified manage and use, it is necessary to coding these items uniformly. If the subsidiary has its own coding, it should be uniformly coding, the rules must be followed GB, department of superscript, the provincial standard, as well as industry standards when building cloud platform, among these standards, enterprise can add their own coding as well as other aspects of coding, which is the basic work of cloud platform.

2. Network Set Up

According to Group enterprise distribution features, the network form can use virtual private network (VPN) to set up which has a higher network security.

7   Conclusion

With the development of enterprises, there will be large amounts of information produced in the production process and effectively using this information in-depth can provide the essential means to improve economic efficiency. This paper discusses the design, implementation also the principle of the Group integrated information management cloud platform, given the specific design ideas and methods from platform technology framework, physical framework, functional design, and coding rules for the enterprise group to build a comprehensive information management cloud platform. Also mention the basic principle in designing and implementing cloud platform.

References

1. Ghemawat, S., Gobioff, H., Leung, S.-T.: The Google file system. In: Proc. of the 19th ACM SOSP, New York (2003)
2. Lammel, R.: Google's MapReduce programming model Revisited (2007), http://www.cs.vu.nl/ralf/MapReduce/paper.pdf
3. Sims, K.: IBM introduces ready-to-use cloud computing collaboration services get clients started with cloud computing (2009), http://www-03.ibm.com/press/us/en/pressrelease/22613.wss
4. IBM. IBM virtualization (2009), http://www-03.ibm.com/systems/virtualization/
5. Hbase Development Team. Hbase: bigtable-like structured storage for hadoop hdfs (2009), http://wiki.apache.org/hadoop/Hbase
6. Zhang, Y.X., Zhou, Y.Z.: 4VP+: a novel meta OS approach for streaming programs in ubiquitous computing. In: Proc. of IEEE the 21st Int'l Conf. on Advanced Information Networking and Applications (AINA 2007), Los Alamitos (2007)
7. Foster, I., Yong, Z., Raicu, I., et al.: Cloud computing and grid computing 360-degree compared. In: Grid Computing Environments Workshop, GCE 2008 (2008)
8. Tim Jones, M.: Cloud computing with Linux (2009), http://download.boulder.ibm.com/ibmdl/pub/software/dw/linux/1-cloud-computing/1-cloud-computing-pdf.pdf
9. Rajkumar, B., Yeo, C.S., Venugopal, S.: Marketoriented cloud computing:vision, hype, and reality for delivering IT services as computing utilities. In: Proc. of the 10th IEEE International Conference on High Performance Computing and Communications (2008)
10. Luis, M.V., Rodero-Merino, L., Caceres, J., Lindner, M.: A break in the clouds: toward a cloud definition. ACM SIGCOMM Computer Communication Review 39(1), 50–55 (2009)
11. Merkle, R.: Protocols for public key cryptosystems. In: Proceedings of IEEE Symposium on Security and Privacy 1980, pp. 122–133. IEEE Computer Society, Oakland (1980)

uDisC: An Ultra-Lightweight and Distributed Scheme for Defending against Data Loss Attack in RFID Networks

Zebo Feng[1,2], Xiaoping Wu[2], Liangli Ma[3], and Wei Ren[4]

[1] Department of Information Security
Naval University of Engineering
Wuhan 430033, P.R. China
[2] Department of Military Affairs
Navy Headquarters
Beijing 100841, P.R. China
[3] Department of Computer Engineering
Naval University of Engineering
Wuhan 430033, P.R. China
[4] School of Computer Science
China University of Geosciences
Wuhan 430074, P.R. China

Abstract. RFID network is applied as a key networking architecture in Internet of Things. It can be largely deployed for object localization and tracing. As the number of RFID tags may be large or the speed of tags may be fast, the reader may misreading some tags. Thus, the data consistence is damaged in central RFID database. Currently, many solutions are proposed for this problem. However, few of them paid much efforts in computation overhead and response delay. Especially, most of them are after-event response, but not in-networking process. In this paper, we make the first attempt to solve the problem on-site. We propose several algorithms for check the missing of tags in real-time. We also propose an adaptive random checking method, called uDisC. Our solution is ultra-lightweight and with the delay bounded by a threshold value that can be manually tuned.

Keywords: RFID, Data Cleaning, Ultra-lightweight, Distributed Intelligence, Internet of Things.

1 Introduction

RFID network is a typical enabling technology for Internet of Things [6,5]. It has been envisioned as a key method to recognize, locate, and trace physical objects, which can be applied in numerous applications such as supply chain management, equipment tracing, and so on.

RFID network usually has three tiers: tags attached to interested objects; network of readers to collect the object information; backend supporting system

Z. Li et al. (Eds.): ISICA 2012, CCIS 316, pp. 66–73, 2012.

to process the collected information. As the number of objects is large, and the readers always have to receive multiple objects simultaneously, certain tags on objects may not be able to read by readers. It is called misreading problem in RFID networks. Although it can be easily avoided by several methods such as adding more readers in one location, or slowing down the speed of reader, or decreasing the number of objects at one-time reading, the performance is inevitably damaged.

To deal with the misreading problem, various solutions are proposed in literature [3,1,7,2,4]. However, those solutions may have some limitations: 1) The solution heavily relies on the global view of the collected information from RFID networks. Thus, the solution is always conducted at central server side. 2) The solution induces a large volume of computational overhead, and thus computation delay is large.

Moreover, currently the data misreading problem is addressed in a un-malicious context. That is, the data misreading is assumed as a system reliability issue (for random data loss), not a system security issue (for intentional data loss). Thus, current solutions may not be able to tackle the malicious attacks intended to data loss or targeting for RFID misreading.

In this paper, we propose a lightweight and distributed data cleaning method. Comparing with the former methods, our scheme can conduct the data cleaning distributively and within much less delay.

The contributions of the paper are listed as follows:

1) we propose an ultra-lightweight and distributed scheme - uDisC, which has much less computation overhead and much less response delay.

2) we point out the data loss attack in malicious context, and our scheme can tackle such kind of attack.

The rest of the paper is organized as follows. In Section 2 we discuss the basic assumption and models used throughout the paper. Section 3 provides the detailed description of our proposed models and analysis. Section 4 gives an overview on relevant prior work. Finally, Section 5 concludes the paper.

2   Problem Formulation

2.1   Network Model

The basic architecture of RFID network has three folds:

1) The tag is attached on the mobile objects.

2) The reader can read the tag data upon the tag moves into the reading scope.

3) The reading data are uploaded to a central server for further processing.

Note that, we hereby propose a connection network between readers. That is, readers construct a network that can share instant reading information. In contrast, the connection in traditional RFID networks is mainly the links between readers and the central server.

The advantages of sharing information between readers can facilitate the on-site processing of reading information. We will propose the details later.

2.2   Attack Model

The previous works concentrate on the un-malicious context, thus attack model is a normal random failure of reading or random loss of reading. Both will result in the misreading problem. In this paper, we consider much stronger adversaries. We assume the existence of malicious context and adversaries intend to make the data loss.

Definition 1. *Misreading Problem. The misreading problem is a reader has not read the tag that in the scope of its reading area.*

In un-malicious context, the misreading problem may be induced by fast mobility of tags or multiple reading of tags simultaneously. In contrast, in malicious context, the misreading problem may be invoked by adversaries' intentionally hiding of tags, for example, shield the tag's signal to remove the trace in database. We formally state such attack as follows:

Definition 2. *Data Loss Attack (DLA). It is an attack to make the misreading at RFID reader intentionally. It can hide the trace of objects although RFID tags are attached.*

2.3   Security Definition and Design Goal

The security for defending against DLA is a reader network can avoid misreading on-site (or in-networking), thus the detection delay is very short and response is in real-time, not after-events.

Therefore, the design goal is to propose a scheme defending against DLA with ultra-lightweight computation overhead, and on-site response within short delay.

3   Proposed Schemes - uDisC

We list certain major notations used in the remainder of the paper in Table 1.

Table 1. Notation

$\mathcal{R}$	reader
*	all neighbors
PS	Pending State
RS	Relief State
WS	Waiting State
NN	Notifying Neighbors
RA	Receive Acknowledgement
RN	Receive Notification
SA	Sending Acknowledgement

3.1   A Basic General Model

Firstly, we propose a basic general model as follows:

1) There exists n readers, denoted as $R_1, R_2, ..., R_n$. The readers have fixed locations. The path for mobile objects between different locations are fixed. Thus, the structures between readers are fixed. For example, visitors walk into a museum. She is attached a tag. The doors are equipped with a reader to monitor the movement of visitors. The paths are thus the routes between different doors.

2) The paths are called links between readers. The links define the neighbors of a given reader. Suppose $N(r)$ is a function that returns the neighbors of the reader r. Here, neighbors are all readers that may read the tag next.

3) Suppose r reads the tag t at time T_1. r will notify all neighbors at time $T_1 + u$, where u is a unavoidable communication delay. If any reader $r' \in N(r)$ reads t later, r' will notify r an acknowledgement. If all reader in $N(r)$ does not acknowledge to r in a threshold time T_{th}, the alert will be generated at r, for example, notify to the central server.

4) After the acknowledgement of r', r' will notify all its neighbors iteratively, similar to what r has done.

3.2   State Transition Diagram

1) State of readers.

The states of readers has three: i) Pending State (PS). After a reader notifies its neighbors, it goes into PS. ii) Relief State (RS). After a reader is acknowledged, it goes into RS. iii) Waiting State (WS). After a reader is notified by another reader, it goes into WS.

2) Events of Readers.

The events of readers has three: i) Notifying Neighbors (NN). ii) Receive Acknowledgement (RA). iii) Receive Notification (RN). iv) Sending Acknowledgement (SA).

3) State Transition Functions.

The state transition function is defined as $s \times e \to s$, where s is a state, e is an event.

$$RS \times NN \to PS,$$

$$PS \times RA \to RS,$$

$$RS \times RN \to WS,$$

$$WS \times SA \to RS.$$

The state is tag-wise. That is, each state is only for one tag.

Some readers at WS state may invoke NN event for other tags. That is,

$$WS \times RN \to WS.$$

Discussions

1) The connections between readers are wired and preloaded, as the paths are fixed.

2) The NN event can be avoided to save traffics. If so, in SA event, acknowledgement message should be sent to all neighbors. Thus, the notification function can be conducted by acknowledgement message. That is, acknowledgement message can be avoided.

3) For each tag, the reader will send one broadcasting notification message and one acknowledgement. If using the previous way, each tag will revoke one broadcasting message.

3.3   Optimal In-Networking Detection Algorithms

We next describe the details in our scheme in the following.

1) One broadcasting for each tag.

Once reading of a tag, the reader broadcasts tag information to its neighbors. That is,

$$\mathcal{R} \to * : \{time, rid, tid\},$$

where $\mathcal{R}$ is a reader; $time$ is current time; rid is the identity of the reader; tid is the identity code of the tag.

The reader will store $< time, rid, tid, state = PS >$ in local database.

If a reader receives a broadcasting message, it will check in the local database whether tag is already stored. If it is stored, it update the record to $< time, rid, tid, state = RS >$.

The reader will check the table for every period time (threshold period). If the threshold period later the $state$ is still PS, the reader will alert.

2) Random Check for tags.

Although the network between readers are wired, the throughput would not be a concern. To further decrease the traffics in the network, the random check can be conducted.

Suppose the probability for random check is p. That is, for each $\lfloor 1/p \rfloor$ tags, the broadcasting will be sent. The traffics are throttled into p as of before. The confidence of the security on DLA becomes p.

3) Adaptive Random Check.

If the number of tags sensed in the area is α, the probability of random check is β. Note that, $\beta = f(\alpha)$. That is, β is a function of α. Thus, the checking traffics can be adaptively throttled.

The adaption strategy depends on the context, and can be selected or tuned on-line. For example, $\beta = k * \alpha$, or $\beta = k/\alpha$.

According to above discussion, we propose several algorithms as follows:

Algorithm 1. Notify()

Require $time, rid$
Ensure $Table$ % $\mathsf{Prvc}_{ABK} = 1, \mathsf{Soundness}_{DATA_u} = 1$.
$tid \Leftarrow ReadTag()$
$Broadcast(time, rid, tid)$
$state \Leftarrow PS$
$Table \Leftarrow < time, rid, tid, state >$

Algorithm 2. ACK()

Require $Null$
Ensure $state$ % $\mathsf{Prvc}_{ABK} = 1, \mathsf{Soundness}_{DATA_u} = 1$.
$tid \Leftarrow ReceiveBroadcast()$
$FIND \Leftarrow Search(tid, Table)$
$state \Leftarrow SearchState(tid, Table)$
if $(FIND .AND. (state == PS))$ then
    $state \Leftarrow RS$
    $Table \Leftarrow < time, rid, tid, state >$
end if

Algorithm 3. Check()

Require $Table$
Ensure $Alert$ % $\mathsf{Prvc}_{ABK} = 1, \mathsf{Soundness}_{DATA_u} = 1$.
$While\{1\}$
{
$time \Leftarrow search(PS, Table)$
if $(time > T_{th})$ then
    $Message("Alert.")$
end if
Wait a period.
}

Algorithm 4. RandomCheck()

Require $time, rid, p$
Ensure $Table$ % $\mathsf{Prvc}_{ABK} = 1, \mathsf{Soundness}_{DATA_u} = 1$.
$tid \Leftarrow ReadTag()$
$prob \Leftarrow random()\%100/100$
if $(prob < p)$ then

    $Broadcast(time, rid, tid)$
    $state \Leftarrow PS$
    $Table \Leftarrow < time, rid, tid, state >$
end if

Algorithm 5. AdaptiveRandomCheck()

```
Require time, rid, α, β
Ensure Table % Prvc_{ABK} = 1, Soundness_{DATA_u} = 1.
timestart ⇐ time()
While{time() − timestart < gap}
{
tid ⇐ ReadTag()
prob ⇐ random()%100/100
if (prob < β) then
    Broadcast(time, rid, tid)
    state ⇐ PS
    Table ⇐< time, rid, tid, state >
end if
count + +
if (count > α) then
    β ⇐ f(α)
    timestart ⇐ time()
    count ⇐ 0
end if
}
```

Analysis

Proposition 1. *The security of uDisC defending against DLA can be guaranteed.*

Proof. As the moving object has to pass one of the current reader's neighbors, the event that the tag is read by a reader among them must happen. Otherwise, the DLA happens.

Claim. Scheme uDisC is ultra-lightweight.

Proof. The induced overhead is traffics between readers. It is a wired network, thus the traffics overhead can be ignored. The probably random check can further mitigate the traffics to p percentage of before.

Proposition 2. *uDisC has T_{th} delay with on-site response.*

Proof. uDisC has the parameterized delay T_{th}, which is an on-site response, not an after-event.

4   Related Work

Massawe, L.V. et al. [3] presented a new adaptive data cleaning scheme called WSTD based on some of the concepts proposed in SMURF but with an improved transition detection mechanism. WSTD compares two window subrange observations or estimated tag counts to detect when transitions occur within a window. Gonzalez, H. et al. [1] proposed a cleaning framework that takes an RFID data set and a collection of cleaning methods, with associated costs, and induces a cleaning plan that optimizes the overall accuracy-adjusted cleaning costs by determining the conditions on prices. Ku, W. et al. [2] designed an n-state detection model and formally prove that their 3-state model can maximize the system performance. They also extended the n-state model to support 2D RFID reader arrays.

5 Conclusions

Different from all current works, we addressed the problem from the viewpoint of in-networking processes. We proposed an ultra-lightweight scheme uDisC to tackle the misreading problem on-site. Several algorithms are proposed for the off-the-shelf implementation. The security and performance of the scheme are analyzed, which justifies the applicability of the scheme.

Acknowledgement. This research was financially supported by National Natural Science Foundation of China (No.61170217) and Fundamental Research Funds for the Central Universities, China University of Geosciences (Wuhan)(No.110109, No.090109).

References

1. Gonzalez, H., Han, J., Shen, X.: Cost-conscious cleaning of massive rfid data sets. In: IEEE 23rd International Conference on Data Engineering, ICDE 2007, pp. 1268–1272 (2007)
2. Ku, W., Chen, H., Wang, H., Sun, M.: A bayesian inference-based framework for rfid data cleansing. IEEE Transactions on Knowledge and Data Engineering (99), 1 (2012)
3. Massawe, L.V., Vermaak, H., Kinyua, J.D.M.: An adaptive data cleaning scheme for reducing false negative reads in rfid data streams. In: 2012 IEEE International Conference on RFID (RFID), pp. 157–164 (April 2012)
4. Nie, Y., Cocci, R., Cao, Z., Diao, Y., Shenoy, P.: Spire: Efficient data inference and compression over rfid streams. IEEE Transactions on Knowledge and Data Engineering 24(1), 141–155 (2012)
5. Tran, T., Sutton, C., Cocci, R., Nie, Y., Diao, Y., Shenoy, P.: Probabilistic inference over rfid streams in mobile environments. In: IEEE 25th International Conference on Data Engineering, ICDE 2009, April 2, pp. 1096–1107 (2009)
6. Vahdati, F., Javidan, R., Farrahi, A.: A new method for data redundancy reduction in rfid middleware. In: 2010 5th International Symposium on Telecommunications (IST), pp. 175–180 (December 2010)
7. Ziekow, H., Ivantysynova, L.: A probabilistic approach for cleaning rfid data. In: IEEE 24th International Conference on Data Engineering Workshop, ICDEW 2008, pp. 106–107 (2008)

A Motion Planning Framework for Simulating Virtual Crowds*

Muzhou Xiong[1], Yunliang Chen[1], Hao Wang[2], and Min Hu[3]

[1] School of Computer Science
China University of Geosciences
Wuhan 430074, P.R. China
mzxiong@gmail.com
[2] School of Material Science and Chemistry
China University of Geosciences,
Wuhan 430074, P.R. China
[3] Chinese PLA Defense Information Academy

Abstract. As crowd simulation technology has becoming an emerging tool for mining and analyzing human crowd movement pattern since last decade, the problem of how to simulate the motion of pedestrian during her/his movement around the simulation environment then becomes one of the core problem for crowd simulation. This paper proposes a motion planning framework for agent-based crowd simulation model. The main idea of motion planning is to direct and simulate the process of agent's selecting velocity to achieve a given desired goal. The framework includes modules of collision detecting and collision avoidance. If agent does not detect any on-coming collision, it moves towards the goal by a so-called *preferred velocity*, based on the given desired goal. However, once a collision is detected, agent needs to make a deviation from its current desired velocity to conduct a collision-free motion. Experiment results from case study shows the proposed framework is able to simulate agent's movement for dynamic environment.

Keywords: Motion Planning, Collision Avoidance, Collision Response, Agent-based Simulation, Crowd Simulation.

1 Introduction

Recently, crowd simulation [1] has become an efficient tool for psychologists, sociologists, computer scientist, and even for architecture designers and urban planners to study the crowd behavior and its movement under diverse what-if scenarios. It has widely applied to transportation systems, sporting and general spectator evacuation, and fire escapes. The existing behavior models for

---

* This study was supported in part by National Natural Science Foundation of China (grant No.61103145), and the Fundamental Research Funds for the Central Universities, (China University of Geosciences (Wuhan), No.CUG100314 and No.CUG120409).

Z. Li et al. (Eds.): ISICA 2012, CCIS 316, pp. 74–83, 2012.
© Springer-Verlag Berlin Heidelberg 2012

crowd simulation can be roughly divided into two categories: *microscopic* and *macroscopic* approach.

As part of a crowd simulation system, motion planning typically decides how an agent chooses velocity by which it avoids the potential collisions and achieves the desired path, and how an agent responds to the collision once it really happens. However, path-planning, as the result of the decision making part, refers to the planning of where the agent desires to get, which is beyond the scope of this paper. In each simulation step, a preferred direction will be generated from the process of path-planning. The agent will move along the direction until the end of the simulation step, if no potential collision is detected; otherwise the velocity will be adjusted based on the predefined trade-off between the preferred direction and the collision-free directions. However, the potential collision will not always be avoided, and in some cases no collision free path exists. The probable reason of this could be: 1) one or both of the two pedestrians involved in the collision do not detect the collision; 2) the velocity adjustment is not enough to avoid it; and 3) one of the two pedestrians is aggressive and expects the other one to respond to the coming collision such that the velocity adjustment from the other pedestrian is not enough to avoid the collision. When collision does occurs, collision response is needed for both of the agents in the subsequent simulation step(s). The process of response is dependent on the type of collision: a slight collision may allow the agents to continue their movement only with a subtle change of original velocity; a heavy collision, in the contrary, will make them stop. After the agent recovers from the collision, it continues to move along its desired trajectory and avoid the potential collisions.

There is a large amount of existing work on collision avoidance for motion planning of multi-agent systems. Some of them [2,3] are aimed at controlling the motion of robotics in dynamic environments, but without considering the motion of the other agent involved in the possible collision. Similar to those methods for the motion planning of robotics, literatures [4,5] focused on the motion of multi-agent also regards the other agents as passively moving obstacle. If both of the agents use such methods, oscillation may happen, as described in [6] where *Velocity Obstacle* (VO) [7] is used.

Most of existing work is towards building a collision avoidance model for a collision-free motion planning. The objective of this paper aims to propose a framework for motion planning so that different collision avoidance can run on the framework for comparison and evaluation.

The rest of the paper is organized as follows. A brief overview of prior work is given in Section 2. The overview of the motion planning framework is described in Section 3. After this, detials of the framework design is introduced in Section 4 and 5, for collision detection and collision avoidance, respectively. Experiment results is reported and analyzed in Section 6. The paper is concluded in Section 7.

2   Related Work

The motion planning is originally from the research field of robotics, where it used for controlling robots in the environments, and avoiding the potential collision amongst other robots or moving obstacle, in the mean time. However, most of the existing work [8,3] do not take into account the fact that the other robot may be affected by the presence and the possible motion of this robot. When the similar method is applied to crowd simulation for collision avoidance [9], the trajectory from the simulation result may be deviated from that of the real pedestrian. A typical example is that oscillation happens when agent avoids the collision with the other one [6] .

Another type of motion planning mechanism for crowd simulation is based on social norms. These methods aim to unveil the principles inside the crowd movement and mimic the behavior of real human in dynamic environments, in which some social norms and psychological factors are combined. It can make the result more realistic to those observed in real life. Generally, there are two kinds of implementation to this: one, like the method in [10], concentrates the principles and features of crowd movement in a macroscopic level, losing the personality of each pedestrian; the other one, like the method proposed by [11], focuses on modeling the individual behaviors at the microscopic level, lacking the consideration of the movement of the whole crowd.

3   Overview of Motion Planning Model

As described above, the motion planning modelling for crowd simulation is to model and direct agent to select velocity in each simulation step to move towards its given destination. When the destination is given, a preferred moving direction can be thereupon determined (i.e., a typical preferred moving direction is moving from the current location to the destination along the shortest path). During its movement around the simulation environment, it may encounter potential collision with other agents or obstacle, collision avoidance model is then needed for simulating agent's movement. Since agent may collide with other agent, the process of collision response is also needed to be considered. Hence, the whole model of motion planning in this paper is composed by two parts: collision avoidance and collision response.

At the beginning of each simulation step, the destination for the simulation time step should be given from some higher level model (for example, a behavior model). Based on the destination, agent's preferred moving direction is also determined. Some environment information, like agents and obstacles around should also be sensed based on the current location of the agent. With this input information, the model helps agent to detect any potential collisions. If there is no such potential collision, the agent will choose preferred velocity (preferred direction as well as a given preferred speed) to move in the current simulation time step. Otherwise, agent should deviate its preferred velocity and choose a proper velocity in order to avoid the on-coming collision. Using this velocity, if the agent really collides with another agent or obstacle, the motion planning model also needs to treat how agent recover from the collision, this refers to the collision response part. The whole process of the motion planning model is illustrated in Figure 1. Detailed process of collision avoidance and response is introduced in the following two sections, respectively.

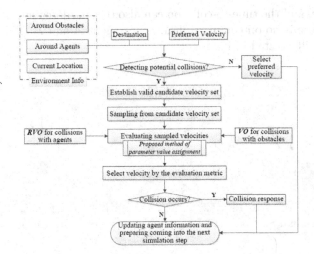

Fig. 1. Overview of the Adaptive Motion Planning Model

4   Coolision Detection

In the following discussion, the shape of agent is assumed as a circle with a specified centre and radius. A collision of agent with another one is defined as the two circles intersects. Similarly, an agent collide with obstacle can be described as the circle for the agent intersects with the obstacle.

The problem of collision detection can be converted into whether moving circles intersect or not. Taking detection of agent colliding with another one as example, the collision detection can be executed as the following steps (illustrated in Figure 2). The initial status (location,shape and location) of the two agents are shown in Figure 2a. Based on these information, the model can detect if there exists a potential collision.

1. As illustrated in Figure 2b, the circle of agent A can be overlaid on agent B, i.e., agent A can be reduced to a point and the circle radius of circle come into $r_A + r_B$; the relative velocity of A respective to B can also be calculated by $\mathbf{v}_A - \mathbf{v}_B$. In this case, agent B can be considered as a static obstacle with radius $r_A + r_B$, and agent A is a moving point.
2. In order to detect the potential collision, it is then needed to figure out the intersection point of the new circle of agent B and the ray for the current location of agent A with the direction of the relative velocity $\mathbf{v}_A - \mathbf{v}_B$. If there exists a intersection point, the distance between A's current location and the intersection point can be calculated. The *time-to-collision*, which means how soon agent A will collide with agent B, can be also figured out (if there exists two intersection points, select the lower one as the time-to-collision). Otherwise, if there is such point, the distance can be considered

as infinite, and the time-to-collision can also be considered as infinite, which means there is no on-coming collision with agent A and B. This modelling process is illustrated in Figure 2c.

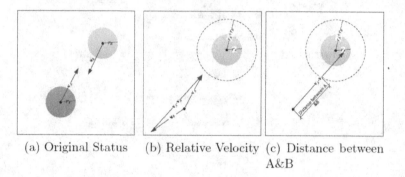

(a) Original Status   (b) Relative Velocity   (c) Distance between
                                                   A&B

Fig. 2. Process of Calculating Time to collision

5   Collision Avoidance

The process of collision avoidance could be divided into four stages:

1. detecting whether there exists an on-coming potential collision with other agent or obstacles; if there is no such collision, just choose the preferred velocity and complete the collision avoidance process; otherwise, go to step 2);
2. establishing the validate set for all candidate velocities based on kinematic and dynamic principles;
3. sampling velocities from the validate velocity;
4. evaluating each sampled velocity by the established evaluation method and choosing the best one as the moving velocity for the current simulation time step.

5.1   Validate Velocity Set and Velocity Sampling

If an agent detects an on-coming collision and decides to avoid the collision, it will deviate from its preferred velocity and choose another to achieve it. A validate velocity set should be first established as the candidate velocities. The validate velocity set is built based on the agent's kinematic and dynamic principles. The validate velocity set is based on agent's maximum speed and maximum acceleration. In Figure 3 the white circle represents all the velocities with the speed less than the maximum speed. The black circle is the velocity set for is all the velocities to which can be accelerated or decelerated from the current selected velocity. The intersection part of the two set (presented in gray color) is then the validate velocity set, in which each velocity can been considered as candidate velocity. However, it is not possible to evaluate every velocity belongs to the validate velocity set, for that the number of velocity is infinite. Hence, typical finite velocities is needed to sample from the set. Thus follows the velocity sampling process. Detailed selection of maximum speed and acceleration will be discussed later.

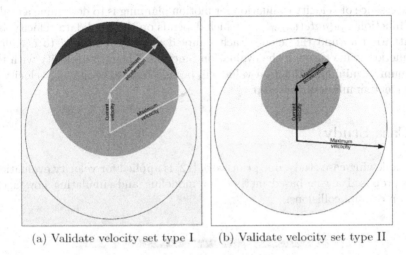

(a) Validate velocity set type I   (b) Validate velocity set type II

Fig. 3. Validate Velocity Set Determination

The main idea of sampling velocity from the validate velocity set is to choose typical velocity to represent all the velocities belonged to the set. Regarding this, the sampling velocity should be uniformly distributed inside the intersection of the two circles (i.e., the validate velocity set). The shape of the validate velocity set falls into two types: 1) the shape illustrated in Figure 3a, in which some velocity satisfied the acceleration requirement is outside the maximum speed set; and 2) the shape of validate set is whole velocity set satisfying the requirements of acceleration (shown in Figure 3b).

For both type of interaction part, a uniformly distributed function is built by Equation 1).

$$P(x, y) = uniformPoint(r_{speed}, r_{acc}, centre_{speed}, centre_{acc}) \qquad (1)$$

where $P(x, y)$ is a point inside the intersection part, and x and y represents the two velocity components along x and y coordinate, respectively; r_{speed} and r_{acc} represents the maximum speed and value of acceleration holding by the agent. $centre_{speed}$ is the center point of the velocity set for those satisfying the maximum speed requirement; accordingly, $center_{acc}$ is the center point of the velocity set for those satisfying the maximum value of acceleration requirement. With a given number of velocity sampling and executing the equation with the same number of time, the results (i.e., a serial of points), uniformly distributed inside the intersection part.

5.2   Velocity Evaluation

Existing motion planning work aims to provide a velocity evaluation for free-collision motion planning. For a general framework of motion planning, it does not need to propose a concert mechanism for the evaluation process, but only a abstract process is necessary.

The essence of velocity evaluation for motion planning is to determine a evaluation function $evaluate(v_{candiate})$, which depends on the candidate velocity and the situation around the agent. Each sampled velocity is needed to calculate the function value through the evaluation function. Then, the velocity with the maximum or minimum function value will be selected as the moving velocity in the current simulation time step.

6   Case Study

In the following case study, our prior work [12] is applied for velocity evaluation, which proposed a rule-based method for modeling and simulating how agent avoid on-coming collisions.

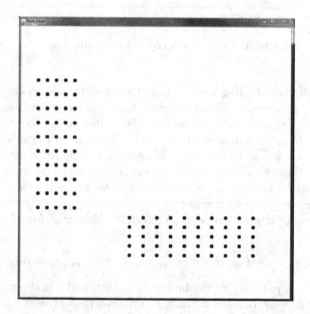

Fig. 4. Agents Initial Positions for Simulation Scenario

The simulation scenario is depicted as follows. There are two groups of agents initially located at the left and bottom side of the environment. For agents located at the left side, their destinations are the right side of the environment. Agents located at the bottom of the environment desire to achieve the positions at the top of the environment. The initial and final positions are shown in Figure 4 and 5, respectively. During the simulation, the desired goal of each agent is set its corresponding destination. Hence, the preferred direction of each agent is along the direction

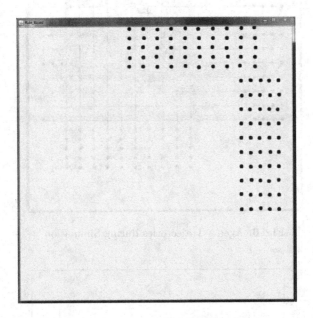

Fig. 5. Agents Final Positions for Simulation Scenario

The trajectory of each agent during the simulation is shown in Figure 6. With a fixed simulation step, the interval between two dots of an agent's trajectory should be equal if agent holding a constant speed. However, the simulation result indicates that the intervals are nonuniform: dots are sparse at the initial and final stage, and dense at the stage when the two groups of agents are gathering at the centre part of the environment. This indicates that agents can hold the desired speed and need to slow down for collision avoidance. This phenomenon is common in real life scenario that pedestrian can hold the desired speed at sparse area and slow down when coming into dense area. The scenario is also applied for the performance evaluation discussed as follows. The performance evaluation result is reported in Figure 7. The main contribution to the simulation cost is during sampling and evaluating velocities from valid set. The average simulation cost per time step is around $9.4ms$ when no agent has arrived at its destination. Since there are 100 agent in the simulation, hence the average simulation cost is around $0.094ms$. This result indicates the proposed motion planning framework is efficient.

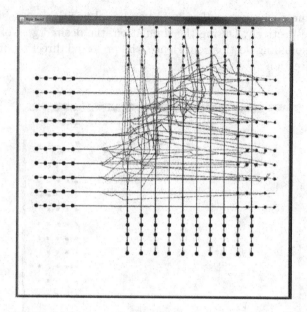

Fig. 6. Agents Trajectories during Simulation

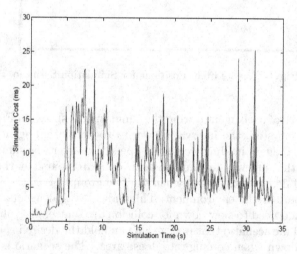

Fig. 7. Performance Evaluation of Motion Planning Framework

7   Conclusion

In this paper, a motion planning framework is proposed for crowd simulation. The framework divides motion planning into several continuous process: collision detection, valid velocity determination, velocity sampling and velocity evaluation. When no collision is detect, agent just follows its preferred velocity directing to its destination. Otherwise, agent must choose a velocity from the valid velocity set to avoid the collision, based on the mechanism of velocity evaluation. Simulation results indicates that the proposed framework is able to simulation agent's motion during its movement towards its destination, and hold a efficient simulation cost.

References

1. Thalmann, D., Musse, S.: Crowd Simulation. Springer (2007)
2. Fox, D., Burgard, W., Thrun, S.: The dynamic window approach to collision avoidance. IEEE Robotics & Automation Magazine 4(1), 23–33 (1997)
3. Jaillet, L., Simeon, T.: A prm-based motion planner for dynamically changing environments. In: Proceedings of IEEE/RSJ Int. Conference on Intellegent Robots and Systems, pp. 1606–1611 (2004)
4. Sud, A., Andersen, E., Curtis, S., Lin, M., Manocha, D.: Real-time path planning for virtual agents in dynamic environments. In: Proceedings of IEEE Virtual Reality, pp. 91–98 (2007)
5. Li, Y., Gupta, K.: Motion planning of multiple agents in virtual environments on parallel architectures. In: Proceedings of IEEE International Conference on Robotics and Automation, pp. 1009–1014 (2007)
6. van den Berg, J., Lin, M., Manocha, D.: Reciprocal velocity obstacles for real-time multi-agent navigation. In: Proceedings of IEEE International Conference on Robotics and Automation, pp. 1928–1935 (2008)
7. Fiorini, P., Shiller, Z.: Motion planning in dynamic environments using velocity obstacles. Int. Journal of Robotics Research 17(7), 23–33 (1998)
8. Gayle, R., Sud, A., Lin, M., Manocha, D.: Reactive deforming roadmaps: Motion planning of multiple robots in dynamic environments. Int. Journal of Robotics Research 21(3), 233–255 (2002)
9. Lamarche, F., Donikian, S.: Crowd of virtual humans: a new approach for real time navigation in complex and structured environements. Computer Graphics Forum 23(3), 509–518 (2004)
10. Stephen, C.: Flow tiles. In: Proceedings of the 2004 ACM SIG-GRAPH/Eurographics Symposium on Computer Animation, San Diego, California, USA, July 26-27, pp. 233–242 (2004)
11. Shao, W., Terzopoulos, D.: Autonomous pedestrians. In: Proceedings of the 2005 ACM SIGGRAPH/Eurographics Symposium on Computer Animation, pp. 19–28 (2005)
12. Xiong, M., Lees, M., Cai, W., Zhou, S., Low, M.Y.H.: The dynamic window approach to collision avoidance. The Visual Computer 26(5), 367–383 (2010)

Personalized Friend Recommendation in Social Network Based on Clustering Method

Zhiwei Deng, Bowei He, Chengchi Yu, and Yuxiang Chen

School of Computer Science
Beijing University of Posts and Telecommunications
Beijing, P.R. China
jobsdzw@gmail.com

Abstract. Personalized recommendation technique has been applied into lots of Electronic Commerce websites. Now some social websites also proposed some application to recommend potential friends for users. However, their algorithms are based on the amounts of common friends, which has some drawbacks in accuracy. We proposed a new recommendation method by combining the existed 'friend of friend' algorithm and content-based recommendation. Through this method we can get more meaningful and apt recommendation outcomes and calculate more quickly in the large amount of data.

Keywords: Forum, Friend Recommendation, Clustering Method.

1 Introduction

Entering an era of data explosion, the network has become an information-sharing platform; it is increasingly difficult to find the information they need in such vast amounts of data. In the condition that people know what information they really need, the search engine provides users with a quick and convenient way to find appropriate items. However, when facing the large amount of data in the network, there are many users who aren't aware of what they are really interested in or what they need. As a result, the recommendation engine provides a good way to deal with the problem and satisfies the users. Recommendation engine helps users filter out the apt information they may need indeed from the flood of information, recommending the users with items, according to the users' behaviors in the past and their information the systems collected. Nowadays, the recommendation engine has been successfully applied to the field of electronic commerce and some music and movies websites, such as Amazon, Netflix, etc.

With the rise of a variety of social networks such as micro blog and forums, many of them have launched an application which can recommend potential friends users may be interested in such as Facebook and twitter. While they both recommend information to the target users, the recommendation engine in the social network is of considerable different from the traditional recommendation system applied in the Electronic Commerce. Compared with the friend

Z. Li et al. (Eds.): ISICA 2012, CCIS 316, pp. 84–91, 2012.
© Springer-Verlag Berlin Heidelberg 2012

recommendation, the traditional-recommendation application has the characters that it has less attributes rarely changed in the items and that the impact factors on the users and items are relatively simple, for example, there can be properties like genre, writer, singer and publication date in the music. However, when considering the relations between people and people, there are many factors related. First, describing a person's interests and hobbies is more complex than the description of items. Secondly, people's interests and hobbies change over time. Thirdly, in the social network, for most of the users, they add people into their friend lists not based on similar interests, but due to geographical location, study or work experience. In this way, recommending potential friends that users may interested in would be influenced by factors above. Importantly, similar people users met through social network are much valuable for themselves. People often gain more help from these weak ties than from strong ties[4]. So research on the social network friend-making method and potential friend recommendation is very necessary.

The idea used in PYMK Friend Recommendation designed by MySpace is based on the user's social relations and personal information profiles, or, the 'friend of friend' method[1]. The similar strategy is applied on Facebook, Renren, twitter and other social networks, in which the recommended results are based on friends in common, similar age, geographic location, school information and etc. However, there are some drawbacks to extend friends through this way. Friends recommended based only on the number of common friends may not necessarily have the same hobbies and habits with the target users. Thus, the user may associate the so-called potential friends and may fall victim into the dilution of 'friendship' based on the above strategy, for the reason that they may share some contents that the target users may not interested[2]. At the same time, this method may cause the waste of lots of useful information in social networks, through which we can obtain more accurate and meaningful friends recommendation results.

As a result, we designed a content-based friend recommendation system. Through the records of the contents users interested in, potential friends who have similar interests and behaviors with target users of users can be presented. There are two differences in our system compared with the existed method.

1. Not like the collaborative filtering method, we recommend potential friends for the target users based on the browsing records and the content users interested in;
2. We apply the clustering algorithm in our system, reducing the system overhead when dealing with a large amount of data. And a lot of work can be carried out offline, which make recommending friends more accurate and faster.

2   The Introduction of Recommendation Method

There are two main methods in the recommended areas to generate recommended.

2.1   Content-Based Recommendation

Content-based filtering method is based on the user's previous behavior, such as users' items ratings, browsing records of goods and their purchasing history, establishing the user model to describe their interests and behaviors. At the same time, the system establishes a model for each item to describe its characteristics. And then match the user models and item models by estimating how much they are related each other, recommending the items users may need. For example, in a music-sharing site, we can establish a user model to describe the users' interests and preferences by collecting user's previous listening history, search history and other information. And then to recommend music's which match users' preferences described by the user model.

2.2   Recommendation Based on Collaborative Filtering

Collaborative filtering and content filtering is dissimilar. Collaborative filtering recommendation recommends items for people based on users who are alike with them. While in content-based filtering algorithm, we calculate the similarity between the items, in the collaborative filtering algorithm, we analyze relationships between users and interdependencies among products to identify new user-item associations. Using collaborative recommendation one identifies users whose tastes are similar to those of the given user and recommends items they have liked[3].

Both information filtering method have a wide application in the field of e-commerce. A major advantage of collaborative filtering is that it can address data aspects that are often elusive and difficult to profile using content filtering and generally speaking, it is more accurate that content-based method. Collaborative filtering suffers from what is called the 'cold start' problem, due to its inability to address the system's new products and users[10]. There are considerable differences between people recommendation and items recommendation and the performance of the two methods remains to be seen when applied into the friend recommendation field. The approach used in nowadays social network, 'friends of friends' method can be seen as a way of adopting the ideas in collaborative filtering. As the preceding discussion, in the social networking, many of the user's friendship are not based on users' interests and hobbies, but by other factors, for example, one may become friend with others for the reason that they come from the same city or country. 'Friends of friends' recommending method may compel users to add some people with whom they are not similar and who may share information the user is not interested in. And new 'friends' may derive from these people, lowering the accuracy of recommendation.

3   Friend Recommendation System

3.1   System Framework Design

In this section, we will make an introduction on the ideas and frameworks of our friend recommendation system. The system is structured as follows:

1. Collecting user information and modeling;
2. User clustering;
3. Friend recommendation.

The framework of our system is showed in the chart.

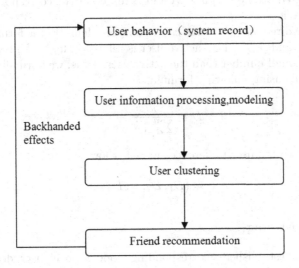

Fig. 1. System Framework

3.2   User Information Collecting and Modeling

We set up a forum using Discuzz and embed the friend recommendation system into the forum. System records the users' behaviors (browsing and posting) in several plates and set different weights on the two kinds of records to character the users' preferences on each plate.

	pid	fid	tid	first	author	authorid	subject	dateline	message	useip	invisible	anonymous
□ ✎ ✗	1	2	1	1	aiku	1	THE YANKEE COMANDANTE	1327926646	For a moment, he was obscured by the Havana night....		0	0
□ ✎ ✗	2	2	2	1	aiku	1	CORY BOOKER: THE DILEMMA OF THE NEW BLACK POLITICL...	1327926646	Reflections are like second marriages—a portion of...		0	0
□ ✎ ✗	3	2	3	1	aiku	1	TO-DO LIST: EGYPTIANS VOTE: GOOGLE BEING EVIL?	1327926646	To know: Egypt held its Presidential election toda...		0	0

Fig. 2. Posting Records on Forum

We adopted the method based on vector space model which is an N-dimensional feature vector to represent users' interest models[6]:

$$n = \{(k_1, w_1), (k_2, w_2), ..., (k_n, w_n)\} \tag{1}$$

Each dimension of vector is composed by users' behaviors on each plate and its corresponding weight. And:

$$k_i = A_{mi} + B_{ti} \tag{2}$$

Where the mi is the number of user's visit on the i_{th} plate, t_i is the number of user's texts posted on the i_{th} plate. A and B represent the corresponding weight of the two behaviors.

However, the direct use of (1) can cause some problems. In the forum, while the amount of visit and post for an active user is relatively high, the low activeness will result to a small number than the active user. Thus, we normalize the users' preference vector using following formula.

$$D = \frac{k_i w_i}{\sum_{i=1}^{n} k_i w_i} \tag{3}$$

Then we can have the user preferences model below:

$$F = (D_1, D_2, ..., D_n) \tag{4}$$

3.3 User Clustering

In section 3.2, we established a corresponding model for each user, describing the features of users on N dimensions. In the method we represented below, we use the cluster model to the system overhead when recommending potential friend. We adopt the cosine similarity to describe the degree of how much they are alike:

$$similarity = \frac{uv}{|u||v|} \tag{5}$$

Where both u and v are vectors established in 3.2 as user models to depict user's interests.

lustering groups people into collections, breaking a whole large package of data into several pieces. And in clusters we can deal with data more convenient and takes less system overload[5]. For the reason that in each cluster the users in it are people who are most similar, the quality of recommendation is not much influenced by the reduction of users. In our system we adopted two algorithms, K-means and DBSCAN. The outcome of clustering in K-means is better. Though DBSCAN do not need the initial set of cluster number, it tends to perform badly in grouping users into several different collections or when number of dimensions goes high.

3.4 Friend Recommendation

After the work in above, we can begin to seek the similar potential friends for users based on the clusters. In our system, we realize the recommendation in four aspects below:

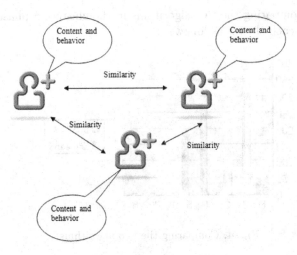

Fig. 3. Clustering users

1. Find people who are most similar with the target user based on clusters. We calculate every people's similarity with target user and find the N-nearest ones as the raw recommendation of similar potential friends;
2. Then we make the outcomes more appropriate and accurate by combining the raw potential friends recommendation list with user's profile and the amount of their common friends, for the reason that recommendations are more persuasive coming from real friends or some common aspects (such as coming from the same school)[2].So we adjust the outcome according to the common information between user and friend recommended;
3. Another aspect that should be taken into count is that users' interests may change with time. To deal with this problem, we recalculate the clusters every one week or two weeks. The update time can be changed over different situation;
4. The new users of our system should be treated differently. They have two problems: a. new users do not have history records which can be used to create a model. b. Which cluster new users belong to is remain unknown. This is similar to the 'cold start' problem in collaborative filtering[7]. To deal with new ones, we use the 'friend of friend' method to recommend potential friend and adjust outcomes by getting direct feedback from users' preferences chart. When people log into a social website at the first time, they tend to add some people they already knew in the real world.

4   Results

1. To test the quality of our method, we get 500 users' data and information from the BYR forum. BYR forum is the official forum of Beijing University of Posts and Telecommunication. It was established in 2003 and has the average of 9000 user visit in a day.

When comparing the two algorithms in the clustering phase, we record the data of each cluster as follows:

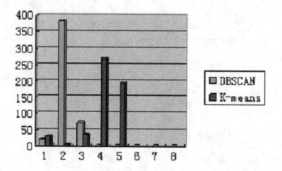

Fig. 4. Comparing the two algorithms

Compared with K-means algorithm, the outcome of DBSCAN has lots of isolated points which is not suitable for recommendation. And DBSCAN is considerably affected by the parameters set due to experience. So the K-means and its variants may be superior.

2. We change the number of clusters presented in K-means from 2 to 6, though we obtain different clusters in the outcomes, we get the same potential friend list recommended by system. So, the clustering method's passive effect on quality of recommendation is low;

3. Our test shows that the pure content-based method produces potential friends that are exactly similar to the target user. And the top-3 friends' distance to target user is less than 0.1. In contrast, two users in the same class has a distance more than 0.8. In the aspect of spiritual friendship, pure content-based method is superior.

5 Related Works

Joonhee Kwon and Sungrim Kim, proposed a method of scoring friendship based on context and friends can be recommended based on the user's physical and social information[8]. Xing Xie, established a friend recommendation machine embedded on a bioscience web, which can recommend users who have similar research fields to each other[9]. Jilin Chen, based on IBM's Beehive dating site compares four recommendation methods including the four friends on their web sites.

References

1. Moricz, M., Dosbayev, Y., Berlyant, M.: PYMK: Friend Recommendation at MySpace. In: Proceedings of the ACM SIGMOD International Conference on Management of Data, pp. 999–1002 (2010)

2. Brzozowski, M.J., Hogg, T., Szabo, G.: Friends and Foes: Ideological social networking. In: Proceedings of the Conference on Human Factors in Computing Systems, pp. 817–820 (2008)
3. Balabanovic, M., Shoham, Y.: Fab:Content-Based, Collaborative Recommendation. Communications of the ACM 3(40), 66–72 (1997)
4. Granovetter, M.: Strength of weak ties. Amer. J. Sociology 78(6), 1360–1380 (1973)
5. Ungar, L., Foster, D.: Clustering Methods for Collaborative Filtering. In: Proc. Workshop on Recommendation Systems, pp. 112–125 (1998)
6. Lin, S.-M., Wang, G.-S., Chen, Y.-Q.: User Modeling and Feature Selection in Personalized Recommender System. Computer Engineering 33(17), 196–198 (2007)
7. Schein, A.I., Popescul, A., Ungar, L.H., Pennock, D.M.: Methods and Metrics for Cold-Start Recommendations. In: Proc. 25th Ann. Int ACM SIGIR Conf., pp. 253–260 (2002)
8. Kwon, J., Kim, S.: Friend Recommendation Method using Physical and Social Context. International Journal of Computer Science and Network Security 10(11), 116–120 (2010)
9. Xie, X.: Potential Friend Recommendation in Online Social Network. In: Proc. IEEE/ACM GREENCOM-CPSCOM 2010, pp. 831–835. IEEE Computer Society (2010)
10. Koren, Y., Bell, R., Volinsky, C.: Matrix factorization techniques for recommender systems. IEEE Comput. 42(8), 30–37 (2009)

The Research of Intrusion Detection Based on Mixed Clustering Algorithm

Nanyan Liu

College of Computer Science and Technology
Xian University of Science and Technology
Xi'an 710054, P.R. China
513318452@QQ.com

Abstract. Nowadays, network security problems are increasing prominent, and how to find intrusion activities quickly and efficiently has become important to the security of system and network resource. we use the feature extraction and feature selection method of rough set and pattern recognition in the feature selection of network intrusion detection and introducing clustering method and genetic algorithm for network intrusion detection. First, we use the feature extraction which based on rough sets theory for the experimental data set. we use a mixed data dissimilarity algorithm and combining it with k-medoids algorithm. Makes the clustering algorithm can deal with a mixed data set which include continuous and discrete data. Last, traditional k-medoids clustering algorithm is difficult to determine the number of existing clustering, sensitive to initial value and easy to fall into local optimal solution. So we present an unsupervised clustering algorithm which combing with genetic algorithm and k-medoids clustering algorithm. All of these methods are efficiently to solve the defects of traditional k-medoids algorithm. And the algorithm can distinguish new attack from already existed attack.

Keywords: Intrusion Detection, Cluster Analysis, Dissimilarity Matrix, Rough Set, Genetic Algorithm.

1 Introduction

Clustering algorithms in data mining is a kind of common, classic analysis method. According to the data mining technology and clustering analysis method of itself, we applied it to the intrusion detection systems[1]. According to the type of data, the characteristic of practical problems and the aim of clustering, the literature presents many clustering algorithm. Distance and dissimilarity measuring method is the most fundamental conception in clustering analysis. The select of measuring method directly influences the quality of clustering results. In this paper, on the basis of the above-mentioned literature combined the dissimilarity matrix of mixed data and K-means algorithm, presents a practical and effective method, with strong versatility, stability. Its of great practical significance for the intrusion detection.

Z. Li et al. (Eds.): ISICA 2012, CCIS 316, pp. 92–100, 2012.

2 Prepare Knowledge

Dissimilarity matrix (or object-by-object structure): This stores a collection of proximities that are available for all pairs of n objects. It is often represented by an n-by-n table:

$$
\begin{bmatrix}
0 & & & \\
d(2,1) & 0 & & \\
M & M & 0 & \\
d(n,1) & d(n,2) & \varLambda & 0
\end{bmatrix}
\tag{1}
$$

Where $d(i,j)$ is the measured difference or dissimilarity between objects i and j. In general, $d(i,j)$ is a nonnegative number that is close to 0 when i and j are highly similar or 'near' each other, and becomes larger the more they differ. Since $d(i,j) = d(j,i)$, and $d(i,i)$, we have the matrix in (1).

Interval-scaled variables are continuous measurements of a roughly linear scale. The measurement unit used can affect the clustering analysis, to help avoid dependence on the choice of measurement units, the data should be standardized. It can be performed as follows.

2.1 Interval-Scaled Varibles

Caculate the mean absolute deviation, s_f.

$$
s_f = \frac{1}{n}(|x_{1f} - m_f| + |x_{2f} - m_f| + \varLambda + |x_{nf} - m_f|)
\tag{2}
$$

Where $x_{1f}, ..., x_{nf}$ f are n measurements of f, and m_f is the mean value of f
Calculate the standardized measurement:

$$
z_{if} = \frac{x_{if} - m_f}{s_f}
\tag{3}
$$

The mean absolute deviation, s_f, is more robust to outliers than the standard deviation, σ_f. When computing the mean absolute deviation, the deviations from the mean are not squared; hence, the effect of outliers is somewhat reduced. The dissimilarity between the objects described by interval-scaled variables is typically computed based on the distance between each pair of objects. The most popular distance measure is Euclidean distance, which is defined as:

$$
d(i,j) = \sqrt{(x_{i1} - x_{j1})^2 + (x_{i2} - x_{j2})^2 + \varLambda + (x_{in} - x_{jn})^2}
\tag{4}
$$

Where $i = (x_{i1}, x_{i2}, \varLambda, x_{in})$ and $j = (x_{j1}, x_{j2}, \varLambda, x_{jn})$ are two n-dimensional data objects.

2.2 Binary Variables

A binary variable has only two states: 0 or 1, where 0 means that the variable is absent, and 1 means that it is present. Its dissimilarity measure, defined in Equation (5).

$$
d(i,j) = \frac{r + s}{q + r + s + t}
\tag{5}
$$

where q is the the number of variables that equal 1 for both objects i and j, r is the number of variables that equal 1 for object i but that are 0 for object j, s is the number of variables that equal 0 for object i but equal 1 for object j , and t is the number of variables that equal 0 for both objects i and j. The total number of variables is p , where $p = q + r + s + t$.

2.3 Categorical Variables

A categorical variable is a generalization of the binary variable in that it can take on more than two states. For example, map color is a categorical variable that may have, say, five states: red, yellow, green, pink, and blue.

Let the number of states of a categorical variable be M. The states can be denoted by letters, symbols, or a set of integers.

The dissimilarity between two objects i and j can be computed based on the ratio of mismatches:

$$d(i,j) = \frac{p - m}{p} \tag{6}$$

Where m is the number of matches, and p is the total number of variables.

2.4 Ordinal Variables

Ordinal variables are very useful for registering subjective assessments of qualities that cannot be measured objectively. For example, professional ranks are often enumerated in a sequential order, such as assistant, associate, and full for professors. The values of an ordinal variable can be mapped to ranks. For example, suppose that an ordinal variable f has M_f states. There ordered states define the ranking $1, \Lambda, M_f$.

The treatment of ordinal variables is quite similar to that of interval-scaled variables when computing the dissimilarity between objects. Suppose that f is a variable from a set of ordinal variables describing n objects. The dissimilarity computation with respect to f involves the following steps:

The value of f for the ith object x_{if}, and f has M_f ordered states, representing the ranking $1, \Lambda, M_f$. Replace each x_{if} by its corresponding rank, $r_{if} \in \{1, \Lambda, M_f\}$.

Since each ordinal variable can have a different number of states, it is often necessary to map the range of each variable onto $[0.0, 1.0]$ so that each variable has equal weight. This can be achieved by replacing the rank r_{if} of the ith object in the fth variable by z_{if}.

$$z_{if} = \frac{r_{if} - 1}{M_{f-1}} \tag{7}$$

2.5 Ratio-Scaled Variables

A ratio-scaled variable makes a positive measurement on a nonlinear scale, such as an exponential scale, approximately following the formula:

$$A^{B1}e \ or \ Ae^{-Bt} \tag{8}$$

Where A and B are are positive constants, and t typically represents time. There are three methods to handle ratio-scaled variables for computing the dissimilarity between objects.

Treat ratio-scaled variables like interval-scaled variables. This, however, is not usually a good choice since it is likely that the scale may be distorted.

Apply logarithmic transformation to a ratio-scaled variable f having value x_{if} for object i by using the formula $y_{if} = log(x_{if})$. The y_{if} values can be treated as interval valued Notice that for some ratio-scaled varibles, log-log or other trasformations may be applied, depending on the variable's definition and the application.

Treat x_{if} as continuous ordinal data and treat their ranks as interval-valued.

2.6 Variables of Mixed Types

A preferable approach is to process all variable types together, performing a single cluster analysis. One such technique combines the different variables into a single dissimilarity matrix, bringing all of the meaningful variables onto a common scale of the interval $[0.0, 1.0]$.

Suppose that the data set contains p variables of mixed type. The dissimilarity $d(i, j)$ between objects i and j is defined as:

$$d(i, j) = \frac{\sum_{f=1}^{p} \delta_{ij}^{(f)} d_{ij}^{(f)}}{\sum_{f=1}^{p} \delta_{ij}^{(f)}} \tag{9}$$

Where the indicator $\delta_{ij}^{(f)} = 0$ if either x_{if} or x_{jf} is missing (i.e., there is no measurement of variable f for object i or j), or $x_{if} = x_{jf} = 0$ and variable f is asymmetric binary; otherwise, $\delta_{ij}^{(f)} = 1$. The contribution of variable f to the dissimilarity between i and j, that is, $d_{ij}^{(f)}$, is computed dependent on its type:

- If f is interval-based: $d_{ij}^{(f)} = \frac{|x_{if} - x_{jf}|}{max_h x_{hf} - min_h x_{hf}}$, where h runs over all non-missing objects for variable f.
- If f is binary or categorical $d_{ij}^{(f)} = 0$ if $x_{if} = x_{jf}$; otherwise $d_{ij}^{(f)} = 1$.
- If f is ordinal: compute the ranks r_{if} and $z_{if} = \frac{r_{if} - 1}{M_f - 1}$, and treat z_{if} as interval-scaled.
- If f is ratio-scaled: either perform logarithmic transformation and treat the transformed data as interval-scaled; or treat f as continuous ordinal data, compute r_{if} and z_{if}, and then treat z_{if} as interval-scaled.

The above steps are identical to what we have already seen for each of the individual variable types. The only difference is for interval-based variables, where here we normalize so that the values map to the interval $[0.0, 1.0]$. Thus, the dissimilarity between objects can be computed even when the variables describing the objects are of different types[8].

2.7   KDDcup99 Data Set

The data set first used at the 3rd International Knowledge Discovery and Data
Mining Tools Competition which is at the same time with KDD99.It contains
variety of simulation data in a military network environment. The data set provides nine weeks of data network connection from a simulated U.S Air Force
LAN. The data set contains about 4,900,000 simulated attack records. Total of
22 kinds of attacks and is divided into four classes: DOS, R2L, U2R, Probing.
This data set have 41 features, 7 classification features, 34 numeric features[9].
Because the entire data set is too large so we often use a subset of 10%(contains
494,020 records, 19.69% is normal, 80.31% is attack) to test the performance of
the algorithm.

2.8   K-Means Algorithm

The original k-means algorithm processing flow is as follows: First, randomly
select k objects, each object represents an initial value or center of clusters. The
remaining each objects, according to its center distance with various clusters,
it will be assigned to a recent cluster. Then, re-calculate the average of each
cluster. This process is repeated until criterion function convergence.

3   Feature Selsection Based on Binary Particle Swarm Optimization(BPSO) and Overlap Information Entropy(OIE)

For the KDD99 data set we used have a higher dimension, large amounts of
data so the clustering need to spend a lot of time making the clustering difficult
to achieve. So we need a feature selection to reduce the original data set and
maintain the integrity of the original data set.

3.1   BOSP&OIE

Binary Particle Swarm Optimization (BPSO) algorithm is an extension of PSO,
it uses binary code. The essence is that the position of particle x_i , which is
generated randomly, is set as 0 or 1. The possibility of the position changing at
next time is judged by sigmoid function (10) of velocity.

$$s(v) = \frac{1}{1 + e^{-v}} \tag{10}$$

Comparing with PSO algorithm, the velocity formula remains unchanged, but
the particles new position is calculated by the following rule.

$$if\ rand() < s(v_{id}^n), then\ x_{id}^{n+1} = 1, else\ x_{id}^{n+1} = 0 \tag{11}$$

Overlap Information Entropy (OIE) proposed by Wang and Shen et al is a
method for analyzing the multi-variables correlation degree quantitatively[10].

The OIE uses $\frac{\lambda_i}{N}$ of correlation matrix to replace the probability of Information Entropy, and its value denotes the correlation degree of multi-variable.

Assuming a multi-variable system has N variables, and time series of the multi-variable system at time $t(t = 1, 2, ..., M)$ is P.

$$P = \{y_i(t)\}_{1 \leq t \leq M, 1 \leq i \leq N}, P \in R_{M \times N} \tag{12}$$

Then, the matrix Q is gotten after doing centralization and standardization to P, and finally the correlation matrix R is obtained by (6).

$$R = Q^T \cdot Q, R \in R^{N \times N} \tag{13}$$

Let $\lambda_i (i = 1, 2, ..., N)$ denotes the eigen-value of R, and H_R representing the OIE can be defined as:

$$H_R = 1 + \sum_{i=1}^{N} \frac{\lambda_i}{N} log_N \frac{\lambda_i}{N} \tag{14}$$

As an uncertainty measurement for multi-variable's correlation, the value of H_R is in the range $[0, 1]$. The more overlap information variables include, the stronger correlation degree variables have.

3.2   Feature Selection Method Description

Using feature subset which OIE is max with class attribute to reflect objects nature is an improvement method of Feature Selection in this thesis, and this method is not dependent on classifier.

In order to get the Optimal Feature Subset by BPSO algorithm, it needs to define fitness function. Given a training sample set DB, A is sample's attribute set and C is class attribute. And for any proper subset A', $A' \subseteq A$, its calculation steps are:

1. Form the training sample DB, get the attribute values, which are all included in A' set, and get class attribute. A new training sample set P' is composed of both of them;
2. Process P' with centralization and standardization. Then, matrix Q' is gotten;
3. Calculate the OIE of P' according to (7). This OIE is fitness value of A'. The attribute subset whose fitness value is max is the Optimal Feature Subset.

The method steps of Optimal Feature Subset Selection are as follows.

1. Initialize particle swarm, $X = (x_1, x_2, ..., x_m)^T$, m is number of particles; particle's position vector of $x_i = (x_{i1}, x_{i2}, ..., x_{iD})$ is set as 1 or 0; D is particle's feature dimension, which is the attribute number of training sample set DB. Each particle x_i is 1, $j = 1, 2, ..., D$, so the first attribute j is selected. Otherwise this attribute is shielded.
2. Set the maximum number of iterations T and every parameter of BPSO algorithm.

3. Calculate each particle's fitness value and local optimal position $pbest_i^*, i = 1, 2, ..., m$, and compare with this particle's $pbest_i$ of last iteration. If $pbest_i^* > pbest_i$, then update $pbest_i$ and its corresponding fitness value. Otherwise, they are both unchanged.

4. From all $pbest_i, i = 1, 2, ..., m$, determine the global optimal position $gbest^*$ of this iteration, and compare with the last $gbest$. It need to make sure whether $gbest$ is requested to update the $gbest^*$ of this iteration and record its corresponding fitness value.

5. According to (1), (3), (4), update the value of each particle' velocity and position vector.

6. Judge whether the iteration is ended, if not, return to step 3, if yes, then output $pbest$ and its corresponding fitness value. The $pbest$ is the Optimization Feature Subset.

The Optimal Feature Subset is recorded as: $A = \{A_1, A_2, ..., A_n\}$. For example, assuming $pbest = 00101$, that is to say feature 3 and feature 5 are selected, and the Optimal Feature Subset is $\{3, 5\}$.

3.3   Feature Selection of KDD99 Data Sets

Parameters of BPSO algorithm are set as: C1 and C2 are 2.0, and the inertia factor W is 0.7, and the particles' number is 40, and the conditions of stopping the algorithm are 100 times. The experimental results on KDD99, the Optimal Feature Subsets OIE is 0.862895. Feature number is 20, and binary string of Optimization Feature Subset is: 00101001001000110100001111011100110101011. It would mean that the Optimal Feature Subset is $\{3, 5, 7, 10, 14, 15, 17, 22, 23, 24, 25, 27, 28, 29, 32, 33, 35, 37, 40, 41\}$. All feature's OIE is 0.274093 and its feature number is 41.

4   K-Means Algorithm Based on Dissimilarity Matrix

We calculate the dissimilarity $d(i, j)$ line by line to cluster center. Then compare the value of $d(i, j)$, judge the variable to the smallest one. At last calculating the mean of each class until no change in.

Algorithm: K-means algorithm based on Dissimilarity matrix
Input: KDD99 data sets matrix D, cluster number k
Output: meet the constraints of the clustering results
Process:

1. Random select k cluster center from D;
2. Begin from $i = 1$(i is the object number ofD), calculate the dissimilarity to each cluster center. Compare these dissimilarity, then classify i to the cluster which has the minimum dissimilarity.
3. Calculated the mean of cluster and update cluster center, turn to step 2.
4. Repeat step 3 until no longer changes.

5   Experiential Results

5.1   Experiential 1

Randomly selected 1000, 2000, 3000 data from KDD99 data sets. Number of attacking data by less than 1%, clustering number k=5.compared with the traditional K-means algorithm, the results are shown in Table 1.

Table 1. Results comparison between the two algorithms

Size of Data Set	Compared Attribute	K-means Algorithm	K-Means Algorithm Based on Dissimilarity Matrix
1000	Detection Rate(%)	45	60
	False Positive Rate(%)	0.75	0.22
2000	Detection Rate(%)	80	85
	False Positive Rate(%)	0.4	0.35
3000	Detection Rate(%)	89	91
	False Positive Rate(%)	0.25	0.21

5.2   Experimental 2

Randomly selected 5 data sets from KDD99 data sets, each data sets contains 3000 data. With each data sets we selected the cluster center number is 5, 8, 10, 12, 15..the results are shown in Table 2.

Table 2. Results comparison between different cluster center numbers

ClusterCenter Number	Detection Rate%	False Positive Rate%
5	91	0.21
8	92	0.19
10	94	0.16
12	95	0.14
15	96	0.13

Can see with the increasing number of clustering center the intrusion detection rate increase.

6   Conclusion

Experimental results denote that the improved algorithm is obviously improved then the previous algorithm in cluster. This illustrates that the improved algorithm is effective, with further research and practical value.

References

1. Hu, C.: Network Intrusion Detection Principle and Technology. Beijing Institute of Press, Beijing (2006)
2. Jiang, S., Li, Q.: Research on dissimilarity for clustering analysis. Computer Engineering and Applications 11(4), 146–149 (2005)
3. Guha, S., Rastogi, R., Shim, K.: ROCK: A robust clustering algorithm for categorical attributes. In: Proceedings of the 15th ICDE, Sydney, Ausralia, pp. 512–521 (1999)
4. He, Z., Xu, X., Deng, S.: Squeezer: an efficient algorithm for clustering categorical data. Journal of Computer Science and Technology 17(5), 611–624 (2002)
5. Guha, S., Meyerson, A., Mishra, N., et al.: Clustering data streams: Theory and practice. IEEE Transactions on Knowledge and Data Engineering 15(3), 515–528 (2003)
6. Portnoy, L., Eskin, L., Stolfo, S.: Intrusion Detection with Unlabeled Data using Clustering. In: Proceedings of ACM CSS Workshop on Data Mining Applied to Security (DMSA 2001), Philadelphia, PA (2001)
7. Eskin, E., Arnold, A., Prerau, M., et al.: A geometric framework for unsupervised anomaly detection: Detecting intrusions in unlabeled data. In: Data Mining for Security Applications (2002)
8. Han, J., Kamber, M.: Data Mining: Concepts and Techniques, 2nd edn. China Machine Press, Beijing (2007)
9. KDD99. KDD99 cup dataset (DB/OL) (1999),
   http://kdd.ics.uci.edu/databases/kddcup99
10. Wang, Q., Shen, Y., Zhang, Y., et al.: Fast quantitative correlation analysis and information deviation analysis for evaluating the performances of image fusion techniques. IEEE Transactions on Instrumentation

Applying Support Vector Machine to Time Series Prediction in Oracle

Xiangning Wu, Xuan Hu, Chengyu Hu, and Guiling Li

School of Computer Science
China University of Geosciences
Wuhan 430074, P.R. China
{wxning,huchengyu}@cug.edu.cn,
huxuancug@gmail.com,
lgldec@yahoo.com.cn

Abstract. Using oracle data mining option(ODM) and the time series stored in oracle database, the SVM (support vector machines) model can be used to predict the future value of the time series. To build SVM model, firstly the trend in time series must be removed, and the target attribute should be normalized. secondly the size of the time window in which include all the lagged values should be determined, thirdly the machine learning method is used to construct SVM prediction model according to the time series data. Comparing with the traditional time series prediction model, SVM prediction models can reveal non-linear, non-stationary and randomness of the time series, and have higher prediction accuracy.

Keywords: Oracle, Time Series, Support Vector Machine, Prediction Model.

1 Introduction

A time series is a sequence of data points, measured at successive time instants spaced at uniform time intervals. The time series prediction is a process to find out the regularity of the data and predict the future values based on the previous values and the current value of the time series.

Traditional linear time series prediction technologies mainly use the theory of mathematical statistics. For example, Box-Jenkins autoregressive model (AR), moving average model (MA), autoregressive integrated moving average model (ARIMA)[1], and Holt-Winter method which is always applied to the time series with seasonal fluctuation. In nearly half a century, this kind of white noise driven linear model has been in a dominant position in the field of time series analysis[2]. However, in many cases, the accuracy of the linear series predicting techniques are not satisfactory because of nonlinear characteristics of some time series, such as irregularity and chaotic.

Therefore, some nonlinear time series predicting techniques appeared. These techniques mainly use the embedding space method and neural network method,

Z. Li et al. (Eds.): ISICA 2012, CCIS 316, pp. 101–110, 2012.
© Springer-Verlag Berlin Heidelberg 2012

especially be represented by chaotic method and neural network based method. However, these non-linear methods are not stable enough. The chaotic systems are sensitive to the initial value and cant be used to predict long-term chaotic time series[3]. For the neural network model, the knowledge obtained from the model is expressed by network structure which is not transparent[4].In addition, the neural network has some drawbacks , such as excessive adaptation and insufficiency of ability to learn.

According to statistical learning theory, Vapnik et al. proposed a learning method, support vector machine(SVM)[5], the SVM embed or map the data of input space into a high dimensional feature space, so as to solve many difficult problems that can not be solved by the linear method in the original sample space. Compared with the neural network algorithm, SVM has simpler training process and better generalization capability. SVM also can be used in field of time series prediction.

In recent years, the Oracle database provided more and better statistical and analysis capabilities, such as Oracle SQL analytical functions and DBMS_STAT_FUNCS program package, which can perform some complex statistical and analysis processing that ordinary SQL statements can not achieve. With the rapid development of data warehouse and online analytical processing(OLAP), Oracle also offers Oracle OLAP option and Oracle Data mining (ODM) option to expand the statistical and analysis capabilities. ODM provide 12 kinds of data-mining algorithms, such as classification, clustering, regression, anomaly detection, attribute importance analysis, association rules, feature extraction, sequence similarity searches and analysis etc.. the SVM algorithm is also included in ODM[6,7].

Using ODM to directly analyze the time series data stored in Oracle database can reduce resources consumption and the time required for data transmission.

2   Landslide Displacement Time Series

Landslide cause huge human and economic losses, people often use global positioning system (GPS) to monitor landslide surface displacements. Landslide displacement monitoring data are collected based on a fixed interval, so it is a typical time series.

Nowaday, some analysis methods are gradually adopted to find the internal regularities in evolution and development of landslides from the historical monitoring time series data, so as to predict future displacement value of the landslides.

Traditional methods of landslide displacement prediction are mostly based on statistical prediction model[8], which use mathematical statistics methods, investigate various internal geological factors and outside environmental factors, create a landslide displacement fitting curve, then predict future landslide displacement value according to the model reflected by the curve. These methods include the gray GM (1,1) model, fuzzy mathematical model, autoregressive model and so on.

However, landslides are nonlinear dynamical systems because their evolution are influenced by geotechnical structure, rainfall, water level and other complex factors. Landslide displacement time series show some complex characteristics, such as non-linear, non-stationarity and randomness, Therefore, the SVM algorithm is suitable for prediction of the landslide displacement time series.

3   Two Methods of Time Series Prediction ODM Provided

3.1   FORECAST Command of the Oracle OLAP Option

The FORECAST command can predict data through one of the following three methods: the linear trend method, exponential growth or Holt-Winter inference method. The former two methods are simple extrapolation methods, while the latter is more complicated, it can be seen as a kind of exponential smoothing or moving average method[6].

FORECAST command support 'single variable time series'. The usage of the FORECAST command related to the Oracle data warehouse and OLAP, need to organize data into a cube with time dimension, and store measurement data in data cubes. FORECAST command is usually applied to linear prediction, it can not capture the complex relationships between the input and output.

3.2   Oracle Data Mining Option (ODM)

ODM is a option module that can help developers to directly mine the data stored in Oracle database. All steps of data mining, such as establishing model, scoring, and metadata management, can be accomplished either through the Oracle Data Mining client, SQL Function, PL/SQL API or Java API interface[7]. We will gradually introduce how to use the ODM SQL Function, PL/SQL API interface to establish a SVM model.

4   Establishing the Time Series Prediction Model

4.1   Lagged Value of the Time Series

ODM SVM regression algorithm build time series prediction model via state-space reconstruction method. In the algorithm, the past value of the target attribute, also be called lagged value, will be used as training inputs, lagged variables can be computed through Oracle's SQL LAG analytic function.

Table 1 illustrates how to use SQL LAG function to compute the lagged value. Y represents the landslide displacement value of current $month$, $LAG(Y, 1)$, $LAG(Y, 2)$, respectively represent the lagged (past) value of a month ago and two months ago.

For example, the target attribute Y and its lagged values are used as the training data, the SVM model obtained can be seen as a predictive function

Table 1. Lagged value computed by SQL LAG analytic function

MONTH	Y	LAG(Y, 1)	LAG(Y, 2)
2006-1	430.8	.	.
2006-2	439.6	430.8	.
2006-3	448.8	439.6	430.8
2006-4	457.1	448.8	439.6

$P()$. the predictive value of November can be inferred from the actual values of October and September, namely, predictive value $Y\_11 = P(Y\_10, Y\_9)$.

4.2 Transforming the Time Series Data

Fig. 1 shows the monthly accumulated displacement monitoring data (unit:mm) of one landslide in Yichang, Hubei Province, China. This time series last from July 2003 to December 2010, time axis scale "04-01" means "January 2004".

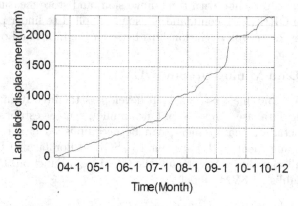

Fig. 1. Monthly accumulated landslide displacement time series

The algorithm requires the time series is stable, however, The series in Fig. 1 has a upward growth trend in the mean. In addition, the variance, size of the swings in the series, is relatively big on both sides of the mean. Therefore, the time series is unstable, the series data needs some transformation prior to modeling.

We use the difference method to remove the trend, that is, the value of time t minus the value of the previous time t-1 get a series of the difference. We no longer use Y, but the difference $D = Y - LAG(Y, 1)$ as the target attribute,and do not use $LAG(Y, 1)$, but the difference $LAG(Y - 1) - LAG(Y, 2)$ as input attributes. If necessary differential processing can be done several times until the trend disappears. We also used the LOG transformation to reduce the range of the variance.

The SQL code of LOG transforming and first order differencing is:

```
CREATE VIEW Landslide_xform AS
SELECT a.*
FROM (SELECT Month, Displacement,
           tp - LAG(tp,1) OVER (ORDER BY Month) tp
           FROM (SELECT Month, Displacement,
                LOG(10, Displacement) tp
                FROM Landslide)) a;
```

In the above statement, table *Landslide* stores the landslide displacement time series, in this table the columns, *Displacement* and *Month*, are value of accumulated landslide displacement and the month of monitoring, the result of data transformation is shown in the view *Landslide_xform*. the internal subquery uses a logarithmic function to stabilize the variance, and outside subquery is used to implements first order differencing.

After stabilizing the series, we use the Z-score method to normalize the series. This process is help for speeding up the convergence of the SVM algorithm, and can make the series have zero mean. The calculation method is to subtract the mean of the series from each sample, then divide the result by the standard deviation of the series, the formula is $(tp - mean(tp))/stddev(tp)$. In view *Landslide_xform*, the average $mean(tp) = 0.0249$, the standard deviation $stddev(tp) = 0.05315$. The result of normalization is shown in *Landslide_norm* view, the following SQL statement implements normalizing :

```
CREATE VIEW Landslide _norm AS
SELECT Month, Displacement, (tp - 0.0249)/ 0.05315 tp
FROM Landslide_xfrm;
```

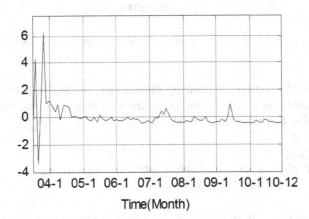

Fig. 2. The time series after log transformation, first order differencing and normalizing

Fig. 2 is the time series which has implemented log transformation, first order differencing and normalizing, and has been transfered to a stationary series with zero mean.

4.3 Selecting Lagged Attributes

Prior to establishing the SVM model, we need to determine which lagged values must be taken as input attributes, in another words, we must determine the size of a delay time window, the window contains all lagged values involved in modeling, the size of the window will directly affect pattern recognition ability of the SVM algorithm. We choose delay window size via analyzing the autocorrelation coefficient of the time series, the window includes the term with the largest autocorrelation coefficient.

Fig. 3 shows the first 16 autocorrelation coefficient (ACF) of the landslide displacement series. The maximum autocorrelation coefficient is at lag 12, which fits well with our expectation that landslide displacement may show annual seasonal effects because of the seasonal rainfall and other climate factors. Based on this autocorrelation analysis, we choose a window of size 12, which include 12 lagged values, $LAG(Y, 1), LAG(Y, 2), ..., LAG(Y, 12)$.

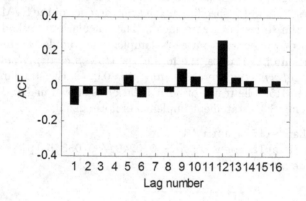

Fig. 3. First 16 autocorrelation coefficients for the time series

To calculate the autocorrelation coefficient, we use SQL analysis functions *CORR* and *LAG*. For example: the correlation coefficient of the series with its first lagged value can be computed as follows:

```
SELECT CORR(ts,tp)
FROM (SELECT tp, LAG(tp,1) OVER (ORDER BY Month) ts
      FROM Landslide _norm);
```

The following PL/SQL procedure xcorr return the cross-correlation coefficient between any two columns (*p_base_col* and *p_lag_col*). *p_in_table* is name of the input time series table, *p_out_table* is name of the output table that will store the correlation coefficients, *p_seq_col* is name of the time column, *p_base_col* is name of the time series column, *p_lag_col* is name of the lagged value column, *p_max_lag* is size of the delay time window.

```
CREATE OR REPLACE PROCEDURE xcorr(
p_in_table    VARCHAR2,
    p_out_table  VARCHAR2,
    p_seq_col    VARCHAR2,
    p_base_col   VARCHAR2,
    p_lag_col    VARCHAR2,
    p_max_lag    NUMBER)
AS
    v_stmt VARCHAR2(4000);
    v_corr NUMBER;
BEGIN
    v_stmt:= 'CREATE TABLE ' || p_out_table ||
        '(lag_num NUMBER, correlation NUMBER)';
    EXECUTE IMMEDIATE v_stmt;

    FOR i IN 1..p_max_lag LOOP
    v_stmt:= 'SELECT CORR(' || p_base_col || ', lag_val) ' ||
        'FROM (SELECT ' || p_base_col || ',' ||
        'LAG(' || p_lag_col || ',' || i || ') ' ||
        'OVER(ORDER BY '||p_seq_col||') lag_val ' ||
        'FROM ' || p_in_table || ')';
    EXECUTE IMMEDIATE v_stmt INTO v_corr;
    v_stmt:='INSERT INTO ' || p_out_table ||
        ' (lag_num, correlation) VALUES(:v1, :v2)';
    EXECUTE IMMEDIATE v_stmt using i, v_corr;
    END LOOP;
END;
```

4.4   Establishing the SVM Prediction Model

Subsequently, we prepare training data set for the SVM model, the following SQL statement is used to create the view *Landslide_lag* which contain 12 lagged values:

```
CREATE VIEW Landslide_lag AS
SELECT a.*
FROM (SELECT Month, Displacement, tp,
                LAG(tp, 1)  OVER (ORDER BY month) L1,
                LAG(tp, 2)  OVER (ORDER BY month) L2,
                LAG(tp, 3)  OVER (ORDER BY month) L3,
                LAG(tp, 4)  OVER (ORDER BY month) L4,
                LAG(tp, 5)  OVER (ORDER BY month) L5,
                LAG(tp, 6)  OVER (ORDER BY month) L6,
```

```
        LAG(tp, 7)  OVER (ORDER BY month) L7,
        LAG(tp, 8)  OVER (ORDER BY month) L8,
        LAG(tp, 9)  OVER (ORDER BY month) L9,
        LAG(tp, 10) OVER (ORDER BY month) L10,
        LAG(tp, 11) OVER (ORDER BY month) L11,
        LAG(tp, 12) OVER (ORDER BY month) L12
    FROM Landslide _norm) a;
```

In view *Landslide_lag*, the records whose serial number is greater than 13 are chosen as the training data. Because the lagged values of the first 13 samples are incomplete, so these 13 records are not selected into the training data set. The view *Landslide_train* created below is the training data set:

```
CREATE VIEW Landslide_train AS
    SELECT Month, tp,
    L1, L2, L3, L4, L5, L6, L7, L8, L9, L10, L11, L12
    FROM  Landslide_lag  a
    WHERE  Month>13;
```

Following PL /SQL code use the view *Landslide_train* as training data set, and create a SVM-based landslide displacement prediction model *Landslide_SVM*:

```
BEGIN
    DBMS_DATA_MINING.CREATE_MODEL(
    model_name          => 'Landslide_SVM',
    mining_function     => dbms_data_mining.regression,
    data_table_name     => 'Landslide_train',
    case_id_column_name => 'Month',
    target_column_name  => 'tp');
END;
```

5   Landslide Displacement Prediction Based on the SVM Regression Model

In order to verify performance of the prediction model, we apply the SVM data mining model on the view *Landslide_lag*, get a predictive value of the next month for each sample.The query below accomplish this process, displacement is the original value and *pred* is the predictive value

```
SELECT Month, Displacement,
       PREDICTION(Landslide_svm USING a.*) pred
FROM Landslide_lag a;
```

However, because the model predictions are derived on the basis of the transformed data, so the predicted values should be reversely transformed back to the original scale of the time series. The reverse transformation includes following steps:

1. Reverse normalization transformation: the predicted values ??multiplied by the standard deviation (0.05315), then add the mean(0.0249);
2. Reverse differential transformation : add the value of the past timepoint to the value of the current timepoint;
3. Reverse logarithmic transformation: the exponentiation computation, 10 raised to power of the result of the step 2.

The following query performs the above three steps of the reverse transformation, the results stored in table *Landslide_pred*:

```
CREATE TABLE Landslide_pred AS
SELECT Month, Displacement, power(10, pred) pred
FROM (SELECT Month, Displacement,
        pred + LAG(lp,1) OVER (ORDER BY Month) pred
FROM (SELECT Month, Displacement, LOG(10, Displacement) lp,
        (PREDICTION(Landslide _SVM USING a.*)*0.05315 + 0.0249) pred
        FROM Landslide _lag a));
```

In this statement, the innermost subquery achieves reverse normalization transformation, the middle layer subquery achieves reverse differential transformation, the outermost layer subquery is exponentiation computation.

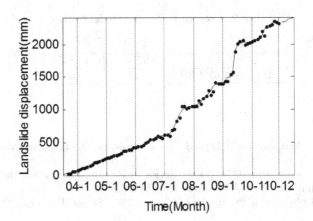

Fig. 4. Actual values, fitted values and predicted values

Fig. 4 shows the actual landslide displacement values (the curve until december 2010), the fitted values of the SVM model (the discrete points until december 2010), the predicted value of the future six months (the curve after december 2010).

Fig. 5 is residuals diagram of the training data. it shows that the residuals are very small, and randomly distributed near zero without trend.

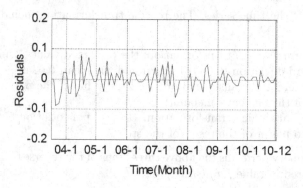

Fig. 5. Residuals diagram of the SVM prediction model

6   Comparison and Analysis

Table 2 presents the Root Mean Squared Error (RMSE) and Mean Absolute Error (MAE) for the training data set and the predicting data set respectively using the SVM model and Autoregressive Integrated Moving Average model(ARIMA).The SVM model is implemented by Oracle's ODM, the ARIMA model is achieved in MATLAB. The result shows that, for the time series of landslide displacement used in current case, the SVM method has better prediction accuracy.

Table 2. Comparison of the model quality

Model	$RMSE$	MAE
SVM	16.743	15.545
ARIMA(0,2,0)(1,0,0)	24.942	16.575

References

1. Box, G.E.P., Jenkins, G.M.: Time Series Analysis, Forecasting and Control. Holden Day, San Francisco (1970)
2. Weigend, A., Gershenfeld, N.: Times Series Prediction: Forecasting the Future and Understanding the Past. Addison-Wesley, MA (1994)
3. Casdagli, M.: Nonlinear Prediction of Chaotic Time Series. J. Physica D 35, 335–356 (1989)
4. Anthony, M., Bartlett, P.L.: Neural Network Learning: Theoretical Foundations. Cambridge University Press, Cambridge (2009)
5. Vapnik, V.: The nature of statistical learning theory. Springer, New York (2000)
6. Oracle USA, Inc.: Oracle OLAP Application Developers Guide, 10g Release 2 (10.2.0.3), B14349-03. Technical report, Redwood City, CA, USA (2006)
7. Oracle USA, Inc.: Oracle Data Mining Application Developer's Guide 11g Release 1 (11.1), B28131-04. Technical report, Redwood City, CA, USA (2008)
8. Glade, T., Anderson, M.G., Crozier, M.J.: Landslide Hazard and Risk. John Wiley and Sons (2006)

Dynamic FP-Tree Pruning for Concurrent Frequent Itemsets Mining

Wei Song, Wenbo Liu, and Jinhong Li

College of Information Engineering
North China University of Technology
Beijing 100144, P.R. China
sgyzfr@yahoo.com.cn

Abstract. To solve the problem of huge memory usage of FP-tree construction and traversal in FP-growth, Dynamic-prune, which is a concurrent frequent itemsets mining algorithm based on FP-tree is proposed. On the one hand, by recording the change of support counts of frequent items during the process of FP-tree construction, dynamic FP-tree pruning is implemented. And the rationality is discussed. On the other hand, by using the concurrency strategy, the construction of FP-tree and the discovery of frequent itemsets can be realized simultaneously. Compared with FP-growth, it is not necessary to mine frequent itemsets after the construction of the whole FP-tree in Dynamic-prune. Experimental results show Dynamic-prune is efficient and scalable.

Keywords: Data Mining, Frequent Itemset, FP-Tree, Concurrency.

1 Introduction

Frequent Itemset Mining (FIM) is a fundamental task in data mining. First introduced by Agrawal et al.[1], several variations to the original Apriori algorithm, and completely different approaches, have been proposed[2][3]. The FP-tree[4] and its variations[5][6] have been shown to be very efficient data structures for mining frequent itemsets. Although FP-tree is often compact, it may not be minimal and still occupy a large memory space. The mining performance of FP-growth approach is closely related to the number of conditional FP-trees constructed during the whole mining process and the construction/traversal cost of each conditional FP-tree. Thus, one of the performance bottlenecks of FP-growth is generating a huge number of conditional FP-trees, which costs a lot of time and space.

To reduce the cost of storage and traversal of huge number of FP-trees, Dynamic-prune, a concurrent FIM algorithm based on dynamic FP-tree pruning, is proposed. Firstly, the F-list structure is used for monitoring the changes of frequent items' counts during the construction of FP-tree. Then, the rationality of pruning FP-trees based on F-list is discussed. And the pruning strategies are presented. Finally, two concurrent processes[7] are designed to discover frequent itemsets over the construction of FP-trees. Compared with FP-growth, it is not

Z. Li et al. (Eds.): ISICA 2012, CCIS 316, pp. 111–120, 2012.

necessary to mine frequent itemsets only after the construction of the whole FP-tree. Experimental results show that Dynamic-prune outperforms FP-growth in both efficiency and scalability.

2   Problem Statement

2.1   Frequent Itemset

Let $I=\{I_1, I_2, ..., I_m\}$ be a finite set of items and D be a dataset containing N transactions, where each transaction $t \in D$ is a list of distinct items $t = \{i_1, i_2, ..., t_{|t|}\}$, $i_j \in I, (1 \leq j \leq |t|)$.

A set $X \subseteq I$ is called an itemset. An itemset with k items is called a k-itemset. The *support count* of an itemset X, denoted as $\sigma(X)$, is defined as the number of transactions in which X occurs as a subset. The support of an itemset X, denoted as $sup(X)$, is the percentage of transactions in D containing X, i.e.: $sup(X) = \sigma(X)/|D|$. Let $minsup$ be the threshold minimum support value specified by user. If $sup(X) \geq minsup$, itemset X is called a frequent itemset.

It is observed in[1] an interesting downward closure property: A k-itemset is frequent only if all of its sub-itemsets are frequent.

2.2   FP-Tree and FP-Growth

The FP-growth method[4] is a depth-first algorithm. In the method, Han et al. proposed a data structure called the FP-tree (frequent pattern tree), which is a compact representation of all relevant frequency information in a database.

An FP-tree T is a tree structure, as constructed below.

1. It consists of one root labeled as 'null', denoted as $T.root$; a set of item-prefix subtrees as the children of the root, denoted as $T.tree$; and a frequent-item-header table, denoted as $T.header$.
2. Each node in the item-prefix subtree consists of five fields: *item, count, node − link, children* and *parent*, where *item* registers which item this node represents; *count* registers the number of transactions represented by the portion of the path reaching this node; *node − link* links to the next node in the FP-tree carrying the same item, or 'null' if there is none; *children* registers the children node of this node, or 'null' if there is none; and *parent* registers the parent node of this node.
3. Each entry in T.header consists of two fields: (1) *item*, item this entry represents and (2)*node − link*, a pointer pointing to the first node in the FP-tree carrying the item.

In an FP-tree T, a *conditional pattern* of X is defined as $CP(X) = (Y : num)$, where YX is the set of items in the path from $T.root$ to X, num is the support count of X in this path. The set of all X's conditional pattern is called the *conditional pattern base* of X, denoted as $CPB(X)$. The FP-tree constructed from the conditional pattern base is called X's *conditional FP-tree*, denoted as $CT(X)$.

Given an item i in $CT(X).header$, by following the linked list starting at i in $CT(X).header$, all branches that contain item i are visited. The portion of these branches from i to the root forms the conditional pattern base of $X \cup \{i\}$, so the traversal obtains all frequent items in this conditional pattern base. The FP-growth method then constructs the conditional FP-tree $CT(X \cup \{i\})$ by first initializing its header table based on the frequent items found, then revisiting the branches of $CT(X)$ along the linked list of i and inserting the corresponding itemsets in $CT(X \cup \{i\})$. The above procedure is applied recursively, and it stops when the resulting new FP-tree contains only one branch. The complete set of frequent itemsets can be generated from all single-branch FP-trees.

3 Dynamic Pruning of FP-Tree

As we can see from Section 2.2, FIM based on FP-growth consists of two steps: FP-tree construction and frequent itemsets discovery. The main work done in the FP-growth method is traversing FP-trees and constructing new conditional FP-trees after the first FP-tree is constructed from the original database. When using FP-growth in large database with low support threshold, the storage and traversal cost of numerous conditional FP-trees are high. Thus, the question is, can we reduce the storage space and traversal time so that the method can be sped up? Furthermore, can we discover frequent itemsets during the process of FP-tree construction, instead of after the construction of the whole FP-tree? The answers are yes, by using a simple additional data structure.

Definition 1. Let T be an FP-tree, $I_T = \{i_1, i_2, ..., i_n\}$ be itemsets in $T.header$, the *frequent item list* (F-list) is an array with size n. Each element of the array corresponds to a tuple $(item, (item))$, where $item \in I_T$; the initial value of $\sigma(item)$ is the support count of $item$ in T, $\sigma(item)$ will be reduced by 1 if a transaction containing $item$ is inserted into T, until $\sigma(item)$ is reduced to 0.

Given the F-list, we first arrange the frequent items in each transaction in conditional database according to the order in F-list, such as frequency-descending order. Then, during the construction of FP-tree, when a transaction t is inserted into the FP-tree, the counts of items contained in t in F-list will subtract 1. If the count of certain item in F-list is reduced to 0, the nodes containing this item and their offspring nodes can be pruned. The rationality of this pruning strategy is based on the following theorem.

Theorem 1. Let i be an item in F-list of FP-tree T, $N = \{n_1, n_2, ..., n_k\}$ be the set of nodes containing i, during the construction of T, if the count of item i in F-list is reduced to 0, nodes in N will not change during the following construction process of T, and the conditional patterns of items contained in the offspring nodes of nodes in N will not change either.

Proof. During the construction of FP-tree, if the count of item i in F-list is reduced to 0, we can see no nodes containing i will be inserted in T, and nodes in N will not change.

For $\forall n \in N$, we have the following two cases:

1. If n is a leaf node, there are no offspring nodes of n.
2. If n is not a leaf node, based on the construction of FP-tree, no items will add into T as offspring nodes of n. Thus, the support counts of items contained in the offspring nodes of n will not change. So the conditional patterns of these nodes will not change either.

To discover frequent itemsets from the pruned branches, we give the following definition.

Definition 2. Let T be an FP-tree, $I_T = \{i_1, i_2, ..., i_n\}$ be itemsets in $T.header$, the *conditional pattern list* of T is an array with size n, denoted as $CPL(T)$. Each element of the array corresponds to a triple (*item, flag, CPB(item)*), where *item* $\in I_T$, *flag* is a Boolean variable whose values are in $\{TRUE, FALSE\}$, $CPB(item)$ is the conditional pattern base of *item*.

According to Theorem 1, during the construction of FP-tree T, once the support count of item i is reduced to 0, we can prune the nodes containing i and their offspring nodes, then store the pruned branches in $CPL(T)$. Set *flag* of i as TRUE, which means we can start the mining process of itemsets containing i. For other items stored in $CPL(T)$ together with i, set their *flags* as FALSE.

4   Concurrent Mining of Frequent Itemsets

To simultaneously prune and discover itemsets during the construction of FP-tree, we propose the Dynamic-prune algorithm based on the concurrent processes. The Dynamic-prune algorithm can be divided into two sub-tasks: One is to prune FP-tree dynamically and store the results into CPL in the shared buffer. The other is to read conditional pattern from CPL and mines frequent itemsets in FP-tree-like method. These two tasks are implemented by the following concurrent Algorithm 1 and Process 2.

Algorithm 1. Process 1

1: **procedure** DYNAMIC PRUNE FP-TREE(*itemset X, CPB(X), minsup*)
2:     Scan database $CPB(X)$ once. Delete infrequent items;
3:     Sort frequent single items in supports descending order. Construct F-list. Sort each pruned transaction in the same order as F-list;
4:     **for** each each transaction $T \in CPB(X)$ **do**
5:         op_P (mutex); //if mutex==1, get CPL in buffer, mutex−; otherwise waiting;
6:         Insert T into FP-tree;
7:         **for** each item $i \in T$ **do**
8:             $\sigma(i) - -$in F-list;
9:             **if** $\sigma(i) == 0$ **then**
10:                 **for** each node n in FP-tree **do**
11:                     **if** $n.item == i$ **then**

```
12:                    Compute CP(i);
13:                    if ∃e ∈ CPL, such that e.item == i then
14:                        CP(i) → e.CPB, e.flag = TRUE;
15:                    elseThere is no element e ∈ CPL, such that e.item == i
16:                        Construct new element m in CPL, m.item = i, CP(i) →
        m.CPB, m.flag = TRUE;
17:                    end if
18:                    Insert(n);//Insert n and its offspring nodes into CPL.
19:                    Delete(n);//Delete n and its offspring nodes from FP-tree.
20:                end if
21:            end for
22:        end if
23:    end for
24:    op_V (mutex); //Release mutex, mutex++.
25:    op_V (signal); // Wake up Process 2 to discover frequent itemsets in CPL
        in buffer.
26:    end for
27: end procedure
28: procedure INSERT(node n)
29:    for each n_c ∈ n.children do
30:        Compute CP(n_c.item);
31:        if ∃e ∈ CPL, such that e.item == n_c.tiem then
32:            CP(n_c.item) → e.CPB;
33:        else//There is no element e ∈ CPL, such that e.item == n_c.item.
34:            Construct new element m in CPL, m.item = n_c.item, CP(n_c.item) →
        m.CPB, m.flag = FALSE;
35:        end if
36:        Insert(n_c);
37:    end for
38: end procedure
39: procedure DELETE(n)
40:    for each n_c ∈ n.children do
41:        Delete(n_c);
42:    end for
43:    if n.children == ∅ then
44:        delete n;
45:    end if
46: end procedure
```

Algorithm 2. Process 2

```
1: procedure FREQUENT ITEMSET MINING(itemsetX, CPB(X), minsup)
2:     op_P (signal); //Waiting for Process 1
3:     op_P (mutex); //if mutex== =1, get CPL in buffer, mutex−; otherwise waiting.
4:     for each e ∈ CPL AND e.flag == TRUE AND e.CPB ≠ ∅ do
5:         Construct conditional FP-tree of e.item, denoted by CT(e.item);
6:         FP-growth (e.item, CT(e.item));
7:         Delete e from CPL;
8:     end for
```

```
 9:      op_V (mutex); // Release mutex, mutex++.
10: end procedure
11: procedure FP-GROWTH(itemset X, conditional frequent pattern treeCT(X))
12:     if CT(X) only contains a single branch B then
13:         for each subset Y of the set of items in B do
14:             Output Y ∪ X, sup(Y ∪ X) equals to the smallest support of nodes in
    Y;
15:         end for
16:     else
17:         for each i ∈ CT(X).header do
18:             FP-growth(X ∪ {i}, CT(X ∪ {i}));
19:         end for
20:     end if
21: end procedure
```

5   Performance Evaluation

We compared the performances of Dynamic-prune with algorithms Apriori[1], FP-growth[4], and Index-BitTableFI[3]. They are all implemented in C++ and compiled with Microsoft Visual Studio 2005.

5.1   Test Environment and Datasets

We chose several real and synthetic datasets for testing the performance of Dynamic-prune. All datasets are taken from the FIMI repository page[8]. The mushroom dataset is described in terms of physical characteristics as nominal-valued attributes as opposed to numerical-values. Some of the important nominal-valued attributes include classification as either poisonous or edible. The chess and connect datasets are derived from their respective game steps. Typically, these real datasets are very dense, i.e., they produce many long frequent item-sets even for very high values of support. We also chose a synthetic dataset T10I4D100K. This dataset mimics the transactions in a retailing environment. Usually the synthetic dataset is sparse when compared to the real sets. Table 1 shows the characteristics of these datasets. The experiments were conducted on a Windows XP PC equipped with a Pentium 2.2GHz CPU and 2GB of RAM memory.

Table 1. Characteristics of datasets used for experiment evaluations

Dataset	Number of Items	Number of Records	Average Length
mushroom	22	8,124	23
chess	75	3,196	37
connect	129	65,557	43
T10I4D100K	870	100,000	11

5.2   The Runtime

Figures 1-4 use total running time as the performance metric. On mushroom, chess and T10I4D100K, the experimental results show that computational time sharply increases with the decreasing of the minsup in Apriori, FP-growth and Index-BitTableFI algorithm, while it increases slowly in Dynamic-prune. On connect, execution time still sharply increases with the decreasing of the minsup in Apriori and Index-BitTableFI. However, both FP-growth and Dynamic-prune illustrate steady trends. In general, Dynamic-prune outperforms Apriori, FP-growth and Index-BitTableFI in execution time on both dense and sparse datasets.

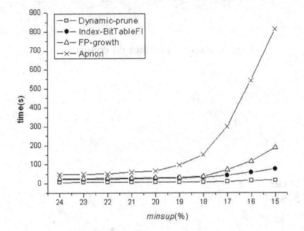

Fig. 1. Execution times on mushroom

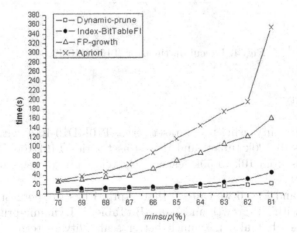

Fig. 2. Execution times on chess

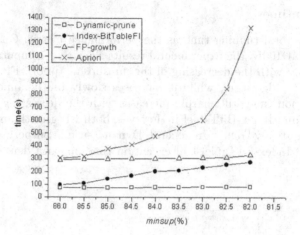

Fig. 3. Execution times on connect

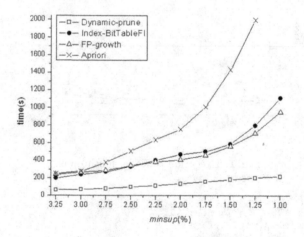

Fig. 4. Execution times on T10I4D100K

5.3   Scalability

We test the scalability using the dataset series T10I4D10-100k with base sizes from 10k tuples to 100k tuples, and the dataset series T10D100kI10-50k with number of items from 10k to 50k respectively. They are all generated by IBM data generator[9].

From Fig. 5 and Fig. 6 we can see that, Apriori has the poorest scalability. In comparison with FP-growth and Index-BitTableFI, Dynamic-prune not only runs much faster, but also has much better scalability in terms of base size and item size: the slope ratio of Dynamic-prune is much lower than those of FP-growth and Index-BitTableFI.

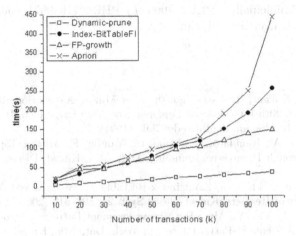

Fig. 5. T10I4D10-100k, minsup=40%

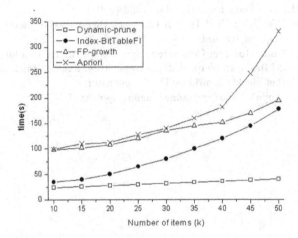

Fig. 6. T10D100kI10-50k, minsup=40%

6 Conclusions

This paper proposes an efficient concurrent frequent itemset mining algorithm Dynamic-prune, which divides the task of FIM into two concurrent subtasks of pruning tree and discovery of frequent itemsets. Dynamic-prune solves the main drawbacks of the FP-growth method, namely, huge storage and traversal cost of conditional FP-trees by pruning FP-tree dynamically. The experimental results showed that Dynamic-prune is efficient and scalable.

Acknowledgment. The work is partly supported by the National Natural Science Foundation of China (61105045), Funding Project for Academic Human Resources Development in Institutions of Higher Learning Under the Jurisdiction of Beijing Municipality (PHR20100509 / PHR201108057), and key project of North China University of Technology.

References

1. Agrawal, R., Srikant, R.: Fast Algorithms for Mining Association Rules in Large Databases. In: 20th International Conference on Very Large Data Bases, pp. 487–499. Morgan Kaufmann, Santiago de Chile (1994)
2. Pietracaprina, A., Riondato, M., Upfal, E., Vandin, F.: Mining Top-K Frequent Itemsets Through Progressive Sampling. Data Min. Knowl. Discov. 21, 310–326 (2010)
3. Song, W., Yang, B.R., Xu, Z.Y.: Index-BitTableFI: An Improved Algorithm for Mining Frequent Itemsets. Knowl.-Based Syst. 21, 507–513 (2008)
4. Han, J., Pei, J., Yin, Y., Mao, R.: Mining Frequent Patterns without Candidate Generation: A Frequent-Pattern Tree Approach. Data Min. Knowl. Discov. 8, 53–87 (2004)
5. Grahne, G., Zhu, J.: Fast Algorithms for Frequent Itemset Mining using FP-Trees. IEEE Trans. Knowl. Data Eng. 17, 1347–1362 (2005)
6. Liu, G., Lu, H., Yu, J.X.: CFP-Tree: A Compact Disk-Based Structure for Storing and Querying Frequent Itemsets. Inform. Syst. 32, 295–319 (2007)
7. Roscoe, A.W.: Understanding Concurrent Systems. Springer, London (2010)
8. Requent Itemset Mining Implementations Repository, http://fimi.ua.ac.be
9. IBM Quest Market-Basket Synthetic Data Generator, http://www.cs.loyola.edu/~cgiannel/assoc_gen.html

MapReduce-Based Bayesian Automatic Text Classifier Used in Digital Library

Zhen Niu, Zelong Yin, and Huayang Cui

Department of Computer Science and Technology
Harbin Institute of Technology, P.R. China
zhenniuhit@gmail.com

Abstract. Bayesian theorem is an effective method for text classification. But it will consume too much time and resource when used in large-scale database such as digital library. How to process data efficiently becomes a vital problem for the further development of digital library. Because Mapreduce model has strong capacity of processing mass data, Mapreduce-based Bayesian classifier can reduce the time that is caused by machine learning and therefore enhance the overall efficiency. The new method can classify documents into different groups according to its subject. In this way, it is helpful for management and storage of information. Thus Mapreduce-based Bayesian text classifier can be used in digital library successfully which will provide better service for people.

Keywords: Mapreduce, Cloud Computing, Bayesian Text Classifier, Machine Learning, Digital Library.

1 Introduction

Nowadays, people can easily get access to numerous documents through digital library. Many countries constructed their national digital libraries which are funded by government, e.g. Korea[1]. Universities and other research institutes establish their own digital library all over the world. Although digital libraries develop very fast, there are still many dilemmas such as an effective method to classify documents (e.g. books, articles, academic journals etc.) automatically according to their subjects. There are many famous algorithms e.g. K-Nearest Neighbor (KNN), Support Vector Machine (SVM), Decision Tree Model, and Neural Network, but those popular algorithms including Bayesian classifier cannot be applied to large database with high speed. So we need find another way to speed up text classification.

In this paper, we introduce Mapreduce to improve Bayesian classifier. Mapreduce programming model has been used widely in different fields and it has many advantages when handling huge database such as a digital library. Nowadays, cloud computing is used widely in different fields especially for large database. Apache Hadoop Mapreduce is an open-source style which is an implementation similar to the Google Mapreduce[2]. Hadoop provides a sophisticated framework for cloud platform which is very useful for processing large-scale date sets[3]. By

Z. Li et al. (Eds.): ISICA 2012, CCIS 316, pp. 121–126, 2012.

parallel computing, Mapreduce distributed processing framework can enhance the efficiency of classifying text in digital library.

2   Background Knowledge

2.1   Digital Library

The definition of digital library varies from time to time. The general idea about it is data repository which is responsible for collecting, storing and managing information[4][5]. With the fast development of digital library, how to provide better management and service becomes one of the toughest problems and classification of document automatically plays a significant role in it. In order to drive the digital library to develop, we use the popular cloud computing to the text classification.

2.2   Mapreduce Used in Data Processing

Mapreduce is a simple programming model for data processing. The main task of Mapreduce is dividing the data in to different blocks. In the model, NameNode divides data into small blocks and then distributes them to DataNodes. DataNodes process small blocks and pass the answer to its NameNode. Fig. 1 shows the use of NameNode and DataNodes.

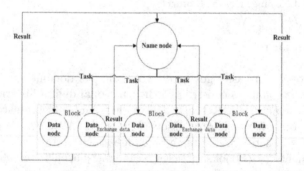

Fig. 1. Name node and data nodes

Map Task. Map process each block separately in parallel. The results are reassembled into aggregate intermediate values.

$Map(in\_key, in\_value) \rightarrow IntermediateResult(out\_key, out\_value)$

Reduce Task. The NameNode then collects the answer and combines them to form the output.

$Reduce(out\_key, out\_value) \rightarrow value\ list$

The process of Mapreduce is showed in Fig. 2.

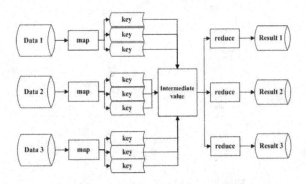

Fig. 2. The process of Mapreduce

3   Mapreduce-Based Bayesian Text Classifier

3.1   Improved Naive Bayesian Algorithm

Based on Naive Bayesian theorem, each document in digital library is represented as a vector $^d(w_1, w_2, ..., w_n)$, where $w_1, w_2, ..., w_n$ are feature items of different subjects. There are n feature items for the documents of a certain subject. We suppose there are m classes $C_1, C_2, ..., C_3$ and then the possibility that d belongs to C_k is:

$$P(C_k|d) = \frac{P(C_k) \times P(d|C_k)}{P(d)} \quad k = 1, 2, ..., m \tag{1}$$

$$P(d) = \sum_{k=1}^{n} P(d|C_k) \quad k = 1, 2, ..., m \tag{2}$$

In order to know which subject a document belongs to, we must calculate the largest $P(C_k|k)$[6]. According to comparison between key words in each document and known feature items of a certain subject, $w_1, w_2, ..., w_n$, we can get the possibility that vector d belongs to C_k. Because key words in each document are random and independent with each other, so we get:

$$P(d|C_k) = P(w_1, w_2, ..., w_n|C_k) = \prod_{i=1, k=1}^{n} P(w_i|C_k) \tag{3}$$

The aim is to classify all documents into different subjects. Before classifying the documents automatically, we must train corpus first and then we can form our own knowledge base. During the training, we can extract key words from corpus as feature items $w_1, w_2, ..., w_n$ and also calculate frequency of each key word $f_1, f_2, ..., f_n$[7][8]. After this process, we can establish a huge and stable knowledge base. The Fig. 3 shows the process of machine learning and this process play a significant role in text classification.

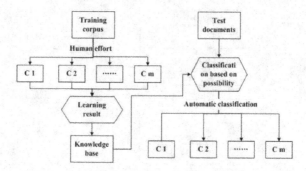

Fig. 3. Machine learning process

3.2   Mapreduce-Based Bayesian Model

Because digital library is a very large database, in order to get precise result through machine learning, we should train large-scale corpus. Only in this way, can Bayesian text classifier show its competitive edge. Thus, we improve Bayesian classifier by using Mapreduce. Mapreduce can reduce the training time and enhance the speed of classification because it has many advantages in data processing.

- Training Model
    - Step 1 Calculate prior possibilities of known documents of different subjects;
    - Step 2 Complete the statistics of key words (feature items) and word frequency;
    - Step 3 Calculate the conditional probability of key words;
  We use the output of step 1 and step 3 to form our knowledge base. Fig. 4 shows how Mapreduce is used in training model.

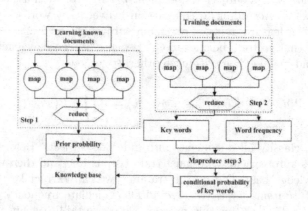

Fig. 4. Training model using Mapreduce

- Classifying Model
  - Step 4 We can use Mapreduce to calculate the $P(C_k|k)$, and then classify the documents into different subjects according to the possibility value. Fig. 5 shows this process;

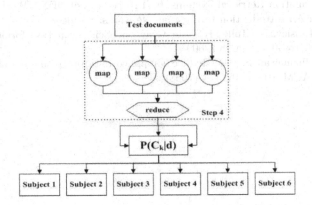

Fig. 5. Test model using Mapreduce

4 Conclusion

Bayesian text classifier is an effective method only when machine learning has large-scale data for training at first. This paper introduces a novel method which uses Mapreduce programming model to process enormous database in order to reduce time and humans effort. Mapreduce-based Bayesian text classifier can not only boost the efficiency, but also enhance the recall, precision and accuracy. Thus it can be used in digital library effectively. Improved Bayesian text classifier can classify documents into different groups according to its subject. Thus it is useful to manage and store large-scale information which means better service of digital library.

References

1. Lee, S., Cho, S.: Digital Libraries in Korea. In: International Conference on Research and Practice, Kyoto, pp. 130–135 (2000)
2. Gunarathne, T., Wu, T.-L., Qiu, J.: MapReduce in the Clouds for Science. In: Cloud Computing Technology and Science (CloudCom), Conference Publication, pp. 565–572 (2010)
3. Yang, G., Huanggang: The Application of MapReduce in the Cloud Computing. In: 2nd International Symposium on Intelligence Information Processing and Trusted Computing (IPTC), Conference Publication, pp. 154–115 (2011)
4. Alias, N.A.R., Noah, S.A., Abdullah, Z., Omar, N., Yusof, M.M.: Application of Semantic Technology in Digital Library. In: 2010 International Symposium in Information Technology (ITSim), vol. 3, pp. 1514–1518 (2010)

5. Warren, P., Thurlow, I., Alsmeyer, D.: Applying semantic technology in digital library: a case study. Emerald, Library Management 26(4/5), 196–205 (2005)
6. Sebastiani, F.: Machine learning in automated text categorization. ACM Computing Surveys 34(1) (2002)
7. de Campos, L.M., Femaindez-Luna, J.M., Huete, J.F.: Building Bayesian Network-Based Information Retrieval Systems. In: Proceedings of DEXA Workshop (2000)
8. Jensen, F.: An Introduction to Bayesian Networks. Springer, New York (1998)
9. Hey, T., Tansley, S., Tolle, K.: Jim Gray on eScience: a transformed scientific method. Microsoft Research (2009)
10. Dean, J., Ghemawat, S.: MapReduce: simplified data processing on large clusters. Commun. ACM 51(1), 107–113 (2008)

On Dependences among Objects and Attributes

Sylvia Encheva

Stord/Haugesund University College
Bjørnsonsg. 45, 5528 Haugesund, Norway
sbe@hsh.no

Abstract. Intensive research on reusable learning objects has been going on for more than three decades. Both content providers and information technology experts work for developing and implementing models that can facilitate the process of developing, storing, classifying and delivering of learning objects to students. Considering the amount of resources required to support that process and the current restrictions on educational budgets there is an obvious need for further research in this area. In this paper we pay special attention on how to select the most suitable learning objects from a database.

Keywords: Rough Sets Approximations, Knowledge, Intelligent Systems.

1 Introduction

Learning objects are defined in various ways in the literature and are popularly understood as small entities that can be used in learning and teaching situations. Animations, Java applets, interactive simulations, and quizzes are among the most applicable ones. Important characteristics of learning objects are discussed in details in [1]. Technical aspects of learning objects are presented in [2], [4], [5], [6], [7], and [12].

Intensive research on reusable learning objects has been going on for more than three decades. Both content providers and information technology experts work for developing and implementing models that can facilitate the process of developing, storing, classifying and delivering of learning objects to students. Considering the amount of resources required to support that process and the current restrictions on educational budgets there is an obvious need for further research in this area. In this paper we pay special attention on how to select the most suitable learning objects from a database.

The rest of the paper is organized as follows. Section 2 contains definitions of terms used later on. Section 3 illustrates our approach. Section 4 contains the conclusion of this work.

2 Background

Let P be a non-empty ordered set. If $sup\{x, y\}$ and $inf\{x, y\}$ exist for all $x, y \in P$, then P is called a *lattice* [3]. In a lattice illustrating partial ordering of

Z. Li et al. (Eds.): ISICA 2012, CCIS 316, pp. 127–133, 2012.

knowledge values, the logical conjunction is identified with the meet operation and the logical disjunction with the join operation.

A *context* is a triple (G, M, I) where G and M are sets and $I \subset G \times M$. The elements of G and M are called *objects* and *attributes* respectively [3], [13].

For $A \subseteq G$ and $B \subseteq M$, define

$$A' = \{m \in M \mid (\forall g \in A) \ gIm\},$$

$$B' = \{g \in G \mid (\forall m \in B) \ gIm\}$$

where A' is the set of attributes common to all the objects in A and B' is the set of objects possessing the attributes in B.

A *concept* of the context (G, M, I) is defined to be a pair (A, B) where $A \subseteq G$, $B \subseteq M$, $A' = B$ and $B' = A$. The *extent* of the concept (A, B) is A while its *intent* is B. A subset A of G is the extent of some concept if and only if $A'' = A$ in which case the unique concept of the which A is an extent is (A, A'). The corresponding statement applies to those subsets $B \in M$ which is the intent of some concepts.

The set of all concepts of the context (G, M, I) is denoted by $\mathfrak{B}(G, M, I)$. $\langle \mathfrak{B}(G, M, I); \leq \rangle$ is a complete lattice and it is known as the *concept lattice* of the context (G, M, I).

2.1 Rough Sets

Rough Sets were originally introduced in [8]. The presented approach provides exact mathematical formulation of the concept of approximative (rough) equality of sets in a given approximation space. An *approximation space* is a pair $\mathcal{A} = (U, R)$, where U is a set called universe, and $R \subset U \times U$ is an indiscernibility relation [9].

Equivalence classes of R are called *elementary sets* (atoms) in $\mathcal{A}$. The equivalence class of R determined by an element $x \in U$ is denoted by $R(x)$. Equivalence classes of R are called *granules* generated by R.

The following definitions are often used while describing a rough set $X, X \subset U$:

- the R-*upper approximation* of X

$$R^\star(x) := \bigcup_{x \in U} \{R(x) : R(x) \cap X \neq \varnothing\}$$

- the R-*lower approximation* of X

$$R_\star(x) := \bigcup_{x \in U} \{R(x) : R(x) \subseteq X\}$$

- the R-*boundary region* of X

$$RN_R(X) := R^\star(X) - R_\star(X)$$

An information system is a pair $S = (U, A)$, where U and A, are non-empty finite sets called the universe, and the set of attributes, respectively such that $a : U \to V_a$, where V_a, is the set of all values of a called the *domain* of a. Any subset B of A determines a binary relation $I(B)$ on U , which will be called an indiscernibility relation, and defined as follows: $(x, y) \in I(B)$ if and only if $a(x) = a(y)$ for every $a \in A$, where $a(x)$ denotes the value of attribute a for element x. Obviously $I(B)$ is an equivalence relation. The family of all equivalence classes of $I(B)$, i.e., a partition determined by B, will be denoted by $U/I(B)$, or simply by U/B; an equivalence class of $I(B)$, i.e., block of the partition U/B, containing x will be denoted by $B(x)$ and called B-granule induced by x, [10].

If (x, y) belongs to $I(B)$ we will say that x and y are B-indiscernible (indiscernible with respect to B). Equivalence classes of the relation $I(B)$ (or blocks of the partition U/B) are referred to as B-elementary sets or B-granules, [10].

Elements in the index set $A = a_1, a_2, ..., a_m$ are the importance degree of attribute set where each index in the system is determined by:

$$S_A(a_i) = \frac{|POS_A(A)| - |POS_{A-a_i}(A)|}{|U|}$$

where $i = 1, 2, 3, ..., m$ and the weight of index a_i is given by

$$w_i = \frac{\cdot \; S_A(a_i)}{\sum_{i=1}^{m} S_A(a_i)}.$$

The institution assessment model is defined by

$$P_j = \sum_{i=1}^{m} w_i f_i$$

where P_j is the comprehensive assessment value of assessed jth object, f_i is the assessment value of ith index a_i according to the comprehensive assessment value, [11].

3   Learning Objects

A set of learning objects is evaluated by educational and IT specialists. The outcome is shown in Table 1, where Lo is a learning object and A is an attribute (criterion). The degree to which a learning object satisfies a criterion varies from the 3 to 1 where 3 is the highest. To avoid confusion we leave empty cells where it should written '1' in Table 1.

3.1   Formal Concept Analysis Approach

The lattice in Fig. 1 illustrates which learning objects have medium and high level of usability with respect to any of the five attributes. Note that f.ex. $Lo7$ and $Lo8$ share four attributes while $Lo1$, $Lo3$, $Lo5$, $Lo5$, $Lo7$ and $Lo8$ share five attributes.

Table 1. Learning objects

	A1	A2	A3	A4	A5
Lo1	2		3	2	
Lo2				2	
Lo3	2		3	2	2
Lo4	2	2		2	3
Lo5	3		3	2	
Lo6			2	2	
Lo7	3	2	3	2	
Lo8	3	2	3	3	
Lo9			3	2	
Lo10		2		2	2

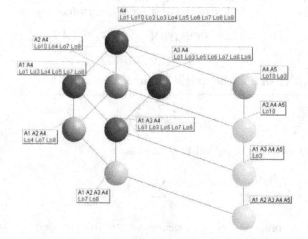

Fig. 1. Lattice for learning objects as in Table 1

Table 2. Learning objects marked as medium

	A1	A2	A3	A4	A5
Lo1	x			x	
Lo2				x	
Lo3	x			x	x
Lo4	x	x		x	
Lo5				x	
Lo6			x	x	
Lo7		x		x	
Lo8		x			
Lo9				x	
Lo10		x		x	x

The lattice in Fig. 2 illustrates which learning objects received medium level of usability with respect to any of the five attributes. It is useful to notice that attribute A4 alone cannot be used to select a subset of particularly useful learning objects, since nearly all of them received the same mark. Learning objects $Lo2, Lo5$ and $Lo9$ are indiscernible if attribute A4 is to be applied alone.

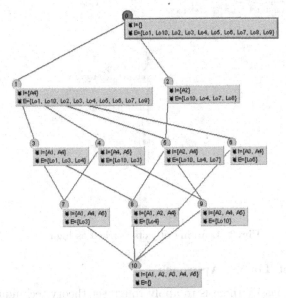

Fig. 2. Lattice for learning objects marked as medium as in Table 2

Table 3. Learning objects

	A1	A2	A3	A4	A5
Lo1			x		
Lo3			x		
Lo4					x
Lo5	x		x		
Lo7	x		x		
Lo8	x		x	x	
Lo9			x		

The lattice in Fig. 3 illustrates which learning objects received high level of usability with respect to any of the five attributes. Furthermore object $Lo8$ is the only one marked with high level with respect to three attributes, while earning objects $Lo2, Lo6$ and $Lo10$ have been marked as medium at highest.

Nodes in a Hasse diagram closer to the lowermost node show concepts where less objects share more attributes, while nodes closer to the uppermost node show concepts where more objects share less attributes. An extent at a node lists all objects in that node and all descendant nodes. An intent at a node lists all attributes possessed by the objects in this node and all ancestor nodes.

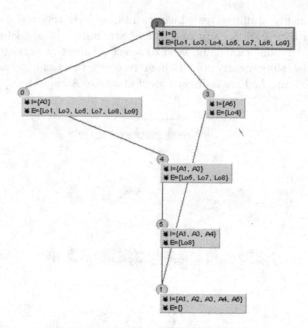

Fig. 3. Learning objects marked as high

3.2 Rough Set Theory Approach

Another way to classify them is to apply rough set theory techniques by calculating indexes weights as described in Section 2.1. The outcome is presented in Table 4.

Table 4. Indexes' weights

Indexes	a_1	a_2	a_3	a_4	a_5
Weights	0.3	0.2	0.3	0.2	0.2

The comprehensive assessment value values for objects $Lo1, ..., Lo10$ are

$$P_1 = 2.3, P_2 = 1.4,$$

$$P_3 = 2.5, P_4 = 2.3,$$

$$P_5 = 2.6, P_6 = 1.7,$$

$$P_7 = 2.8, P_8 = 3.0,$$

$$P_9 = 2.5, P_{10} = 1.8$$

The best three learning objects are $Lo8, Lo7$, and $Lo5$ with corresponding assessment values $3, 2.8$, and 2.6. The least desirable is object $Lo2$ with assessment value 1.4.

4   Conclusion

Providing students with personalized assistance in a form of learning objects is an essential part of modern education. Our work presents an structured way of filtering the best learning objects from a pool of already existing ones. This approach also supports their reusability. Future work is needed for selecting an optimal set of attributes to be applied in the evaluation process of each learning object.

References

1. Barritt, C., Alderman Jr., F.L.: Creating a Reusable Learning Objects Strategy: Leveraging Information and Learning in a Knowledge Economy. Wiley Books, San Francisco (2004)
2. Cisco Systems Inc. Enhancing the Learner Experience (v1.1), Internet (2003), http://www.apan.net/meetings/busan03/materials/ws/education/articles/EnhancingLearnerExp.pdf
3. Davey, B.A., Priestley, H.A.: Introduction to lattices and order. Cambridge University Press, Cambridge (2005)
4. Eap, T.M., Hatala, M., Gasevik, D.: Technologies for enabling the sharing of learning objects. International Journal of Advanced Media and Communication Archive 2(1), 1–19 (2008) ISSN: 1462-4613
5. Encheva, S., Tumin, S.: Cooperative Learning Objects in an Intelligent Web-Based Tutoring System. In: Advanced Industrial Conference on Telecommunications, Service Assurance with Partial and Intermittent Resources Conference, E-Learning on Telecommunications Workshop (AICT / SAPIR / ELETE 2005), Lisbon, Portugal, July 17-22, IEEE Computer Society (2005) ISBN 0-7695-2388-9504-508
6. Encheva, S., Tumin, S.: On facilitating automated evaluation. In: IADIS International Conference e-Learning, pp. 223–225 (2008) ISBN: 978-972-8924-58-4
7. Gehringer, E., Ehresman, L., Conger, S., Wagle, P.: Reusable Learning Objects Through Peer Review: The Expertiza approach. Innovate 3(5) (2007)
8. Pawlak, Z.: Rough Sets. International Journal of Computer and Information Sciences 11, 341–356 (1982)
9. Pawlak, Z.: Rough Sets: Theoretical Aspects of Reasoning About Data. Kluwer Academic Publishing, Dordrecht (1991)
10. Pawlak, Z.: In Pursuit of Patterns in Data Reasoning from Data - The Rough Set Way. In: Alpigini, J.J., Peters, J.F., Skowron, A., Zhong, N. (eds.) RSCTC 2002. LNCS (LNAI), vol. 2475, pp. 1–9. Springer, Heidelberg (2002)
11. Ramasubramanian, P., Banu Priya, S., Dhanalakshmi, T.: Study on assessment method for technical institutions using rough sets. International Journal of Research and Reviews in Information Sciences (IJRRIS) 1(4), 99–103 (2011) ISSN: 2046-6439
12. Valderrama, R., Ocana, L., Sheremetov, L.: Development of intelligent reusable learning objects for web-based education systems. Expert Systems with Applications 28(2), 273–283 (2005) ISSN: 09574174
13. Wille, R.: Concept lattices and conceptual knowledge systems. Computers Math. Applications 23(6-9), 493–515 (1992)

An Improved Bayesian Inference Method
for Data-Intensive Computing[*]

Feng Ma[1,2] and Weiyi Liu[1]

[1] School of Information Science and Engineering
Yunnan University
Kunming, P.R. China
[2] School of Computer and Information
Yunnan University of Finance and Economics
Kunming, P.R. China
Mafeng2009@foxmail.com

Abstract. Recent years, data-intensive computing has become a research hotspot. It also proposed a new challenge to traditional Bayesian inference methods. It is known that, traditional Bayesian inference methods could do a good job when all data stay in a single station. However, when come to data-intensive computing, it would hard to apply to new situation directly because data often been distributed in multi stations under data intensive computing. Among different Bayesian inference methods, random algorithm often been regarded as a common and effective one. And the sampling method adopted in random algorithm would largely influence the efficiency of this random algorithm. Gibbs sampling method often been used in random algorithm for Bayesian inference. Taking all of this into consideration, an improved Bayesian inference method for data-intensive computing is developed in this paper, which first use improved Gibbs sampling method in each station to gain the suitable information, then union them together to infer the final result. The validity of this method is discussed in theory and illustrated by experiment.

Keywords: Data-intensive Computing, Bayesian Inference, Gibbs sampling.

1 Introduction

Nowadays, there are more and more data needed to be processed in all kinds of application, and data-intensive computing[1] has become a research hotspot. Bayesian network, a probability graph model, can effectively present and infer uncertain knowledge in data when all of it stays in a single station. However, under data-intensive computing environment, data often been distributed into

---

[*] This paper is sponsored by: The National Natural Science Foundation of China(No.61163003), the Yunnan Provincial Department of Education Foundation(No.2011Y029).

Z. Li et al. (Eds.): ISICA 2012, CCIS 316, pp. 134–144, 2012.

multi stations. So traditional Bayesian inference methods can not been used in data-intensive computing directly, improvement has to be made to make it suitable for the new situation.

In traditional Bayesian inference methods, random algorithm is a frequently used approximate method, and the sampling method adopted in the random algorithm often largely influences its efficiency. Gibbs sampling method is a common and effective sampling method, which has been used in many fields[2–5]. Therefore, this article proposed a Bayesian inference method under Data-Intensive Computing, which adopts improved Gibbs sampling method as the core of the random algorithm in inference process. It first use improved Gibbs sampling method in each station to gain the suitable information independently, then union them together to infer the final result. The validity of this method is discussed in theory and illustrated by experiment.

The main contributions of this paper are:

- The proposed Bayesian inference method under data-intensive computing extends the applied scope of the traditional Gibbs sampling method, so that it can process the data which distributed in multi stations;
- We proved the validity of this method. That is to say, assuming that we put all these distributed data into a single station and use the traditional Bayesian inference method in it, the final result would be equal with that we got by Bayesian inference method under data-intensive computing in multi stations.

The remainder of this paper is organized as follows: Section 2 gives some background knowledge, Section 3 discuss about the Bayesian inference method under data-intensive computing, Section 4 shows the experimentation result and gives the analysis, and Section 5 gives the conclusions and future work.

2 Data-Intensive Computing and Bayesian Inference

Due to the rapid development of Information collection technology, there are more and more data needed to be processed in all kinds of field, and data-intensive computing has become a important research point. Data-intensive computing is the combination of traditional data management high performance computing, data analysis and data mining. The main research area of it is mass data automatically analysis and intelligent process to find out the potential knowledge. And it has important theory and practical value[1, 6, 7]. To meet the need of these auto analysis and intelligent process, valid knowledge presentation is the important base, effective knowledge inference is the important guaranteed.

Probability graph model as an effective model of knowledge presentation and inference is the research hotspots in Artificial Intelligent. Bayesian Network (BN)[8], is a kind of graph model which could present the dependence between variables. It is a effective tool in dealing with uncertain reasoning and data analysis[9–11]. BN also can be seen as a DAG. The nodes stand for the random variables, and the arcs represent the dependent relationship between the variables.

Bayesian networks are directed acyclic graphs (DAGs), in which nodes represent random variables and the lack of an arc between two nodes represents the conditional probabilistic independence of the two unlinked variables. The networks structure is a factorization of a joint probability distribution[19]. A node without parents is called a root node and a node without children is a leaf node. Consider an example concerning O_3 State related variables in the atmosphere. As shown in Fig. 1, the O_3 state of the previous day (*PreO3*, non-methane hydrocarbons(NMHC), sulfur dioxide(SO_2), O_3 and mitric oxide (NO) are root nodes and the O_3 state is a leaf node. Root nodes, which have no causal node, may only be associated with unconditional probabilities, other nodes. That is, *Pollutants* and O_3 state must have own conditional probability table. Table 1 shows conditional probability table that is related to the *Pollutants* and its parent nodes. We assume $NMHC$, SO_2, O_3 and NO have their own States whether *ture*(T) or *false*(F), respectively. If States of $NMHC$, SO_2, O_3 and NO are all 'T', the true state probability of *Pollutants* is about 0.5.

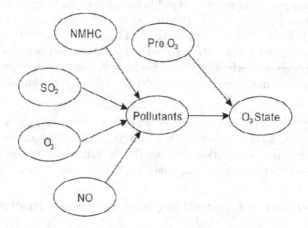

Fig. 1. Example of Bayesian Network

There are many kinds of Bayesian inference methods[12–15], and MCMC(Markov Chain Monte Carlo) method[9] is one of them. Furthermore, Gibbs sampling method is the special form of MCMC method. The traditional Gibbs Sampling process can be described as follows[16]:

Assuming that $\psi(x) = \psi(x_1, ..., x_m)$ is a m-dimension joint distribution, $\psi(x_i|x_{-i})$ is the conditional distribution, $x_{-i} \overset{\triangle}{=} \{x_j, j \neq i\}$, $\vec{e} = \{e_1, e_2, ..., e_n\}$ is called evidence variables.

1. Taking $x^{(0)} = (x_1^{(0)}, ..., x_m^{(0)}$ as initial valuethen sampling data according to $\psi(x_i|x_{-i})$.
2. Got $x_i^{(0)}$ from $\psi(x_i|x_{-i}^{(0)})$, $i = 1, 2, ..., m$.
3. Finish one transition from $x^{(0)}$ to $x^{(1)}$, that is $q(x^{(0)} \to x^{(1)})$.
4. Repeat (1)-(3), until $x^{(t)}, t = 1, 2, ...$, reach a stationary distribution.

Table 1. Characteristics of datasets used for experiment evaluations

$NMHC$	SO_2	O_3	NO	$Pollutants$
T	T	T	T	0.95
T	T	T	F	0.80
T	T	F	T	0.85
T	T	F	F	0.50
T	F	T	T	0.85
T	F	T	F	0.50
T	F	F	T	0.50
T	F	F	F	0.20
F	T	T	T	0.85
F	T	T	F	0.50
F	T	F	T	0.50
F	T	F	F	0.20
F	F	T	T	0.50
F	F	T	F	0.20
F	F	F	T	0.20
F	F	F	F	0.05

3 Bayesian Inference Method under Data-Intensive Computing

In this section, we will discuss about the detail of the Bayesian Inference Method under Data-Intensive Computing(BIMDIC), which take the improved Gibbs sampling method as the core algorithm. Compared with traditional Gibbs sampling method, BIMDIC has two main issues need to be considered for data distributed in multi stations. One is how to make sure that each station has the same joint distribution $\psi(x)$, and the other is that the whole evidence variables may be divided into several groups and reach different situations, how to make the final infer result suitable for the whole data. To make it simple, we put a Bayesian network, which could present the joint distribution $\psi(x)$ in every station. So the following part will focus on how to resolve the second problem.

3.1 Description of the BIMDIC

For the convenience of description, we assume that $S_1, S_2, ..., S_n$ are the n stations, g is the Bayesian network in each of them. $\vec{X} = \{X_1, X_2, ..., X_n\}$ is the variable in g. To make it easy to follow, we would like to give some definitions and propositions before describe the detail of the BIMDIC.

Definition 1. *If S_m stands for one of the n stations, evidence variables $\vec{e_m} = \{X_i = x_i, X_j = x_j, ..., X_k = x_k\}$ reaches on S_m, $1 \leq m \leq n$, $1 \leq i \leq j \leq k \leq n$, then it called $\vec{e_m} \in S_m$.*

Definition 2. *If $\vec{e_m} \in S_m$, $1 \leq m \leq n$, and $\vec{E} = \vec{e_1} \cup \vec{e_2} \cup ... \cup \vec{e_n}$, then $\vec{E}$ is called entirely evidence variables for whole data distributed in n stations.*

Definition 3. *During taking traditional Gibbs sampling in S_m with $\overrightarrow{e_m} \in S_m$, $1 \leq m \leq n$, a Markov chain will be produced, noted as $MC(\overrightarrow{e_m})$, when stationary distribution is reached in this station. If we take out the special statures in $MC(\overrightarrow{e_m})$, which suitable for the $\overrightarrow{E}$, and put them together, then it called the sub-sample of $MC(\overrightarrow{e_m})$ in S_m, noted as $MC'(\overrightarrow{e_m})$.*

Therefore, we would firstly using traditional Gibbs sampling method in every station with its own stationary distribution and produce a Markov chain during this process. In order to got the reasonable result in the final step, here we imporve the Gibbs sampling, but taking out the $MC'(\overrightarrow{e_m})$ form each station, and put them together to do the final infer.

Based on the above discussion, the Bayesian Inference Method under Data-Intensive Computing, noted as BIMDIC, can be described as follows:

Input:
$S_1, S_2, ..., S_n$: the n stations
g: the Bayesian network contains joint distribution $\psi(x)$
$\overrightarrow{X} = \{X_1, X_2, ..., X_n\}$: the variables in g
$\overrightarrow{e_m} = \{X_i = x_i, X_j = x_j, ..., X_k = x_k$: the evidence variables in S_m, $1 \leq i \leq j \leq k \leq n$
Output:
$P(\overline{X}|\overrightarrow{E})$: the conditional probability of $\overline{X}$, here $\overline{X} = \{X_i | X_i \notin \overrightarrow{E}\}$
Variables:
$SD_m, 1 \leq m \leq n$: the stationary distribution in S_m
$MC(\overrightarrow{e_m})$: the Markov chain under SD_m
$MC'(\overrightarrow{e_m})$: the sub-sample of $MC(\overrightarrow{e_m})$ in S_m
Steps:
Begin

1. Using traditional Gibbs sampling method on $S_m, 1 \leq m \leq n$, until it reach SD_m, and $MC(\overrightarrow{e_m})$ will be got;
2. Figure out the $\overrightarrow{E}$, $\overrightarrow{E} = \overrightarrow{e_1} \cup \overrightarrow{e_2} \cup ... \cup \overrightarrow{e_n}$;
3. As $\overrightarrow{e_m} \in S_m$, find out the reasonable $MC'(\overrightarrow{e_m})$, which suitable for the $\overrightarrow{E}$, from $MC(\overrightarrow{e_m})$ in SD_m;
4. $MC'(\overrightarrow{E}) = MC'(\overrightarrow{e_1}) \cup MC'(\overrightarrow{e_2}) \cup ... \cup MC'(\overrightarrow{e_n})$;
5. Calculate the $P(\overline{X}|\overrightarrow{E})$ on the basic of $MC'(\overrightarrow{E})$

End

3.2   Validity of the BIMDIC

In the issue of many data-intensive computing methods, an important standard is that, whether the final result has validity. As it come to BIMDIC, which means how to prove the final result is suitable for the whole data in all stations.

Theorem 1. *Assuming that S is a station which includes all the data in $S_1, S_2, ...,$ S_n, $\overrightarrow{E} \in S$, and $P'(\overline{X}|\overrightarrow{E})$ is the result got from the S by traditional Gibbs* *sampling method;* $P(\overline{X}|\overrightarrow{E})$ *is the result got by BIMDIC form* $S_1, S_2, ..., S_n$, $\overrightarrow{e_m} \in S_m$, $1 \le m \le n$, *and* $\overrightarrow{E} = \overrightarrow{e_1} \cup \overrightarrow{e_2} \cup ... \cup \overrightarrow{e_n}$, *then* $P'(\overline{X}|\overrightarrow{E}) = P(\overline{X}|\overrightarrow{E})$.

In other words, we use BIMDIC to get the final result from distributed stations, and can prove that it is equal with one got by traditional Gibbs sampling method in a single station.

Proof. For a single station S, according to the proof of Stuart Russell in[12], the sampling process will reach a Detailed Balance when stationary distribution appears. At this moment, the statistical result of the passed Markov chain would be the correct value of variables conditional probability. Assume that $\{X_i = x_i, X_j = x_j\} \subseteq \overrightarrow{E}$, X_k is the sampling variables, $\overline{X} = \{X_q|X_q \ne X_k\}$, $\pi(X_k)$ is the posterior probability when $X_k = x_k$. To make it easy to follow, we first proof the situation of $n = 2$, while when $n > 2$, it could be proofed by the same way. As $q(X_k \to X_k') = q((x_k, \overline{x}) \to (x_k', \overline{x}))=p(x_k'|\overline{x}) = p(x_k'|e_{ij}, \overline{x} - e_{ij})$, $e_{ij} = (X_i = x_i, X_j = x_j)$. So for S_1, $q(X_{1k} \to X_{1k}') = q((x_{1k}, \overline{x}) \to (x_{1k}', \overline{x}))=p(x_{1k}'|\overline{x}) = p(x_{1k}'|e_i, \overline{x} - e_i)$, $e_i = (X_i = x_i)$. For S_2, $q(X_{2k} \to X_{2k}') = q((x_{2k}, \overline{x}) \to (x_{2k}', \overline{x}))=p(x_{2k}'|\overline{x}) = p(x_{2k}'|e_j, \overline{x} - e_j)$, $e_j = (X_j = x_j)$. As for any statue in $MC'(\overrightarrow{e_i})$, there is always exit $X_j = x_j$, so $q(X_k \to X_k')=q((x_k, \overline{x}) \to (x_k', \overline{x}))=p(x_k'|\overline{x})=p(x_i'|e_i, X_j = x_j, \overline{x} - e - (X_j = x_j))=p(x_i'|e_{ij}, \overline{x} - e_{ij})$. By the same reason, for any statue in $MC'(\overrightarrow{e_j})$, $q(X_k \to X_k')=p(x_j'|e_{ij}, \overline{x} - e_{ij})$. That is to say, $\pi(X_k)q(X_k \to X_k)=p(x_k, \overline{x} - e_{ij}|e_{ij}), p(x_k|e_{ij}, \overline{x} - e_{ij})=p(x_k, \overline{x} - e_{ij}, e_{ij})p(\overline{x} - e_{ij}|e_{ij})p(x_k|e_{ij}, \overline{x} - e_{ij}) = p(x_k|e_{ij}, \overline{x} - e_{ij}|e_{ij}) = q(X_k \to X_k)\pi(X_k) = \pi(X_k)q(X_k \to X_k)$, which means the detailed balance is also exit in $MC(\overrightarrow{E})$. Taking $mb(X_k)$ as the Markov blanket of X_k, then $p(x_k|mb(X_k)) = p(x_k|e_{ij}, \overline{x} - e_{ij})$ will be got. As $p(x_k|mb(X_k)) = p'(\overline{X}|\overrightarrow{E})$, and $p(x_k|e_{ij}, \overline{x} - e_{ij}) = p(\overline{X}|\overrightarrow{E})$, so $p'(\overline{X}|\overrightarrow{E}) = p(\overline{X}|\overrightarrow{E})$.

4   Experimentation and Analysis

4.1   Experimentation

The experimentation environment contains a computer, which has an Intel Core i5-2300 2.8G CPU, 8G memory and 2T hard disk. In order to build a multi stations environment, four virtual machines were created with VMware Workstation software in this computer and each of them has 1G memory and 20G hard disk. The operating system in virtual machines is Ubuntu, and and BIMDIC was implemented by C and MPI language. For the convenience of description, four virtual machines are noted as S_0, S_1, S_2 and S_3.

As mentioned before, a Bayesian network g is needed in every station. Here, we chose a classical one, Chest Clinic[17, 18], which is shown in Fig. 2. Chest Clinic is a small Bayesian network, which is mainly to judge whether a patient has tuberculosis, lung cancer or bronchitis by different factors, such as smoking, visited to Asia, dyspnea and so on. Since its structure is not very complex, but very typical, it often been regarded as the first choice for many experiments when a Bayesian network is needed. In this experiment, we let the entirely evidence variables

$\vec{E} = \{Tuberculosis = absent, Bronchitis = present\}$. The main purpose of this experiment is to compare the final results got by BIMDIC with that got by the traditional Bayesian inference method. If they almost come to the same value, the validity of BIMDIC can be proofed. So the four stations are divided into two groups. The first group just has S_0 with Chest Clinic and $\vec{E}$ in it, and traditional Bayesian inference method is used to fetch final result. The second group has S_1, S_2 and S_3. Besides Chest Clinic, S_1 has $\vec{e_1} = \{Tuberculosis = absent\}$ and S_2 has $\vec{e_2} = \{Bronchitis = present\}$. The final results are shown in Fig. 3 and Fig. 4.

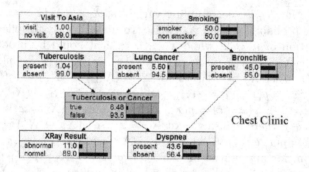

Fig. 2. Chest Clinic

Time performance of both methods are also involved in the experimentation. Here, TBIM stands for the traditional Bayesian inference method and the 4 in BIMDIC 4 means there are 4 stations in this situation. For the same reason, BIMDIC 6 and BIMDIC 8 means there are 6 and 8 stations in these kind of situation respectively. The result is shown in Fig. 5.

4.2   Analysis

From Fig. 3, we can see that in the final results of traditional Bayesian inference method, variable Smoking, Tuberculosis or Cancer and Lung Canner began converge to 0.63, 0.07 and 0.08 when sampling records reach 9000, 8000 and 10000 respectively. The rest non-evidence variables began to converge while sampling records reach 11000. When that come to Fig.4, all the non-evidence variables are converge to the same value compared with that in Fig.3, but the initial point for which began to converge are not all the same. In Fig. 4, Smoking, Tuberculosis or Cancer and Lung Canner began converge when sampling records reach about 12000. The rest non-evidence variables began to converge while sampling records reach 11000 to 13000. That is because, in the BIMDIC we take statues from multi stations and union them together to be a new Markov chain. But the evidence in the sub stations may not contain all the information in entirely

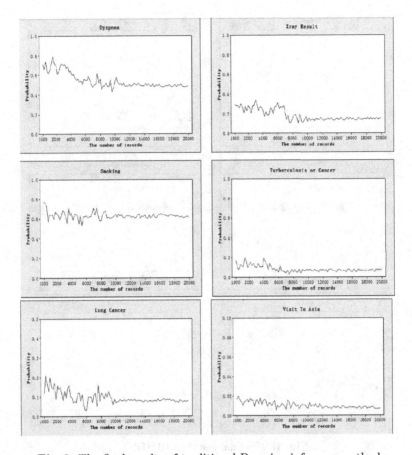

Fig. 3. The final results of traditional Bayesian inference method

evidence variables. So when we take statues which suitable for the entirely evidence variables from each station, we can hardly predicate how many we could get. Therefore the converge speed of every non-evidence variables are also hard to predicate. However, converge speed of these variables doesn't influence its final converged value. So the final results got by BIMDIC are equal to that got by the traditional Bayesian inference method.

From the Fig5, we can see that the TBIM's time performance is very sensitive to the number of records. Compare with it, BIMDIC has a better time performance on the same data sets and the more stations the better.

But when the data stations reach a certain level, the improvement degree is growing slowly. That is because when the number of data sets is not very large, the more data stations may make the data sets too scattered, which will cause even more data movements during the application executed. However, when the number of data sets is continue to grow, the time performance become better and better.

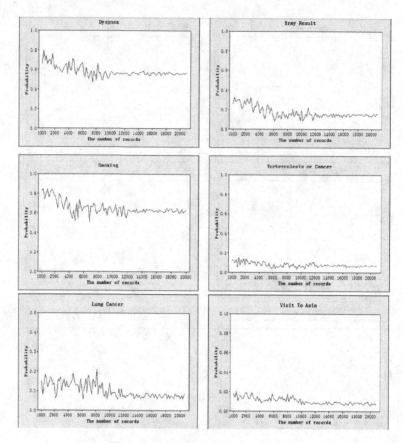

Fig. 4. The final result of BIMDIC

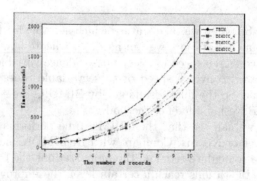

Fig. 5. The time performance of BIMDIC and TBIM

5 Conclusion

In this paper, we proposed a Bayesian inference method under data-intensive computing to deal with the situation when data is distributed in multi stations. Compared with the traditional Bayesian inference method, this method could get the equal final results when every station has the same Bayesian network, and the union set of all the evidence variables in them is the same as that in traditional Bayesian inference method. The validity of this method has been discussed in theory, and illustrated in real data.

However, after experiment in a lot of data sets, we note that converge speed of variables in this method is very sensitive to the structure of Bayesian network and can hardly predicated. This may cause disturbance if we try to prove the efficiency of this method. So, how to improve this situation would become our future work.

References

1. Perkins, A.: Data intensive computing. In: Proceedings of the 2006 ACM/IEEE Conference on Supercomputing (2006)
2. Chen, K., Jiang, W., Martin: A note on some algorithms for the Gibbs posterior. Statistics and Probability Letters 80, 1234–1241 (2010)
3. van Hasselt, M.: Bayesian inference in a sample selection model. Journal of Econometrics, 221–232 (2011)
4. Omori, Y.: Efficient Gibbs sampler for Bayesian analysis of a sample selection model. Statistics and Probability Letters, 1300–1311 (2007)
5. Kottegoda, N.T., Natale, L., Raiteri, E.: Gibbs sampling of climatic trends and periodicities. Journal of Hydrology, 54–64 (2007)
6. Eastern forum of science and technology. What is data-intensive computing (2010), http://www.efst.sh.cn/showKnowledge.do?id=577
7. Agrawal, D., El Abbadi, A., Antony, S., Das, S.: Data Management Challenges in Cloud Computing Infrastructures. In: Kikuchi, S., Sachdeva, S., Bhalla, S. (eds.) DNIS 2010. LNCS, vol. 5999, pp. 1–10. Springer, Heidelberg (2010)
8. Pearl: Probabilistic reasoning in intelligent systems: networks of plausible inference, pp. 116–131. Morgen Daufmann Publishers Inc., California (1998)
9. Russell, S.: Artificial IntelligenceA modern method, 2nd edn. The People's Posts and Telecommunications Press (2010)
10. Liu, W., Li, W., Yue, K.: Intelligent data analysis. Science Press, Beijing (2007)
11. Deshpande, S.: Probabilistic graphical models and their role in database. In: Proc. of VLDB 2007, pp. 1435–1436 (2007)
12. Tuncozgur, B., Elbeyli, L., Gungor, A., Isik, F., Akay, H.: Chest wall reconstruction with autologas rib grafts in dogs and report of a clinic case. European Journal of Cardio-Thoracic Surgery 16(3), 292–295 (1999)
13. Pericchi, L.R.: Model Selection and Hypothesis Testing based on Objective Probabilities and Bayes Factors. In: Handbook of Statistics, vol. 25, pp. 115–149 (2005)
14. Tsukuma, H.: Generalized Bayes minimax estimation of the normal mean matrix with unknown covariance matrix. Journal of Multivariate Analysis, 2296–2304 (2009)

15. Chen, L., Yang, M.: Empirical Bayes testing for equivalence. Journal of Statistical Planning and Inference, 2670–2681 (2011)
16. Liu, W., Han, C., Shi, Y.: Unsupervised Learning for Finite Mixture Models Via Modi ed Gibbs Sampling. Journal of Xi'an Jiaotong University, 15–19 (2009)
17. Lauritzen, S.L., Spiegelhalter, D.J.: Local computations with probabilities on graphical structures and their application to expert systems. Royal Statistics Society B, 157–194 (1988)
18. Zhang, L., Guo, H.: Introduction to Bayesian Networks. Science Press, Beijing (2006)
19. Lauritzen, S.: Local computation with probabilities on graphical structures and their application to expert systems. Roy. Stat. B, 157–224 (1988)

Preferences Predictions of Learning Objects Supported by Collaborative Recommendations

Sylvia Encheva

Stord/Haugesund University College
Bjønsonsg. 45, 5528, Haugesund, Norway
sbe@hsh.no

Abstract. A numerous amount of models supporting automated provision of personalized recommendations to students have been developed lately. Students usually receive recommendations based on outcomes of tests they have already taken. In this paper we explore methods from formal concept analysis for predicting students' preferences with respect to learning objects going to be suggested to them before they have taken relevant tests.

Keywords: Intelligent Tutoring Systems, Formal Concept Analysis, Education.

1 Introduction

Latest economic developments often result in restrictions on educational budgets among other things. Thus, the role of intelligent tutoring systems in education is getting more and more important every day. Intelligent tutoring systems as a research field have been of a particular interest to educators and researchers for a number of years. One of the main problems in this field is how to provide automated advices to each student tailored to her needs. Some of the attempts in that direction are based on students' tests results, [5] and [6]. Students receive recommendations based on a test outcomes and are usually invited to take a new test with similar questions and thus find out whether the learning level is increasing and or certain skills are obtained.

In this work we look at the process of providing automated assistance to students from a different angle. Suppose we know students preferences about learning objects from several topics due to their responses to Web-based questionnaires. How can we predict their preferences with respect to learning objects belonging to consecutive topics? Our approach is using methods from formal concept analysis and concept lattices, [9].

The rest of the paper is organized as follows. Selected theory used for the model development is presented in Section 2. The model description can be found in Section 3. The paper ends with a conclusion placed in Section 4.

Z. Li et al. (Eds.): ISICA 2012, CCIS 316, pp. 145–151, 2012.
© Springer-Verlag Berlin Heidelberg 2012

2   Preliminaries

Let P be a non-empty ordered set. If $sup\{x, y\}$ and $inf\{x, y\}$ exist for all $x, y \in P$, then P is called a *lattice* [3]. In a lattice illustrating partial ordering of knowledge values, the logical conjunction is identified with the meet operation and the logical disjunction with the join operation.

A *context* is a triple (G, M, I) where G and M are sets and $I \subset G \times M$. The elements of G and M are called *objects* and *attributes* respectively [3], [9], and [2].

For $A \subseteq G$ and $B \subseteq M$, define

$$A' = \{m \in M \mid (\forall g \in A) \ gIm\},$$
$$B' = \{g \in G \mid (\forall m \in B) \ gIm\}$$

where A' is the set of attributes common to all the objects in A and B' is the set of objects possessing the attributes in B.

A *concept* of the context (G, M, I) is defined to be a pair (A, B) where $A \subseteq G$, $B \subseteq M$, $A' = B$ and $B' = A$. The *extent* of the concept (A, B) is A while its *intent* is B. A subset A of G is the extent of some concept if and only if $A'' = A$ in which case the unique concept of the which A is an extent is (A, A'). The corresponding statement applies to those subsets $B \in M$ which is the intent of some concepts.

The set of all concepts of the context (G, M, I) is denoted by $\mathfrak{B}(G, M, I)$. $\langle \mathfrak{B}(G, M, I); \leq \rangle$ is a complete lattice and it is known as the *concept lattice* of the context (G, M, I).

The entry-level concept for an object and the entry-level concept of an attribute are defined in [1] as follows. The entry-level concept of an object is the unique concept for which the object is a member of the extent and is not a member of the extent of any sub-concept. The entry-level concept of an attribute is the unique concept for which the attribute is a member of the intent and is not a member of the intent of any super-concept. Note that in [1] the authors use entity and feature instead of object and attribute which are much more common in the field of formal concept analysis.

We apply the following hypothesis presented in [1] - *Users who rate the same items tend to rate items the same.*

A generator of a closed set Z is a minimal (w.r.t. inclusion) set X such that $X\dot{\phi} = Z$, [8].

Various aspects of learning and instructions are taken in [4] and [7].

3   The Model

Suppose we have information from a group of students about their preferences related to learning objects in N topics. The question is - if we know the preferences of another group of students concerning the first $N - l$ topics can we predict their preferences for the last l topics? We are assuming a subject where the last l topics require knowledge and skills from the first $N - l$ topics. In our case $N = 6$ and $l = 2$.

For the purpose of our study we engage two groups of students, A with ten students denoted $1, ..., 10$ and B with five students denoted $11, ..., 15$ in Table 1 and Table 2. Students in group A have taken tests related to all six topics while students in group B have taken tests related to the first four topics. All students are invited to rate the suggested learning objects stating that they have found a particular learning object useful or not useful.

Attributes in Table 1 and Table 2 represent topics, where each topic is supported by two learning objects. Learning objects under the first four topics are denoted by a, b, c, d, e, f, g, h and for the last two topics by $D1, D2, I1, I2$.

A nonempty cell in Table 1 and Table 2 shows that a student has found that learning object useful if denoted by u and not useful if denoted by n. An empty cell means that a learning object has not been rated. Negative ratings are not involved in the predictions of the current investigation. The prediction is based only on learning objects rated as useful.

Table 1. Responses from the first ten participants with respect to six attributes

	Fractions		Polynomials		Trig. functions		Log. functions		Derivation		Integration	
	a	b	c	d	e	f	g	h	D1	D2	I1	I2
1	u			u	u		u		n	u	u	
2					u	u	u		u		n	
3		u	u				u	u	n	u		u
4	u			u	u		u		u	n	u	
5			n		u		n			u		n
6		u		u		u	u		u		u	
7		u	u		u		u			u		n
8	u		u		u			u	n	n	u	
9	u			u	u		u			u		u
10				u		u		u	n	n		u

A Hasse diagram depicting the concept lattice for the context of Table 1 can be seen in Fig 1. Nodes closer to the lowermost node show concepts where less objects share more attributes, while nodes closer to the uppermost node show concepts where more objects share less attributes. An extent at a node lists all objects in that node and all descendant nodes. An intent at a node lists all attributes possessed by the objects in this node and all ancestor nodes.

Our goal is to predict which objects among D1, D2, I1, and I2 are going to be on the preference list of students 11, 12, 13, 14, and 15. From the concept lattice in Fig. 2 we extract the entry-level concepts for students 11, 12, 13, 14, and 15. If an entry-level concept is insufficient we consider the nearest relevant concept, provided such a concept is placed on the next level in the concept lattice, otherwise we apply majority voting. We then turn to the concept lattice in Fig. 1 to extract information about the preferences regarding attributes D1, D2, I1, I2 of students 1, 2, ..., 10 that appear in the entry-level concepts for students 11, 12, 13, 14, and 15.

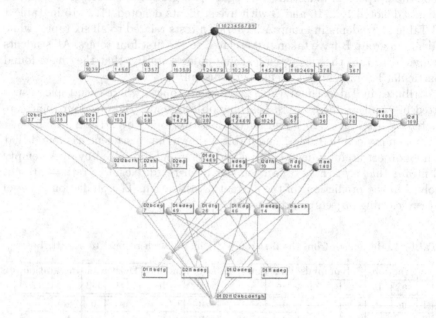

Fig. 1. Lattice of the context in Table 1

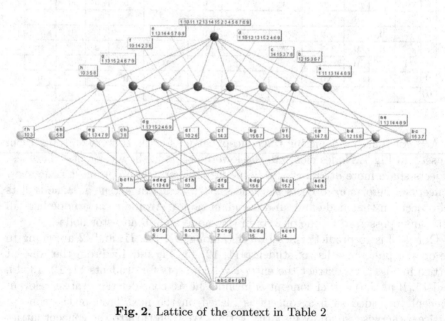

Fig. 2. Lattice of the context in Table 2

Table 2. Responses from all participants with respect to the first four attributes

	Fractions		Polynomials		Trig. functions		Log. functions	
	a	b	c	d	e	f	g	h
1	u			u	u		u	
2			n		u	u	u	n
3		u	u		n	u		u
4	u			u	u	n	u	
5	n		n		u	n		n
6		u		u		u	u	n
7		u	u		u	n	u	
8	u		u		u	n	n	u
9	u		n	u	u		u	
10		n		u		u		u
11	u				n		n	
12		u		u		n		n
13	u	n		u	u		u	
14	u		u		u	u		n
15		u	u	u		n		u

The entry-level concept for student 11 contains also students 1, 4, 8, and 9. In order to make a prediction we have to consider two subsets of the set of students 1, 4, 8, and 9, because the data derived from this set is inconclusive. Students 4 and 9 share more attributes than students 1 and 9. Since students 4 and 9 prefer D1 we assign D1 to student 11 as well. The second preference for student 11 is I1 since students 1, 4 and 9 prefer I1, i.e. majority voting.

The entry-level concept for student 12 contains student 6. Thus the assigned preferences for student 12 follow the preferences for student 6, D1 and I1.

The entry-level concept for student 13 contains also students 1, 4, and 9. Since students 1, 4, and 9 do not generate a single concept we choose the preferences for student 13 by using majority voting. Students 4 and 9 prefer D1 while students 4 and 9 prefer I1. Thus the assigned preferences for student 13 are D1 and I1.

The entry-level concept for student 14 does not contain other students. The nearest concept including student 14 and another student contains actually student 8. Thus one of the preferences for student 14 is I1. In order to determine the other preference for student 14 we again use majority voting because the student appears in two different concepts that are on equal distance from the concept including student 8. Student 14 appears in one concept with students 7 and 8 and in another concept with students 1, 4, 8, and 9. The second preference for student 14 is I1 since students 1, 4 and 9 prefer I1.

The entry-level concept for student 15 does not contain other students. The nearest concepts including student 15 include students 7 and 8 but this is not useful since students 7 and 8 do not share preferences regarding attributes D1, D2, I1, and I2. The nearest concept including student 15 and another student contains students 3 and 7. Thus one of the preferences for student 15 is D2. The other preference is I1 since student 15 appear in another concept with students 1, 4 and 9 and they prefer I1.

3.1 Discussions

So far we have applied two gradings useful and not useful. Finer grading looks more accurate but some students find it a bit difficult to use a larger grading scale and definitely time consuming. As a result they skip the rating part.

While the authors of [1] settled for the nearest neighbour concept when a entry concept is inconclusive we apply majority voting most of the time since the nearest neighbour concept can some times be too far away in order to draw reasonable conclusions.

In this work we have been using data from four topics to predict preferences in two new topics. It is of interest to develop two different predictions based on the four topics again but focusing on one new topic at a time. The results can be then compared with the ones described above.

4 Conclusion and Future Work

Conclusions obtained in this way can be used to re-evaluate the quality of available learning materials and consecutively make adjustments.

There are often students who belong to a certain group but have not rated some of the learning materials. In case lack of knowledge and or skills is detected when they take later tests the group preferences from the first test can be used as a base for automated selection of necessary recommendations.

Faculty members with limited experience in teaching a particular subject can benefit from considering students preferences and improve the effectiveness of their teaching.

In a future work it is of interest to investigate to which extend generators can be involved in the process of predicting students preferences.

References

1. Boucher-Ryan, P., Bridge, D.: Collaborative Recommending using Formal Concept Analysis. Knowledge-Based Systems 19, 309–315 (2006)
2. Carpineto, C., Romano, G.: Concept Data Analysis: Theory and Applications. John Wiley and Sons, Ltd. (2004)
3. Davey, B.A., Priestley, H.A.: Introduction to lattices and order. Cambridge University Press, Cambridge (2005)
4. Dummer, P., Ifenthaler, D.: Automated Knowledge Visualization and Assessment. Computer-Based Diagnostics and Systematic Analysis of Knowledge, Part 2, 77–115 (2010)
5. Encheva, S.: Evaluation of Learning Outcomes. In: Luo, X., Spaniol, M., Wang, L., Li, Q., Nejdl, W., Zhang, W. (eds.) ICWL 2010. LNCS, vol. 6483, pp. 72–80. Springer, Heidelberg (2010)

6. Encheva, S.: On Indiscernibility in Assessments. In: Kim, T.-h., Ma, J., Fang, W.-c., Park, B., Kang, B.-H., Ślęzak, D. (eds.) UNESST 2010. CCIS, vol. 124, pp. 39–47. Springer, Heidelberg (2010)

7. Ifenthaler, D.: Learning and instruction in the digital age. In: Spector, J.M., Ifenthaler, D., Isaas, P., Sampson, D.G. (eds.) Learning and Instruction in the Digital Age: Making a Difference Through Cognitive Approaches, Technology-Facilitated Collaboration and Assessment, and Personalized Communications, pp. 3–11. Springer, New York (2010)

8. Pfaltz, J.L., Taylor, C.M.: Scientific Discovery Through Iterative Transformations of Concept Lattices. In: Workshop on Discrete Mathematics and Data Mining (at 2nd SIAM Conf. on Data Mining, Arlington VA), pp. 65–74 (April 2002)

9. Wille, R.: Concept lattices and conceptual knowledge systems. Computers Math. Applications 23(6-9), 493–515 (1992)

A Method of Face Detection
Based on Skin Color Model in Fixed Scene

Yan Rao and Ruliang Zhang

College of Science
Guizhou Minzu University
Guiyang, P.R. China
gr_yan@sina.com

Abstract. This paper presents a rapid detection method for face detection in stationary scenes. First of all, we obtain the binary image based on background picture subtraction and extract the moving target area with SHEN filter. Then, we select the YCbCr color model, based on skin-color clustering model, we can do color segmentation on the sub-image of moving target. Following that, we use elliptical template to detecting the face region and marking the face positioning. Lastly, we implement the algorithm with Matlab7.0, and do experiments in our face testing set. The experimental data manifest the good robustness of the suggested method for face detection in the stationary scenes, and its strong adaptability to attitude, expression and age. Also, the results show that this method is definitely practical in real time. In a word, the accuracy is above 84 percent.

Keywords: Face Detection, Shen Filter, Skin Color Segmentation, Ellipse Template.

1 Introduction

With the development of intelligent IT, people are eager to achieve fast and effective authentication on video surveillance, distance education, human-computer interaction technology and security, and so on. The face detection issue originated from face recognition, which refers to determine the existence of the face region in the inputting image, and then determine the location of the face and facial features. The application of face detection is now far beyond the scope of the face recognition system with its more and more academic and practical value. In the last few years, some achievements in face detection [1-9] have been obtained. These achievements have distinguished effects in special conditions. This article presents a rapid method to detect the face in stationary scenes.

The advantages of the method are the fixed area which is similar to face skin color can be deleted in the image of motion information sequence, and on the target tracking and motion detection fields, it not only can effectively suppresses background noise and diminishes false detection rate, but also can narrow the face detection range and accelerate the detection speed.

Z. Li et al. (Eds.): ISICA 2012, CCIS 316, pp. 152–158, 2012.
© Springer-Verlag Berlin Heidelberg 2012

The outline of this paper is as follow: section 1 presents the introduction, section 2 introduces basic steps of the algorithm, section 3 gives experimental results of the algorithm and the section 4 gives the conclusion.

2 The Basic Steps of the Algorithm

2.1 Extracting Moving Target (person) from the Video Images

Building Background Chart and Getting the Initial Target. In order to locate the face in the video image, interference of background image should be reduced as far as possible, which can make it more accurate and faster to locate the face. The moving objects are extracted with the background subtraction due to its fixed scene.

$$Df_i f_j(x, y, z) = f(x, y, z, t_i) - f(x, y, z, t_j) \tag{1}$$

$f(x, y, z, t_i)$, $f(x, y, z, t_j)$ is the brightness values in the (x, y, z) position of pixel at the moment of i and j, $0 \leq f(x, y, z, t) \leq 255$. In this paper, $j = 0$, thus is the fixed background. The result is shown in Fig. 1.

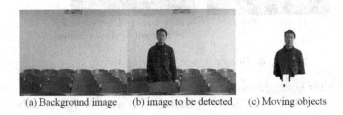

(a) Background image   (b) image to be detected   (c) Moving objects

Fig. 1. Gets the moving target

Split a Clearing Moving Target by the Threshold of Bimodal Method. The initial moving target is not very clear because here are still some backgrounds. By observing the histogram, it's found that the histogram is typical two peaks distribution, as shown in Fig.2. Through analyzing, the left side of the peak is the background, and the right side of the peak is the motion target.

It is necessary to find out the valley point between two peaks of the histogram in order to separate the moving object from the image. In this paper, SHEN filter[4] will be used to calculate the $f'(x)$ and $f''(x)$ of the discrete images, so as to find out the minimum point (if $f'(x) = 0$ and $f''(x) > 0$, then the point is the minimum one), as shown in Fig. 2. Therefore the moving target can be gotten with this threshold. And then the clear and complete moving target also can be gotten by using mathematical morphology to removing interference, as shown in Fig. 3.

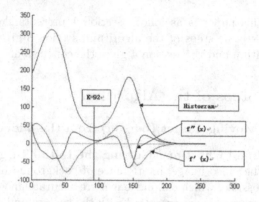

Fig. 2. Histogram, $f'(x)$ and $f''(x)$ of target image

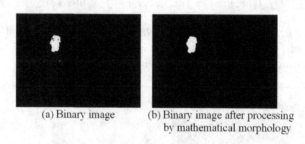

(a) Binary image   (b) Binary image after processing
                           by mathematical morphology

Fig. 3. Binary image of moving target

The SHEN filter is exponential filter which can smooth the image. It includes left filter and right filter. It is assumed that $I(x)$ is a signal functionthe original image is I_0, then[4].

Left filter:

$$f_L(i) : I_1(0) = I_0(0), i = 1, 2, ..., N - 1 \tag{2}$$

$$I_1 i = a_0 \cdot I_0(i) + (1 - a_0) \cdot I_1(i - 1) \tag{3}$$

Right filter:

$$f_R(i) : I_2(N - 1) = I_1(N - 1), i = N - 2, N - 3, ..., 1, 0 \tag{4}$$

$$I_2 i = a_0 \cdot I_1(i) + (1 - a_0) \cdot I_2(i + 1) \tag{5}$$

With $0 < a_0 < 1$, they have relations with:

$$I_2(x) - I_1(x) \approx C_1 \cdot D_1(x) \tag{6}$$

$$I_2(x) - I_0(x) \approx C_2 \cdot D_2(x) \tag{7}$$

Then

$$D_1(x) = \frac{d}{dx}(I_0 * f), D_2(x) = \frac{d^2}{dx^2}(I_0 * f) \tag{8}$$

So we can use them to calculate $f'(x)$ and $f''(x)$ of the image.

2.2   Face Detection Based on YCbCr Color Space

Skin Color Segmentation with the Determinate Threshold of Cb, Cr.
Firstly, it's essential to transform the image from RGB to YCbCr color space (as
shown in Fig. 4). After getting the CbCr, it's found that Cb has some interference
between 73 and 127 by observing, that's, there is one point between 120 and 127
suddenly increasing to form a leap .What's more, Cr has some interference on
the range between 133 and 173, that's, here is one point between 140 and 133
suddenly increasing to form a leap (shown in Fig. 5). In this paper, the leap point
is found because $f''(x)$ is sensitive to the leap, as shown in Fig. 6. The skin color
segmentation threshold can be gained by combining Cr=[133,173], Cb=[77,127]
[6]. And the binary image can be obtained by using this threshold.

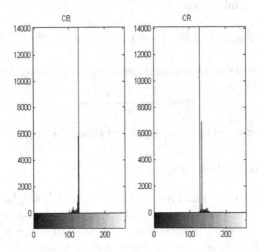

Fig. 4. Target image of YCbCr

Fig. 5. Histogram of Cb, Cr

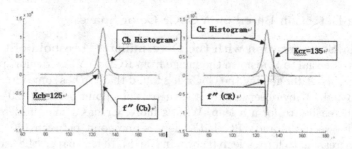

Fig. 6. $f''(Cb)$ and $f''(Cr)$

The Face Region Can Be Extracted by the Combination of Oval Template after Using Mathematical Morphology to Remove Interference. After the segmentation of skin color, the generated color binary image has the noise, and possible similar skin regions which can't be get rid of with the mathematical morphology. Thus, the verifying of face region is required. Because the shape of face is similar to the oval, the face region can be found by finding elliptical approximation in the detection image[6].

The center of the ellipse (x_0, y_0):

$$x_0 = \frac{\sum_{(x,y)\in B} x}{S}, y_0 = \frac{\sum_{(x,y)\in B} y}{S} \tag{9}$$

S is the number of pixels in each connected region B.

The angle of dip θ:

$$\theta = \frac{1}{2} arctan(\frac{2\mu_{11}}{\mu_{20} - \mu_{02}}) \tag{10}$$

Major axis a and the minor axis b:

$$a = \frac{\sum_{i=1}^{S} |y' - y_0|}{S} \times 1.5, b = \frac{\sum_{i=1}^{S} |x' - x_0|}{S} \times 1.5 \tag{11}$$

Then we can calculate connected component S and the similarity of elliptical.

$$similar = \frac{4 \times S}{\pi \times a \times b} \tag{12}$$

If the similar> 0.7, that is the face region, as shown in Fig. 7.

3   Experimental Results

In this paper, some experimental results of the algorithm will be presented here. In order to validate the performances of the algorithm, 120 various photos were selected to be tested. People in these photos are from different ages, with various poses and facial expressions. These photos are in different lighting condition, different backgrounds and different sizes. Some experimental results are shown in Fig. 8.

Fig. 7. The result of Face detection

4   Conclusion

The experimental results show that this algorithm has less limitations of the human face and has a better detection effect for glass-weared as well as nonpositive upright face. Through testing the image library, this paper proves that

Fig. 8. The experimental results

the proposed face detection algorithm has good detection results for the frontal face. The detection rate of glass-weared as well as non-positive face is also high. The detection rate of the algorithm in single detection is above 90%, while the detection rate in group detection is above 78%. However, the rate of side face detection is relatively low, which should be improved in the future work of the algorithm proposed in this paper.

References

1. Du, Y., Yang, N.: Research of Face Detection in Color Image Based on Skin Color. Energy Procedia 13, 9395–9401 (2011)
2. Liang, Y., Ma, L., Zhang, L., Miao, Q.: Face Localization Based on Edge Information of Skin Color And Eye. Energy Procedia 13, 3678–3683 (2011)
3. Chaves-González, J.M., Vega-Rodríguez, M.A., Gómez-Pulido, J.A., Sánchez-Pérez, J.M.: Detecting skin in face recognition systems: A colour spaces study. Digital Signal Processing 20(3), 806–823 (2010)
4. Shen, J., Castan, S.: An optimal linear operator for step edge detection. CVGIP: Graphical Model and Image Processing 54(2), 112–133 (1992)
5. Sun, H.-M.: Skin detection for single images using dynamic skin color modeling. Pattern Recognition 43(4), 1413–1420 (2010)
6. Khan, R., Hanbury, A., Stöttinger, J., Bais, A.: Color based skin classification. Pattern Recognition Letters 33(2), 157–163 (2012)
7. Yang, J., Ling, X., Zhu, Y., Zheng, Z.: A face detection and recognition system in color image series. Mathematics and Computers in Simulation 77(5-6), 531–539 (2008)
8. Lee, K.-M.: Component-based face detection and veri-fication. Pattern Recognition Letters 9(3), 200–214 (2008)
9. Huang, T.-H., Yu, Y.-M., Qin, X.-G.: A High-Performance Skin Segmentation Method. Procedia Engineering 15, 608–612 (2011)

Parallel Remote Sensing Image Processing: Taking Image Classification as an Example

Xiaoyue Wang[1], Zhenhua Li[1,*], and Song Gao[2]

[1] School of Computer Science
China University of Geosciences
Wuhan, Hubei 430074, P.R. China
xiaoxiaoyue.wang@gmail.com
zhli@cug.edu.cn
[2] Computer Science and Software Engineering
Auburn University
Auburn, AL 36849, United States
song.gao@auburn.edu

Abstract. This paper introduces an architecture of parallel remote sensing image processing software, with advantages including high scalability, platform-independence, language-independence, and so on. It helps achieve high-performance computing in this field. MPI is used as the fundamental distributed message passing protocol. An object-oriented wrapper, Boost.MPI library, is used in the software to manipulate MPI. Open Source libraries such as GDAL and Open-CV are studied in this paper to help develop detailed image processing programs and implement classification algorithms. A number of experiments are conducted to test the parallel classification programs. The results indicate that in most cases the performance is significantly improved, especially for multi-spectral remote sensing image classification, in which a highest speed-up of 3.92 is reached.

Keywords: Remote Sensing Image Processing, Parallel Computing.

1 Introduction

Remote sensing is one of the fastest developing techniques in the 20th century. With help of spaceborne sensors, remote sensing can be used to get information of observed objects, regions or phenomenons without any direct contact. It provides macro-observations on changes of epidemic topographic, distribution of mineral sources, and so on in a wide range[1]. Figure 1 shows how remote sensing images are produced.

Since remote sensing techniques keep developing, spectral resolution ranges from $10^{-3}\lambda$ to $10^{-1}\lambda$[2]. The size of the image and the number of the bands increase very fast, thus lead to unprecedent mass of data and tremendous calculation requirement.

---

* Corresponding author.

Z. Li et al. (Eds.): ISICA 2012, CCIS 316, pp. 159–169, 2012.
© Springer-Verlag Berlin Heidelberg 2012

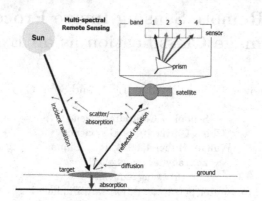

Fig. 1. Producing Remote Sensing Image

Since remote sensing image processing is facing challenges of complexity and speed-up, high-performance computing techniques, especially parallel processing, can play an extremely important role in this field[3]. Under this circumstance, the combination of remote sensing image processing and parallel processing is becoming natural and urgent[4]. The subject of this paper is the product of this combination.

This paper develops an available parallel processing system, in which MPI is adopted as the fundamental distributed platform, GDAL and OpenCV are employed to implement detailed procedures and algorithms.

In this system, UI (*User Interface*) and computation module are separate, thus ensures platform-independence. On the other hand, to make better use of the computing capability of high-performance clusters, communications between different modules are implemented by concise and language-independent protocol, `Protobuf`. Experiments are conducted to test the system. Detailed results will be discussed in the Section 4.

2   Methods

In order to develop such a parallel system, the following questions should be taken into consideration:

- Which framework to choose to construct the fundamental distributed components?
- How to display and process such high dimensional image?
- Which classifiers to integrate in the classification module?

These questions are discussed in this section.

2.1   Fundamental Framework–MPI

MPI (*Message Passing Interface*) is a language-independent communication protocol that keeps prevailing in high-performance computation[5]. Supported functions like point-to-point communication, collective communication, essential virtual topology, synchronization are all provided in a language-independent way.

The basic unit of MPI framework is process, which is also addressed by processor. Typically, to achieve better performance, each CPU is assigned a single process once the agent starts the MPI program.

In this paper, the MPI framework is manipulated through Boost.MPI library, which is one of the most popular object-oriented interface libraries for MPI. The detailed implementation is discussed in Section 3.

2.2   Image Processing–GDAL

Since remote sensing contains much more bands than ordinary images, and that variety of remote sensing image formats exist, GDAL is used to handle remote sensing image file I/O.

GDAL (*Geospatial Data Abstraction Library*) is a translator library for raster geospatial data formats released under an X/MIT style open-source license by *Open Source Geospatial Foundation*[6].

Although different spaceborne sensors have different data formats (such as HDF, TM, BIP, etc.), GDAL provides an uniform abstract module available for most of the translations between remote sensing image formats and common raster image formats.

The abstract data model of GDAL including *Dataset, Coordinate System, Affine Geo-transform, Raster Band, Color-Table, Sub-Dataset Domain, Image Structure Domain*, etc. [7]. A data set (represented by GDALDataset) contains particular information of each pixel as well as some relevant details, including size of the image and total number of the bands.

In image processing, the first step is open a image, the following code is to read a image.

```
GDALDataset *poDataset;
poDataset=(GDALDataset *)GDALOpen(argv[1],GA_ReadOnly);
```

No matter which image format it is, it can be read into in this poDataset. After that, detailed information can be retrieved through functions provided by GDAL. For example, GetRasterBand() function is used to get a single chosen band, GetXSize() and GetYSize() get the width and length of the image.

2.3   Image Classification–OpenCV

Image Classification. Remote sensing image classification approaches can be classified into two classes: spectral-space based classification and feature-space based classification. In this paper, feature-space based classifier is used. For a remote sensing image with N bands, a pixel(x,y) in band i is denoted as x_i. Since there are N bands, this pixel can be represented by a vector $X = (x_1, x_2, ..., x_n)$. The corresponding N dimensional Euclidean space is called feature space[8].

Classification methods are mostly depended on the distribution of these pixels in feature space.

Based on whether to use training set or not, these methods can be divided into supervised classification and unsupervised classification. The former one means there are some samples to train the classifier before it is used to do the classification, on the other hand, the latter need no training.

In this paper, two supervised classifiers, Naive Bayesian [9]and SVM are used.

OpenCV. OpenCV library is used to implement the classification algorithms and image formats transformation.

OpenCV (*Open Source Computer Vision Library*) is a library developed by *Inter* and now supported by *Willow Garage*[10]. Composed by a series of C functions and few C++ classes, it contains many general algorithms in image processing and computer vision.

OpenCV's principal modules are:

- **CV**: core function library;
- **CVAUX**: auxiliary function library;
- **ML**: machine learning learning library;
- **HIGHGUI**: GUI function library.

Algorithms in ML are used in this paper, specifically `CvNormalBayesClassifier` and `CvSVM`. These two classifiers are implementations of the algorithm *Naive Bayesian* and *Support Vector Machine*. Some interfaces for these two classifiers are shown as follows.

- training classifier :

```
virtual bool train
(const CvMat* _train_data, const CvMat* _responses);
```

- predicting:

```
virtual float predict (const CvMat* _sample) const;
```

In addition, the following two functions are used to serialize and de-serialize classifiers in order to broadcast them over MPI:

- Serialization:

```
virtual void save
(const char*filename, const char* name=0);
```

- De-serialization:

```
virtual void load
(const char *filename, const char* name=0);
```

Since `ML` module implements many useful machine learning algorithms, the parallel classification programming is simplified, and the overall reliability of the programs is improved as well.

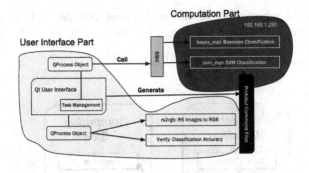

Fig. 2. System Structure

3 System Design

This section gives detailed descriptions of this parallel system.

3.1 Overview of System Structure

To take advantages of high-performance computer clusters that contain lots of parallel processing nodes, UI is not necessary for these computers. A very loosely coupling structure is used in order to separate computation module and UI module from each other. Therefore, this kind of structure makes cross-platform and cross-language development possible in the future.

The overview of system structure is shown in Figure 2. The specific design can be summarized as follows.

- The system is mainly divided into UI part and computing end, the former takes responsibility for user interaction, such as parameter settings and image display, while the latter is responsible for actual computing tasks;
- Modules exist as stand-alone binary files, which act as separate processes to support the whole system. In this way, the separation between UI and computing part is achieved;
- The management and control of UI modules is supported by QProcess, which is provided by the compiler *QtCreator*, it supports the communication requirements by redirecting input/output flows;
- Communications between UI modules are mainly through Protobuf messages (for large data) and standard output (for small data);
- SSH and a set of SSH-based commands are used to accomplish the exchanges and management between UI and computation nodes.

3.2 System Communication Flow

As mentioned above, supervised classification is divided into training and classifying. Experiments show that training is indivisible and cost much less time, therefore, a mixed process –serial training and parallel classifying – is adopted. UI is responsible for training, while the computing end is responsible for classifying.

The whole system flow is shown in Figure 3.

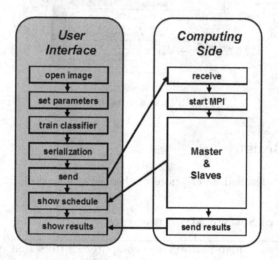

Fig. 3. Communication Flow

The steps are as follows:

1. Open a image, choose three bands to display;
2. Choose two sample areas for class 1 and class 0. The areas will be set by coordinates;
3. Set classification parameters, such as classifier type, serial or parallel, and how much processes to choose;
4. After choose sample areas, the chosen classifier is trained by sample instances and then serialized;
5. The serialized classifier, along with whole image and other parallel parameters is sent to the computing part;
6. The computing end does the classifying and then sends the result back, shown to the user.

3.3 Computing End Design

Based on MPI framework, the computing end is divided into two kinds of nodes: one is *Master* and the other one is *Slave*. There is only one *Master* but several *Slave*. *Master* takes responsibility for management, and *Slave* is responsible for computing.

After receiving parameters and image from UI part, computing end starts the MPI agent. A *Master* and several *Slaves* are built. *Master* divides the whole

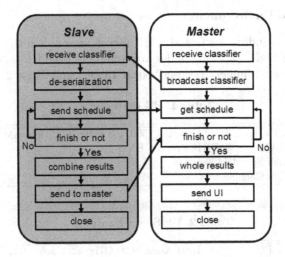

Fig. 4. Flow Chart of *Master* and A *Slave*

classification tasks into several pieces, broadcasts serialized classifier and related information to *Slaves*.

When the *Slaves* receive, they de-sterilized the information and start their own classification tasks according to the commands parallelling.

Question comes how to deal with the results. Generally, there are two ways:

- Once a pixel classification result comes out, the *Slave* sends it back to *Master*;
- Each node will write its results continuously into a file, and then send the file back to *Master*.

Experiments show that the first method leads to delay and serious CPU occupation, thus decreases the classification speed. The second method can bring prodigious speed acceleration. Under this circumstance, when a *Slave* finishes its tasks, it will send back all the classification results together.

In order to show the classification schedule to the users, a progress bar is placed on the bottom of the interface, thus leads to the requirement of real-time communications between UI and computing side.

The complete process is shown in Figure 4.

4   Experiments and Results

This section is the description of detailed experiments, including hardware environment, software environment and tested images. Finally, the results table is given.

4.1 Hardware Environment

Two computers including a PC and a HP server are used in the experiments. PC is used as UI and computing nodes are running on the HP server.

The hardware environment is described in Table 1 and Table 2.

Table 1. PC Hardware

CPU	Intel Core 2 T7500, 2.2 GHz, L2 Cache 4MB
RAM	DDRII 667, 3 GB
Harddisk Speed	5400 rpm
Harddisk Cache	8 MB

Table 2. HP Server Hardware

CPU	2 * [Intel Xeon, 3.20 GHz, L2 Cache 2MB]
RAM	2 GB
Harddisk Cache	128 MB

Both PC and HP server are connected to the same wired LAN. During the experiments, no other hosts are connected to the same LAN, so the connection and bandwidth are considered to be stable. No other programs other than the tested ones that require more resources are running on the hosts, therefore it can be considered that the hardware resource is fully used by the tested programs.

4.2 Software Environment

Both hosts use 64bit Linux operating system. The server uses *Ubuntu Server 10.10*, while the PC uses *Archlinux 2.6.39*. Open-source libraries such as GDAL and OpenCV are from repositories. Library versions may differ between PC and HP server, as a result, the modules are compiled separately in each testing environment. MPI version is *OpenMPI 1.4.1*.

4.3 Results

Figure 5 is a classification experiment conducted on a TM remote sensing image. In this experiment, Naive Bayesian algorithm is used. 1500 samples, including 600 liked points and 900 disliked points, are in the training samples of the Bayesian classifier. The image displayed on the left part is generated by the 4th, 3rd, 2nd band of the remote sensing image.

Because of the lack of standard data set, pseudo-classification accuracy is used to verify the classification accuracy. 20% of the samples are not used for training, instead, they are used to verify the classification result. Based on this method, the pseudo-classification accuracy of figure 5 is 97.1445%.

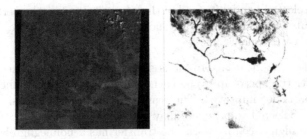

Fig. 5. A TM RS Image and Classification Result

SVM and *Naive Bayesian* classifiers are used to classify two remote sensing images. One is a 7348*5968 TM image with seven raster bands (*multi-spectral*), and the other one is a 256*3400 H5 image with 242 raster bands (*hyper-spectral*). Serial mode, 5 processes mode, 9 processes mode have been tested in the experiments. The Speed-up S is calculated based on 1.

$$S = T_{serial}/T_{parallel} \qquad (1)$$

Clearly $S > 0$, when $S > 1$, the speed is enhanced; When $S < 1$ the speed is reduced.

Image Format	Width	Height	Bands Count	Selected Transfer	Tran. Samp.	Classif. Algor.	Process Count	Time (s)	speedup
TM	7348	5968	7	No	1500	Bayes	Serial	221.8	-
							5	114.1	1.94
							9	107.4	2.07
						SVM	Serial	2380.3	-
							5	682.4	3.49
							9	606.6	3.92
h5	256	3400	242	No	1700	Bayes	Serial	6	-
							5	42.6	0.14
							9	40.8	0.15
						SVM	Serial	106.7	-
							5	70	1.52
							9	66.6	1.60
				Yes	2000	Bayes	Serial	4.2	-
							5	4.2	1
							9	4.1	1.02
						SVM	Serial	109.7	-
							5	34.6	3.17
							9	30.4	3.61

Fig. 6. Classification Results of A TM and A H5 RS Image

4.4   Discussion

During the experiments, it is suspected although wired network is used, it still takes a very long time to transfer files. This is more obvious when a fast algorithm is tested on large size files. This problem is possibly due to the limitation of hard-disk speed. After modifying the programs making it transfer only useful raster bands (selected bands) instead of the whole image file, the performance is significantly improved. The results are shown in Figure 6. From the result table, we can reach several conclusions:

- The parallel classification programs significantly speed up the processing. In one test, the speed-up even reaches 3.92. However, in some cases, the acceleration is not improved. The reason is that file transmission takes too much time. Selected transferring may help;
- Despite the high-speed network, file transfer has a nonnegligible impact on the speed, especially for fast classification algorithms;
- For fast algorithms and quite large images, parallel processing is a good choice according to the experiments conducted in this paper. *SVM* is more suitable than *Naive Bayesian* to be parallelized. However, for some large size images, bayesian classifier is also helpful;
- The computing end contains two hyper-thread Xeon Processors, which afford 4 threads. From the chart we can see that 9 processes is a little faster than 5 processes in many cases, so it is appropriate to set the number of processes slightly bigger than the biggest number of parallel treads, which can make better use of the CPU.

5   Conclusion

An architecture of parallel remote sensing image processing is designed in this paper. With the use of open-source libraries, a group of programs are developed. The architecture has a variety of advantages including platform-independent, language-independent and high scalability.

The classification experiments of the two remote sensing images indicated that parallel system brings significantly speed improvement in most cases.

Since we realize that file transfer somehow impact on the performance, it's more practical to select certain bands of the image to transfer to the computing end for classification. On the other hand, classification algorithms also impact the speed-up. This system is mainly useful for slow algorithms and multi-spectral images (with less bands).

The future work will concentrate on implementing more algorithms in this system, making it suitable in different conditions.

References

1. Mei, A., Peng, W., Qin, Q., Liu, H.: Introduction to Remote Sensing, 1st edn. (2001)
2. Chen, P., Tong, Q., Guo, H.: Incentive Research of Remote Sensing Information, 1st edn. (1998)
3. Tong, Q., Zhang, B., Zhen, F.: Theory, Techniques and Application of Hyperspectral Remote Sensing, 3rd edn. (2006)
4. Shen, Z., Luo, J., Huang, G., et al.: Distributed computing model for processing remotely sensed images based on grid computing. Information Science (2007)

5. Boost. MPI Document, http://www.boost.org
6. GDAL, 201x, GDAL-Geospatial Data Abstraction Library: Version 1.8.0, http://gdal.osgeo.org
7. GDAL Raster Formats, http://gdal.org/formats_list.html
8. Zhang, L., Zhang, L.: Hyper-spectral Remote Sensing, 1st edn. (2005)
9. John, G.H., Langley, P.: Estimating Continuous Distributions in Bayesian Classifiers. In: Proceedings of the Eleventh Conference on Uncertainty in Artificial Intelligence (1995)
10. Bradski.G, The OpenCV Library, Dr.Dobb's Journal of Software Tools

CNAR-M: A Model for Mining Critical Negative Association Rules

Tutut Herawan[1] and Zailani Abdullah[2]

[1] Faculty of Computer System and Software Engineering
Universiti Malaysia Pahang
Lebuhraya Tun Razak, 26300 Kuantan Pahang, Malaysia
tutut@ump.edu.my
[2] Department of Computer Science
Universiti Malaysia Terengganu
21030 Kuala Terengganu, Terengganu, Malaysia
zailania@umt.edu.my

Abstract. Association rules mining has been extensively studied in various multidiscipline applications. One of the important categories in association rule is known as Negative Association Rule (NAR). Significant NAR is very useful in certain domain applications; however it is hardly to be captured and discriminated. Therefore, in this paper we proposed a model called Critical Negative Association Rule Model (CNAR-M) to extract the Critical Negative Association Rule (CNAR) with higher Critical Relative Support (CRS) values. The result shows that the CNAR-M can mine CNAR from the benchmarked and real datasets. Moreover, it also can discriminate the CNAR with others association rules.

Keywords: Data Mining, Negative, Association Rules, Critical Relative Support.

1 Introduction

Data mining[1] emerged as a rapidly growing interdisciplinary field that merges together statistics, databases, machine learning, and others related area. The main aim of data mining is to find interesting patterns in database but is hidden among the massive of data. It is a part of knowledge discovery in database (KDD) process. Its primary goals are to be prediction and description. The first part is more on using some variables in database to predict unknown values of other variable. For the second part, it focuses on interpreting the patterns of data. Until this recent, data mining has been successfully applied in various multi-domain many applications.

Association rule mining is one of the most important and widespread techniques in data mining[2]. It falls under the second part of data mining. The main objective of association rule mining is to find the correlations, association or casual structure among set of items in data repository. The problem of association rule mining was first introduced by Agrawal for market-basket analysis[3–5].

Z. Li et al. (Eds.): ISICA 2012, CCIS 316, pp. 170–179, 2012.

Typically, two main stages are involved before generating the association rules (ARs). First, find all frequent items from transactional database. Second, generate the common ARs from the frequent items. An item is said to be frequent if it appears more than a minimum support threshold. These frequent items are then used to produce the ARs. Besides that, confidence is another measure that always used in pair with the minimum support threshold.

Infrequent itemset is a set of item that is rarely appeared in the transactional database. It is also known as non-frequent, unusual, exceptional, abnormal, inbalance or sporadic itemset. These itemsets are sometimes very important because it can provide a significant information for certain domain applications such as detection for air pollution, network intruders, machine critical faulty[6], abnormal learning problems[7], and many more. From the previous research developments, most of the tradition association rules mining algorithms[8–15] are still have a limitation in term of efficiency and evaluating the real datasets. Usually, the low minimum support is used to extract capture the least itemset. However, it may drives into producing the abundant of unnecessary association rules. As a result, it creates another challenge in comprehensively identifying which association rules are really useful and significant. Moreover, the low minimum support will also proportionally increase the computational resources or memory consumption. Since the complexity of study, difficulties in algorithms[11] and it may require excessive computational cost, there are very limited attentions have been paid to discover least association rules.

Negative association rule is a contrast of the most regular or positive association rules. This rule describes the relationship between the itemset and the absence of others. The most popular negative association rule measure was introduced by Brin et al.[16]. Extracting a complete set of negative association rules[17, 18] is very useful and important. However the formation of this rule always involved with infrequent items. The infrequent itemset is a set of items that is rarely found in the database but may produce an interesting result. Typically, it can only be captured by lowering the minimum support threshold. Therefore, in this paper we propose a model called Critical Negative Association Rules Model (CNAR-M) to extract three types of rules; Critical Negative Association Rules, Negative Association Rules and Positive Association Rules. Critical Relative Support (CRS) measure[19] is embedded in the model to ensure CNAR can be captured.

The contributions of this paper are as follows. First, we develop a model named CNAR-M that can be used to mine three types of association rules. Second, we embed the scalable CRS measure to extract the meaningful set of negative rules known as CNAR. Third, we evaluate the performance of CNAR-M against the benchmarked and real datasets. We believe that, the information obtained from the experiment can be further studied by the respectively experts domain.

The reminder of this paper is organized as follows. Section 2 describes the related work. Section 3 explains fundamental definitions of association rules and measures. Section 4 discusses the proposed method. The result and discussion will be shown in Section 5. Finally, conclusion and future direction are reported in section 6.

2   Related Work

Association rule was first introduced by Agrawal et al. [3] and still attracts many attentions from knowledge discovery community[20-29]. Numerous studies have been done in term of infrequent and negative correlation of itemset. Zhou et al. [9] suggested an approach to mine the ARs by considering only infrequent itemset. They proposed two algorithms called Matrix-based Scheme (MBS) and Hash-based scheme (HBS) to discovery the ARs among infrequent items. The limitation is, both algorithms are not a good solution for mining ARs with any length among frequent items due to the expensive cost of hash collision. Ding[15] discussed association rule mining among rare items. He designed a new disk-based data structure called Transactional Co-occurrence Matrix (TCOM). However, the implementation of this algorithm is too costly.

Yun et al. [11] proposed the Relative Support Apriori Algorithm (RSAA) to generate rare itemsets. Relative support is determined by selecting the largest among the confidence values for the candidate itemset against each data item. The challenge is if the minimum allowable relative support is set close to zero, it takes similar time taken as performed by Apriori. Koh et al. [10] introduced Apriori-Inverse algorithm to mine infrequent itemsets without generating any frequent rules. It captures the sporadic rules using maximum support and minimum confidence thresholds. However, the main constraints are it suffers from too many candidate generations and time consumptions during generating the rare ARs. Liu et al. [12] proposed Multiple Support Apriori (MSApriori) algorithm to extract the rare ARs. The minimum support for each item is calculated using the "support difference".

Brin et al. [16] mentioned for the first time about the notion of negative relationship. It is based on the chi-square and lift. The statistical test is employed to verify the independence between two variables. Indeed, a correlation metric is used to determine either it is positive and negative relationships between items. Tan et al. [30] suggested a novel algorithm for deriving indirectly associated itempairs known as indirect rules. The extraction of the indirect rules is basically closely related to the existing items that have a negative association. Hamano et al. [31] introduced a new method to mine both positive and negative indirect association rules similar to Tan et al. (2000) which is based on special measure.

Wu et al. [32] proposed a new algorithm to generate both positive and negative association rules. On top of support-confidence framework, an extra measure called mininterest is introduced for a better pruning on the existing frequent itemsets. Antonie et al. [33] also proposed a new algorithm by extending and modifying the supportCconfidence framework with a sliding correlation coefficient threshold. This algorithm can extract both positive and negative association rules with the strong correlation among them.

Hahsler [34] proposed a stochastic mixture model known as Negative-Binomial distribution. This model utilizes the process of generating transaction data and finally generate Negative-Bionamial frequent itemsets. Zhao et al. [9] also suggested two simple practical and effective schemes to mine AR among the rare item which are MatrixCBased Scheme (MBS) and Hash-Based Scheme (HBS).

These two schemes only consider infrequent items in generating the association rules. Tsai et al. [35] proposed an algorithm named Generalized Negative Association Rules (GNAR) which improved the traditional Apriori algorithm by reducing the computational cost during mining negative association rules. It also reduces the number of nonCinteresting negative association rules.

3   Preliminaries

In this part, important definition pertinent to the proposed model will be discussed.

3.1   Definition

Definition 1. (Association Rules) *An association rule is a form of $A \Rightarrow B$, where $A, B \sqsubset I$ such that $A \neq \varnothing, B \neq \varnothing$ and $A \cap B \neq \varnothing$. The set A is called antecedent of the rule and the set B is called consequent of the rule.*

Definition 2. (Frequent Items) *An itemset is said to be frequent if the support count satisfies a minimum support count (minsupp). The set of frequent itemsets is denoted as L_k.*

Definition 3. (Confidence) *The confidence of the ARs is the ratio of transactions in D contains A that also contains B. The confidence also can be considered as conditional probability $P(B|A)$.*

$$Conf(A, B) = \frac{Supp(A \cup B)}{Supp(A)}$$

Definition 4. (Lift Correlation) *The occurrence of itemset A is independence of the occurrence of itemset B if $P(A \cup B) = P(A)P(B)$; otherwise itemset A and B are dependence and correlated. The lift between occurrence of itemset A and B can be defined as:*

$$lift(A, B) = \frac{P(A \cap B)}{P(A)P(B)}$$

or

$$lift(A, B) = \frac{Conf(A \Rightarrow B)}{Supp(B)}$$

The strength of correlation is measure from the lift value. If $lift(A, B) = 1$ or $P(B|A) = P(B)$(or $P(A|B) = P(B)$) then B and A are independent and there is no correlation between them. If $lift(A, B) > 1$ or $P(B|A) > P(B)$(or $P(A|B) > P(B)$), then A and B are positively correlated, meaning the occurrence of one implies the occurrence of the other. If $lift(A, B) < 1$ or $P(B|A) < P(B)$(or $P(A|B) < P(B)$), then A and B are negatively correlated, meaning the occurrence of one discourage the occurrence of the other. Since lift measure is not down-ward closed, it definitely will not suffer from the least item problem. Thus, least itemsets with low counts which per chance occur a few times (or only once) together can produce enormous lift values.

Definition 5.(Critical Relative Support) *A Critical Relative Support (CRS) is a formulation of maximizing relative frequency between itemset and their Jaccard similarity coefficient.*

The value of Critical Relative Support denoted as CRS and

$$CRS(A, B) = max\left(\left(\frac{Supp(A)}{Supp(B)}\right), \left(\frac{Supp(B)}{Supp(A)}\right)\right)$$
$$\times \left(\frac{Supp(A \Rightarrow B)}{Supp(A) + Supp(B) - Supp(A \Rightarrow B)}\right)$$

CRS value is between 0 and 1, and is determined by multiplying the highest value either supports of antecedent divide by consequence or in another way around with their Jaccard similarity coefficient. It is a measurement to show the level of CRS between combination of the both Least Items and Frequent Items either as antecedent or consequence, respectively.

4   Proposed Method

4.1   Model Overview

There are several main components involved in critical negative association rules model. Fig. 1 shows an overview model of CNAR. The discussions about the main components are as follows.

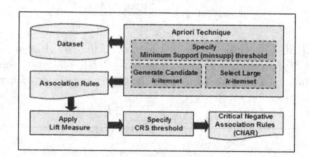

Fig. 1. An overview of CNAR-M

Dataset. Dataset is a set of data which is stored in a flat file format. All data (itemsets) in datasets are segregated by one extra space. The data are "uploaded" into certain objects such array, sortedlist, hashes etc. before further processing.

Apriori Technique. There are three sub-processes. First, specify minimum support (*minsupp*) threshold. It is an initial stage to extract the reasonable itemset from dataset. Second, generate candidate *k−itemsets*. Several candidate *k − itemsets* will be produced and *minsupp* will not take into account. Third, select large *k − itemset*. Only items that fulfill the minsupp will be extracted and became large $k-itemsets$. Sub-process of second and third will be repeated several times until $n-itemsets$ or no more large itemset can be produced.

AssociationRules. All itemsets from the pervious process stage will be converted into association rules. The formation of association rules is in term of one-or-many items as antecedent and only one item as consequence.

Apply Lift Measure. Lift formula (Definition 4) will be applied into generated association rules to discriminate between positive and negative association rules. Only negative association rules will be captured and forwarded into the next stage.

Specify CRS threshold. Minimum CRS (min-CRS) will be specified as a threshold. CRS measure (Definition 5) will be applied into all negative association rules. The total critical negative rules based on predefined different min-CRS will be computed.

4.2 Pseudo-Code Development

The procedure of NCAR-M in term of pseudo-code is presented below. This pseudo-code is conformed to the implementation of CNAR-M as explained in the previous an overview model.

Algorithm 1. NCAR-M Pseudo-code

1: Input: Dataset(D), minsupp(α), min-CRS(φ)
2: Output: Critical Negative Association Rules (CNAR)
3: Load dataset
4: Generate candidate itemsets, CI_i
5: **if** $Supp(CI_i) > \alpha$ **then**
6:     Generate large itemsets, $LI_j \leftarrow CI_i$
7: **end if**
8: **for** $LI_j \neq eof$ **do**
9:     Generate Association Rules, AR_m
10: **end for**
11: **for** $AR_m \neq eof$ **do**
12:     Calculate Lift
13:     **if** $Lift(AR_m) > 1$ **then**
14:         Generate Positive Association Rules, PAR_a
15:         else If $Lift(AR_m) = 1$ Then
16:         Generate Positive Association Rules, $NeuAR_b$
17:         else If $Lift(AR_m) < 1$ Then
18:         Generate Positive Association Rules, NAR_c
19:     **end if**
20: **end for**
21: **for** $NAR_c \neq eof$ **do**
22:     Calculate CRS
23:     **if** $CRS(NAR_c) \leq \varphi$ **then**
24:         Generate Critical Negative Association Rules, CNAR
25:     **end if**
26: **end for**

---

5   Experimental Test

In this section, we do comparison tests between different types of rules. The performance analysis is made by comparing the total number of rules being produced. We used two benchmarked datasets and one real dataset. These experiments have been conducted on Intel Core^{TM} 2 Quad CPU at 2.33GHz speed with 4GB main memory, running on Microsoft Windows Vista. All algorithms have been developed using C# as a programming language.

The benchmarked datasets were the Car Evaluation dataset and Breast Cancer dataset. Both datasets were taken from the UCI Machine Learning Repository[36]. The Car Evaluation dataset was created by Marko Bohanec with 1,729 transactions and 6 attributes. While for the Breast Cancer dataset was prepared by Dr. William H. Wolberg, from the University Wisconsin Hospitals Madison, Wisconsin, USA. This dataset contains 699 transactions and 13 attributes. The real dataset is Kuala Lumpur Air Pollution dataset. The data were taken in Kuala Lumpur on July 2002 as presented and used in[19]. The items involved were i.e. $\{CO_2, O_3, PM_{10}, SO_2, NO_2\}$. The value of each item is with the unit of part per million (ppm) except PM_{10} is with the unit of micrograms (μgm). The data were taken for every one-hour every day. The actual data is presented as the average amount of each data item per day. For brevity, each data item is mapped to parameters and 5 respectively. The Minimum Support (minsupp=20%) and minimum Critical Relative Support (min-CRS=0.6) were specified as threshold values for all experimented datasets. The result from the experiments is shown as in Fig. 2.

For Car Evaluation dataset, the total number of Negative Association Rules (NAR) is the highest as compared to others. Critical Negative Association Rules (CNAR) appeared about 33% from the rest association rules. For Breast Cancer dataset, CNAR occurred 36% from the whole association rules. The highest number of association rules is conquered by NAR. For the last dataset (KL Air Pollution), total number of NAR and CNAR are the same, which is 47% from the rest association rules.

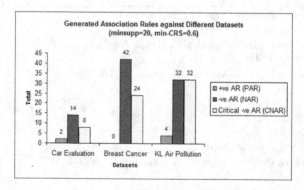

Fig. 2. The Different Types of Association Rules Being Generated by Diverse Datasets

6 Conclusion

Association rule mining is one of the well-known techniques in data mining that have been widely applied in multidiscipline applications. It aims at finding the hidden and interesting association rule from database. One of the important categories in association rules is negative association rule (NAR). However, not all NAR are really critical or significant. Indeed, the significances of negative association rules are nearly not discussed in all literatures. Therefore, in this paper we proposed a model called Critical Negative Association Rule Model (CNAR-M) to extract NAR with the higher Critical Relative Support (CRS) values. The result shows that the CNAR-M can mine and discriminate Critical Negative Association Rule (CNAR) from the benchmarked and real datasets. In average, the total number of Critical Negative Association Rules (CNAR) is less than the standard NAR.

In a near future, we planned to apply CNAR-M into several benchmarked datasets and real datasets as an attempt to evaluate it scalabilities. Moreover, we also interested to convert the implementation from Apriori technique into Trie-based technique.

Acknowledgement. This work is supported by Research Management and Innovation Centre, Universiti Malaysia Pahang.

References

1. Fayyad, U.M., Piatetsky-Shapiro, G., Smyth, P., Uthurusamy, R.: From Data Mining To Knowledge Discovery: An Overview. In: Advanced in Knowledge Discovery and Data Mining, pp. 1–34. AAAI Press (1996)
2. Morzy, T., Zakrzewicz: Data mining. In: Handbook on Data Management in Information Systems, pp. 487–565. Springer, Heidelberg (2003)
3. Agrawal, R., Imielinski, T., Swami, A.: Database mining: A Performance Perspective. IEEE Transactions on Knowledge and Data Engineering 5(6), 914–925 (1993)
4. Agrawal, R., Imielinski, T., Swami, A.: Mining Association Rules Between Sets Of Items In Large Databases. In: Proceedings of the ACM SIGMOD International Conference on the Management of Data, pp. 207–216 (1993)
5. Agrawal, R., Srikant, R.: Fast Algorithms For Mining Association Rules. In: Proceedings of the 20th International Conference on Very Large Databases (VLDB 1994), pp. 487–499 (1994)
6. Abdullah, Z., Herawan, T., Deris, M.M.: Mining Significant Least Association Rules Using Fast SLP-Growth Algorithm. In: Kim, T.-h., Adeli, H. (eds.) AST/UCMA/ISA/ACN 2010. LNCS, vol. 6059, pp. 324–336. Springer, Heidelberg (2010)
7. Romero, C., Romero, J.R., Luna, J.M., Ventura, S.: Mining Rare Association Rules From E-Learning Data. In: Proceeding of The Third International Conference of Education Data Mining, Pittsburgh, USA, pp. 171–180 (2010)
8. Kiran, R.U., Reddy, P.K.: An Improved Multiple Minimum Support Based Approach To Mine Rare Association Rules. In: Proceeding of IEEE Symposium on Computational Intelligence and Data Mining, pp. 340–347 (2009)

9. Zhou, L., Yau, S.: Association Rule and Quantitative Association Rule Mining among Infrequent Items. In: Proceeding of ACM SIGKDD 2007, Article No. 9 (2007)
10. Koh, Y.S., Rountree, N.: Finding Sporadic Rules Using Apriori-Inverse. In: Ho, T.-B., Cheung, D., Liu, H. (eds.) PAKDD 2005. LNCS (LNAI), vol. 3518, pp. 97–106. Springer, Heidelberg (2005)
11. Yun, H., Ha, D., Hwang, B., Ryu, K.H.: Mining Association Rules on Significant Rare Data Using Relative Support. The Journal of Systems and Software 67(3), 181–191 (2003)
12. Liu, B., Hsu, W., Ma, Y.: Mining Association Rules with Multiple Minimum Supports. In: Proceeding of ACM SIGKDD 2007, pp. 337–341 (1999)
13. Wang, K., Hee, Y., Han, J.: Pushing Support Constraints Into Association Rules Mining. IEEE Transaction on Knowledge and Data Engineering 15(3), 642–658 (2003)
14. Tao, F., Murtagh, F., Farid, M.: Weighted Association Rule Mining using Weighted Support and Significant Framework. In: Proceeding of ACM SIGKDD 2003, pp. 661–666 (2003)
15. Ding, J.: Efficient Association Rule Mining Among Infrequent Items. Ph.D. Thesis, University of Illinois at Chicago (2005)
16. Brin, S., Motwani, R., Silverstein, C.: Beyond Market Basket: Generalizing Ars To Correlations. Special Interest Group on Management of Data (SIGMOD), pp. 265–276 (1997)
17. Tsai, L.M., Lin, S.J., Yang, D.L.: Efficient Mining Of Generalized Negative Association Rules. ACM Digital Library (2010)
18. Wu, X., Zhang, C., Zhang, S.: Efficient Mining For Both Positive And Negative Association Rules. ACM Digital Library (2004)
19. Abdullah, Z., Herawan, T., Deris, M.M.: Scalable Model for Mining Critical Least Association Rules. In: Zhu, R., Zhang, Y., Liu, B., Liu, C. (eds.) ICICA 2010. LNCS, vol. 6377, pp. 509–516. Springer, Heidelberg (2010)
20. Abdullah, Z., Herawan, T., Deris, M.M.: Mining Significant Least Association Rules Using Fast SLP-Growth Algorithm. In: Kim, T.-h., Adeli, H. (eds.) AST/UCMA/ISA/ACN 2010. LNCS, vol. 6059, pp. 324–336. Springer, Heidelberg (2010)
21. Abdullah, Z., Herawan, T., Noraziah, A., Deris, M.M.: Extracting Highly Positive Association Rules from Students' Enrollment Data. Procedia Social and Behavioral Sciences 28, 107–111 (2011)
22. Abdullah, Z., Herawan, T., Noraziah, A., Deris, M.M.: Mining Significant Association Rules from Educational Data using Critical Relative Support Approach. Procedia Social and Behavioral Sciences 28, 97–101 (2011)
23. Abdullah, Z., Herawan, T., Deris, M.M.: An Alternative Measure for Mining Weighted Least Association Rule and Its Framework. In: Zain, J.M., Wan Mohd, W.M., El-Qawasmeh, E., et al. (eds.) ICSECS 2011, Part II. CCIS, vol. 180, pp. 480–494. Springer, Heidelberg (2011)
24. Abdullah, Z., Herawan, T., Deris, M.M.: Visualizing the Construction of Incremental Disorder Trie Itemset Data Structure (DOSTrieIT) for Frequent Pattern Tree (FP-Tree). In: Badioze Zaman, H., Robinson, P., Petrou, M., Olivier, P., Shih, T.K., Velastin, S., Nyström, I. (eds.) IVIC 2011, Part I. LNCS, vol. 7066, pp. 183–195. Springer, Heidelberg (2011)
25. Herawan, T., Yanto, I.T.R., Deris, M.M.: Soft Set Approach for Maximal Association Rules Mining. In: Ślęzak, D., Kim, T.-h., Zhang, Y., Ma, J., Chung, K.-i. (eds.) DTA 2009. CCIS, vol. 64, pp. 163–170. Springer, Heidelberg (2009)

26. Herawan, T., Yanto, I.T.R., Deris, M.M.: SMARViz: Soft Maximal Association Rules Visualization. In: Badioze Zaman, H., Robinson, P., Petrou, M., Olivier, P., Schröder, H., Shih, T.K. (eds.) IVIC 2009. LNCS, vol. 5857, pp. 664–674. Springer, Heidelberg (2009)

27. Herawan, T., Deris, M.M.: A soft set approach for association rules mining. Knowledge Based Systems 24(1), 186–195 (2011)

28. Herawan, T., Vitasari, P., Abdullah, Z.: Mining Interesting Association Rules of Student Suffering Mathematics Anxiety. In: Zain, J.M., Wan Mohd, W.M.b., El-Qawasmeh, E., et al. (eds.) ICSECS 2011, Part II. CCIS, vol. 180, pp. 495–508. Springer, Heidelberg (2011)

29. Abdullah, Z., Herawan, T., Deris, M.M.: Mining Significant Least Association Rules Using Fast SLP-Growth Algorithm. In: Kim, T.-h., Adeli, H. (eds.) AST/UCMA/ISA/ACN 2010. LNCS, vol. 6059, pp. 324–336. Springer, Heidelberg (2010)

30. Tan, P.N., Kumar, V.: Discovery of Indirect Associations In Web Usage Data. In: Web Intelligence, pp. 128–152. Springer, Heidelberg (2003)

31. Hamano, S., Sato, M.: Mining Indirect Association Rules. In: Perner, P. (ed.) ICDM 2004. LNCS (LNAI), vol. 3275, pp. 106–116. Springer, Heidelberg (2004)

32. Wu, X., Zhang, C., Zhang, S.: Mining Both Positive And Negative Association Rules. In: Proceedings of 19th International Conference on Machine Learning, Sydney, Australia, pp. 558–665 (2002)

33. Antonie, M.L., Zaiane, O.: Mining Positive and Negative Association Rules for an Approach for Confined Rules. Technical Report TR04-07, University of Alberta (2004)

34. Hahsler, M.: A Model Based Frequency Constraint For Mining Associations From Transaction Data. In: DMKD, pp. 137–166. Springer Science and Business Media (2006)

35. Tsai, L.M., Lin, S.J., Yang, D.L.: Efficient Mining of Generalized Negative Association Rules. ACM Digital Library (2010)

36. UCI Machine Learning Repository, http://archive.ics.uci.edu/ml/datasets.html (accessed on June 01, 2012)

MT2Way: A Novel Strategy for Pair-Wise Test Data Generation

Khandakar Fazley Rabbi, Abul Hashem Beg, and Tutut Herawan

Faculty of Computer Systems & Software Engineering
University Malaysia Pahang
Lebuhraya Tun Razak, Gambang 26300, Kuantan, Pahang, Malaysia
fazley.rabbi@ymail.com
ahbeg_diu@yahoo.com
tutut@ump.edu.my

Abstract. Reducing the number of test cases by utilizing minimum possible amount of time during the testing process of software and hardware is highly desirable. For ensuring the reliability of the method the combination of a complete set of available inputs is recommended to be executed. But generally an exhaustive numbers of test cases are hard to execute. Besides, test data generation is an NP-hard (non-deterministic polynomial-time hard) problem. This is likely to present considerable difficulties in defining the best possible method for generating the test data. The reduction of test cases depends on the interaction level, 2-way interaction or pair-wise test data can reduce high number of test cases and it efficiently addresses most of the software errors. This paper presents MT2Way, an effective 2-way interaction algorithm to generate the test data which is more acceptable in terms of the number of test cases and execution time. The performance tests show that MT2Way achieve better results in terms of system configuration, generated test size, and executing time as compared to other techniques.

Keywords: Combinatorial Interaction Testing, Software Testing, Pair-wise Testing, Test Case Generation.

1 Introduction

Software testing and debugging is one of the integral part of software development life cycle in software engineering but this process is still very labor-intensive and expensive [1]. The most practical and realistic project management is characterized by an effective and well balanced allocation of project resources. But this is generally observed that half of the project resources are consumed in efforts for introducing highly reliable software applications and hardware devices. All the researcher and practitioners focus on finding automatic cost-effective software testing and debugging techniques which can maintain the high error detection rate and ability to ensure high quality of software product release [1, 2]. Nowadays, software engineering research on software testing focus on test coverage criterion design, test generation problem, test oracle problem, regression testing

Z. Li et al. (Eds.): ISICA 2012, CCIS 316, pp. 180–191, 2012.

problem and fault localization problem [1]. Among this problem test generation problem is an important issue in producing error free software [1]. Just to realize the volume of the problem a system which is constituted by considering only six parameters with ten values is expected to generate 106 numbers of test cases. Obviously this much number of test cases is un-realistic as it is expected to miserably fail to keep itself well within the boundaries of time and resource constraints. Therefore it is very important to reduce the number of test cases keeping the effectiveness of detecting errors [3]. To solve this problem pair-wise interaction technique is one of the most useful as it keeps good balance between the quality and effectiveness of combinations. To understand the problem more briefly, the follow the option dialog in Microsoft excel (Figure 1) view tab option, there are 20 possible configurations needed to be tested along with the Gridline colors (which takes 56 possible values). Each other configuration takes two values (checked or unchecked). So to test this view tab exhaustively there is $2^{20} \times 56$ i.e. 58,720,256 (2 is the values and 20 is the number of configurations) number of test cases need to be executed. Assuming each test case may consume 5 minutes to execute; results around 559 years to complete the exhaustive test of this view tab [4].

This is also similar for hardware products as well. A product which has 20 on/off switches will take $2^{20} = 1,048,576$ test cases. It will take 10 years if each single test case needs around 5 minutes to execute [4]. Nowadays, research work in combinatorial testing aims to generate least possible test cases [5]. The solution of this problem is NP-hard (non-deterministic polynomial-time hard) [6].

This is likely to present considerable difficulties in defining the best possible method for generating the test data. The reduction of test cases depends on the interaction level. 2-way interaction or pair wise test data can reduce high number of test cases. This efficiently addresses most of the software errors. So far many approaches have been proposed and also many tools have been developed to find out the least possible test suit in polynomial time [5,12]. This paper represents an effective 2-way interaction algorithm to generate the test data which is more acceptable in terms of the number of test cases and execution time.

The rest of this paper is organized as follow. Section 2 describes the related work. Section 3 describes proposed MT2Way and its software developed. Section 4 describes results and comparison test. Finally, the conclusion of this work is described in Section 5.

2 Related Work

There are some strong empirical facts [7,8] that lack of testing of both functional and nonfunctional is one of major source of software and systems bug/errors. National Institute of Standard and Technology (NIST) estimated that the cost of software failure to the US economy at 6×1010, which was the 0.6 percent of GDP [9,10]. Another report found that more than one-third of this cost can be reduce by improve software testing structure. To improve software testing structure automatic testing is a critical concern [10, 11]. In the course of automation, software can become more practical and scalable. However, the automated

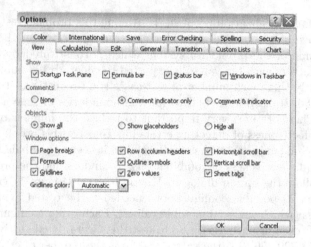

Fig. 1. Option Dialog for Microsoft Excel

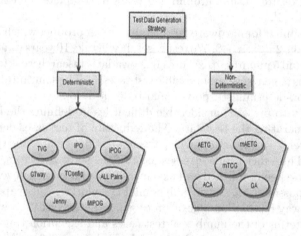

Fig. 2. Test case generation categories

generation of test case still represents challenges [10]. The underlying problem is known to be un-decidable and NP-hard thus researchers have focus on the techniques that search to identify near optimal test sets in a reasonable time [12-27]. Basically, each test suite constructed by N×k array where N represents the number of test cases. Each of the single test cases is a combination of the k parameter values [17]. In test data generation, it is normally focused on all t-way combinations of inputs and the values occur in the test suite. In the t-way t determines the interaction strength of test suite. Thus, the single setting of the system that needs to be tested is defined by the single test case. In a nut shell, test data generation strategies can generate test cases in non-deterministic or deterministic way. Figure 2 shows the most available test data generation strategies.

The Automatic Efficient Test Generator or AETG [28, 29] and its deviation mAETG [30] generate test data using computational approach. This approach uses the Greedy technique to build test cases based on covering as much as possible uncovered pairs. AETG uses a random search algorithm [30]. Genetic Algorithm, Ant Colony Algorithm [30] is the variants of AETG. Genetic algorithm [31] creates an initial population of individuals (test cases) and then the fitness of those individuals is calculated. Then it starts discarding the unfit individuals by the individual selection methods. The genetic operators such as crossover and mutation are applied on the selected individuals and this continues until a set of best individuals found. Ant Colony Algorithm [31, 18] candidate solution is associated with the start and end point. When an ant chooses one edge among the different edges, it would choose the edge with a large amount of pheromone which gives the better result with the higher probability. The In-Parameter-Order [32] or IPO Strategy starts with an empty test set and adds one test at a time. It creates the test cases by combination of the first two parameters, and then add third and calculate how many pair is been covered then add fourth and calculate and then fifth and calculate until all the values of each parameter is checked. This approach is deterministic approach. G2Way [31] is based on backtracking algorithm uses computational deterministic strategy. G2Way uses customized markup language to describe base data. The G2Way backtracking algorithm tries to combine generated pairs so that it covers highest pairs. Finally when it covers all the pairs, the test case treats as a final test suite.

3   Proposed MT2Way Algorithm

In this section, the proposed MT2Way is described. The proposed MT2Way algorithm can be divided into 3 steps, i.e.

1. Pair generation
2. Tuple generation
3. Test case construction

3.1   Pair Generation

Pair wise algorithm first takes the parameters and corresponding values, makes the 2 way combinations. The combinations stored into the memory. As an example

1. Parameter A has 3 values a1, a2, a3
2. Parameter B has 2 values b1, b2
3. Parameter C has 4 values c1, c2, c3, c4

The pair generation algorithm first generates header pairs. The header pair from this example is:

Table 1. Generated header and value pairs

Pairs	Value
AB	((a1, b1), (a1, b2), (a2, b1), (a2, b2), (a3, b1), (a3, b2)
AC	(a1, c1), (a1, c2), (a1, c3), (a1, c4), (a2, c1), (a2, c2) (a2, c3), (a2, c4), (a3, c1), (a3, c2), (a3, c3), (a3, c4)
BC	(b1, c1), (b1, c2), (b1, c3), (b1, c4), (b2, c1), (b2, c2), (b2, c3), (b2, c4)

Table 2. Generated N-Tuple from Table 1

Pairs	N-Tuple from Pairs
AB	(a1, b1, 0)
	(a1, b2, 0)
	(a2, b1, 0)
	(a2, b2, 0)
	(a3, b1, 0)
	(a3, b2, 0)
AC	(a1, 0, c1)
	(a1, 0, c2)
	(a1, 0, c3)
	(a1, 0, c4)
	(a2, 0, c1)
	(a2, 0, c2)
	(a2, 0, c3)
	(a2, 0, c4)
	(a3, 0, c1)
	(a3, 0, c2)
	(a3, 0, c3)
	(a3, 0, c4)
BC	(0, b1, c1)
	(0, b1, c2)
	(0, b1, c3)
	(0, b1, c4)
	(0, b2, c1)
	(0, b2, c2)
	(0, b2, c3)
	(0, b2, c4)

After creating the header pairs, it calculates the possible values into all the pairs. In this example, AB contains $3 \times 2 = 6$ values (pairs). The expression can be written as follows:

$N_p = [V_1 * V_2]$, where N_p=number of Pairs, V_1=Value of Param 1, V_2=Value of Param 2.

So, in the pair generation algorithm stores all the value pairs into the memory which are 6, 12, and 8 respectively for AB, AC, and BC.

3.2 Tuple Generation

Tuple generation algorithm first generates the N-tuple based on the pairs, where N is the number of the parameters. To make a complete N-tuple from a pair, the values are replaced by 0 which is not exists in that pair. As an example, the generated 3-tuples can from Table 1, which can make the algorithm more understandable. Table 2 illustrates the N-tuple.

From Table 2, all the pair values converted to a set of test case i, e, N-tuples and the missing parameter is replaced by 0.

3.3 Test Case Construction

In this part of the algorithm, the missing parameters try to adjust with the all possible values. It selects the value from the missing parameter position which gives the highest possible pair coverage. From N-Tuple the 0 indicates the missing parameter and it's positing. The pseudo code of the algorithm has been shown below.

Algorithm 1. Test Case Construction

1: Begin
2: Let $C_T = \{\}$ as empty set, where CT represents the Tuples Set
3: Let $C_{TT} = \{\}$ as empty set, where CTT represents the single Tuple in the Tuple Set.
4: **for** e **doach** value tuple in C_T
5:     C_{TT} = tuple
6:     **for** e **doach** value val in C_{TT}
7:         **if** v **thenal** is equals to 0
8:             read val position as P
9:             Find parameter position from P
10:            **for** e **doach** values in P
11:                replace 0 with the values
12:                create test case set as C_{TS}
13:                get pair coverage of C_{TS} = PC
14:                **if** P **thenC** is the highest pair Coverage
15:                    select C_{TS} as the final test case
16:                Else if
17:                    Search the value which generates highest pair coverage = C_{FS}.
18:                    select C_{FS} as the final test case
19:                **end if**
20:            **end for**
21:        **end if**
22:    **end for**
23: **end for**
24: End

3.4   Complete Flow Chart

Fig. 3, shows the work flow of details of the algorithms.

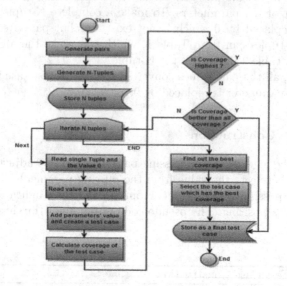

Fig. 3. Flowchart of the Algorithm

3.5   Software Development

From the Fig. 4, "Add Parameter" button adds the parameters and "Add Values" button add values of the selected parameter. From the input wizard section the user can be able to select an XML file from where the parameter and value

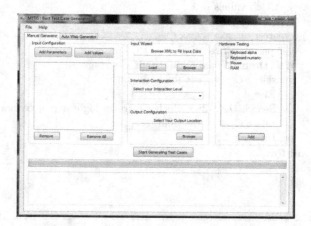

Fig. 4. The initial view of application

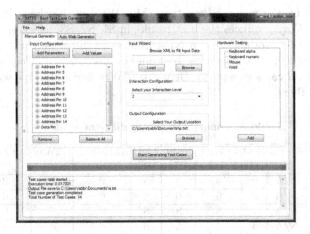

Fig. 5. The view after generating test cases

can be loaded. As this is the pair-wise testing, the interaction configuration section will select 2 always. In the output configuration user can able to select the file as a XML file or as a TEXT file, where the test cases will store.

Fig. 5 shows the output of the system when user select RAM from the hardware testing section. On the left side, a tree is showing the possible parameters and values of a RAM. Interaction configuration section showing the 2 which means it is a pair-wise interaction. Output configuration is showing the location of output data. After clicking on the "Start Generating Test Cases" generates the test cases which save to the output location. In the text area, a summary of process is shown.

4   Results and Discussion

To evaluate the efficiency of proposed algorithm (MT2Way), for pair-wise test data generation, the paper evaluates 6 different configurations. Among those the first 3 are non-uniform parameterized values and the rest are uniform as follows:

1. S1: 3 parameters with 3, 2 and 3 values respectively.
2. S2: 3 parameters with 2, 1 and 3 values respectively.
3. S3: 5 parameters with 3, 2, 1, 2 and 2 values respectively.
4. S4: 3 2-valued parameters
5. S5: 3 3-valued parameters
6. S6: 4 3-valued parameters.

The consideration of the parameters and assumptions are according to the following existing strategies that supports pair-wise testing: AETG [7, 8], IPO [10], AllPairs [14], TConfig [15], Jenny [16], TVG [17], G2Way [9] tools. This is done to compare results with those tools. Table 3 shows the comparison of test suite size generated by the proposed strategy (MT2Way) with others. It shows that proposed MT2Way produces the best results in S1, S2, S3, S4 except S5 and S6. However test case production is a NP-complete problem and it is well known that no strategy may perform the best for all cases. Also TConfig shows the best result for all cases except S3. Hence MT2Way is comparable with TConfig outperforms others.

Table 3. Comparison among different strategy

Config	AETG [8]	IPO [11]	TConfig [16]	Jenny [17]	TVG [18]	ALL Pairs [10]	G2Way	MT2Way
S1	N/A	9	9	9	9	9	9	9
S2	N/A	6	6	6	6	6	6	6
S3	N/A	7	8	8	8	9	7	6
S4	N/A	4	4	5	6	4	4	4
S5	N/A	10	9	9	10	10	10	10
S6	9	10	9	13	12	10	10	10

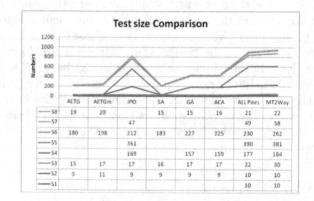

Fig. 6. Comparison Based on the Test Size

Table 4. Comparison Based on Execution Time (in seconds)

System	AETG	AETGm	IPO	SA	GA	ACA	ALL Pairs	MT2Way
S1	NA	NA	NA	NA	NA	NA	0.08	0.003
S2	NA	NA	NA	NA	NA	NA	0.23	0.004
S3	NA	NA	NA	NA	NA	NA	0.45	0.02
S4	NA	NA	0.3	NA	866	1180	5.03	0.08
S5	NA	NA	0.72	NA	NA	NA	10.36	0.225
S6	NA	6,001	NA	10,833	6,365	7,083	23.3	0.474
S7	NA	NA	0.05	NA	NA	NA	1.02	0.042
S8	NA	58	NA	214	22	31	0.35	0.019

Moreover, eight system configurations have been tested with the proposed strategy. The following systems are:

- S1: 3 3-valued parameters,
- S2: 4 3-valued parameters,
- S3: 13 3-valued parameters,
- S4: 10 10-valued parameters,
- S5: 10 15-valued parameters,
- S6: 20 10-valued parameters,
- S7: 10 5-valued parameters,
- S8: 1 5-valued parameters, 8 3-valued parameters and 2 2-valued parameters.

Fig. 6, shows the generated test size by each algorithm, and Table 4 shows the execution time for each system. The darker cell entries row-by-row indicates the best performance (i.e. in term of execution time).

Referring to the Fig. 6, MT2Way generates test cases each time which is competitive to other well-known strategies.

Concerning execution time, MT2Way gives the best result in all systems. As oppose to S6 where every strategy takes too much long time except ALL Pairs. In all the system, MT2Way generates less than 1 sec. Thus compare to execution time MT2Way is much more acceptable than other strategies.

5   Conclusion and Further Work

This paper proposed a Tuple based algorithm MT2Way for test case generation for pair-wise testing. The algorithm is able to reduce test suites effectively compared to other algorithms. The proposed algorithms could be further extended to support higher t-way interaction testing which is under investigation in University Malaysia Pahang.

Acknowledgement. This work is supported by Research Management and Innovation Centre, Universiti Malaysia Pahang.

References

1. Lei, Y., Kacker, R., Kuhn, D.R., Okun, V., Lawrence, J.: IPOG: A General Strategy for T-Way Software Testing. In: Proceedings of the 14th Annual IEEE International Conference and Workshops on Engineering of Computer-Based Systems (ECBS 2007), pp. 549–556 (2007)
2. Cui, Y., Li, L., Yao, S.: A New strategy for pairwise test case generation. In: Third International Symposium on Intelligent Information Technology Application (IITA), pp. 303–306 (2009)
3. Younis, M.I., Zamli, K.Z., Isa, N.A.M.: Algebraic Strategy to Generate Pairwise Test Set for Prime Number Parameters and Variables. In: Proceedings of the IEEE International Conference on Computer and Information Technology (ITSim), pp. 1–4 (2008)

4. Chen, X., Gu, Q., Qi, J., Chen, D.: Applying Particle Swarm optimization to Pairwise Testing. In: Proceedings of the 34th Annual IEEE Computer Software and Application Conference (COMPSAC), pp. 107–116 (2010)
5. Klaib, M.F.J., Muthuraman, S., Ahmad, N., Sidek, R.M.: A Tree Based Strategy for Test Data Generation and Cost Calculation for Uniform and Non-Uniform Parametric Values. In: Proceedings of the 10th IEEE International Conference on Computer and Information Technology (CIT), pp. 1376–1383 (2010)
6. Harman, M., Jones, B.F.: Search-based Software Engineering. Information and Software Technology 43, 833–839 (2001)
7. Leffingwell, D., Widrig, D.: Managing Software Requirements: A Use Case Approach. Addison-Wesley (2003)
8. Glass, R.L.: Facts and Fallacies of Software Engineering. Addison Wesley (2002)
9. National Institute of Standards and Technology: The Economic Impacts of Inadequate Infrastructure for Software Testing. Planning Report 02-3 (May 2002)
10. Harman, M., McMinn, P.: A Theoretical and Empirical Study of Search-Based Testing: Local, Global, and Hybrid Search. IEEE Transaction on Software Engineering 36(2), 226–247 (2010)
11. McMinn, P.: Search-Based Software Test Data Generation: A Survey. Journal of Software Testing, Verification and Reliability 14(2), 105–156 (2004)
12. Gong, D., Yao, X.: Automatic detection of infeasible paths in software testing. IET Software 4(5), 361–370 (2010)
13. Samuel, P., Mall, R., Bothra, A.K.: Automatic test case generation using unified modeling language (UML) state diagrams. IET Software 2(2), 79–93 (2008)
14. Pomeranz, I., Reddy, S.M.: On Test Generation with Test Vector Improvement. IEEE Transactions on Computer-Aided Design of Integrated Circuits and Systems 29(3) (2010)
15. Younis, M.I., Zamli, K.Z., Isa, N.A.M.: A Strategy for Grid Based T-Way Test Data Generation. In: Proceedings of the First International Conference IEEE Conference on Distributed Framework and Applications (DFmA), pp. 73–78 (2008)
16. McCaffrey, J.D.: An Empirical Study of Pairwise Test Set Generation using a Genetic. In: Proceedings of the IEEE Seventh International Conference International Conference on Information Technology: New Generations (ITNG), pp. 992–997 (2010)
17. Chen, X., Gu, Q., Qi, J., Chen, D.: Applying Particle Swarm optimization to Pairwise Testing. In: Proceedings of the 34th Annual IEEE Computer Software And Application Conference (COMPSAC), pp. 107–116 (2010)
18. Chen, X., Gu, Q., Zhang, X., Chen, D.: Building Prioritized Pairwise Interaction Test Suites with Ant Colony Optimization. In: Proceedings of the IEEE 9th International Conference on Quality Software (QSIC 2009), pp. 347–352 (2009)
19. Calvagna, A., Gargantini, A., Tramontana, E.: Building T-wise Combinatorial Interaction Test Suites by means of Grid computing. In: Proceedings of the 18th IEEE International Workshops on Enabling Technologies: Infrastructures for Collaborative Enterprises (WETICE 2009), pp. 213–218 (2009)
20. McCaffrey, J.D.: Generation of Pairwise Test Sets using a Simulated Bee Colony Algorithm. In: Proceedings of the IEEE International Conference on Information Reuse & Integration (IRI 2009), pp. 115–119 (2009)
21. Yuan, J., Jiang, C., Jiang, Z.: Improved Extremal Optimization for Constrained Pairwise Testing. In: Proceedings of the IEEE International Conference on Research Challenges in Computer Science (ICRCCS 2009), pp. 108–111 (2009)

22. Zamli, K.Z., Younis, M.I.: Interaction Testing: From Pairwise to Variable Strength Interaction. In: Proceedings of the 2010 Fourth IEEE Asia International Conference on Mathematical/Analytical Modeling and Computer Simulation (AMS), pp. 6–11 (2010)
23. Lei, Y., Kacker, R., Kuhn, D.R., Okun, V., Lawrence, J.: IPOG: A General Strategy for T-Way Software Testing. In: Proceedings of the 14th Annual IEEE International Conference and Workshops on Engineering of Computer-Based Systems (ECBS 2007), pp. 549–556 (2007)
24. Kimoto, S., Tsuchiya, T., Kikuno, T.: Pairwise Testing in the Presence of Configuration Change Cost. In: Proceedings of the Second IEEE International Conference on Secure System Integration and Reliability Improvement (SSIRI 2008), pp. 32–38 (2008)
25. Kuhn, D.R., Okun, V.: Pseudo-Exhaustive Testing for Software. In: Proceedings of the 30th Annual IEEE/NASA Software Engineering Workshop (SEW 2006), pp. 153–158 (2006)
26. Ahmed, B.S., Zamli, K.Z.: PSTG: A T-Way Strategy Adopting Particle Swarm Optimization. In: Proceedings of the Fourth IEEE Asia International Conference on Mathematical/Analytical Modeling and Computer Simulation (AMS), pp. 1–5 (2010)
27. Kim, J., Choi, K., Hoffman, D.M., Jung, G.: White Box Pairwise Test Case Generation. In: Proceedings of the Seventh IEEE International Conference on Quality Software (QSIC 2007), pp. 286–291 (2007)
28. Cohen, D.M., Dalal, S.R., Fredman, M.L., Patton, G.C.: The AETG System: An Approach to Testing Based on Combinatorial Design. IEEE Transactions on Software Engineering 23, 437–444 (1997)
29. Cohen, D.M., Dalal, S.R., Kajla, A., Patton, G.C.: The Automatic Efficient Test Generator (AETG) System. In: Proceedings of the 5th International Symposium on Software Reliability Engineering, pp. 303–309 (1994)
30. Younis, M.I., Zamli, K.Z., Mat Isa, N.A.: IRPS – An Efficient Test Data Generation Strategy for Pairwise Testing. In: Lovrek, I., Howlett, R.J., Jain, L.C. (eds.) KES 2008, Part I. LNCS (LNAI), vol. 5177, pp. 493–500. Springer, Heidelberg (2008)
31. Shiba, T., Tsuchiya, T., Kikuno, T.: Using Artificial Life Techniques to Generate Test Cases for Combinatorial Testing. In: Proceedings of the 28th Annual International Computer Software and Applications Conference (COMPSAC 2004), pp. 72–77 (2004)
32. Lei, Y., Tai, K.C.: In-Parameter-Order: A Test Generation Strategy for Pairwise Testing. In: Proceedings of the 3rd IEEE International High-Assurance Systems Engineering Symposium, pp. 254–261 (1998)

MaxD K-Means: A Clustering Algorithm for Auto-generation of Centroids and Distance of Data Points in Clusters

Wan Maseri Wan Mohd, Abul Hashem Beg,
Tutut Herawan, and Khandakar Fazley Rabbi

Faculty of Computer Systems & Software Engineering
University Malaysia Pahang
Lebuhraya Tun Razak, Gambang 26300, Kuantan, Pahang, Malaysia
{maseri,tutut}@ump.edu.my,
ahbeg_diu@yahoo.com,
fazley.rabbi@ymail.com

Abstract. K-Means is one of the unsupervised learning and partitioning clustering algorithms. It is very popular and widely used for its simplicity and fastness. The main drawback of this algorithm is that user should specify the number of cluster in advance. As an iterative clustering strategy, K-Means algorithm is very sensitive to the initial starting conditions. In this paper, we propose a clustering technique called MaxD K-Means clustering algorithm. MaxD K-Means algorithm auto generates initial k (the desired number of cluster) without asking for input from the user. MaxD K-means also used a novel strategy of setting the initial centroids. The experiment of the Max-D means has been conducted using synthetic data, which is taken from the Llyod's K-Means experiments. The results from the new algorithm show that the number of iteration improves tremendously, and the number of iterations is reduced by confirming an improvement rate is up to 78%.

Keywords: K-means Algorithm, Partitioning Algorithm, Clustering, MaxD K-means, Data Mining.

1 Introduction

The K-means is one of the classical and well-researched algorithms for unsupervised learning to solve the essential clustering problem. It tries to find the possible classes of data objects, structured categories, whose associates are identical in some way. The cluster therefore corresponds to a collection of objects that are "equivalent" to each other and are "different" and objects belonging to other groups. The K-means can be considered as the most important unsupervised learning approach. K-means method has the following prospective benefits: (i) covering different types of attributes, (ii) to discover clusters of arbitrary shape, (iii) the minimum requirements for domain knowledge to determine input parameters, (iv) it can be uses with noise and outliers, and (v)

Z. Li et al. (Eds.): ISICA 2012, CCIS 316, pp. 192–199, 2012.
© Springer-Verlag Berlin Heidelberg 2012

to minimize the difference between the data. Therefore, it relates to many fields such as marketing, biology, and image recognition [1].

Clustering is an essential technique used unsupervised classification to recognize some of the structures involved in the use of objects. The purpose of cluster analysis is to recognize objects into subsets that have aspect in the viewpoint of a particular problem. In particular, the clustering, a set of patterns, usually vectors in a multidimensional place, are organized into clusters so that patterns in the same cluster are similar in some sense and patterns in different categories are different in the sense. In some clustering concerns, the number of clusters, K, this is known in advance. In such conditions, clustering can be developed as a distribution model n in N dimensions statistic locations between groups of K so that the goals of a group are more similar to each other than trends in different clusters. This contains the minimization of an optimization specification extrinsic. K-Means algorithm is very well-known and widely used clustering technique appropriate in such conditions [2]. Clustering is often the very first steps in data analysis. It can be used to recognize natural categories in data sets and to recognize very subjective elements that might reside there, without having any basic knowledge on characteristics of data. Therefore, many classification methods have been developed such as hierarchical clustering [3-8], the mixture densities [9,10], graph partitioning [11], and spectral classification [12] and these methods have been used in a wide range of areas such as computer vision, data mining, bio-informatics and information retrieval, to name a few [13]. The Pseudo-code of the Lloyds K-Means algorithm [14] is shown in the following Figure.

```
Input:
    D= {t1, t2, .... Tn   } // Set of elements
    K               // Number of desired clusters
Output:
    K               // Set of clusters
K-Means algorithm:
    Assign initial values for m1, m2,.... mk
    repeat
        assign each item ti     to the clusters which has the closest mean;
        calculate new mean for each cluster;
    until convergence criteria is met;
```

Fig. 1. Pseudo-code of the Lloyd's K-Means algorithm

K-Means is a simple algorithm that has adapted to areas with many problems. Similar to other algorithm, K-Means clustering has some limitations [15,16,17]. To solve the existing K-meanss problem is the main vision of this research.

Hence, this paper proposes a new approach to overcome existing problems mentioned above, called MaxD K-means. The new clustering algorithm provides a technique to define the initial parameter of K-means through the auto generation of the number of clusters using the maximum distance of data points and a novel approach of defining initial centroid for effective and efficient clustering process. The algorithm helps the user in estimating the number of clusters which is highly dependent on the domain knowledge, which is not so desirable.

The rest of this paper is organized as follow. Section 2 describes the related work. Section 3 describes proposed Max D-Kmeans algorithm. Section 4 describes results and comparison test. Finally, the conclusion of this work is described in Section 5.

2   Related Works

There are different method has been proposed to improve the efficiency of k-means algorithm [18,19,20]. Among them it can find the k-means algorithm is one of the more common ones. But we know that K-means algorithm is sensitive to the initial cluster centers and easy to get immovable in local optimal solutions [21]. Moreover, when the number of data points is large, it takes a tremendous amount of time to find a global optimal solution [22,23].

Bandyopadhyay and Maulik [2] described a Genetic Algorithm-based clustering algorithm. In their strategy the chromosome encodes the centers of the clusters instead of a possible partition of the data points. The algorithm attempts to develop appropriate cluster centers, while optimizing a given clustering metric. In addition, the usefulness of KGA-clustering algorithm for classification of pixels of a satellite image to distinguish between the different areas of land has been designated. Note that even if the GAs is usually done with binary strings, they have implemented the encoding of floating point chromosome. Otsubo et al. [24] presented a computerized the identification of the clusters by using the k-means clustering technique. In their research they present a computerized technique to recognize clusters separately to determine the tensor representing a reduction of stress and the spread of tensors. To this end, uses a technique called k-means for the purpose of the division to reduce the stress tensor obtained by inversion methods into multiple clusters. Currently, the number of clusters, k, must be specified by the user. The k-means requires a well-defined distance between the objects to classify. The stress difference defined by Orif and Lisle [19] is a useful distance between the tensors of stress reduction. The parameter space is adequate, since the Euclidean distance between points in the parameter space is equal to the stress difference between the stresses that are represented by points. They tested the technique by artificial data sets. It has been shown that the resolution of visual identification of the clusters was often insufficient, and that the present technique correctly detected highlights from artificial data were generated with known stress. Kalyani and Swarup [25] presented a modified K-means algorithm (PSOKM) using particle swarm optimization technique for the evaluation of static security, transient. Training set of vectors generated from offline simulations are presented as input to the PSO algorithm based K-means classification using supervised active learning to adjust its weight vectors (cluster centers). The proposed algorithm was implemented in IEEE 30 bus, 57 bus, 118 bus and 300 bus standard of test cases, and its performance was compared with other K-means algorithm. Their results showed that the high-accuracy classifiers with lower rate of misclassification can be exchanged with the classification PSOKM.

Bagirov, et al. [26] have developed a new version of the modified global K-means algorithm. This algorithm computes step by step through the clusters k-1 cluster centers from the previous iteration to solve the problem of k-partitions. An important step in the calculation of this algorithm is a starting point for the center of the cluster k-th. This starting point was calculated by minimizing the additional function known as clusters. The results of their numerical experiments show that in most cases, the proposed algorithm is faster and more accurate than the global k-means algorithm. At the same time, similar results the proposed algorithm requires much less evaluations and CPU time than changing the global k-means algorithm. Therefore, the proposed algorithm is a significant improvement in changing the global K-means algorithm. Moreover, this improvement is even more important that all size of the data set increases.

3 Max D-Kmeans Algorithm

To calculate the centroid and build cluster has been developed the MaxD-Kmeans algorithms are shown in Figures 2 and 3. Figure 2 describes the algorithm for calculating centroids. At the first step, data set has been inserted

Input: X is the set { x_1, x_2, x_3......................x_n}, where n represents the number of input values
Output: Y is the set {} of clusters
1. Let x_s= {} represent   the sorted values
2. Let k= number of total centroids
3. Let C= {} represents set of the number of total centroids
4. Let X_{max} represents the maximum value in X
5. Let X_{min} represents the minimum value in X
6. Let C_t represents total centroids
7. X_s = sort (x);
8. X_{min}= read first value of (x_s);
9. X_{max}= read last value of (X_s);
10. C_t []= $\underline{Xmin}$;
11. For each X_f in X_s
12. If ($X_f > X_{min}$)
13. {
14. Ct = (X_f-X_{min})/2 + 2
15. }
16. Else
17. {
18. (C_f=X_{min})/ 2+ X_{min}
19. }
20. K= count (t);
21. }

Fig. 2. Max-D k-means algorithm (Calculate_centroid)

Input:	Y = { } represents the cluster
	X = { } represents the input values
Output:	Cm= {} {} represents the cluster members
1.	For each (Y_f in Y)
2.	{
3.	P = [Y_f];
4.	If([Y_f]> =[x_f] and [Y_f]<= [X_f+1])
5.	P= [Y_f];
6.	Cm= P[];
7.	}

Fig. 3. Max-D k-means algorithm (Build_Cluster)

and arranged in the ascending order. After that mean has been calculated for each centroid which has been stored in the centroid table. On the other hand, Figure 3 describes the algorithm for build cluster based on the centroid table.

4   Result and Discussion

This section describes experimental results of MaxD K-means and its performance comparison with Llyod's K-Means.

4.1   Performance on Synthetic Dataset[15]

In order to study the effectiveness of MaxD-Kmeans Algorithm experiment has been conducted using synthetic data which is taken from the Llyods K-Means experiments [15]. To show the significant improvement of the new algorithm, the experiment was divided into several cycles. Table 1 shows the result of Max-D k-means algorithm where input was k defined the number of desired cluster and the initial values for centroids $c_1, c_2, c_3, ..., c_k$ were used to defined centroids. In addition, Table 2, Shows the comparative result between Max-D K-means and Llyods K-Means algorithm. The comparative results show that, with the Max-D k-means algorithm, the number of iteration improves tremendously when N is larger. It shows that the number of iterations is reduced from 18 to 4, which is an improvement of 78%.

Table 1. Result of Max-D K-Means Algorithm

Number of Iteration	Number of Clusters	Cluster's Members
1	1	{3,5,11,13,4,12}
	2	{21,26,31}
2	1	{3,5,11,13,4,12}
	2	{21,26,31}

Table 2. A Comparison between Max-D K-Means and Llyod's K-Means using N=80

Comparative Algorithm	Total Clusters	Cluster Member
	1	2
	2	9
	3	9
Max-D K-Means	4	8
N= 80	5	6
K = 17	6	8
Number of iteration= 4	7	16
Total Cluster=10	8	18
	11	3
	17	2
	1	2
	2	12
	3	0
Llyod's K-Means	4	12
N=80	5	11
K=10	6	8
Number of Iteration=18	7	10
Total Cluster =10	8	13
	9	3
	10	9

5   Conclusion

In this paper, we have proposed a parameter less data clustering technique based on maximum distance of data and Llyod K-means algorithm, which requires a number of clusters, k, must be determined beforehand, which is not desirable, since the number of cluster configuration needs domain knowledge. In order to study the effectiveness of the proposed approach for setting the parameters of K-Means algorithm, the experiment has been done using synthetic data, which has been taken from the Llyods K-Means experiments. The experimental results show that the use of new approach to defining the centroids, the number of iterations has been reduced where the improvement was 78%.

Acknowledgement. This work was supported by Fundamental Research Grant Scheme (FRGS- RDU110104), University Malaysia Pahang under the project "A new Design of Multiple Dimensions Parameter less Data Clustering Technique (Max D-K means) based on Maximum Distance of Data point and Lloyd k-means Algorithm".

References

1. Zhou, H., Liu, Y.: Accurate integration of multi-viewrange images using k-means clustering. Pattern Recognition 41, 152–175 (2008)
2. Bandyopadhyay, S., Maulik, U.: An evolutionary technique based on K-Means algorithm for optimal clustering. Information Sciences 146, 221–237 (2002)
3. Herawan, T., Yanto, I.T.R., Deris, M.M.: Rough Set Approach for Categorical Data Clustering. In: Ślęzak, D., Kim, T.-h., Zhang, Y., Ma, J., Chung, K.-i. (eds.) DTA 2009. CCIS, vol. 64, pp. 179–186. Springer, Heidelberg (2009)
4. Yanto, I.T.R., Herawan, T., Deris, M.M.: Data clustering using Variable Precision Rough Set. Intelligent Data Analysis 15(4), 465–482 (2011)
5. Yanto, I.T.R., Vitasari, P., Herawan, T., Deris, M.M.: Applying Variable Precision Rough Set Model for Clustering Student Suffering Study's Anxiety. Expert System with Applications 39(1), 452–459 (2012)
6. Herawan, T., Yanto, I.T.R., Deris, M.M.: ROSMAN: ROugh Set approach for clustering supplier chain MANagement. International Journal of Biomedical and Human Sciences 16(2), 105–114 (2010)
7. Herawan, T., Deris, M.M., Abawajy, J.H.: A rough set approach for selecting clustering attribute. Knowledge Based Systems 23(3), 220–231 (2010)
8. Duda, R., Hart, P., Stork, D.: Pattern Classification, 2nd edn. John Wiley and Sons, New York (2001)
9. Dempster, A.P., Laird, N.M., Rubin, D.B.: Maximum likelihood from incomplete data via the EM algorithm. J. Roy. Statist. Spc. 39, 1–38 (1977)
10. McLachlan, G.L., Basford, K.E.: Mixture Models: Inference and Application to clustering. Marcel Dekker (1987)
11. Jiambo, S., Jitendra, M.: Normalized cuts and image segmentation. IEEE Trans. Pattern Anal. Machine Intell. 22, 288–905 (2000)
12. Stella, Y., Jianbo, S.: Multiclass spectral clustering. In: Proc. Internat. Conf. on Computer Vision, pp. 313–319 (2003)
13. Murino, L., Angelini, C., Feis, I.D., Raiconi, G., Tagliaferri, R.: Beyond classical consensus clustering: The least squares approach to multiple solutions. Pattern Recognition Letters 32, 1604–1612 (2011)
14. Dunham, M.: Data Mining: Introductory and Advance Topics. N.J. Prentice Hall (2003)
15. Chiang, M., Tsai, C., Yang, C.: A time-efficient pattern reduction algorithm for k-means clustering. Information Sciences 181, 716–731 (2011)
16. Xu, R., Wunsch, D.: Survey of clustering algorithms. IEEE Transaction on Neural Netowrks 16(3), 645–678 (2005)
17. Jain, A.K., Murty, M.N., Flynn, P.J.: Data clustering: a review. ACM Computing Surveys 31(3) (1999)
18. Kanungo, T., Mount, D., Netanyahu, N.S., Piatko, C., Silverman, R., Wu, A.: An efficient K-means clustering algorithm: analysis and implementation. IEEE Trans. Pattern Anal. Mach. Intell. 24(7), 881–892 (2002)
19. Likas, A., Vlassis, N., Verbeek, J.J.: The global K-means clustering algorithm. Pattern Recognition 36, 452–461 (2003)
20. Charalampidis, D.: A modified K-means algorithm for circular invariant clustering. IEEE Trans. Pattern Anal. Mach. Intell. 27(12), 1856–1865 (2005)
21. Selim, S.Z., Ismail, M.A.: K-means type algorithms: a generalized convergence theorem and characterization of local optimality. IEEE Trans. Pattern Anal. Mach. Intell. 6, 81–87 (1984)

22. Spath, H.: Cluster Analysis Algorithms. Ellis Horwood, Chichester (1989)
23. Chang, D., Xian, D., Chang, W.: A genetic algorithm with gene rearrangement for K-means clustering. Pattern Recognition 42, 1210–1222 (2009)
24. Otsubo, M., Sato, K., Yamaji, A.: Computerized identification of stress tensors determined from heterogeneous fault-slip data by combining the multiple inverse method and k-means clustering. Journal of Structural Geology 28, 991–997 (2006)
25. Kalyani, S., Swarup, K.S.: Particle swarm optimization based K-means clustering approach for security assessment in power systems. Expert Systems with Applications 38, 10839–10846 (2011)
26. Bagirov, A.M., Ugon, J., Webb, D.: Fast modified global k-means algorithm for incremental cluster construction. Pattern Recognition 44, 866–876 (2011)

Real-Time and Automatic Vehicle Type Recognition System Design and Its Application

Wei Zhan[1,2] and Qiong Wan[3]

[1] College of Computer Science
Yangtze University
Jingzhou, Hubei, China
[2] School of Computer Science
China University of Geosciences
Wuhan 430074, P.R. China
[3] Jingzhou Yangtze River Bridge
Jingzhou, Hubei, China

Abstract. Via a fixed camera, real-time video including moving vehicles of a highway toll station is collected, with technology of digital image processing and recognition, all frames include vehicles can be detected automatically from the video, and vehicle type will be recognized automatically. The system includes four modules: reading video and decomposing it into frames; moving vehicle detection; vehicle image processing and vehicle type recognition from image. Tests show that the system design is simple and effective. Vehicle image processing algorithm in the system is simpler than that of references and vehicle type recognition algorithm through counting black pixels number including in vehicle body contour is a new idea.

Keywords: Image Processing, Image Recognition, Vehicle Type Recognition.

1 Introduction

Motion detection[1,2] from complex video has become a hot research area in recent years. Motion detection is a fundamental and important part of many visual detecting and tracking systems. Motion detection is commonly used in the field of video surveillance[3], optical motion capture[4], multimedia application[5] and video object segmentation[6]. At present, common motion detection method are mainly temporal difference between two consecutive frames, image subtraction with background, Support Vector Machine (SVM)[7], Self-Organizing Map[8] and optical flow estimation[9].

Moving object recognition is an important branch in research on computer vision, the goal of vehicle recognition is to separate moving object from background and recognition. With rapid development of digital image processing and recognition technology, vehicle type recognition system[10,11] has become an important part of Intelligent Transportation System[12,13]. In recent references, there are a great many research on vehicle type recognition based on video,

Z. Li et al. (Eds.): ISICA 2012, CCIS 316, pp. 200–208, 2012.
© Springer-Verlag Berlin Heidelberg 2012

the primary method[14,15] of vehicle type recognition are: radio wave or infrared contour scanning, radar detection, vehicle weight, annular coil[16,17] and laser sensor measurement. Because vehicle type recognition result can be used in road traffic monitoring, vehicle type classification, automatic license plate recognition, and so vehicle type recognition based on video is a hot research direction. In this paper, a real time vehicle type recognition system is designed for moving vehicle detection and recognition automatically from video.

2   System Architecture

The goal of system design includes vehicle type recognition automatically, traffic flow and charge statistics. As Fig. 1 shows: the system consists of four main modules: reading video capture card and decomposing video into digital image frames; moving vehicles detection; vehicle digital image processing module and vehicle type classification module. Video collected via camera is be digitized by video capture card, at the same time, real time video displays on screen and the system decompose this video into digital image frames which will be stored into memory for further processing.

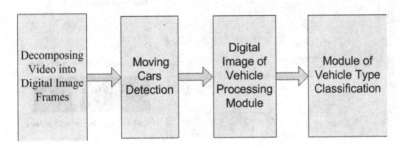

Fig. 1. System Architecture

3   Detailed System Design

3.1   Reading Video Capture Card and Decomposing Video into Frames

Program interface of video capture card is be packaged in a named DSStream.h head file, through this Software Development Kit(SDK) of video capture card, attribute of video from camera can be controlled, such as: video input port, video standard, video file ratio of image frame, video file format, and so on. It is particularly worth mentioning is that current image frame's Device-Independent Bitmaps(DIB)[18] data of real time video can be get through SDK interface function DSStream_GetCurrentDiB include in head file DSStream.h, so a video file can be decomposed into one by one digital image frames which is so called decomposing video into frames[19].

3.2   Moving Vehicle Detection Algorithm

For finding cars in video, the first step is moving vehicle detection, first a certain selected area in image should be compared with corresponding area of the current AVI frame, through comparing pixels changing result in the detection area, moving vehicles can be found, the area is so called monitoring area which setting will directly influence the following image processing algorithm. In this system, a rectangular detection area is used. Fig. 2 shows a sample how to select rectangular detection area, on the left, an black rectangular is the detection area, on the right there are three little rectangular image: the top is image cut from detection area, the middle is the background, the bottom is pixels' changing image in detection area. After computing, if pixels changing ratio in detection area exceeds the threshold value, system thinks there is a car in detection area, at the same time, current cars image will be stored in memory and be as input parameter of digital image processing module.

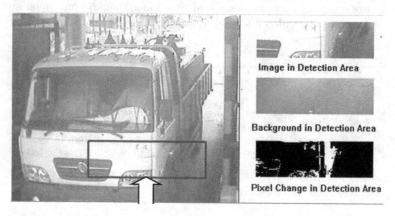

Fig. 2. Rectangular Detection Area

Before image processing, a background must be selected. There two kind of background choosing method: statically and dynamic. In this system, background is determined manually and statically, when system is not running or there are no cars passing in the rectangular detection, a static image is selected as background template.

How to judge weather a car in or out of the detection area: a threshold value is be set when system initializes, such as 5% or 10%, if change ratio in detection area of the image be is more than this threshold value, system thinks there is a moving car in detection area, otherwise, there is no car.

3.3   Module of Vehicle Image Processing

The following is algorithm flow of vehicle image processing, Fig. 3 illustrates the vehicle image processing in detail.

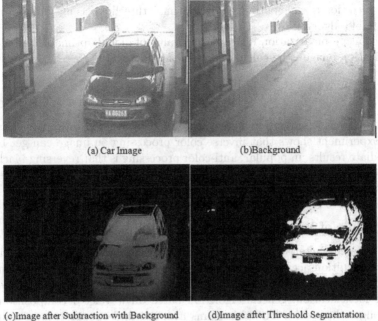

(a) Car Image	(b)Background

(c)Image after Subtraction with Background	(d)Image after Threshold Segmentation

(e)Image after Inverse Color)

Fig. 3. Vehicle Image Processing

Processing Flow of Vehicle Digital Image:

1. Input specified frame include vehicle image;
2. Background Subtraction;
3. Threshold Segmentation;
4. Inverse Color Processing on Image;
5. Counting the number of black pixels include in vehicle body contour;
6. Output Vehicle Type Parameters;

Algorithm of Threshold segmentation[20] is consisting of two steps: first is calculating global optimal threshold, second is calculating binarization image through global optimum threshold value.

1. Calculating global optimal threshold: RGB Data of every pixel is be read out in order to calculating global optimal threshold using global optimal threshold algorithm.
2. Calculating binarization image: Using value of global optimal threshold, binarization image is be calculated out, as follows:

$$g(x,y) \begin{cases} 0, & |d(x,y)| < Threshold \\ 255, & |d(x,y)| > Threshold \end{cases} \qquad (1)$$

Where $g(x,y)$ is binarization image, threshold is a global optimal numerical value. Experiment shows that inverse color processing on image can get better intermediate results, in popular, anti-color processing is that reversing black and white pixels each other on binarization image. After above processing, vehicle body contour image will be get.

3.4   Module of Vehicle Type Recognition

Vehicle type characteristic data and vehicle type recognition algorithm will influence recognition accuracy, common vehicle type classification algorithm are: vehicle recognition based on neural network, models based on support vector machine identification, genetic algorithm based on wavelet decomposition and genetic algorithm (GA). The algorithms mentioned above are complex, in this paper, a simple and effective algorithm of extracting vehicle type parameters is designed, that is via counting number of black pixels include in vehicle body contour.

Repeated experiments show that for the car image after inverse color, the bigger car the more black pixels include in vehicle body contour, so the black pixels' total number include in vehicle body contour is be analyzed as vehicle type characteristic data. Fig. 4 shows vehicle classification algorithm including two parts: Algorithm of Extracting Vehicle Type Characteristic Data and Vehicle Type Classification Algorithm.

4   System Test

4.1   Test on Non-motor Vehicle

Fig. 5 shows when a person goes into monitor area, system has not recognized the person as a car, system flag of vehicle in and out has not changed, which shows that the setting of the rectangular image detection region and the threshold value of segmentation are reasonable and effective.

4.2   System Test on Cars Fleet

As shows in Fig. 6, there is a cars fleet, system has not response timely, cars in fleet has not be separated timely by system which leads to recognition error. More effective algorithm need be designed to solve this difficult recognition problem, at the same time, it is a key question in the future research.

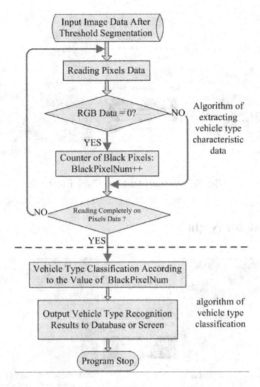

Fig. 4. Algorithm of Vehicle Type Recognition

Fig. 5. Test on Non-motor Vehicle

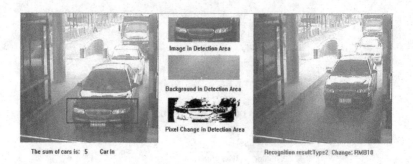

Fig. 6. System Test on Cars Fleet

4.3 System Test on Night

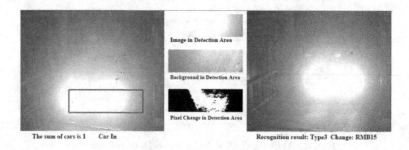

Fig. 7. System Test On night. On night, the head lamp light of cars is too light makes the system failure.

5 Conclusion

Motion detection is foundation of motion tracking. The normal method of motion object detection is by image subtraction which is easy to be affected by light change of surroundings. In this paper, motion object detection and tracking system via image pyramids optical flow is designed. The system framework includes five main modules: Decomposing AVI files into frames; Optical flow computation on every frame; Computing threshold of optical flow image and segmentation; Morphological transformation on image and Extracting moving object by rectangle division.

Vehicle type recognition technology can be widely used in statistics of traffic and toll automatic. The system design is simple and effective, algorithm of vehicle image processing is simpler than that of references, there are only three steps: first is background image subtraction, second is threshold segmentation, the last is inverse color processing on image, large number of experiments show that the algorithm of vehicle image processing in this system is simply to achieve. Especially to say, compared with complex vehicle type recognition algorithm mentioned in references, algorithm of counting the number of black pixels included in vehicle body contour is new method for vehicle recognition.

Through system test on AVI files including a large number of real vehicle image, we find three key points may be improved: first, background changing leads to fault recognition results, so in the future research, a background model and eliminating algorithm will be add into system algorithm as a auxiliary way in order to detach moving objects from background more precisely; second, shadow of targets make incorrect detection results too, the reason is there is no shadow detection and elimination module in the system; third, optical flow computation on every frame of the AVI file is very time-consuming, so we hope that an improved optical flow algorithm will be designed for higher processing speed and accuracy in the future work.

References

1. Zhan, W.: Moving Object Detection from Video with Optical Flow Computation. Informationan 15, 4157–4164 (2012)
2. Zhan, W., Luo, Z.: Research of Vehicle Type Recognition System Based on Audio Video Interleaved Flow for Toll Station. Journal of Computer 7, 741–744 (2012)
3. Cheung, S., Kamath, C.: Robust Background Subtraction With Foreground Validation for Urban Traffic Video. Journal of Applied Signal Processing, 2330–2340 (2005)
4. Carranza, J., Magnor, M.: Free-Viewpoint Video of Human Actors. ACM Transactions on Graphics 22, 569–577 (2003)
5. El Baf, F., Bouwmans, T.: Comparison of Background Subtraction Methods for a Multimedia Learning Space. In: International Conference on Signal Processing and Multimedia (2007)
6. Colombari, Fusiello, A., Murino, V.: Video Objects Segmentation by Robust Background Modeling. In: ICIAP, pp. 155–164 (2007)
7. Choi, B., Han, S., Lim, J., Chung, B., Ryou, J.: Design and Performance Evaluation of Temporal Motion and Color Energy Features for Video Rating System. IJACT: International Journal of Advancements in Computing Technology 3, 8–15 (2011)
8. Wu, Y., Zhou, G., Wu, J.: A Monitoring System for Supermarket Based on Trajectory of Palm. IJACT: International Journal of Advancements in Computing Technology 2, 7–15 (2010)
9. Shi, X.: Research on Moving Object Detection Based on Optical Flow Mechanism, pp. 6–14. University of Science and Technology of China (2010)
10. Zhan, W., Luo, Z.: System Design of Real Time Vehicle Type Recognition Based on Video for Windows (AVI) Files. In: Chen, R. (ed.) ICICIS 2011 Part II. CCIS, vol. 135, pp. 681–686. Springer, Heidelberg (2011)
11. Zhan, W.: Research of Vehicle Recognition System for the Road Toll Station Based on AVI Video Flow. China University of Geosciences, Wuhan (2006)
12. Tian, B.: Research on Automatic Vehicle Recognition Technology in Intelligent Transportation system, pp. 23–26. XiDian University (2008)
13. Xia, W.: Video-Based Vehicle Classification Method Research, pp. 1–2,7,16, 24–29. HuaZhong University of Science & Technology (2007)
14. Ji, C.: Vehicle Type Recognition Based on Video Sequences. Journal of LiaoNing University of Technology (Natural Science Edition) 30, 5–7 (2006)

15. Cao, Z.: Vehicle Detection and Classification Based on Video Sequence, pp. 20–46. ZheJiang University (2004)
16. Cao, Z., Tang, H.: Vehicle Type Recognition in Video. Computer Engineering and Applications, 226–228 (2004)
17. Xiong, S.: Research of Automobile classifying method based on Inductive Loop, pp. 2–5. ChangSha University of Science and Technology (2009)
18. Xiao, Y.: Fast Decomposition of Avi Data Stream by Using Functions in VFW Library. Journal of Information Engineering University, 2–3 (2002)
19. Lang, R.: Digital Image Processing with Visual C++. BeiJing Hope Electronic Press, Beijing (2003)
20. Gonzalez, R.C.: Digital Image Processing, 2nd edn. Publishing House of Electronics Industry, Beijing (2005)

Comparison of Three Multi-objective Optimization Algorithms for Hydrological Model

Xiaomin Huang[1,2], Xiaohui Lei[2], and Yunzhong Jiang[2]

[1] School of Environmental Science and Engineering
Donghua University, Shanghai 201620, P.R. China
huangxiaomin123123@126.com
[2] China Institute of Water Resources and Hydropower Research
Beijing 100038, P.R. China

Abstract. In our research of this article, the efficiency of Multi-objective Particle Swarm Optimization (MOPSO), Non-dominated Sorting Genetic Algorithm (NSGA-II), and Multi-objective Shuffled Complex Evolution Metropolis (MOSCEM-UA) algorithms were compared by implementing the Hydrological Model (HYMOD) and the related observed daily precipitation, evaporation and runoff data. High flow Nash-Sutcliffe efficiency and Low flow Nash-Sutcliffe efficiency were used to optimize the model parameters as two criterions; the time consumption, the dominating rate and, the quality of Pareto set (distance, distribution, and extent) were used to analyze the performance of the three algorithms. Compared with NSGA-II and MOSCEM-UA, the MOPSO algorithm performed most efficiently to complete each trail. The non-dominant solutions derived from MOPSO algorithm were seldom dominated by those from the other two algorithms, while a high proportion of solutions drawn from NSGA-II are dominated by the other two algorithms. When we come to the three optimization goal of multi-objective optimization, there is a complication. The shortest distance of the resulting non-dominated set to the Pareto-optimal front was from the NSGA-II algorithmthe most uniform distribution of the solutions was derived from the MOSCEM-UA algorithm; and the maximal extent of the obtained non-dominated front was stemmed from the MOPSO algorithm. The results demonstrated that all three algorithms were able to find a good approximation of the Pareto set of solutions, but differed in the rate of convergence to the optimal solutions.

Keywords: Comparison of Multi-objective Optimization, Hydrological Model, MOPSO, NSGA-I, MOSCEM-UA.

1 Introduction

Rainfall-runoff process in heterogeneous real world can be commonly simplified and represented by various hydrological models[1], These models are conversion and simplification of reality, thus no matter how spatially sophisticated and accurate they may be, those models only represent aspects of conceptualization

Z. Li et al. (Eds.): ISICA 2012, CCIS 316, pp. 209–216, 2012.

or empiricism of modelers (or hydrologists). Accordingly, their outputs are as reliable as hypothesis, structure of models, and quantity and quality of input data, and parameter estimates[3][4]. While hydrologic models often contain parameters that cannot be measured directly but which can only be inferred by a trial-and-error (calibration) or auto optimization process which can help adjust the parameter values to closely match the input-output behavior of the model to the real system it represents. Traditional calibration procedures, which involve manual adjustment of the parameter values, are labor-intensive, and their successes are strongly dependent on the experience of the modelers. Automatic model calibration methods, which seek to take advantage of the speed and power of computers that are objective and relatively easy to implement, have therefore become more popular[5]. A lot of popular options have been introduced to optimize hydrological model and successfully applied to complex multi-objective optimization problems, such as NSGA (Non-dominating Sorting Genetic Algorithm)[6], NSGA-II[7], MOPSO[8] and MOSCEM-UA[9]. In recent years, some researchers have made efforts on theory and application on those algorithms. But there are still few comprehensive studies that compare the performance and different aspect of those algorithms. For better understanding the specific advantages and drawbacks of each algorithm, this article compared the efficiency, the dominating rate and the quality of Pareto-optimal set (distance, distribution, and extent) of the MOPSO, NSGA-II, MOSCEM-UA in this paper.

2   Multi-objective Optimization Algorithms

MOPSO, NSGA-II and MOSCEM-UA all seek the Pareto optimal set instead of a single solution. Although these algorithms employ different methodologies, ultimately they all try to balance rapid convergence to the Pareto front with maintaining a diverse set of solutions along the full extent of an applications tradeoffs. As initial sampling distribution, a common approach of these three algorithms is to assign upper and lower bounds for each parameters, and to use uniform sampling in this space to create the initial population of points to be iteratively improved with the optimization algorithm.

Regarding the whole evolution process, MOSCEM-UA is significantly different from MOPSO and NSGA-II, although their search populations are randomly initialize. MOSCEM-UA uses the complex shuffling method and the Metropolis-Hastings algorithm to conduct search. Offspring are generated using a multivariate normal distribution developed utilizing information from the current draw of the parallel sequence within a complex. The acceptance of a new generated candidate solution is decided according to the scaled ratio of candidate solutions fitness to current draws fitness of the sequence. Complex shuffling helps communication between different complexes and promotes solution diversity[10].

Binary, SBX crossover, and mutation operators are implemented to generate the child population in NSGA-II algorithm. Pareto rank and crowding distance are the two criteria for parents selection. And then these parents generate children (offspring) using SBX operator for crossover and polynomial mutation. This is the elitism strategy of NSGA-II algorithm.

MOPSO algorithm is population-based, and uses a geographically-based approach to maintain diversity. As it introducing external archiving, crowding distance and mutation operator, MOPSO uses a measure of performance similar to the fitness value used with evolutionary algorithms. And the adjustments of individuals are analogous to the use of a crossover operator. The algorithm updates the velocity and the position of each particle according to global bests of particle gained from last iteration. And mutation is carried out among individuals. Then non-dominated sorting is conducted on all these new individuals with the individuals in the archive. Comparison to NSGA-II, MOPSO multiplies the chances to keep individuals' changes and make it easier to maintain diversity.

3 Case Study

3.1 Model and Materials

HYMOD model is a five-parameter conceptual model raised by British scholar R.J. Moore[11], which is based on saturation excess runoff yield mechanism. The model consists of a relatively simple rainfall excess model, connected with two series of linear reservoirs (three identical quick and a single for the slow response)[12](Fig. 1). Fvie parameters in HYMOD model require sensitivity analysis: the maximum storage capacity in the catchment (C_{max}), the degree of spatial variability of the soil moisture capacity within the catchment (B_{exp}), the factor distributing the flow between the two series of reservoirs (Alpha), and the residence time of the linear quick and slow reservoirs (R_q, R_s). The parameter ranges are defined in Table 1. The objective functions adopted in this study are high flow Nash-Sutcliffe efficiency[13] (1) and low flow Nash-Sutcliffe efficiency (2).

$$F_1(x) = \frac{\sum_{i=1}^{n}(Q_{o,i} - Q_{s,i})^2}{\sum_{i=1}^{n}(Q_{o,i} - \overline{Q_o})^2} \tag{1}$$

$$F_2(x) = \frac{\sum_{i=1}^{n}(log(Q_{o,i}) - log(Q_{s,i}))^2}{\sum_{i=1}^{n}(log(Q_{o,i} - log(\overline{Q_o})))^2} \tag{2}$$

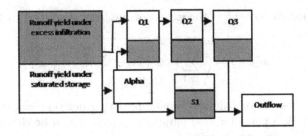

Fig. 1. Schematic representation of the HYMOD model

Table 1. Parameter and parameter ranges used in optimization

Parameter Name	C_{max}	B_{exp}	Alpha	R_s	R_q
Ranges	[200,500]	[0.1, 2.0]	[0.1, 0.99]	[0.001, 0.1]	[0.3, 0.7]

Where, Q_s is simulated runoff; Q_o is the observed runoff; $\overline{Q_o}$ is the mean value of observed runoff; x is the parameter set corresponding to the objective.

Coverage, one of metrics of performance proposed by Zitzler and Thiele[14], was use in this study. This metric is used to show that the outcomes of an algorithm dominate the outcomes of the other. The equation can be expressed as follows:

$$C_{x',x''} = \frac{|\{a'' \in x''; \exists a' \in x' : a' \leqslant a''\}|}{|x''|} \times 100\% \quad (3)$$

Where $x', x'' \subseteq x$ are two sets of decision vectors. The function C maps the ordered pair (x', x'') to the interval $[0, 1]$. When the value $C_{x',x''} = 1$ means that all points in x' are dominated by or equal to points in x''.

Because the optimization goal of multi-objective optimization consists of multiple objectives: i) The distance of the resulting non-dominated set to the Pareto-optimal front should be minimized; ii) An good (in most cases uniform) distribution of the solutions found is desirable; iii) The extent of the obtained non-dominated front should be maximized, i.e., for each objective a wide range of values should be covered by the non-dominated solutions. We adopt the following metrics which allow assessment of each of the criteria[15].

1. The The function M1 gives the average distance to the Pareto-optimal set $\overline{x} \subseteq x$:

$$M_1(x') = \frac{1}{|x'|} \sum_{a' \in x'} min\{\|a' - \overline{a}\| ; \overline{a} \in \overline{x}\} \quad (4)$$

2. The function M_2 takes the distribution in combination with the number of non-dominated solutions found into account:

$$M_2(x') = \frac{1}{|x' - 1|} \sum_{a' \in x'} |\{b' \in x'; \|a' - b'\| > \delta\}| \quad (5)$$

3. The function M_3 considers the extent of the front described by x':

$$M_3(x') = \sqrt{\sum_{i=1}^{m} max\{\|a'_i - b'_i\| ; a', b' \in x'\}} \quad (6)$$

Where $x' \in x$ is a set of pair wise non-dominating decision vectors, δ(usually greater than 0), is a neighborhood parameter which should be chosen appropriately, and $\|.\|$ is a distance metric.

While M_1 is intuitive, M_2 and M_3 need further explanation. The distribution metrics give a value within the interval $[0,1]$; it reflects the average fraction of members of x' which lie outside the neighborhood of an arbitrary solution in x'. Obviously, the higher the value the better the distribution for an appropriate neighborhood parameter. The functions M_3 use the maximum extent in each dimension to estimate the range to which the fronts spreads out.

3.2 Study Area

In this study, a test case on the Xiangjiaping hydrological station located on Hanjiang River was conducted. The catchment area is about 6448 square kilometers. A 65-day warm-up period was used in this study based on the methodological recommendations of Vrgt[8] et al. A calibration period was used in this study from January 1, 1981 to March 11, 1983. The validation period was used from March 12, 1983 to March 9, 1984.

3.3 Results and Discussions

For application in this study, we use initial particle population $s = 200$ and a total of 40,000 model evaluations for all those three algorithms. For computational efficiency, we decide to run MOSCEM-UA with the number of complexes $q = 5$. In MOPSO algorithm, based on the methodological recommendations of Raquel[15] et al. The inertia weight(w), cognitive(c_1), and social parameter(c_2) are 0.4, 1, and 1, respectively; and the mutation probability is 0.5.

Table 2 shows the optimization results. Considering the randomness of optimization methods, the computational cost was calculated as the mean value of trails of each algorithm that is implemented for five times. And the mean values of running time, Pareto sets coverage rate and the Pareto front spacing rate were also analyzed. It can be seen that MOPSO algorithm consumed much less time, only 14.8055s, and gets relatively smaller value for both high flow and low flow Nash-Sutcliffe efficiency. Compared with NSGA-II and MOSCEM-UA, the MOPSO algorithm was the most efficient one to complete each trail. The non-dominated solutions derived from MOPSO algorithm were seldom dominated by those from the other two algorithms. While a high proportion of solutions drawn from NSGA-II were dominated by the ones from MOPSO and MOSECM-UA algorithms.

When we come to the three optimization goal of multi-objective optimization, there is a complication. The shortest distance of the resulting non-dominated set to the Pareto-optimal front was from the NSGA-II algorithm with an average value of function M_1 equal to 0.0419. The most uniform distribution of the solutions was derived from the average value of function M_2 equal to 0.0310. And the maximal extent of the non-dominated front was stemmed from the MOPSO algorithm with the average value of function M_3 equal to 0.5061. They all have their own merits, which are determined by their search strategies. The results of first repetition of every algorithm were used to analyze their performances in the following parts.

The convergence behaviors of the three algorithms were analyzed in this study by the Pareto set. Fig. 2 shows the evolution of F1 Pareto extremes with

Table 2. Result comparison on LN data

Algorithm	Trials	Time used(s)	$C_{1,2}\%$		M_1	M_2	M_3
MOPSO	1	14.5313	$C_{pso,ga}=0$	$C_{pso,sce}=6$	0.1078	0.0011	0.5074
	2	15.6797	$C_{pso,ga}=0$	$C_{pso,sce}=1.5$	0.1030	0.0327	0.5004
	3	14.6797	$C_{pso,ga}=0$	$C_{pso,sce}=0$	0.1070	0.0344	0.5101
	4	15.1484	$C_{pso,ga}=0$	$C_{pso,sce}=0.5$	0.0917	0.0332	0.5088
	5	13.9883	$C_{pso,ga}=0$	$C_{pso,sce}=0.5$	0.0971	0.0328	0.5039
	average	14.8055	$\overline{C}_{pso,ga}=0$	$\overline{C}_{pso,sce}=1.7$	0.1013	0.0268	0.5061
NSGA-II	1	125.3984	$C_{ga,pso}=100$	$C_{ga,sce}=100$	0.0628	0.0325	0.4645
	2	116.2188	$C_{ga,pso}=100$	$C_{ga,sce}=100$	0.0242	0.0269	0.3971
	3	118.6602	$C_{ga,pso}=100$	$C_{ga,sce}=98$	0.0765	0.0305	0.4699
	4	116.5312	$C_{ga,pso}=100$	$C_{ga,sce}=59.5$	0.0194	0.0324	0.4886
	5	117.1641	$C_{ga,pso}=100$	$C_{ga,sce}=100$	0.0264	0.0191	0.4328
	average	118.7945	$\overline{C}_{ga,pso}=100$	$\overline{C}_{ga,sce}=91.5$	0.0419	0.0283	0.4506
MOSCEM-UA	1	133.7500	$C_{sce,pso}=73$	$C_{sce,ga}=0$	0.0650	0.0336	0.4928
	2	143.9688	$C_{sce,pso}=71.5$	$C_{sce,ga}=0$	0.0792	0.0352	0.5084
	3	135.9297	$C_{sce,pso}=93.5$	$C_{sce,ga}=0$	0.0621	0.0348	0.5165
	4	146.6172	$C_{sce,pso}=86.5$	$C_{sce,ga}=0$	0.0269	0.0302	0.4401
	5	152.0625	$C_{sce,pso}=79.5$	$C_{sce,ga}=0$	0.0487	0.0342	0.4910
	average	143.2391	$\overline{C}_{sce,pso}=80.8$	$\overline{C}_{sce,ga}=0$	0.0620	0.0310	0.4787

increasing number of HYMOD evaluations. MOPSO, NSGA-II, and MOSCEM-UA all converge fast to function value $F1 < 0.26$ at about 1600 model evaluations. While MOPSO could continually converge to a smaller function value $F1 < 0.2380$ at about 30000 model evaluations. The searches of NSGA-II and MOSCEM-UA algorithm stopped at a relative local optimum.

The Pareto fronts of the multi-objective optimization with MOPSO, NSGA-II, and MOSCEM-UA algorithm are presented in Fig. 3. It seems that considerable trade-off exists between objectives F1 and F2. The trade-off was largest for the NSGA-II Pareto solutions (purple line) and smallest for MOPSO (red line), indicating that the latter one finds better solutions for objective F2, exhibiting less trade-off. Overall, these three algorithms all have sampled dense and uniform Pareto front. To compare the results, it should be considered that the number of

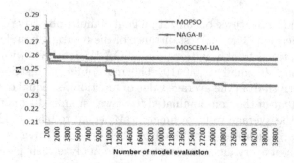

Fig. 2. Evolution of the Pareto extremes (F1) values as a function of the number of HYMOD model evaluations with the MOPSO, NSGA-II, and MOSCEM-UA algorithms

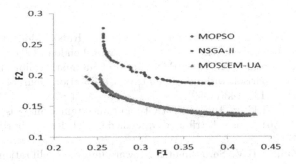

Fig. 3. Pareto Fronts determined by the MOPSO algorithm, NSGA-II algorithm, and MOSCEM-UA algorithm

MOSCEM-UA's Rank 1 Pareto points was 186. In contrast, all 200 points of the last NSGA-II generation were Pareto Rank 1. That was caused by a difference in search strategy.

4 Conclusion

In this study, three different multi-objective optimization algorithms (MOPSO, NSGA-II, MOSCEM-UA) and HYMOD model are introduced and testified by an application on the Xiangjiaping hydrological station located on Hanjiang River. In this multi-objective framework, High flow Nash-Sutcliffe efficiency and Low flow Nash-Sutcliffe efficiency were used to optimize the Hydrological model parameters. And the consuming time, the dominating rate and the quality of Pareto-optimal set (distance, distribution, and extent) were three criteria employed to analyze the performance of the three algorithms.

The MOPSO algorithm was the efficient one to complete each trial, and got relatively smaller value for both high flow and low flow Nash-Sutcliffe efficiency. The non-dominated solutions derived from MOPSO algorithm are seldom dominated by those from the other two algorithms. A high proportion of solutions drawn from NSGA-II were dominated by the other two algorithms. A complication appears in terms of Pareto-optimal sets quality. The shortest distance of the resulting non-dominated set to the Pareto-optimal front was from the NSGA-II algorithm; the most uniform distribution of the solutions was derived from the MOSCEM-UA algorithm; and the maximal extent of the obtained non-dominated front was stemmed from the MOPSO algorithm.

Besides, we analyzed convergence behaviors and the Pareto Fronts. The results demonstrate that all three algorithms are able to find a good approximation of the Pareto set of solutions, but differ in the rate of convergence to the optimal solutions.

References

1. Wagener, T., Wheater, H.S., Gupta, H.V.: Rainfall-Runoff Modeling in Gauged and Ungauged Catchments. Imperial College Press, London (2004)
2. Gupta, H.V., Sorroshian, S., Yapo, P.O.: Status of automatic calibration for hydrologic models: comparison with multilevel expert calibration. Journal of Hydrologic Engineering 4(2), 135–143 (1999)
3. Muletha, M.K., Nicklow, J.W.: Sensitivity and uncertainty analysis coupled with automatic calibration for a distributed watershed model. Journal of Hydrology 306, 127–145 (2005)
4. Boyle, D.P., Gupta, H.V., Sorooshian, S.: Toward improved calibration of hydrological models: Combining the strengths of manual and automatic methods. Water Resources Research 36(12), 3663–3674 (2000)
5. Srinivas, N., Deb, K.: Multi-objective function optimization using non-dominated sorting genetic algorithms. Evolutionary Computation 2, 221–248 (1995)
6. Deb, K., Pratap, A., Agarwal, S., et al.: A fast and elitist multiobjective genetic algorithm: NSGA-II. IEEE Transactions on Evolutionary Computation 6(2), 182–197 (2002)
7. Coello Coello, C.A., Lechuga, M.S.: MOPSO: A proposal for multiple objective particle swarm optimization. In: Proceedings of international Conference on Evolutionary Computation, pp. 1051–1056 (2002)
8. Vrugt, J.A., Gupta, H.V., Bastidas, L.A., et al.: Effective and effi cient algorithm for multiobjective optimization of hydrologic models. Water Resources Research 39, 1–19 (2003)
9. Tang, Y., Reed, P., Wagener, T.: How effective and efficient are multiobjective evolutionary algorithms at hydrologic model calibration. Hydrology and Earth System Sciences Discussions 2, 2465–2520 (2005)
10. Moore, R.J.: The probability-distributed principle and runoff production at point and basin scale. Hydrological Sciences Journal 30(2), 273–297 (1985)
11. Bos, A., Vreng, A.: Parameter optimization of the HYMOD model using SCEM-UA and MOSCEM-UA. University of Amsterdam, Amsterdam (2006)
12. Smakhtin, V.Y., Sami, K., Hughes, D.A.: Evaluating the performance of a deterministic daily rainfall-runoff model in a low flow context. Hydrology Process 12(5), 797–811 (1998)
13. Zitzler, E., Thiele, L.: An evolutionary algorithm for multiobjective optimization: The strength pareto approach. Techinical Report 43. Computer Engineering and Communication Networks Laboratory (TIK), Swiss Federal Institute of Technology (ETH) Zurich, Gloriastrasse 35, CH-8092 Zurich, Switzerland. TIK-Report, No.43 (1998)
14. Zitzler, E., Deb, K., Thielel, L.: Comparison of multi-objective evolutionary algorithms: empirical results. IEEE Transaction on Evolutionary Computation 18(2), 173–195 (2008)
15. Raquel, C.R., Naval Jr., P.C.: An effective use of crowding distance in multiobjective particle swarm optimization. In: Proceedings of the Genetic and Evolutionary Computation (GECCO 2005), Washington, DC, USA (2005)

The Application Study
of Dynamic Pricing Decision System
Based on Multi-objective Optimization

Qing Zhou* and Qinlan Yuan

School of Business Administration
China University of Petroleum-Beijing
Changping 18 Fuxue Road, Beijing, P.R. China
zhouqing00@hotmail.com

Abstract. As the recent volatility of international prices of sulfur indicates, and how to make a reasonable price of sulfur is a matter of concern for the sulfur production enterprises. A dynamic pricing model is
presented in this paper, and the improved fuzzy decision based on multi-
objective optimization is provided. While the key issue is to construct
the matrix based on the volatility factors, to analyze historical pricing
data to predict future price direction, and to give the final pricing scheme
of sulfur based on fuzzy decision-making.

Keywords: Fuzzy Decision, Fuzzy AHP, Dynamic Pricing Decision System, Multi-Objective Optimization.

1 Introduction

There is almost no nature sulfur in our nation, most sulfur productions are affiliated to petroleum industry, and sulfur is mainly recycled from petroleum refining
and natural gas purification. Therefore, our countrys sulfur supply mainly depends on import, which accounts for more than 60% of national total sulfur
consumption. This result in that National Sulfur Price is subjected to Import
Sulfur Price, that is, sulfur pricing of domestic sulfur-product industries is based
on sulfur CIF. Due to the fierce variation of international sulfur price, the sulfur
pricing of domestic sulfur-product industries has attracted our attention.

According to statistics, in 2011, China had imported about 7,850 thousand
ton sulfur from January to October, and has been one of the biggest sulfur
import countries in the world. Although it is predicted that domestic sulfur
industries will output about 4,000 thousand ton sulfur, the sulfur market will
still be dominated by import sulfur. A comparison of the international sulfur
price in those nearly 4 years indicates that the sulfur price has been stepping up
after plunged in 2008. The average price, in 2008, was $490, and it slowed down to
$45 in 2009. In 2010, the average price rise to $140. And in this year, the average

---

* This work is partially supported by NSFC Grant #71101153 to Q. Zhou.

Z. Li et al. (Eds.): ISICA 2012, CCIS 316, pp. 217–227, 2012.
© Springer-Verlag Berlin Heidelberg 2012

price of sulfur has climbed to \$216 from January to November. Meanwhile, the sulfur price in domestic market has being in a high level this year.

Since the Domestic Sulfur Price is mainly based on the Import Sulfur Price, it often can't reflect the domestic market, which causes serious information lag and dull sale. Besides, because of the rigid sulfur pricing mechanism which is always lagged behind the market, and the complex price adjustment procedure, Domestic Sulfur Price often cant be adjusted timely.

So, pricing decision of sulfur industries must consider these three cardinal principles as below:

1. **Reasonable Pricing**
   Pricing of sulfur industries should be market-oriented, as well as keep in step with the import price. In principle, In-factory price should be roughly the same between domestic sulfur users and import sulfur users, and smooth sales also should be guaranteed.
2. **Ensure Safety Stock**
   The storage should be rigidly kept in safe range, which on one hand it should guarantee normal order. On other hand the storage should be limited to its capacity.
3. **Reasonable Economic Goals**
   Ensuring the production and the sale being successful, the sulfur manufacturers should pursue the maximization of the economic interests.

The sulfur price of domestic sulfur industries mainly based on the import price because the demands of domestic sulfur market seriously depend on import. In addition, pricing of the sulfur manufacturers should also consider both its output and the market. In general, the short-term pricing decision should be taken serious attention to the relevant factors as below:

1. International sulfur price and total import: As one of the biggest sulfur-import countries, China is highly depended on import sulfur, which accounts for over 50% of the national sulfur demand. Owing to there will be 3000-4000 thousand ton potential surplus in the global sulfur market in nearly future, the situation of sulfur oversupply can hardly be avoided, which in turn will control the trend of domestic sulfur price.
2. International crude oil price: As a secondary product of refining oil and natural gas, sulfur price is directly influenced by the variable price of the global crude oil.
3. The supply and demand in domestic sulfur market: Because the amount of sulfur consumptions are most use for producing phosphate, about 70%, so the demand of sulfur in the downstream mainly depends on the phosphate market. In 2010, the domestic sulfur industries had outputted about 3,200 thousand ton, however, the amount of sulfur consumption had reached to 13,670 thousand ton, with the self-supply rate only stayed at about 23.4%. In 2011, the external dependence was still in high rate, about 56%.

4. The demand of target market: Variation of the demand in the target market direct influence the pricing decision of the sulfur industries.
5. The special of sulfur production: Pricing decision of sulfur industries must consider its special production. Contrast to other chemical products, sulfur production is affiliate to oil production, whose production is involuntary and supply is rigid.
6. Logistics costs: Due to the domestic sulfur demand mainly depends on import, the variation of international logistics costs will influence domestic sulfur price.
7. The policy: the policy also affects the sulfur price decision, such as, since January 1st, 2010, the government recovered to collect sulfur import tariff, this caused much higher sulfur import price, and to a certain degree, it restrained the sulfur import.

In recent years, the price of global sulfur waves heavily, and there are many factors influence the domestic sulfur market, so it is worth for sulfur industries to pay close attention to how to make a right judgment and decide a reasonable price in the competitive market. Strategies for how to pricing for sulfur sale mainly include cost-plus pricing, pricing by expected value, expected value pricing, competing-oriented pricing, demand-oriented pricing and so on.

2   Dynamic Pricing Model

The dynamic pricing model is structured base on the current situation of domestic sulfur market. It is constructed to ensure that sulfur pricing can keep up with the dynamic market and the sulfur industries can supply timely and smooth, which finally help industries to reach their fixed benefit goals. By detail analysis of the factors that influence pricing of sulfur industries, we decided to apply the Fuzzy Analytic Hierarchy Process method (FAHP) and case based reasoning method to make a sulfur dynamic pricing program.

2.1   The Hierarchical Structure of Sulfur Pricing Decision

Due to the special of sulfur production and the current situation of import dependence of sulfur demand, the sulfur industries mainly base on the import price and the stock capacity to make sale decision, rather than seeking for maximum profit. The principle of pricing is firstly to guarantee a smooth sale and then pursue economic interest. Therefore, considering the current situation of sulfur industries and the impact factors of pricing, we can draw a hierarchical chart that based on Analytic Hierarchy Process method (AHP) as below:

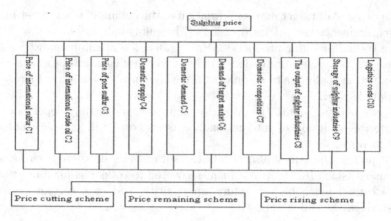

Fig. 1. Impact Factors of Pricing

2.2   Structure A Judgment Matrix

The core concern of the application of Analytic Hierarchy Process (AHP) is how to structure, examine and revise the judgment matrix, and how to set weight for every element of the judgment matrix.

Table 1. Initial weights of the indices table (A)

Evaluating Factors	IndicesC_i	Weights	Fluctuation Sensitivity $V_i(U, N, D)$ (rising, remaining, dropping)	Plus-minus Relationship of Fluctuation
Price	International sulfur price(C1)	Great important	(5, 1, 4)	+
	International crude oil price(C2)	Very important	(4, 4, 1)	+
	Port sulfur price(C3)	Great important	(5, 1, 4)	+
Market Supply and Demand	Domestic supply(C4)	Important	(3, 1, 4)	-
	Domestic demand(C5)	More important	(1, 1, 2)	+
	Demand of the target market(C6)	More important	(1, 1, 3)	+
	Domestic competitors(C7)	general	(1, 1, 2)	-
The Production of Enterprise	The output of sulfur industries(C8)	Very important	(2, 1, 5)	-
	Storage of sulfur industires(C9)	Great important	(2, 1, 5)	-
	Logistics costs(C10)	general	(1, 1, 1)	+

Table 2. Indices weights of expected scheme table (B)

Evaluating Facotrs	Indices(temCi)	Weights	Fluctuation Range(Fi)
Price	International sulfur price(temC1)	Great important	0.04
	International crude oil price(temC2)	Very important	0.03
	Port sulfur price(temC3)	Great important	0.04
Market Supply and Demand	Domestic supply(temC4)	Important	-0.03
	Domestic demand(temC5)	More important	0
	Demand of the target market(temC6)	More important	0
	Domestic competitors(temC7)	general	0
The Production of Enterprise	The output of sulfur industries(temC8)	Very important	0.05

This model aims to solve the problem of dynamic decision of sulfur price. So when constructing the judgment matrix model, it is rather not enough to consider the Impact Factors Weights only which were set by experts when in evaluating, the variation of the all the impact factors also should be considered. Moreover, we should focus on how far the variable index factors influence the pricing decision. Hence, the judgment matrix of this model involves two relevant indices parameter table: initial weights table (Table 1) and indices weights of expected scheme table (Table 2). The weight value of both the Table 1 and Table 2 may be the same, or may be different. As the changing of the market and the adjustment of the pricing scheme, the weight value will be updated intermittently.

Table 1 demonstrates sensitivity value of which the Impact Factors react to the fluctuation of sulphur price, and the table displays the Plus-minus relationship of the fluctuation of sulphur price.

Table 2 demonstrates the indices weights of impact factors and the fluctuation range of the factors' indices during decision-making period.

Structure a Judgment Matrix with Criterion Layer. According to the fundamental principles of the Analytic Hierarchy Process (AHP), through pairwise comparison method among the factors, we can get the criterion layer of judgment matrix A_{ij},

$$A_{ij} = (C_i/C_j + temC_i/temC_j)/2$$

And according to the fluctuation range of the factors' indices during decision-making period in the expected scheme table (table 2), we set up a revised matrix R_{ij},

$$R_{ij} = F_i/F_j/n$$

Revise the criteria judgment matrix: $A_{ij} = A_{ij} + R_{ij}$;
Then compute the coincident indicator of the criterion layer C_I,

$$C_I = \frac{\lambda_{max} - n}{n - 1}$$

And compare the random consistency of the matrix: $C_R = C_I/R_I$, R_I is the average random coincident indicator, whose relevant parameter can be got by using the Analytic Hierarchy Process(AHP). Generally, when $C_R < 0.1$, the consistency of the judgement matrix is satisfied; otherwise, when $C_R > 0.1$, we need to adjust the revised matrix until it is satisfied.

Structure a Judgment Matrix with Scheme Layer. The key point of creating the scheme layer judgment matrix is the sensitivity values which are displayed by the Impact Factors that react to the fluctuation of sulfur price. Base on the change of the Impact Factors, we should revise the sensitivity values, the judgment matrix can be structured as below:

1. Compute the sensitivity value V_0 and the weighted average of the fluctuation range F_0:

$$V_0 = \sum_{t=1}^{n}(U_i, N_i, D_i)/n; \quad F_0 = \sum_{i=1}^{n} F_i/n;$$

2. Revise the sensitivity values $V_i(U, N, D)$;

$$E_i = \begin{cases} V_i(U)/V_0(U); & IF((U[i]/V_0[U]) > (D[i]/V_0(D))) \\ V_i(D)/V_0(D); & Others \end{cases}$$

```
IF Fi <> 0 then    Vi (U,N,D)
    U [i] =  F [i]*100 + (U [i]/ VO(U))*F [i]/FO;
    D [i] =  F [i]*100 + (D [i]/ VO(D))*F [i]/FO;
ELSE  Vi (U,N,D)
    U [i] = U [i]/ VO(U);
    D [i] = D [i]/ VO(D);
    N [i] = E[i] + VO(N)*FO;
ENDIF
```

3. Structure a judgment matrix:

$$B_{ij} = V_i(U, N, D)/V_j(U, N, D)$$

Comparing the random consistency of the scheme layer judgment matrix until it is satisfied.

2.3   Dynamic Pricing Decision Process

In order to make a wise pricing decision, we should apply both the fuzzy analytic hierarchy process method and the case-based reasoning method, and need to figure out the variation of the impact factors base on the market information. In particular, it consists of three steps: Firstly, to figure out the expected scheme of the sale price, which can offer as a reference for the competent department. Secondly, to compute errors depends on the final decision scheme. Finally, according to the errors, we should decide whether it is need to revise some indices value or not, and update the database. Depending on those steps, we make the future decision. The algorithm process is presented as the follow:

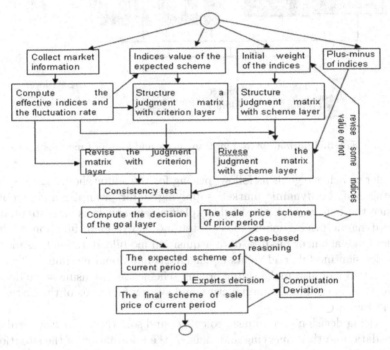

Fig. 2. Dynamic pricing decision process

3   Developing a Pricing Decision System

The sulfur pricing decision system is designed to support the sulfur pricing decision being semi-structured, and mainly aim to improve the effective of the sulfur price decision. Sulfur pricing decision system is a multiplexed system which is consists of series subsystems, with diversified structure. It is structured mainly based on the sulfur pricing problems that need to be solved. The main components of the system are man-machine interactive system, database, model base, knowledge base and inference engine.

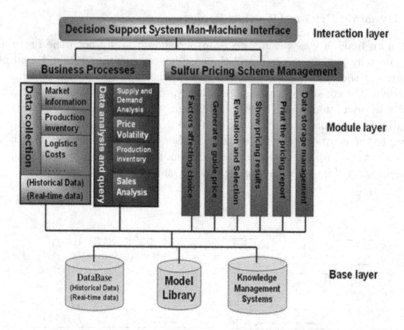

Fig. 3. Information system framework and system functions

In order to achieve goals as follow: pricing for the sulfur should keep up with the changes of the dynamic market, supplying sulfur promptly and smoothly, and reach the benefit goal, we should exploit a relevant decision system on the basis of dynamic pricing model. The system needs five main function modules: basic data management module, data acquisition module, data analysis module, pricing decision module and rights management functional module.

The system can also make sulfur pricing schemes report using some decision support techniques. The frame structure and system functions of the system are shown in Fig. 3.

The pricing decision system need to collect and sort the sulfur historical data and live data, and then querying and analysis the information of the situation of demand, storage, sale and the fluctuation of price of the sulfur market. Based on this, through relevant fuzzy-decision algorithm and the supported technology, we can get the sulfur pricing-scheme report. Based on the final decision scheme and case-based reasoning technology, we should revise the indices weights.

1. Basic data management module

   The main function of this module is to collect the basic data of the factors that influence the sulfur price, including international sulfur producing area, international petroleum production area, port, competitor, target market, domestic supply market, domestic demand market and the transportation.

2. Data collecting module

   The main function of this module is sorting the basic data, querying data by time, and preliminary analyzing. This task can support the later data analysis and pricing decision. Besides, this module can also query and call the basic data and data in management module.

3. Data analysis module

This module depends on collecting all kinds of data. When new data emerges or data obsolesces, the manager need to update the database in order to guarantee the data used are all new, thus can do reasonable analysis and make sure the pricing decision accuracy.

4. Pricing decision module

To structure this module depends on three modules, include data management module, data collecting module and data analysis module. Supporting by those modules, and considering the importance degree of factors, we can create several schemes. By comparing with each scheme, we can finally select the optimal pricing scheme. As the Fig. 4 displayed:

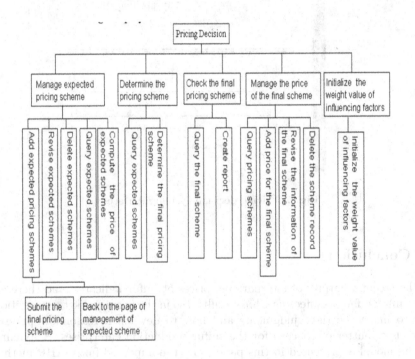

Fig. 4. Pricing Decision Module

In order to build programs with strong cohesion and loose coupling for web application system development, the MVC (Model View Controller) design pattern is used on a Three Tier Client/Server Architecture. Then this web application system is developed with the struts2 framework, in order to achieve loose coupling between the layers, the separation of business logic and view logic, and to improve system flexibility, reusability, maintainability and scalability, while to simplify its tasks through the Analysis, Design, Test and Release Phases.

Finally, according to customer requirements, system development is based on ORACLE database. From the comprehensive consideration of the technical maturity, safety and reliability, the system uses the Apache Tomcat 6.0 as a Web Server and the established the directory structure for the sulfur pricing system. While the finished Pricing Decision System Interface is as follow (Fig. 5).

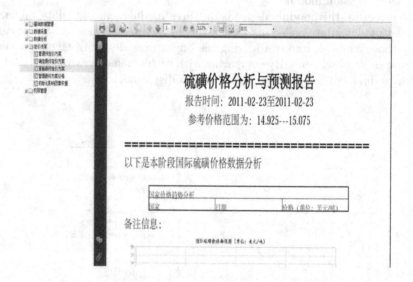

Fig. 5. Pricing decision system interface

4   Conclusion

As the recent volatility of international prices of sulfur indicates, and there are many market factors affecting China's sulfur, so in the fierce market competition, how to make the right judgments and how to develop a reasonable price of sulfur is a matter of concern for the sulfur production enterprises. A dynamic pricing model is presented in this paper, and the improved fuzzy AHP method is adopted. While the key issue is to construct the matrix based on the volatility factors, to analyze historical pricing data to predict future price direction, and to give the final pricing scheme of sulfur based on fuzzy decision-making for multi-objective optimization.

In order to improve the effectiveness of pricing and to make reasonable pricing decisions, it is necessary to develop a computer information system and to achieve the purpose of scientific decision-making.

References

1. Satty, T.L.: Axiomatic foundation of the analytic hierarchy process. Management Science 32(7), 841–855 (1986)
2. Satty, T.L.: The analytic hierarchy process. MC Graw-Hill, New York (1980)
3. Olcer, A.I., Odabasi, A.Y.: A new fuzzy att ributive group decision making met hodology and it s application to propulsion/manoeuvring system selection problem. Journal of Operational Research 166, 93–114 (2005)

Twisted Helical Antenna for Satellite-Mobile Handset Using Dynamic Multi-objective Self-adapting Differential Evolution Algorithm

Lian Zhang[1], Sanyou Zeng[1], Zhu Liu[1],
Steven Gao[2], Zhengjun Li[3], and Hongyong Jing[3]

[1] School of Computer Science
China University of GeoSciences
430074 Wuhan, Hubei, P.R. China
{lianzhang1023,sanyouzeng,evolution.cpp}@gmail.com
[2] Surrey Space Center University of Surrey GU2 7XH
Guildford, Surrey, UK
s.gao@surrey.ac.uk
[3] China Academy of Space Technology 710000 Xi'an
Shaanxi, P.R. China
lzj@cast504.com

Abstract. This paper presents a twisted helical antenna, which operates at 2.4GHz and is applied in satellite mobile communications. It is designed using dynamic multi-objective self-adaptive differential evolution (SaDE) algorithm which combines SaDE and dynamically controlling constraint technique. The main reason for adopting dynamic multi-objective SaDE in this paper is that few literatures have used dynamic multi-objective differential evolution (DE) for solving antenna optimization problem so far. A small twisted helical antenna optimized on a perfect ground plane. Results show that while our optimized antenna met the design requirements, its performance is not inferior to the quadriflilar helical antenna as seen in reference. The design shows the capability of evolution algorithm (EA) as an efficient optimization tool for searching globally optimal solutions for antenna design.

Keywords: twisted helical antenna, dynamic multi-objective optimization, self-adaptive differential evolution, evolutionary antenna.

1 Introduction

With evolution algorithm (EA) being widely applied, an approach of using EA in conjunction with an electromagnetic simulator has been introduced for antenna designs. For instance, EA was applied to design wire antennas [1,2] and quadrifilar helical antennas [3,4]. NASA firstly designed an antenna which meets NASA's Space Technology 5 (ST5) mission spacecraft requirements using EA [5].

Twisted helical antenna is an attractive candidate antenna for satellite mobile handsets because of the simpleness of their geometry, and the capability of providing circular polarization over a broad angular region [6].

Z. Li et al. (Eds.): ISICA 2012, CCIS 316, pp. 228–239, 2012.

Differential evolution(DE) is a heuristic random search algorithm which was introduced by Price and Storn [7] to solve global optimization problems. Because of the effectiveness and high efficiency, DE has been successfully used in number of practical cases. However, choosing suitable parameter values is a frequently problem-dependent task and demands user's previous experience. Despite its crucial importance, there is no consistent methodology for determining the control parameters. So self-adaptive DE (SaDE) was proposed with the idea of "evolution of the evolution" [8]. In this paper, a self-adaptive method is adopted to determine the values of control parameters F and CR in order to produce a flexible DE to solve problems more efficiently.

EA includes multi-objective EA and dynamic adaptive techniques. As to dynamic technique, representative work on dynamic evolutionary computation work could be seen in Shengexiang Yang'book [9] and Yaochu Jin's literature [10,11]. There are few literatures employing dynamic constrained optimization techniques to solve constrained optimization problem (COP). A method of using dynamic multi-objective optimization technique to solve COP was proposed in [12]. Dynamic multi-objective technique is used to improve the SaDE in solving COP. Multi-objective technique makes a balance between objectives and constraints optimization, and dynamically controlling constraint technique makes the constraints converge gradually.

This paper is organized as follows: In section 2, we introduce the dynamically controlling constraint technique. Section 3 gives a briefly explanation of the original SaDE algorithm and illustrates the dynamic multi-objective SaDE in detail. We discuss the antenna design using dynamic multi-objective SaDE in section 4. A design example and its performance analysis are presented. Finally, section 5 concludes the paper.

2 Dynamic Controlling Constraint Technique

The general COP is concerned with finding a solution in the search space satisfies requirements to minimize the function objective value. The expression of a constrained optimization problem is:

$$
\begin{aligned}
min \quad & y = f(\boldsymbol{x}) \\
st: \quad & g_i(\boldsymbol{x}) \leq 0, i = 1, 2, \ldots, q \\
& h_i(\boldsymbol{x}) = 0, i = q+1, q+2, \ldots, m \\
where \quad & \boldsymbol{x} = (x_1, x_2, \ldots, x_n) \in X \\
& X = \{(x_1, x_2, \ldots, x_n) | l_i \leq x_i \leq u_i, i = 1, 2, \ldots, n\} \\
& \boldsymbol{l} = (l_1, l_2, \ldots, l_n) \\
& \boldsymbol{u} = (u_1, u_2, \ldots, u_n)
\end{aligned}
\tag{1}
$$

where $\boldsymbol{x}$ is decision vector, X is decision space, $\boldsymbol{l}$ and $\boldsymbol{u}$ are the lower bound and upper bound of decision space. $f(\boldsymbol{x})$ is an objective function, $g_i(\boldsymbol{x}) \leq 0$ and $h_i(\boldsymbol{x}) = 0$ are q inequality constraints and $m - q$ equality constraints.

To handle constraints well by evolutionary algorithms, researchers have done much work. Michalewicz and Schoenauer [13] classify these methods into four categories and Coello also concludes these techniques in [15].

In the selection operator of a multi-objective evolutionary algorithm, that one solution dominates another helps fast convergence of the population while the both solutions are indifferent benefits wide distribution of the population. Therefore, Pareto domination can be used to achieve a good trade-off between convergency and diversity [12]. That is why multi-objective evolutionary algorithms are widely exploited to be used to solve COP problems. In the section, we construct a dynamic constrained multi-objective problem which converges to the COP in $Eq(1)$.

Generally we prefer the solutions which are with small objective values and with small violation values when solving COPs. It's natural to regard two items as two objectives to form a multi-objective problem. The first item is actually the original objective $f(x)$ (see $Eq(1)$). The second objective determined by the second item is to treat constraints, which is called violation objective.

The violations of inequality constraints and equality constraints in $Eq(1)$ are often defined as:

$$G_i(x) = max\{g_i(x), 0\}, i = 1, 2, \ldots, q, \tag{2}$$

$$H_i(x) = |h_i(x)|, i = q+1, q+2, \ldots, m. \tag{3}$$

The violation objective is usually constructed as:

$$\varphi(x) = \frac{\sum\limits_{i=1}^{q} \frac{G_i(x)}{\max\limits_{x \in P(0)} \{G_i(x)\}} + \sum\limits_{i=q+1}^{m} \frac{H_i(x)}{\max\limits_{x \in P(0)} \{H_i(x)\}}}{m} \tag{4}$$

P is for population, and $P(0)$ is for initial population.

Then, a two-objective optimization problem can be constructed from the COP in $Eq(1)$:

$$\begin{aligned} min \quad & y = (f(x), \varphi(x)) \\ where \quad & x = (x_1, x_2, \ldots, x_n) \in X \\ & X = \{(x_1, x_2, \ldots, x_n) | l_i \le x_i \le u_i, i = 1, 2, \ldots, n\} \\ & l = (l_1, l_2, \ldots, l_n) \\ & u = (u_1, u_2, \ldots, u_n) \end{aligned} \tag{5}$$

About searching global optimum many research has been done by collecting information in feasible solutions. In practice, some infeasible solutions contain more information about global optimum than some feasible ones especially when global optimum is on the boundary feasible region and infeasible region. Searching global optimum need the help of some infeasible solutions, so a relaxing technique which handles both inequality and equality constraints for collecting information from these helpful infeasible solutions is used [14]. The relaxing process of multi-objective problem (MOP) described as follows:

$$\begin{aligned}
min \quad & \mathbf{y} = (f(\mathbf{x}), \varphi(\mathbf{x})) \\
st: \quad & g_i(\mathbf{x}) \le e_i^{(k)}, i = 1, 2, \ldots, q \\
& h_i(\mathbf{x}) = e_i^{(k)}, i = q+1, q+2, \ldots, m \\
where \quad & \mathbf{x} = (x_1, x_2, \ldots, x_n) \in X \\
& X = \{(x_1, x_2, \ldots, x_n) | l_i \le x_i \le u_i, i = 1, 2, \ldots, n\} \\
& e_i^{(k)} = (e_1^{(k)}, e_2^{(k)}, \ldots, e_m^{(k)}), e_i^{(k)} > 0 \\
& e^{(k+1)} \le e^{(k)} \\
& \lim_{k \to \infty} e^{(k)} = 0
\end{aligned}$$

$$(6)$$

Here, $e^{(k)}$ is an evolution environment concerning constraints. At the beginning of evolution, in environment 0, all constraints boundaries of dynamic multi-objective optimization problem (DMOP) in $Eq(6)$ are broadly relaxed, Then, the boundaries are tightened and environment 0 turns to environment 1. With evolution, when the all individuals in population are feasible under current boundaries, constraints boundaries will be tightened again to environment 2. The rest can be done in the same rule until constraints boundaries turn down to 0 when DMOP can converge to COP. In this way, the methods of solving DMOP can be used to solve COP. $e = (e_1, e_2, \ldots, e_m)$ is the **relax-boundary vector**. A monotonically decreasing function is used. Three typical phases are listed here for inequality constraints:

$$\begin{cases}
e_i^{(0)} = \max_{\mathbf{x} \in P(0)} \{G_i(\mathbf{x})\} \\
\quad \vdots \\
e_i^{(k)} = \max_{\mathbf{x} \in P(0)} \{G_i(\mathbf{x})\} e^{-\left(\frac{k}{c_i}\right)^2} \\
\quad \vdots \\
e_i^{(\infty)} = \max_{\mathbf{x} \in P(0)} \{G_i(\mathbf{x})\} e^{-\left(\frac{\infty}{c_i}\right)^2} = 0, i = 1, 2, \ldots, q
\end{cases}$$

$$(7)$$

In the same way, equality constraints ones are listed:

$$\begin{cases}
e_i^{(0)} = \max_{\mathbf{x} \in P(0)} \{H_i(\mathbf{x})\} \\
\quad \vdots \\
e_i^{(k)} = \max_{\mathbf{x} \in P(0)} \{H_i(\mathbf{x})\} e^{-\left(\frac{k}{c_i}\right)^2} \\
\quad \vdots \\
e_i^{(\infty)} = \max_{\mathbf{x} \in P(0)} \{H_i(\mathbf{x})\} e^{-\left(\frac{\infty}{c_i}\right)^2} = 0, i = 1, 2, \ldots, q
\end{cases}$$

$$(8)$$

In practice, after environment changes K times, e can be equal to ε and evolution will stop. $\varepsilon = (\varepsilon_1, \varepsilon_2, \ldots, \varepsilon_m)$ meets a certain high precision requirement, i.e. $\varepsilon_i = 1E-12, i = 1, 2, \ldots, m$. Because of environment changing count k approaches to generation g, max generation G value is used to replace K value. The following formulas can be got:

$$\begin{cases} \max_{\boldsymbol{x} \in P(0)} \{G_i(\boldsymbol{x})\} e^{-\left(\frac{K}{c_i}\right)^2} = \varepsilon_i, i = 1, 2, \ldots, q \\ \max_{\boldsymbol{x} \in P(0)} \{H_i(\boldsymbol{x})\} e^{-\left(\frac{K}{c_i}\right)^2} = \varepsilon_i, i = q+1, q+2, \ldots, m \end{cases} \quad (9)$$

Therefore, the vector c for each constraints can be derived from $Eq(9)$:

$$\begin{cases} c_i = \dfrac{K}{\sqrt{-\ln \frac{\varepsilon_i}{\max\limits_{\boldsymbol{x} \in P(0)} \{G_i(\boldsymbol{x})\}}}}, i = 1, 2, \ldots, q \\ c_i = \dfrac{K}{\sqrt{-\ln \frac{\varepsilon_i}{\max\limits_{\boldsymbol{x} \in P(0)} \{H_i(\boldsymbol{x})\}}}}, i = q+1, q+2, \ldots, m \end{cases} \quad (10)$$

At environment k, If $\boldsymbol{x} \in X, G_i(\boldsymbol{x}) \leq e_i^{(k)}, i = 1, 2, \ldots, q$ and $H_i(\boldsymbol{x}) = e_i^{(k)}, i = q+1, q+2, \ldots, m$, this solution $\boldsymbol{x}$ is said to be relax-feasible, otherwise, $\boldsymbol{x}$ is said to be relax-infeasible.

To obtain all helpful information in infeasible solutions, new constrained Pareto-dominance rules are needed by modifying the traditional dominance rule [15]:

1. A relax-feasible solution is always preferred over a relax-infeasible one
2. Between two relax-feasible or two relax-infeasible solution, traditional Pareto-dominance rules are used.

3   Dynamic Multi-objective Self-adapting Differential Evolution Algorithm

3.1   Differential Evolution Algorithm

There are several variants of DE [7]. In this paper, we use the DE scheme which can be classified using notation [7] as $DE/rand/1/bin$ strategy. This strategy is the most often used in practice and can be described as follows.

An individual is represented by a D-dimensional vector. A population consists of NP individuals $x_{i,G}, i = 1, 2, \ldots, NP$. G denotes one generation. NP is the population size. The initial population is generated stochastically with uniform distribution.

The crucial idea behind DE is a new scheme for generating trail parameter vectors. Mutation and crossover are used to generate trail vectors, and selection determines which vectors in current population will survive into the next generation [7].

Mutation Operation

$$v_{i,G+1} = x_{r_1,G} + F * (x_{r_2,G} - x_{r_3,G}) \quad (11)$$

where r_1, r_2, r_3 are random integers generated from the interval $[1, NP]$, and $i \neq r_1 \neq r_2 \neq r_3$. F is a real number ($F \in [0, 2]$) for scaling differential variation $(x_{r_2,G} - x_{r_3,G})$.

Crossover Operation. After mutation, the crossover operation is applied to each pair of the mutated vector $V_{i,G+1}$ and the target vector $X_{i,G}$ to generate a trail vector $T_{i,G+1}(t_{1i,G+1}, t_{2i,G+1}, \ldots, t_{Di,G+1})$.

$$t_{ji,G+1} = \begin{cases} v_{ji,G+1}, & if\, r(j) \leq CR\, or\, j = rand(i) \\ x_{ji,G}, & if\, r(j) > CR\, and\, j \neq rand(i) \end{cases}$$

For $j = 1, 2, \ldots, D$. CR is the crossover constant $[0, 1]$ which is specified by the user. $r(j)$ is the jth evaluation of a uniform random number generator with outcome $\in [0, 1]$. $rand(i)$ is a randomly chosen index belong $1, 2, \ldots, D$ which ensure that the trail vector containing at least one parameter from the mutant vector.

Selection Operation. If the trail vector $T_{i,G+1}$ yields a better objective function value than $X_{i,G}$, then $X_{i,G+1}$ is set to $T_{i,G+1}$, otherwise, the old value $X_{i,G}$ is retained.

3.2   Self-adapting Differential Evolution Algorithm

It's a problem-dependent work to choose suitable control parameter values frequently. Some researches propose a self-adapting method for control parameters [8, 16], and the mechanism is F and CR are applied at individual levels. In Table 1, Each individual in current population has a set of control parameters(F and CR) which will adjusted by the means of evolution. The better values of these control parameters lead to better individuals which, in turn, are more likely to survive and generate offspring, hence, produce the better control parameter values. New control parameters $F_{i,G+1}$ and $CR_{i,G+1}$ are calculated as follows:

Table 1. Self-adapting: encoding aspect

$x_{1,G}$	$F_{1,G}$	$CR_{1,G}$
$x_{2,G}$	$F_{2,G}$	$CR_{2,G}$
...	...	...
$x_{NP,G}$	$F_{NP,G}$	$CR_{NP,G}$

$$F_{i,G+1} = \begin{cases} F_l + rand_1 * F_u, & if\, rand_2 < \tau_1 \\ F_{i,G}, & if\, rand_2 \geq \tau_1 \end{cases} \tag{12}$$

$$CR_{i,G+1} = \begin{cases} rand_3, & if\, rand_4 < \tau_2 \\ CR_{i,G}, & if\, rand_4 \geq \tau_2 \end{cases} \tag{13}$$

$rand_j, j \in 1, 2, 3, 4$ are uniform random value $\in [0, 1]$. τ_1 and τ_2 represent probabilities to adjust factors F and CR, respectively. $F_{i,G+1}$ and $CR_{i,G+1}$ are produced before mutation operation. So, they influence the mutation, crossover, and selection operations of the new vector $x_{i,G+1}$.

3.3   Dynamic Multi-objective Self-adapting DE Algorithm

The framework of dynamic multi-objective SaDE is below.

Algorithm 1: Framework of dynamic multi-objective SaDE
  Step 1: Initialization. Randomly produce population $P(0)$ with size NP. Set evolution generation counter $g = 0$ and environment parameter $k = 0$. Initialize the algorithm-related parameter F_l, F_u, τ_1, τ_2 and calculate F_0, CR_0 using $Eq(12)$ and $Eq(13)$. Initialize relax-boundary $e = e^{(0)}$. Calculate vector, c using $Eq(10)$ $\varphi(x)$ using $Eq(4)$ and $f(x)$.
  Step 2: Generate an offspring population $T(g)$ of size NP from $P(g)$. (Algorithm 2)
  Step 3: For each $x \in T(g)$, calculate two objective values: $\varphi(x)$ and $f(x)$.
  Step 4: Merge the parents $P(g)$ and the offsprings $T(g)$ into $U(g)$: $U(g) = P(g) \cup T(g)$.
  Step 5: Select NP solutions from $U(g)$ into the next population $P(g + 1)$ using the non-dominated sorting technique.
  Step 6: If all $x \in P(t)$ are relax-feasible, then set $k = k + 1, e = e^{(k+1)}$ and update the relax-feasibility of all $x \in P(g)$.
  Step 7: $g = g + 1$. If $g = T(\text{max generation})$, goto **Step 8**, otherwise goto **Step 2**.
  Step 8: Output $P(g)$.

Algorithm 2: Generating offsprings using SaDE
  Step 1: Empty $T(g)$.
    FOR $i = 1$ TO NP
      Step 2: Randomly generate three integers from the interval $[1, NP]$: r_1, r_2, r_3 and $i \neq r_1 \neq r_2 \neq r_3$. Create $v = (v_1, v_2, \ldots, v_D)$.
      FOR j TO D
      Step 3: Generate random value $r(j)$ uniformly distributed on $(0, 1)$ and randomly chose a index $rand(i)$ belong $1, 2, \ldots, D$.
        IF $(r(j) \leq CR || j == rand(i))$
        $v_i = x_{r_1j} + F * (x_{r_2j} - x_{r_3j})$
        ELSE
        $v_i = x_{ij}$
      END FOR j
    END FOR i
  Step 4 Insert v into $T(g)$: $T(g) = T(g) \cup v$.
  Step 5 Update F and CR.

4   Design Antenna Using Dynamic Multi-objective Self-adapting DE Algorithm

4.1   Antenna Structure Coding

A large amount of results on solving numerical optimization problems indicate that the real-coded can promote search effectively by designing appropriate variable, so real-coded is used to make mathematical model for antenna problems.

As shown in Fig.1 and Table 2, there are seven wires including the feeder.

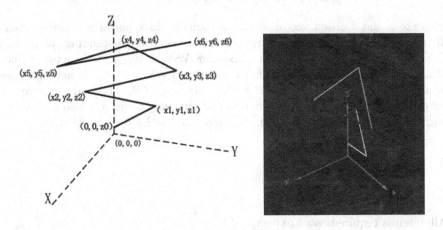

Fig. 1. An model of antenna structure   **Fig. 2.** Simulation antenna structure

The feeder is a very short conductor which starts from $(0,0,0)$ to $(0,0,z_0)$ and extends along the positive z-axis. Each wire is determined by starting coordinate and they are combined end to end. The wire radius is r. The chromosome of antenna structure is $(r, z_0, x_1, y_1, z_1, x_2, y_2, z_2 \ldots, x_6, y_6, z_6)$, all of the 20 numbers are real-coded.

Table 2. Coding of Antenna

Wire	Start Point	End Point	Radius
Feeder	$(0,0,0)$	$(0,0,z_0)$	r
Wire 1	$(0,0,z_0)$	(x_1,y_1,z_1)	r
Wire 2	(x_1,y_1,z_1)	(x_2,y_2,z_2)	r
Wire 3	(x_2,y_2,z_2)	(x_3,y_3,z_3)	r
Wire 4	(x_3,y_3,z_3)	(x_4,y_4,z_4)	r
Wire 5	(x_4,y_4,z_4)	(x_5,y_5,z_5)	r
Wire 6	(x_5,y_5,z_5)	(x_6,y_6,z_6)	r

4.2 Modeling of Antenna Optimization Problem

The model is related to a certain kind of antenna optimization problem which will be mentioned in Section 4, the problem can be described as following.

$$\min \ f(\boldsymbol{x}) = VSWR + \sum_{\varphi=0°}^{355°} \sum_{\theta=0°}^{70°} (-Gain_{(\varphi,\theta)})$$

$$st: \quad g_{VSWR}(\boldsymbol{x}) - T \arg VSWR \le \boldsymbol{e}_i^{(k)} \ i = 1, 2, \ldots, q$$

$$T \arg Gain - g_{Gain_{(\varphi,\theta)}}(\boldsymbol{x}) \le \boldsymbol{e}_i^{(k)} \ i = q+1, q+2, \ldots, n$$

$$T \arg AR - g_{AR_{(\varphi,\theta)}}(\boldsymbol{x}) \le \boldsymbol{e}_i^{(k)} i = n+1, n+2, \ldots, m$$

$$(0° \le \theta \le 70°, 0° \le \varphi \le 355°)$$

where $\boldsymbol{x}$ is the decision vector. θ and φ indicate direction in a spherical coordinate system. g_{VSWR} is the actual value of the voltage standing wave ratio (VSWR) and $T \arg VSWR$ is the maximum VSWR. $g_{Gain_{(\varphi,\theta)}}$ is actual values of antenna gain and $T \arg Gain$ is the minimum gain. $g_{AR_{(\varphi,\theta)}}$ is actual values of axial ratio (AR) and $T \arg AR$ is the maximum AR. There are m inequality constraints totally. k is the environment changing count and determines the numerical change of $\boldsymbol{e}_i^{(k)}$ (k and $\boldsymbol{e}_i^{(k)}$ are defined in $Eq(6)$ in Section 2).

5   Results and Analysis

5.1 Requirements

All antenna requirements are provided below.

Table 3. Requirements of the Single-armed Antenna

Property	Requirement
Frequency	$2.4GHz$
VSWR	1.5
Gain Pattern	$> 0dBic, 0° \le \theta \le 70°, 0° \le \varphi \le 360°$
Axial Ratio	$< 3dB, 0° \le \theta \le 70°, 0° \le \varphi \le 360°$
Polarization	Right-hand circular polarization
Input Impedance	50Ω
Volume	Cube with x-axis 50mm,y-axis 50mm,z-axis 100mm
Reflection Surface	Perfect ground plane

5.2 Parameters Setting

For dynamic multi-objective SaDE, the parameters we token are as follows:
  (1) Population Size: $NP = 100$
  (2) Max Generation: $G = 5000$
  (3) Algorithm-related parameters: $F_l = 0.1, F_u = 0.9, \tau_1 = 0.1, \tau_2 = 0.1$

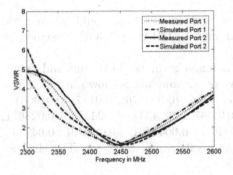

Fig. 3. VSWRs for quadrifilar helical antenna

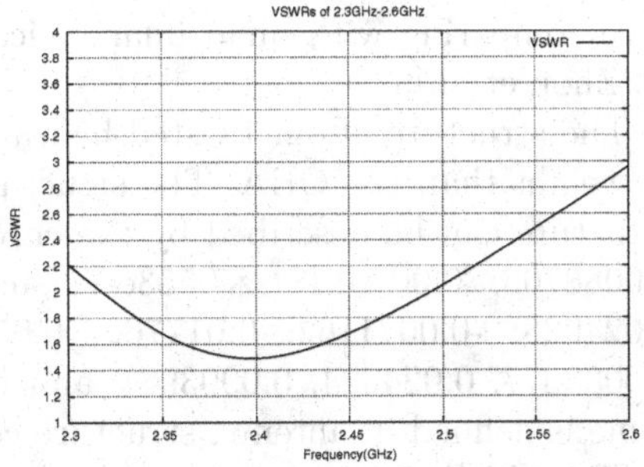

Fig. 4. VSWRs for twisted helical antenna

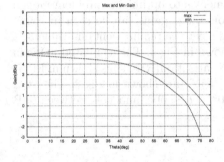

Fig. 5. Maximum and minimum gain

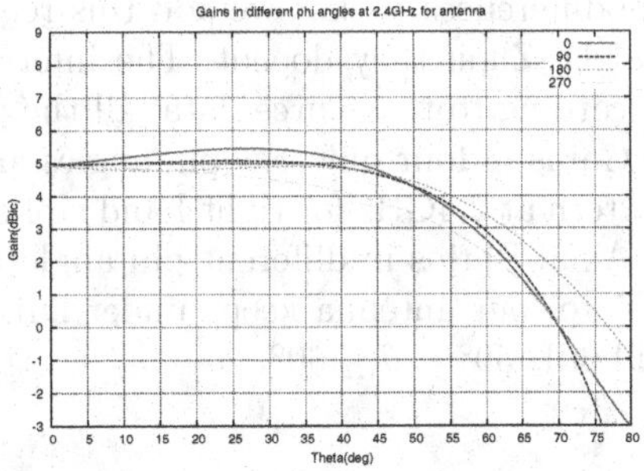

Fig. 6. Gains in different phi angles

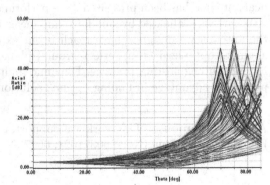

Fig. 7. Simulated 3D pattern

Fig. 8. Axial Ratios in different phi angles

5.3 Result of Experiment

We got some results after several tests, and one of them would be analyzed here comparing with quadrifilar helical antenna(QHA) at 2.4GHz designed by D. Zhou et al [3].

The structure of our twisted helical antenna is smaller in volume and lighter in weight than the QHA. The structure of our antenna is shown in Fig.2. The structure can be described by the chromosome (0.0005769, 0.0187, -0.010081, -0.0088706, 0.0046148, 8.0553e-7, -0.013587, 0.024345, 0.024456, -0.023671, 0.024988, -0.0071162, 0.017762, 0.050267, -0.0048481, -0.024999, 0.047943, 0.0083677, 0.024524, 0.090398), and these values are in meters. The chromosome is defined in antenna structure coding.

The VSWR of our antenna is lower than the VSWR of QHA around 2.4GHz. The frequency sweep of VSWRs can be seen respectively in Fig.3 and Fig.4.

The gain of QHA begins below 0dBic and becomes above 0dBic when $\theta \geq 40°$, but the gain of our antenna keeps above 0dBic in $0° \leq \theta \leq 70°$. The reason for the differences is that we add this requirement into optimizing problem model, and D. Zhou they do not. The information about gains can be seen in Fig.5. According to the source data, all the gains meet the requirements in Table 3.

Gains in four different phi angles are shown in Fig.6. The simulated 3D gain pattern at 2.4GHz for right-hand circular polarization(RHCP) is shown in Fig.7.

Axial Ratios in different phi angles($0° - 355°$) are shown in Fig.8. The axial ratio of our antenna keeps under 3dB in $0° \leq \theta \leq 50°$, and it is between 3dB and 6dB $50° \leq \theta \leq 60°$.

6 Conclusion

An efficient dynamic multi-objective SaDE algorithm, for designing a twisted helical antenna, has been presented. The performance of the best selected antenna structure was validated by simulator and measurement. The results confirm that the antenna we designed have reached our expected completely. However, the antenna ,optimized on a perfect ground plane, should be further modified to be used in engineering practice. Dynamic multi-objective SaDE algorithm has proven its advantage in finding optimal solutions for antenna designs.

Acknowledgments. This work was supported by the National Natural Science Foundation of China (No.s: 60871021, 60473037).

References

1. Altshuler, E.E., Linden, D.S.: Wire-antenna designs using genetic algorithms. IEEE Antennas Propag. Mag. 39, 33–43 (2000)
2. Jones, E.A., Joines, W.T.: Design of Yagi-Uda antennas using genetic algorithms. IEEE Trans. Antennas Propag. 45, 1386–1392 (1997)

3. Zhou, D., et al.: Quadrifilar helical antenna design for satellite-mobile handsets using genetic algorithms. Microwave and Optical Technology Letters 51, 2668–2671 (2009)
4. Lohn, J.D., Kraus, W.F., Linden, D.S.: Evolutionary Optimization of a Quadrifilar Helical Antenna. Proc. IEEE Antenna and Propagation Soc. Mtg. 3, 814–817 (2002)
5. Hornby, G.S., Lohn, J.D., Linden, D.S.: Evolutionary design of an X-band antenna for NASA space technology 5 mission. Evolutionary Computation 19(1), 1–23 (2011)
6. Abd-Alhameed, R.A., Mangoud, M., Excell, P.S., Khalil, K.: Investigations of polarization purity and specific absorption rate for two dual-band antennas for satellite-mobile handsets. IEEE Trans. Antennas Propag. 53, 2108–2110 (2005)
7. Storn, R., Price, K.V.: Differential evolution-A simple and efficient adaptive scheme for global optimization over continuous spaces. Institute of Company Secretaries of India, Chennai, Tamil Nadu. Tech.Report TR-95-012 (1995)
8. Brest, J., et al.: Self-Adapting Control Parameters in Differential Evolution: A Comparative Study on Numerical Benchmark Problems. IEEE Transactions on Evolutionary Computation 10(6), 646–657 (2006)
9. Yang, S., et al.: Evolutionary Computation in Dynamic and Uncertain Environments. SCI, vol. 51. Springer (2007)
10. Jin, Y., Branke, J.: Evolutionary optimization in uncertain environments-a survey. IEEE Transactions on Evolutionary Computation 9(3), 303–317 (2005)
11. Jin, Y., Sendhoff, B.: Constructing Dynamic Optimization Test Problems Using the Multi-objective Optimization Concept. In: Raidl, G.R., Cagnoni, S., Branke, J., Corne, D.W., Drechsler, R., Jin, Y., Johnson, C.G., Machado, P., Marchiori, E., Rothlauf, F., Smith, G.D., Squillero, G. (eds.) EvoWorkshops 2004. LNCS, vol. 3005, pp. 525–536. Springer, Heidelberg (2004)
12. Zeng, S., Chen, S., Zhao, J., Zhou, A., Li, Z., Jing, H.: Dynamic Constrained Multi-objective Model for Solving Constrained Optimization Problem. In: IEEE Congress on Evolutionary Computation, pp. 2041–2046 (2011)
13. Michalewicz, Z., Schoenauer, M.: Evolutionary Algorithms for Constrained Parameter Optimization Problems. Evol. Comput. 4(1), 1–32 (1996)
14. Jia, L., Zeng, S., et al.: Dynamic Multi-objective Differential Evolution for Solving Constrained Optimization Problem. In: IEEE Congress on Evolutionary Computation, pp. 2649–2654 (2011)
15. Coello Coello, C.A.: Theoretical and Numerical Constraint-Handling Techniques used with Evolutionary Algorithms: A Survey of the State of the Art. Computer Methods in Applied Mechanics and Engineering 191(11-12), 1245–1287 (2002)
16. Qin, A.K., Suganthan, P.N.: Self-adaptive Differential Evolution Algorithm for Numerical Optimization. In: IEEE Congress on Evolutionary Computation, vol. 2, pp. 1785–1791 (2005)

Task Scheduling
for Imaging Reconnaissance Satellites
Using Multiobjective Scatter Search Algorithm

Zilong Shen, Huanxin Zou*, and Hao Sun

School of Electronic Science and Engineering
National University of Defense Technology
Changsha 410073, P.R. China
{zlongshen,hxzou2008}@163.com,
clhaosun@gmail.com

Abstract. Through analysis of constraint characteristics and modeling of the imaging reconnaissance satellites (IRS) task scheduling problem, a novel multiobjective scatter search algorithm is proposed to find the pareto-optimal set of solutions. Three new components, including an adaptive probability mutation operator based searching strategy, a constrained-dominance comparator based on number of the constraint violations and a solution combination method based on dual crossover operators are incorporated into the standard Archive-Based hYbrid Scatter Search (AbYSS) algorithm. Experimental results demonstrate the proposed scatter search algorithm is valid and effective.

Keywords: Multiobjective Scatter Search, Multiple Satellites Task Scheduling, Imaging Reconnaissance Satellites.

1 Introduction

Imaging reconnaissance satellites are important instruments for military purpose to gather information from outer space. In order to avoid insufficiency of the reconnaissance capability within single reconnaissance satellite, it is necessary to construct an integrated system within which multiple satellites carry out the imaging task cooperatively[1]. The scheme of united mission scheduling[2] proposed by Wang Jun is an effective way to tackle the problem of multiple satellites task scheduling. In addition to the satellites task scheduling problem for point targets survey, the satellite reconnaissance task scheduling problems for area targets survey have been studied by Chongyou Wu[3].

As a population based global-optimization algorithm, scatter search, such as the M-scatter search algorithm[4], the MOSS-II algorithm[5] and the SSPMO algorithm[6], have been widely used to solve multiobjective optimization problems in the literature.

---

* Corresponding author.

Z. Li et al. (Eds.): ISICA 2012, CCIS 316, pp. 240–249, 2012.

AbYSS (Archive-Based hYbrid Scatter Search)[7] is a hybrid algorithm which enhances scatter search algorithm with randomized operators, and has been widely used in many EAs. It has also been used to deal with many practicable multiobjective optimization problems. For example, AbYSS algorithm is applied in the satellite constellation design domain[8]. In order to resolve the IRS task scheduling problem for point targets survey, improvements and modifications of the AbYSS algorithm are needed.

The rest of this paper is organized as follows. The next section gives some basic definitions and describes the mathematical model of task scheduling problem for multiple imaging reconnaissance satellites. Outlined and detailed description of the task scheduling algorithm for multiple imaging reconnaissance satellites are presented in section 3. Analysis of the experiments and validation of the task scheduling algorithm are discussed in section 4. Conclusion and future work are given in section 5.

2 Description and Modeling of Task Scheduling Problem for Multiple Imaging Reconnaissance Satellites

Task planning and scheduling technology for multiple imaging reconnaissance satellites is one of the foundations on which cooperative reconnaissance means could be guaranteed to succeed. Due to the fact that order of complexity for this new kind of problem is very high, heuristic search algorithms[9] and some related approximation algorithms, such as greedy algorithm[10], tabu search algorithm[11] and simulated annealing algorithm[12], become the dominant methods for the problems. To fully exploit the algorithm, improvements and modifications should be made to guarantee its correctness and effectiveness while considering the characteristics of this task planning and scheduling problem.

2.1 Basic Hypotheses

There is only one imaging sensor on each satellite. Different side-looking angles will always be needed when different imaging reconnaissance tasks are executed by the same satellite, thus, transition of the side-looking angle will be taken into consideration during the interval between successive imaging reconnaissance tasks. For the short-term planning and scheduling problem, revisiting event between satellites and targets will not occur, at the same time, the effectiveness of every imaging reconnaissance task in respect of time will not be lost.

Given the point targets, their precise location cannot be predicted without loss of accuracy, consequently, an approximate time interval during which a point target is illuminated by the sensor onboard will need to be estimated. Conditions of cloudiness, solar altitude and resolution of the imaging sensor will be taken into account at the stage of preprocessing. The overlap between successive imaging reconnaissance tasks will not be taken into account in our model for the sake of convenience.

2.2 Definitions of the Related Sets, Parameters and Decision Variables

Set Sat contains all the satellites involved in the cooperative imaging reconnaissance task scheduling. Set $Task$ contains the entire imaging reconnaissance tasks that will be carried out by the satellites in set Sat. Set $Task_j$ contains the entire imaging reconnaissance tasks that can be performed by the satellite with index j in set Sat. Variable $Value_j$ is the profit gained by executing the task with index j in set $Task$. Variable $Energy_j$ and $Memory_j$ are respectively the initial amount of energy and memory capacity onboard with satellite j. Variable $Time_{i,j}^k$ and $Energy_{i,j}^k$ are respectively the amount of storage and energy consumed during execution of the task with index i, by contrast, Variable and are respectively the amount of time and energy consumed during the transition process between successive imaging reconnaissance tasks performed by the same satellite with index k. Variable $Task_\alpha^s$ and $Task_\beta^s$ are respectively the first and last virtual task of the satellite with index s, both of which are with length of execution time 0. Variable $duration_j^s$ is the minimum length of execution time for the task j carried out by satellite s while $[\alpha_i^s, \beta_i^s], s \in Sat, i \in Task_s$ represents the time window resource for task i carried by satellite s while variable α_i^s is the begining of this time interval and variable β_i^s is the ending of this time interval. Two decision variable are constructed in the model of this task planning and scheduling problem. The first one $D_{i,j}^s$ is an indicator whose value is assigned true if task i and task j are two successive imaging reconnaissance tasks carried out by the same satellite s or is assigned false otherwise. The second one $Time_j^s$ is the begining of the execution time interval of the task j carried out by satellite s.

2.3 Modeling of the Imaging Reconnaissance Task Scheduling Process

The formulas that describe the objectives and constraints of this mathematical model are listed as follows:

$$Target_1 = max \sum_{s \in Sat} \sum_{i \in Task_s \cup Task_\alpha^s/\{j\}, j \in Task_s} Value_j \cdot D_{i,j}^s \qquad (1)$$

$$Target_2 = min \sum_{s \in Sat} \sum_{s \in Sat \cup Task_\alpha^s/\{j\}, j \in Task_s} (Energy_j^s + Energy_{i,j}^s) \cdot D_{i,j}^s \quad (2)$$

$$\sum_{s \in Sat} \sum_{i \in Task_s \cup Task_\alpha^s/\{j\}, j \in Task_s} D_{i,j}^s \leq 1 \qquad (3)$$

$$\sum_{j \in Task_s \cup Task_\beta^s} D_{Task_\alpha^s, j}^s = 1, \forall s \in Sat \qquad (4)$$

$$\sum_{i \in Task_s \cup Task_\alpha^s} D_{i, Task_\beta^s}^s = 1, \forall s \in Sat \qquad (5)$$

$$D_{i,j}^s(Time_i^s - \alpha_i^s) \geq 0, \forall s \in Sat, i \in Task_s \cup Task_\alpha^s/\{j\}, j \in Task_s \quad (6)$$

$$D_{i,j}^s(Time_i^s + duration_i^s - \beta_i^s) \leq 0,$$
$$\forall s \in Sat, i \in Task_s \cup Task_\alpha^s/\{j\}, j \in Task_s \quad (7)$$

$$D_{i,j}^s(Time_i^s + duration_i^s + Time_{i,j}^s - \alpha_j^s) \leq 0,$$
$$\forall s \in Sat, i \in Task_s \cup Task_\alpha^s/\{j\}, j \in Task_s \quad (8)$$

$$\sum_{i,j \in Task_s, i \neq j} D_{i,j}^s \cdot (Energy_{i,j}^s + Energy_i^s) \leq Energy_s, \forall s \in Sat \quad (9)$$

$$\sum_{i,j \in Task_s, i \neq j} D_{i,j}^s \cdot Memory_j^s \leq Memory_s, \forall s \in Sat \quad (10)$$

$$D_{i,j} \in \{0,1\}, s \in Sat, i \in Task_s \cup \{Task_\alpha^s\}, j \in Task_s \cup \{Task_\beta^s\} \quad (11)$$

Formula (1) declares the first objective of this model which maximizes overall profits of every imaging reconnaissance task which is actually executed according to the solutions of this problem. The second objective of this model described by formula (2) is to minimize overall consumption of energy onboard in order to provide adequate remaining resources for the next short-term of task planning and scheduling. Formula (3) illustrates that every task should be executed at most one time. The virtual tasks for every satellite must be executed according to Formula (4), (5). Formula (6), (7), (8) are the temporal constraints that must be complied with during transition process and observation process. Formula (9), (10) are the resources constraints that memory capacity and energy consumed along the whole process should not exceed the initial amounts of these resources onboard. Evaluation of the violations and evaluation of the multiple objectives are constructed according to the formulas above. Ranges of the related variables in the model are explained by Formula (11).

The feasible solutions obtained by the scheduling system should satisfy the requirements posed by Formulas (3) till (11). Solutions in the nondominated solution set are selected and sorted according to Formula (1), (2).

3   Design and Analysis of the Task Scheduling Algorithm for Multiple Imaging Reconnaissance Satellite

The algorithm proposed in this paper is constructed on the foundation of the AbYSS algorithm which transforms the scatter search template proposed by F. Glover[13] to an effective multiobjective optimization algorithm. However, original AbYSS algorithm[7] cannot fit the problem of task scheduling for IRS well due to the fact that this problem needs special treatment of the local constraints and global constraints as well as accurate evaluations of constraint violations. Outline of the task scheduling algorithm proposed by us is given in Fig. 1.

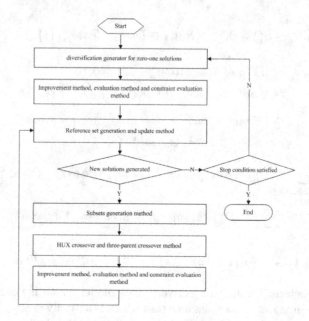

Fig. 1. Outline of the task scheduling algorithm

3.1 Genetic Encoding Scheme and Generation Method of the Initial Set

Length-fixed genetic encoding is adopted as the solution encoding scheme used in our algorithm. All the decision variables $D_{i,j}^s$ of Boolean type corresponding to the same satellite are encoded as a substring which is then linked together with substrings corresponding to other satellites to produce the final complete chromosome.

According to this encoding scheme of binary string, the first diversification generator for zero-one solutions in the template for scatter search and path relinking[13] proposed by F. Glover is adopted as the initial set of solutions generation method whose input is a vector as a seed solution to produce a collection of solutions as output. Each of the solutions in the initial set is improved by a local search method which repeatedly mutates the solutions to look for better solutions in the neighborhood.

3.2 Evaluation Method

The evaluation method of overall constraint violations adopted in original AbYSS algorithm[7] is based on the magnitude of physical quantities for different types of constraint violations which are then summed to get the overall constraint violations. However, it fits the problem we face not well due to the fact that different types of physical quantities cannot be summed together simply without considering the dimension of physical quantity. Instead, evaluation method based on number of the overall violated constraints is used as measurement of the overall constraint violations without difficulty in considering the problem of dimension.

3.3 Genetic Operators

The adaptive probability mutation operator is used in local search process to search for 'better solution' in the neighborhood that mutates every bit in an individual solution with adaptive probability. As a consequence, an individual solution is mutated with larger probability in the bit position which has a greater number of local constraint violations. It is more likely that the components with higher quality are reserved in an individual solution in comparison with other components under the adaptive probability mutation operator.

There are two subsets in the reference set, RefSet1 and RefSet2. RefSet1 is composed of solutions with high quality in the initial set while RefSet2 contains solutions promoting diversity. In order to promote diversification in RefSet1, dual crossover operators including a HUX crossover operator[14] and a three-parent crossover operator are introduced as the solution combination method in RefSet1. The HUX crossover operator generates combinations of two individuals from RefSet1 while the three-parent crossover operator combines two individuals from RefSet1 and one individual selected randomly from RefSet2 as the third parent. The three-parent crossover operator is constructed based on principles of multi-parent recombination[15] proposed by A. E. Eiben, under which each bit of the first parent from RefSet1 is checked with bit of the second parent from RefSet1 to see whether they are the same. If same then the bit is taken for the offspring solution to reserve the components with high quality, otherwise, the bit of the third parent from RefSet2 is taken for the offspring solution with the purpose of introducing components promoting diversity. Therefore, the three-parent crossover operator promotes exchanges between RefSet1 and RefSet2.

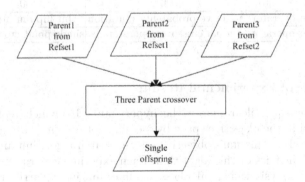

Fig. 2. Three-parent crossover

4   Experiments

Four simulation experiments are conducted to verify correctness and effectiveness of this task scheduling algorithm. There are 6 imaging reconnaissance satellites operating cooperatively in each of the simulation experiments with simulation period of 24 hours to execute 35, 72, 120, and 268 imaging reconnaissance tasks respectively. These simulation experiments are conducted on the assumption that both IRS in orbit and onboard payload and imaging sensors function well during the simulation period. Data of imaging reconnaissance satellites involve point targets with the fixed imaging reconnaissance income randomly generated in the interval [1,100].

Parameter settings of our algorithm are given as follows. The external archive contains an elite group of individuals with the maximum allowable capacity of 100. The initial set of solutions contains diverse solutions with size of 20. Both the size of RefSet1 and RefSet2 are 10. The maximum number of iterations in each local search process is 1. Base mutation probability of the adaptive probability mutation operator is 0.01. The crossover probability of HUX operator is 0.9 and the crossover probability of three-parent crossover operator is 0.99. A series of experiments using NSGA-II algorithm[17] are also conducted to solve the task scheduling problems as the scatter search algorithm. The comparisons between scatter search and NSGA-II in the simulation experiments are listed in Table 1.

These simulation experiments are constructed in programming language java of version 1.6.0_29 with integrated development environment of NetBeans IDE 7.0.1, running on the personal computer with configuration of Intel(R) Core(TM) i3-2100 CPU 3.10GHz, 4GB RAM.

Table 1. Settings for the experiments

Algorithm	Multiobjective scatter search	NSGA-II
Initial set generator	diversification generator	random generator
Solutions set size	20	100
Mutation operator	adaptive probability mutation	bit flip mutation
Crossover operator	dual crossover operators	single point crossover

4.1   Analysis of Experimental Results

Median and interquartile range of the hypervolume[16] which represents the volume covered in the objective space by members of the final elite group of solutions generated by this multiobjective scatter search algorithm are evaluated as performance metrics of this algorithm. Each experiment is running 16 times and statistical analysis which calculates median and interquartile range of all the corresponding hypervolume belonging to the same experiment is conducted afterwards. Pareto front which reflects the multiobjective characteristics of optimization algorithm is utilized as an important indicator to evaluate the results of this algorithm program.

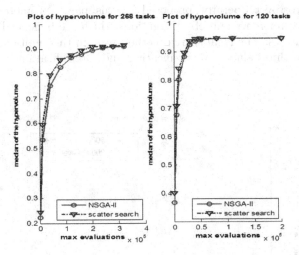

Fig. 3. Median of the hypervolume with 120 and 268 imaging reconnaissance tasks

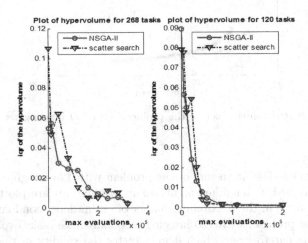

Fig. 4. IQR of the hypervolume with 120 and 268 imaging reconnaissance tasks

Median and interquartile range of the hypervolume derived by this multiobjective scatter search algorithm under different maximum numbers of evaluations with 120 and 268 imaging reconnaissance tasks are plotted in Fig. 3 and Fig. 4. With the increase in the maximum number of evaluations, both of the performance metrics converge to a relatively stable value. Compared with the NSGA-II algorithm[17], this multiobjective scatter search algorithm converges more quickly with a smaller size of population. It can be shown that more contributions in the global search process are made by diversification generators for the initial solution set of this scatter search algorithm than the random generators for the initial solution set of the NSGA-II algorithm. Furthermore, the three-parent crossover operator proposed in this paper promotes exchanges between the two reference subsets of solutions which are beneficial to generating diverse solutions. The local search method used in this multiobjective scatter search algorithm balances diversification and intensification of this search process.

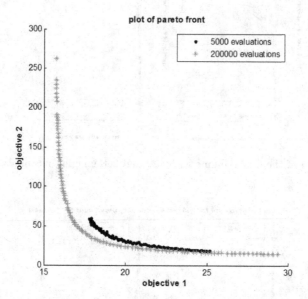

Fig. 5. Nondominated solution set of the problem with 120 imaging reconnaissance tasks

The nondominated solution sets of the problem with 120 imaging reconnaissance tasks obtained under different maximum evaluations are plotted in Fig. 5. With the increase in the maximum number of evaluations, nondominated solution set of the problem with 120 imaging reconnaissance tasks evolves in the direction of true pareto front which demonstrates the validity of this multiobjective scatter search algorithm.

5   Conclusion and Future Works

The task scheduling system for multiple imaging reconnaissance satellites is a critical part of the integrated reconnaissance system, which will be widely used in future reconnaissance and surveillance network. Based on the original AbYSS algorithm, solution encoding scheme, adaptive probability mutation operator, dual crossover operators and violation evaluation method for individual solution are constructed in our proposal. Performance of this algorithm is verified in the simulation experiment.

References

1. Morris, R.A., Dungan, J.L., Bresina, J.L.: An Information Infrastructure for Coordinating Earth Science Observations. In: Proceedings of the 2nd IEEE International Conference on Space Mission Challenges for Information Technology, Washington, D.C., pp. 397–404 (2006)
2. Wang, J.: A New Model and Algorithm of Multi-Objective United Imaging Scheduling. Ph.D. thesis, National University of Defense Technology, Changsha, China (2007) (in Chinese)
3. Wu, C.: Research on Satellite Scheduling Problem for Area Targets Survey. M.S. dissertation, National University of Defense Technology, Changsha, China (2006) (in Chinese)
4. Vasconcelos, J.A., Maciel, J.H.R.D., Parreiras, R.O.: Scatter Search Techniques Applied to Electromagnetic Problems. IEEE Trans. Magn. 4, 1804–1807 (2005)
5. Beausoleil, R.P.: "MOSS-II" Tabu/Scatter Search for Non-linear Multiobjective Optimization. In: Siarry, P., Michalewicz, Z. (eds.) Advances in Metaheuristics for Hard Optimization. Natural Computing Series, pp. 39–67. Springer, Heidelberg (2008)
6. Molina, J., Laguna, M., Martí, R., Caballero, R.: SSPMO: A Scatter Tabu Search Procedure for Non-linear Multiobjective Optimization. Inf. J. Comput. 19(1), 91–100 (2007)
7. Nebro, A.J., Luna, F., Alba, E., Dorronsoro, B., Durillo, J.J., Beham, A.: AbYSS: Adapting Scatter Search for Multiobjective Optimization. IEEE Transactions on Evolutionary Computation 12(4), 439–457 (2008)
8. Xie, B.: AbYSS Algorithm Research and Its Application on Constellation Optimization Design. M.S. dissertation, China University of Geosciences, Wuhan, China (2008) (in Chinese)
9. Bonet, B., Geffner, H.: Planning as Heuristic Search. Artificial Intelligence 129, 5–33 (2001)
10. Frank, J., Jónsson, A., Morris, R., Smith, D.E.: Planning and Scheduling for Fleets of Earth Observing Satellites. In: Proceedings of Sixth Int. Symp. on Artificial Intelligence, Robotics, Automation and Space 2002, Montreal (2002)
11. He, R.: Research on Imaging Reconnaissance Satellite Scheduling Problem. Ph.D. thesis, National University of Defense Technology, Changsha, China (2004) (in Chinese)
12. Bai, J., Sun, K., Yang, G.: Mathematical Model and Hybrid Scatter Search for Cost Driven Job-shop Scheduling Problem. Journal of Networks 6(7), 974–981 (2011)
13. Glover, F.: A Template for Scatter Search and Path Relinking. In: Hao, J.-K., Lutton, E., Ronald, E., Schoenauer, M., Snyers, D. (eds.) AE 1997. LNCS, vol. 1363, pp. 13–54. Springer, Heidelberg (1998)
14. Eshelman, L.: The CHC Adaptive Search Algorithm: How to Have Safe Search When Engaging in Nontraditional Genetic Recombination. In: Rawlings, G.J.E. (ed.) Foundations of Genetic Algorithms, pp. 265–283. Morgan Kaufmann (1991)
15. Eiben, A.E., Raué, P.-E., Ruttkay, Z.: Genetic Algorithms with Multi-Parent Recombination. In: Davidor, Y., Männer, R., Schwefel, H.-P. (eds.) PPSN 1994. LNCS, vol. 866, pp. 78–87. Springer, Heidelberg (1994)
16. Zitzler, E., Thiele, L.: Multiobjective Evolutionary Algorithms: A Comparative Case Study and the Strength Pareto Approach. IEEE Trans. Evol. Comput. 3, 257–271 (1999)
17. Deb, K., Pratap, A., Agarwal, S., Meyarivan, T.: A Fast and Elitist Multiobjective Genetic Algorithm: NSGA-II. IEEE Trans. Evol. Comput. 6, 182–197 (2002)

A Multi-objective Differential Evolutionary Algorithm Applied in Antenna Optimal Problem

Yuanyuan Fan, Qingzhong Liang, and Sanyou Zeng

School of Computer Science
China University of Geosciences, China
yyfan@cug.edu.cn

Abstract. To solve the difficulties of evolutionary antenna, a new multi-objective differential algorithm NS-TADE is proposed, which has twins-crossover and adaptive parameters and integrates with Pareto dominating sorting introduced in NSGA-II. The algorithm aims at reducing antenna simulation times meanwhile satisfying and conciliating multiple objectives. The algorithm has been applied to a practical antenna design problem. The result of the experiment shows that the problem has been solved effectively, and multiple non-inferior solution have obtained in some cases to provide more choice for users.

Keywords: Multi-objective Differential Algorithm, Evolutionary Antenna.

1 Introduction

Antennas are widely used in both civil and military use, such as broadcasting, television, satellite communication, radar and radio astronomy. The relationship between the excitation and directional characteristics is described in Maxwell's equations, which is highly non-linear, multi-modal and non-differentiable. As a consequence, soundly designing an antenna is an extremely complicated huge problem.

With the rapid development of intelligent computing and computational electromagnetism, automation and Intelligentization has become one of the mainstream research areas. More and more researchers have applied evolutionary computation onto intelligent antenna design research since 1990s[1–3]. Considering that evolutionary algorithms are efficient, simply and universal, existing evolutionary algorithms can be easily applied to evolutionary antenna[4–6].

However, it would have required special evolutionary algorithms to solve the difficulties of evolutionary antenna. Firstly, the complicated relationship between antenna structure and its direction characteristics requires evolutionary algorithms to be simpler and more efficient, so a differential evolutionary algorithm would be a good choice owing to its characteristics. Secondly, evaluating an antenna solution is computational valuable, so the algorithms should reduce the evaluation times as more as possible. Thirdly, it is common that multiple conflicted requirements should be satisfied in practical antenna designs. Therefor a

Z. Li et al. (Eds.): ISICA 2012, CCIS 316, pp. 250–257, 2012.

multi-objective differential evolutionary algorithm named as NS-TADE is proposed, which has twins-crossover and adaptive parameters, and integrates with Pareto dominating sorting introduced in NSGA-II[7]. The algorithm aims at reducing antenna simulation times meanwhile satisfying and conciliating multiple objectives. We apply this algorithm to design antennas for NASA ST-5 mission requirements[8], and successfully designed some novel antennas satisfying the requirements. The results also show that the multi-objective algorithm is able to find multiple non-dominated solutions satisfying the requirements, providing more options for users.

2   NS-TADE

On account of the evolutionary antenna's characteristics mentioned above, a multi-objective algorithm NS-TADE (Non–dominated Sorting Twins–crossover Adaptive–parameters Differential Evolution) specifically is proposed. In the algorithm some improvements have been made on the crossover, the adaptive adjustment of the control parameters and the selection. In the meantime the algorithm integrates with Pareto dominating sorting introduced in NSGA-II to deal with multiple conflicted requirements.

2.1   Crossover

Only one of the children generated by the crossover is used in the traditional differential evolutionary algorithm, which will inevitably lose some useful genetic information and make the algorithm convergence slower. So the other child is preserved in NS-TADE.

The formulas that generate the two children $v_i(t + 1)$ and $u_i(t + 1)$ is as formula 1 and 2.

$$v_{ij}(t + 1) = \begin{cases} h_{ij}(t + 1), & if(rand1_{ij} \leq CR_i) \ or \ (j = rand(i)) \\ x_{ij}(t), & otherwise \end{cases} \quad (1)$$

$$u_{ij}(t + 1) = \begin{cases} x_{ij}(t), & if(rand1_{ij} \leq CR_i) \ or \ (j = rand(i)) \\ h_{ij}(t + 1), & otherwise \end{cases} \quad (2)$$

Where t is the number of iterations; $h_{ij}(t + 1)$ denotes the component of the individual $h_i(t+1)$ generated from the parent $x_i(t)$ in the mutation; $CR_i \in [0, 1]$ is the crossover probability of $x_i(t)$; $rand1_{ij} \in [0, 1]$ and $rand(i) \in [1, D]$ are random numbers.

2.2   Parameter Adjustment

Every individual $x_i(t)$ has its own F_i and CR_i, which are adjusted adaptively in NS-TADE according to the result of the comparison with its child $x_i(t+1)$. The adjustment will happen when a individual is not placed by its child, that is, the child can not matches the parent. Otherwise both of the two parameters would be assumed to be good and maintained. The adaptive adjustment is shown as formula 3 and 4.

$$F_i(t+1) = \begin{cases} F_l + rand_1 \times F_u, \, if(x_i(t) == x_i(t+1) \text{ and } rand_2 < \tau_1)) \\ F_i(t), \hspace{3.5cm} otherwise \end{cases} \quad (3)$$

$$CR_i(t+1) = \begin{cases} rand_3, \quad if(x_i(t) == x_i(t+1) \text{ and } rand_4 < \tau_2)) \\ CR_i(t), \hspace{2.7cm} otherwise \end{cases} \quad (4)$$

Where $rand_j$ is a random decimal in [0,1], $j \in$1,2,3,4; τ_1 and τ_2 denote the probability to maintain F_i and CR_i, $\tau_1 = \tau_2 = 0.1$; $F_l = 0.1$, $F_u = 0.9$, so the scopes of F_i and CR_i are in [0.1, 1.0] and [0, 1] respectively.

2.3   Selection

Definition 1 (constrained multi-objective optimization problem)

$$Minimize \; F(x) = Minimize[f_1(x), \dots, f_m(x)]$$

$$s.t. \begin{cases} l_i \leq x_i \leq u_i \,, \quad i = 1, \dots, D \\ g_i(x) \leq 0 \,, \quad\quad j = 1, \dots, J \\ h_k(x) = 0 \,, \quad\quad k = 1, \dots, K \end{cases} \quad (5)$$

Where $f_1(x), \dots, f_m(x)$ are a set of objectives to be minimized; $[l_i, u_i]$ limits the scope of each component x_i of the solution x , D is the dimension of the problem, $g_j(x)$ is inequality constraint and $h_k(x)$ is equality constraint. Only minimization problem is discussed here for convenience.

The violations of the inequal constraints and the equal constraints are given respectively in the formula 6 and 7, while the total violation is defined as the formula 8.

$$V_j(x) = max\{0, g_j(x)\}, \; j = 1, \dots, J \quad (6)$$

$$V_{J+k}(x) = |h_k(x)|, \; k = 1, \dots, K \quad (7)$$

$$V(x) = \sum_{m=1}^{J+K} V_m(x) \quad (8)$$

To solve a constrained multi-objective optimization problem, we propose a partial order which substitute for the Pareto dominance $\prec$ in NSGA-II.

Definition 2 (constrained Pareto dominant). Let x_u and x_v are two solution of the optimization problem defined in Definition 1. If $(V(x_u) < V(x_v)$ or $V(x_u) = V(x_v)$ and $x_u \prec x_v)$, x_u is said to be constrained Pareto dominant over x_v, which is written as $x_u \prec_{pBetter} x_v$.

There are three cases that $v_i(t+1)$ competes with $x_i(t)$ in NS-TADE. If $v_i(t+1)$ dominates $x_i(t)$, $v_i(t+1)$ will substitute for $x_i(t)$ to be preserved to the next generation. $v_i(t+1)$ will be joined to the population directly if they do not dominate each other. Only when $x_i(t)$ dominates $v_i(t+1)$, $u_i(t+1)$ will be evaluated and compete with $x_i(t)$. The cases of $u_i(t+1)$ are similar to those of $v_i(t+1)$. The newly generated population may be larger than the old one, so the next population will be selected according to the constrained Pareto dominance defined above and the elitist non-dominated sorting approach proposed in NSGA-II. This selection will maintain more useful genetic information but make fewer evaluations, which could overcome the difficulty of a long time in evaluation.

2.4 Main Loop

The main loop of the algorithm NS-TADE is shown in Algorithm 1.

Algorithm 1. NS-TADE

1: t=0;
2: Initialize and evaluate P_0, $P_t = \{x_i(t), i = 1, 2, \ldots, N\}$;
3: **while** termination criteria is not satisfied **do**
4:     **for** each individual $x_i(t) \in P_t$ **do**
5:         Generate offspring individuals $v_i(t+1)$ and $u_i(t+1)$;
6:         evaluate $v_i(t+1)$;
7:         **if** $v_i(t+1) \prec_{pBetter} x_i(t)$ **then**
8:             $x_i(t+1) = v_i(t+1)$;
9:         **end if**
10:         **if** $!v_i(t+1) \prec_{pBetter} x_i(t)$ and $!x_i(t) \prec_{pBetter} v_{i+1}(t+1)$ **then**
11:             $P_t = P_t \bigcup \{v_{i+1}(t+1)\}$;
12:         **end if**
13:         **if** $x_i(t) \prec_{pBetter} v_i(t+1)$ **then**
14:             evaluate $u_i(t+1)$;
15:             **if** $u_i(t+1) \prec_{pBetter} x_i(t)$ **then**
16:                 $x_i(t+1) = u_i(t+1)$;
17:             **else if** $x_i(t) \prec_{pBetter} u_i(t+1)$ **then**
18:                 $x_i(t+1) = x_i(t)$;
19:             **else**
20:                 $P_t = P_t \bigcup \{u_{i+1}(t+1)\}$;
21:             **end if**
22:         **end if**
23:     **end for**
24:     Select P_{t+1} from P_t with $\prec_{pBetter}$ and non-dominated sorting approach of NSGA-II;
25:     $t = t + 1$;
26: **end while**
27: output the best individual

---

3   Experiments

3.1   Antenna Design Problem

The antenna optimization problem in the experiment comes from the antenna requirements in ST-5 mission of NASA, and the specification are summarized in table 1.

Table 1. Statistics of NS-TADE

Property	Specification
Transmit Frequency	8470 MHz
Receive Frequency	7209.125 MHz
VSWR	$< 1.2 : 1$ at Transmit Freq
	$< 1.5 : 1$ at Receive Freq
Gain Pattern	$\geq 0\ dBic(40° \leq \theta \leq 80°, 40° \leq \phi \leq 80°)$
Input Impedance	50Ω
Diameter	$< 15.24cm$
Height	$< 15.24cm$
Antenna Mass	$< 165g$

The gain pattern and VSWR are two optimization objectives which must be achieved simultaneously, while part of gain pattern is frequently sacrificed to lower VSWR for general engineering purpose. So the antenna design requirements can be turned into a constrained multi-objective optimization problem[9].

3.2   Experiment Setup

The experiments had been conducted on a PC with Quad–core CPU and a 2GB memory. Given the large cost of evaluation, the population size is set to 50 which is half the size of normal multi-objective evolutionary algorithms . The termination criteria are that the algorithm has run for 5000 generations, or the best individual has satisfied the design requirements. The scheme of mutation used is the DE/current to best/1/bin[10].

3.3   Results and Discussion

We performed 30 independent trial runs from different random initial populations.

The statistics of the experiments are listed in table 2. The average number of iterations is 167.167 and the average time consuming is 42.3 minutes in which the solution can be found to satisfy the requirements. But the average variance of the last generation is 0.166366441 and the algorithm fell into a local optimum for three times. The reason could be two aspects that the variance rapidly becomes smaller. One reason may be that NS-TADE has a fast convergence and prone to precocity. The other reason may be due to the small population size which may affect the population diversity.

Table 2. Statistics of NS-TADE

No.	Number of iterations	Time consuming (minute)	Variance of the last generation
01	128	29	0.0988784829316754
02	128	27	0.143774275741805
03	117	22	0.305895607356397
04	132	27	0.0874711238839937
05	111	30	0.113864624865353
06	139	32	0.0390047734499672
07	142	32	0.107990047584353
08	125	22	0.152022417693447
09	130	29	0.0431616871600022
10	137	35	0.167539308937145
11	500	207	0
12	162	35	0.0404115453694257
13	183	40	0.0754597374252975
14	450	163	0
15	96	19	0.186578071564499
16	93	18	0.139289721468905
17	75	17	0.162257164693847
18	96	20	0.150213835637656
19	147	27	0.0968404707501741
20	221	32	1.91785090098348
21	216	43	0.0635407856892287
22	147	33	0.0594548764993128
23	121	24	0.0974060019195735
24	118	21	0.0859936165426504
25	273	67	0.0464507896394217
26	73	15	0.159730866066606
27	89	19	0.175342619376329
28	410	132	0
29	129	27	0.0636400853842187
30	127	25	0.210926992066998
average	167.167	42.3	0.166366441

The specifications obtained in each run are listed in table 3. Owing to the Pareto optimal solutions adopted by NS-TADE, more than one non-inferior solution has obtained in some of the runs (in tenth, eighteenth and thirtieth run). The situation did not happen in each run, that may be due to the small size of population.

Table 3. Specification obtained

No.	VSWR at Trans-Freq	VSWR at Rec-Freq	Minimum Gain at Trans-Freq	Minimum Gain at Rec-Freq
01	1.155651689	1.478553653	0.35045257	0.375703245
02	1.148641944	1.492819667	0.06037815	1.380067348
03	1.175721645	1.494766474	0.09670328	1.761479497
04	1.187683821	1.49743402	0.02688995	1.559838414
05	1.152550817	1.262093663	0.35063630	0.248139426
06	1.191099644	1.49678123	0.00245581	1.433038116
07	1.191697478	1.485808492	0.04357645	0.041236814
08	1.156418085	1.495791912	0.21815574	1.758759379
09	1.199119806	1.494859457	0.16470265	1.720188975
10	1.194608569	1.463027358	0.00312720	0.161768034
	1.191075683	1.470767498	0.0252945	0.296198934
11	♯	♯	♯	♯
12	1.199327111	1.499386668	0.03480975	1.848783255
13	1.192058802	1.375122547	0.37450638	0.715737522
14	♯	♯	♯	♯
15	1.187982798	1.485482812	0.18965465	1.907765388
16	1.187973142	1.485945344	0.07002274	1.916835904
17	1.19959414	1.478772879	0.01253684	1.824256778
18	1.186358929	1.48028028	0.19988418	1.607199192
	1.177996159	1.496033669	0.34289497	1.729623914
19	1.185702324	1.479666352	0.05325047	1.534772873
20	1.196490288	1.4762851	0.23235147	0.877423763
21	1.199985981	1.499936342	0.00046317	0.565959036
22	1.196048737	1.495703936	0.15421803	1.855134606
23	1.1810776	1.499699473	0.09733643	1.779580593
24	1.199787617	1.496070385	0.15338507	1.709859013
25	1.197395563	1.489474058	0.00709312	0.917584777
26	1.170994043	1.487927556	0.30815193	1.704694629
27	1.199498415	1.489861369	0.32537252	1.769364119
28	♯	♯	♯	♯
29	1.188233733	1.499853969	0.01677709	1.797950268
30	1.189654827	1.478783488	0.0994084	1.200440288
	1.19189167	1.462366343	0.22456411	1.388506770
	1.191607237	1.454189777	0.31750432	1.365190387

4 Conclusion

In this work a new multi-objective algorithm NS-TADE is proposed for the difficulties of evolutionary antenna problems. In the algorithm, several improvements are made in the respects of crossover, selection and control parameters adjustment, while the algorithm integrates with Pareto dominating sorting introduced in NSGA-II.

The result of the experiment shows that with NS-TADE a practical multi-objective optimization problem of antenna design has been solved effectively. In the meantime, more than one non-inferior solution has been obtained in some cases, which can provide more choice for users.

Acknowledgement. This work was supported by the Fundamental Research Founds for National University(China University of Geosciences (Wuhan), No. CUGL120284), and National Civil Aerospace Pre-research Project.

References

1. Haupt, R.L.: Thinned arrays using genetic algorithms. IEEE Transactions on Antennas and Propagation 42, 993–999 (1994)
2. Linden, D.S.: Automated Design and Optimization of Wire Antennas using Genetic Algorithms. MIT (1997)
3. Lohn, J.D., Linden, D.S., Hornby, G.S., et al.: Evolutionary design of an X-band antenna for NASA's space technology 5 mission, pp. 155–163 (2003)
4. Jin, N., Rahmat-Samii, Y.: Advances in Particle Swarm Optimization for Antenna Designs: Real-Number, Binary, Single-Objective and Multiobjective Implementations. Antennas and Propagation 55(3), 556–567 (2007)
5. Goudos, S.K., Siakavara, K., Vafiadis, E.E., et al.: Pareto Optimal Yagi-Uda Antenna Design Using Multi-Objective Differential Evolution. Progress in Electromagnetics Research-Pier 105, 231–251 (2010)
6. Lu, J., Ireland, D., Lewis, A.: Multi-Objective Optimization in High Frequency Electromagnetics-An Effective Technique for Smart Mobile Terminal Antenna (SMTA) Design, pp. 1072–1075. IEEE, Athens (2009)
7. Deb, K., Pratap, A., Agarwal, S., et al.: A Fast and Elitist Multi-Objective Genetic Algorithm: NSGA-II. IEEE Transactions on Evolutionary Computation 6(2), 182–197 (2002)
8. Lohn, J.D., Hornby, G.S., Linden, D.S.: An Evolved Antenna for Deployment on Nasa's Space Technology 5 Mission. Genetic Programming Theory and Practice II, vol. 8, pp. 301–315. Springer US (2005)
9. Fan, Y., Liang, Q., Zeng, S.: A Study on Modeling Evolutionary Antenna Based on ST-5 Antenna and NEC2. In: Cai, Z., Hu, C., Kang, Z., Liu, Y. (eds.) ISICA 2010. LNCS, vol. 6382, pp. 474–486. Springer, Heidelberg (2010)
10. Storn, R., Price, K.: Differential evolution–a simple and efficient heuristic for global optimization over continuous spaces. Journal of Global Optimization 11(4), 341–359 (1997)

A Complete On-chip Evolvable Hardware Technique Based on Pareto Dominance

Qingzhong Liang, Yuanyuan Fan, and Sanyou Zeng

School of Computer Science
China University of Geosciences, China
qzliang@cug.edu.cn

Abstract. To increase the speed of evolvable hardware, a complete on-chip evolvable hardware technique is adopted, where both hardware evaluation and evolutionary algorithm itself are configured on chip. At the same time, a multi-objective evolutionary algorithm based on Pareto dominance is proposed to satisfy and conciliate multiple objectives in many combinational circuits design. This method is applied to the design of a 1-bit full adder and its feasibility is validated by the result of the experiment. The data of result also shows that the speed of evolvable hardware is dramatically increased.

Keywords: Intrinsic Evolvable Hardware, Multi-objective Algorithm.

1 Introduction

The human's activities to explore and transform nature gradually go to some extreme environments such as outer space, deep sea and so on. Human can not directly arrive such places because of technology or danger. In such circumstances, the circuits concerned can not designed in advance, or disability in some device will destroy the normal function of the whole hardware. Therefore, one of the most promising solutions is to use the hardware which could adaptively change its configuration and function through the interaction with the environment, i.e., evolvable hardware.

Evolvable hardware(EHW) is hardware which is built on programmable logic devices (PLD) and whose architecture can be reconfigured by using genetic learning to adapt to the new environment[1], Namely evolutionary algorithms(EAs) and PLD form the hardware and the software of evolvable hardware respectively.

Since the conception is proposed for the first time in the early 1990s[2, 3], evolvable hardware has become a focus topic in the researches on design of hardware systems. In many fields evolvable hardware has been attracted much attention and a large number of achievements have been obtained, such as in Electronic Design Automation(EDA), artificial intelligence(AI), space technology, military control, etc[4–9].

There are two difficulties that hinder the development of evolvable hardware technology, which are slow evolutionary speed and multi-objectives to be satisfied simultaneously. Therefore, we try to overcome the two difficulties from the

Z. Li et al. (Eds.): ISICA 2012, CCIS 316, pp. 258–266, 2012.

aspects of hardware and software respectively. Firstly, both hardware evaluation and evolutionary algorithm itself are configured on chip for speedup, which is a kind of complete on-chip evolvable hardware technique. Secondly, a multi-objective evolutionary algorithm based on Pareto dominance is proposed for multiple outputs or other requirements in combinational circuits design. Finally, we apply this method to design a 1-bit full adder and validate its feasibility. And the data of the experiment result shows that the speed of evolvable hardware is dramatically increased.

2   A Complete On-chip Evolvable Hardware Technique

Numerous advantages of using Field Programmable Gate Arrays(FPGAs) in evolvable hardware have been identified in many researches, including reconfiguration capability, the ability to change after launch, the potential to accommodate on-chip and off-chip failures, etc[10].

The In-System Programming(ISP) and dynamic reconfiguration technology of FPGAs is adopted in evolvable hardware, so that the configuration information of hardware is converted into chromosomes and optimized by evolutionary algorithm. With the method innovative hardware would be obtained to meet the design requirement which always break through the conventional circuit model. Furthermore the hardware could keep updating its internal structure automatically to adapt to a changing environment.

Evolvable hardware can be divided into extrinsic and intrinsic EHW according to the evaluation methods[11]. Extrinsic EHW, also known as off-line EHW, is evaluated by means of software simulation. Its advantage is flexible and not confined to specific hardware platforms, while the disadvantage lies in the computational cost and deviations. Intrinsic EHW, also known as on-line EHW, is evaluated based on real-time configurations and behaviors of practical circuits. The evaluation results are reliable and the evaluation speed is fast. Therefore, intrinsic EHW represents the development trend of evolvable hardware.

To realise the complete on-chip evolvable hardware, both the evolutionary algorithm and the hardware individuals should be configure onto the hardware. The structure is shown as figure 1. The evolutionary algorithm(EA) pipeline module is used to implement the evolution process. The evolving design module

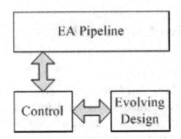

Fig. 1. The structure for complete on-chip evolvable hardware

represents the phenotype of the hardware individuals.The two modules transfers data through the control module, and the process is shown as figure 2.

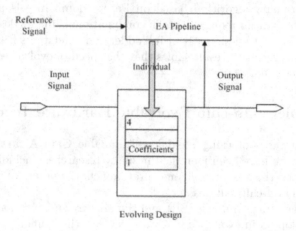

Fig. 2. The process of data transfer

The chromosomes are decoded into the configuration information of FP-GAs(phenotypes) and transferred to the evolving design module. The config-uration information is downloaded onto FPGAs to configure circuits and the output signals are fed back to the EA pipeline module. Then the hardware indi-viduals are evaluated according to the output signal and evolved adaptively in the EA pipeline module.

Handle-C is a kind of hardware design programming language proposed by Celoxica which originates from ISO/ANSI-C[12, 13], so that with it we can easily reuse most of classic algorithms coded in C language. With the development toolkit DK offered by Celoxica, the source codes in Handle-c can be compiled into netlists which directly target FPGAs. And then the netlists will be finally download onto hardware with the routing tool of FPGAs. By the thought that hardware is designed using software methods, evolutionary algorithm on-chip is easier to implement.So in our work the evolutionary algorithm was programmed with Handle-C language, and then simulated and debugged in DK3.

3   A Multi-objective EA for EHW

3.1   Encoding

The encoding method of Cartesian Genetic Programming(CGP) is adopted[14], in which the circuits array is made of several logic cells. One of examples is shown in figure 3.The external input signals are labeled by 0-2. The rectangles labeled by 3-11 represents logic cells whose functions is to be determined. There are two outputs of the whole circuit.

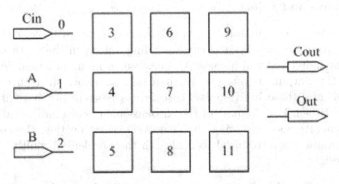

Fig. 3. The logic cell array of CGP encoding

A logic gate or a multiplexer both with two input signals and one output signal can be implemented in a logic cell. A triple code $(A_i\ B_i\ C_i)$ is adopted to represent the logic cell i. Where A_i and B_i indicate the first and second input signal source respectively. C_i identifies the function of the logic cell, which represents one of logic function shown in table 1 if it is negative, or a multiplexer if positive. Four basic gates are used here:-1 for AND, -2 for XOR, -3 for OR, -4 and -5 for NOT on first and second input respectively.

Table 1. The Encoding for Gate Function

Gate Value	Gate Function
-1	$A\&B$
-2	$A \oplus B$
-3	$A \mid B$
-4	$!A$
-5	$!B$

If the code of a logic cell is (2 8 -1), for example, it means that its logic function is AND and its first input comes from the external input signal 2 and the second input from the output of logic cell 8. Another example is (2 8 1), which means that its logic function is multiplexer, its input comes from the same sources, and its select signal comes from external input 1.

The code of the whole circuit also include the code for the circuit outputs in addition to the triple codes for every gate. So a total code for a individual is like:1 2 0 0 2 -2 1 0 -2 0 5 5 2 5 -2 2 0 5 6 7 5 2 0 -1 7 6 5 7 8. The first twenty-seven digits describe the nine gates of the circuit, and the last two digits indicate where the two outputs come from.

3.2   Evaluate and Select

Normally evaluation of digital circuits is based on its truth table, that is if an output of the circuit matches the corresponding output in the truth table, the fitness of the individual will increase by one. But in many practical design, the number of the outputs is often more than one, therefore such an evaluation method will miss those individuals whose fitness is small but part of outputs have been satisfied. Usually part of the chromosome of such individuals is good and may generate good offsprings by crossover. Based on this, the concept of Pareto domination is introduced to deal with the problem of multiple outputs in circuit designs.

Definition 1 (Feasible Region). *Let the circuit codes mentioned above is the decision variables of circuit design problem and its dimension is D. All of the feasible codes of the circuit would form the feasible region of the circuit design problem, shown as formula 1.*

$$\Omega = \{x \in Z^D | l_i \le x_i \le u_i, i = 1, 2, \ldots, D\} \tag{1}$$

Definition 2 (Multi-objective Optimization Problem). *Take the m outputs of the circuit as m objectives to be optimized. For every output, the number of the its values matching the truth table is the value of its corresponding objective function. Therefore, the circuit design problem has been turn to a multi-objective optimization problem:*

$$MaximizeF(x) = Maxmize[f_1(x), f_2(x), \ldots, f_m(x)], x \in \Omega \tag{2}$$

It will take more hardware resources to save and lookup the values with the traditional method of truth table. So the Bit operation in Handle-C language is adopted on the unique characteristics of hardware. The values of the objective functions are obtained by operate directly on the bits of outputs. With the method much of hardware resource would be conserved.

Definition 3 (Pareto Dominance). *Let $x_u \in \Omega$ and $x_v \in \Omega$ are two of decision variables. x_u is Pareto dominant over x_v, recorded as $x_u \succ x_v$, if and only if:*
a)$\forall i \in \{1, 2, \ldots, n\}, f_i(x_u \ge f_i(x_v))$; and
b)$\exists j \in \{1, 2, \ldots, n\}, f_j(x_u > f_j(x_v))$.

If there is no decision variable Pareto dominant over x_u, it is called non-inferior solution.

The evaluation of the multi-objective evolutionary algorithm proposed here is based on the concept of Pareto Dominance defined in Definition 3. Take a 1-bit full adder for example. There are three input signals, which amount to eight input combinations. While there are two output signals which are summary and carry respectively. So the scope of every objective function is $[0, 8]$. According to Definition 3, we have: $(3, 3) = (3, 3)$, $(7, 4) \succ (4, 4)$, $(4, 5) \prec (7, 5)$, $(2, 5) \not\succ (6, 3)$ and $(2, 5) \not\prec (6, 3)$.

The elitist non-dominated sorting approach from NSGA-II[15] is adopted as the selection policy in our evolutionary algorithm. The selection operator creates a mating pool by combining the parent and offspring populations and selects the best N solutions(N is the size of population) according to the relation of Pareto Dominance between the individuals.

3.3 Crossover and Mutation

With the triple encoding approach, the genes in every position have the same meanings. So neither the point crossover nor uniform crossover will result in an error offspring. And the mutation on any gene position will not result error either when the new gene value is selected in its feasible range. The uniform crossover and uniform mutation is used here.

3.4 Main Loop

The main loop of the algorithm NS-TADE is shown in Algorithm 1.

Algorithm 1. NS-TADE

1: t=0;
2: Initialize and evaluate P_0, $P_t = \{x_i(t), i = 1, 2, \ldots, N\}$;
3: **while** termination criteria is not satisfied **do**
4:    Generate offsprings Q_t by uniform crossover and uniform mutation;
5:    Evaluate Q_t;
6:    Select P_{t+1} from $P_t \bigcup Q_t$ with the non-dominated sorting approach of NSGA-II;

7:    $t = t + 1$;
8: **end while**
9: output the best individual

---

4   Experiments

To validate the feasibility of the complete on-chip evolvable hardware technique based on Pareto Dominance, a 1-bit full adder is designed. The population size is set to 30. The termination criteria are that the algorithm has run for 500 generations.

The first set of experiments were conducted on a Celoxica RC203 board on which both evolutionary algorithm and evolving designs(1-bit full adders) are integrated. One of the evolutionary result is shown as figure 4, which is output through the monitor of RC203.

The three individuals shown in the figure 4 have met the design requirements, that is for every input condition the summary and the carry match the corresponding values in the truth table. In the first line 500 represents the number of

Fig. 4. One of the evolutionary result through the monitor of RC203

iterations. The following lines are information of three individuals: 3 means the number of gates used in the circuit; 8 8 means the fitness of the two objective function; and last twenty-nine digits represent the structure of the circuit which is represented by 8-bit complement. The whole circuit matrix includes three external inputs(the input carry C_{in}, the augend A, the addend B), two external outputs(the summary Sum and the output carry C_{out}), and 3×3 logic cells whose function is to be determined shown as figure 3.

In the case of the first adder, it can be seen from the codes that only three gates (No. 5, No.7, No.8) were used in the whole circuit. The codes for the three gates are (1 0 254)(2 5 254)(2 0 5). 254 is the 8-bit complement of -2, represents the exclusive-or gate. So the circuit is made of two exclusive-or gates and one multiplexer, shown as figure 5.

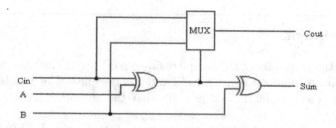

Fig. 5. The circuit diagram of a 1-bit full adder

At the same time the evolutionary optimization of the adder was conducted on a general-purpose computer with 2.4GHz Pentium 4 processor and 512M memory in extrinsic way, that is all the hardware individuals are evaluated by

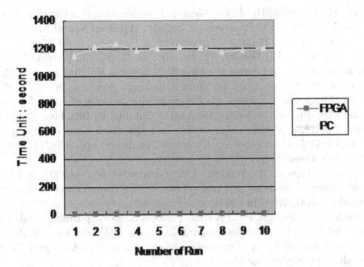

Fig. 6. The comparison of the time consuming

software simulation based on truth table. We performed ten independent trial runs in the two ways respectively and compare the time consuming. The results is shown in figure 6.

As can be seen from the figure 6, it takes an average of twenty-two minutes in extrinsic way to evolve an adder while it takes only an average of six seconds in intrinsic way, whose speed has been increased almost 200 times.

5   Conclusion

In this work a complete on-chip evolvable hardware technique based on Pareto Dominance is proposed to overcome the difficulties in increasing evolutionary speed and handling multi-objectives. The results of the experiment have proven the technique's feasibility and shows an important advantage for its speed. If the objective function is redesigned, the technique can be easily applied to other related evolved hardware areas.

Acknowledgement. This work was supported by the Fundamental Research Founds for National University(China University of Geosciences (Wuhan), No.CUGL120289), and National Civil Aerospace Pre-research Project.

References

1. Higuchi, T., et al.: Evolvable hardware with genetic learning. In: 1996 IEEE International Symposium on Circuits and Systems, ISCAS 1996, Connecting the World, May 12-15, vol. 4, pp. 29–32 (1996)

2. de Garis, H.: Evolvable Hardware: Genetic Programming of a Darwin Machine. In: Albrecht, R.F., Reeves, C.R., Steele, N.C. (eds.) Artificial Neural Nets and Genetic Algorithms. Springer (1993)
3. Higuchi, T., Niwa, T., Tanaka, T., Iba, H., de Garis, H., Furuya, T.: Evolving Hardware with Genetic Learning: a First Step Towards Building a Darwin Machine. In: Proc. of the 2nd Int. Conf. on From Animals to Animats, pp. 417–424 (1993)
4. Koza, J., Al-Sakran, S., Jones, L.: Cross-domain features of runs of genetic programming Used to evolve designs for Analog Circuits, Optical Lens Systems. Controllers, Antennas, Mechanical Systems, and Quantum Computing Circuits. In: 2005 NASA/DoD Conference on Evolvable Hardware, pp. 205–214. IEEE Computer Society Press (2005)
5. Hornby, G.S., Lipson, H., Pollack, J.B.: Generative representations for the automatic design of modular physical robots. IEEE Transactions on Robotics and Automation 19(4), 703–719 (2003)
6. Bergmann, N.W., Sutton, P.R.: A High-Performance Computing Module for a Low Earth Orbit Satellite using Reconfigurable Logic. In: Proceedings of Military and Aerospace Applications of Programmable Devices and Technologies Conference, Greenbelt, MD, September 15-16 (1998)
7. Wells, E.B., Loo, S.M.: On the Use of Distributed Reconfigurable Hardware in Launch Control Avionics. In: Proceedings of Digital Avionics Systems Conference, TBD day/month, TBD location (2001)
8. Lohn, J.D., Linden, D.S., Hornby, G., Kraus, W.F., Rodriguez, A., Seufert, S.: Evolutionary Design of an X-Band Antenna for NASA's Space Technology 5 Mission. In: Proc. 2004 IEEE Antenna and Propagation Society International Symposium and USNC/URSI National Radio Science Meeting, vol. 3, pp. 2313–2316 (2004)
9. Lin, Y.-H., Chen, J.-C.: A Hardware Design of Neuromolecular Network with Enhanced Evolvability: A Bioinspired Approach. Journal of Electrical and Computer Engineering 2012, Article ID 278735, 11 pages (2012)
10. Lambert, C., Kalganova, T., Stomeo, E.: FPGA-based systems for evolvable hardware. In: Proc. of World Academy of Science, Engineering and Technology, Vienna, Austria, pp. 123–129 (2006)
11. Gordon, T.G.W., Bentley, P.J.: On evolvable hardware. STUDFUZZ, vol. 101, pp. 279–323 (2002)
12. Guo, Z., Buyukkurt, B., Najjar, W., Vissers, K.: Optimized generation of data-path from C codes for FPGAs. In: Proceedings of the Conference on Design, Automation and Test in Europe, vol. 1, pp. 112–117 (2005)
13. Handel-C Language Overview. Celoxica, Inc. (2004), http://www.celoxica.com
14. Miller, J.F., Harding, S.L.: Cartesian Genetic Programming. In: Proceedings of the 11th Annual Conference Companion on Genetic and Evolutionary Computation Conference, pp. 3489–3512 (2009)
15. Deb, K., Pratap, A., Agarwal, S., et al.: A Fast and Elitist Multi-Objective Genetic Algorithm: NSGA-II. IEEE Transactions on Evolutionary Computation 6(2), 182–197 (2002)

An Effective Particle Swarm Optimization for Global Optimization

Mahdiyeh Eslami[1,*], Hussain Shareef[1],
Mohammad Khajehzadeh[2], and Azah Mohamed[1]

[1] Electrical, Electronic and Systems Engineering Department
National University of Malaysia
Selangor, Malaysia
m.eslami@eng.ukm.my
[2] Civil Engineering Department, Anar Branch,
Islamic Azad University, Anar, Iran

Abstract. In this paper, a novel chaotic particle swarm optimization with nonlinear time varying acceleration coefficient is introduced. The proposed modified particle swarm optimization algorithm (MPSO) greatly elevates global and local search abilities and overcomes the premature convergence of the original algorithm. This study aims to investigate the performance of the new algorithm, as an effective global optimization method, on a suite of some well-known benchmark functions and provides comparisons with the standard version of the algorithm. The simulated results illustrate that the proposed MPSO has the potential to converge faster, while improving the quality of solution. Experimental results confirm superior performance of the new method compared with standard PSO.

Keywords: Particle Swarm Optimization, Chaotic Sequence, Nonlinear Acceleration Coefficient, Global Optimization.

1 Introduction

Particle swarm optimization (PSO) is a subset of the heuristic algorithm has been developed and successfully applied to a number of benchmarks and real-world optimization problems[1]. Compared with other heuristic algorithms, PSO has less parameter to be adjusted. It has the ability to escape from local minima, it is easy for computer implementation and coding and it has more effective memory capability. Although the standard version of PSO (SPSO) has many advantages and several attractive features, it is also observed that this technique does not always perform as per expectations, and they will smoothly slip into the local near-optimal solutions when the optimization problem is relatively complex, and it cannot jump over the obstruction. Recently, several investigations have been undertaken to improve the performance of the SPSO[2-16]. A survey of the art on the different version of the PSO has been presented by the author

---

* Corresponding author.

Z. Li et al. (Eds.): ISICA 2012, CCIS 316, pp. 267–274, 2012.
© Springer-Verlag Berlin Heidelberg 2012

in[17]. Hence, in this paper, research is concentrated on the development of a modified PSO algorithm (MPSO) with the purpose of enhancing the feasibility, solution quality, and solution speed. In this paper, a novel chaotic particle swarm optimization with nonlinear time varying acceleration coefficient is introduced. The proposed MPSO greatly elevates global and local search abilities and overcomes the premature convergence of the original algorithm. In this study, the performance of the new algorithm, as an effective global optimization method, is investigated on a suite of some well-known benchmark functions. The simulated results illustrate that the proposed MPSO has the potential to converge faster, while improving the quality of solution.

2   Particle Swarm Optimization (PSO)

In a PSO system, multiple candidate solutions coexist and collaborate simultaneously. Each solution called a particle, flies in the problem search space looking for the optimal position to land. A particle, during the generations, adjusts its position according to its own experience as well as the experience of neighboring particles. A particle status on the search space is characterized by two factors: its position (X_i) and velocity (V_i). The new velocity and position of the particle will be updated according to the following equations[18]:

$$V_i(k+1) = (w \times V_i(k) + c_1 \times Rand(.) \times [p_i(k) - X_i(k)]$$
$$+ c_2 \times rand(.) \times [p_g(k) - X_i(k)]) \tag{1}$$

$$X_i(k+1) = X_i(k) + V_i(k+1) \tag{2}$$

where w is an inertia weight and is a scaling factor controlling the influence of the old velocity on the new one. c_1 and c_2 are two positive constants known as cognitive and social coefficients, explaining the weight of the acceleration terms that guide each particle toward the individual best (pbest) and the swarm best positions (gbest), respectively. Rand(.) and rand(.) are two independent random numbers selected in each step according to a uniform distribution in a given interval [0, 1]. The inertia weight, w, is usually evaluated by[18]:

$$w = w_{max} - \frac{w_{max} - w_{min}}{k_{max}} k \tag{3}$$

where w_{max} and w_{min} are maximum and minimum value of w, k_{max} is the maximum number of iteration and k is the current iteration number.

3   Modified Particle Swarm Optimization

As a member of stochastic search algorithms, PSO has a major drawback. Although PSO constitutes a huge success and converges to an optimum much faster than other evolutionary algorithms, it usually cannot improve the quality of the solutions as the number of iterations is increased[19]. PSO usually suffers from premature convergence in the early stages of the search and henceforth, it is unable to locate the global optimum, especially when high multimodal problems are being optimized. In the current research, to enhance the performance, prevent the premature convergence, and provide a good balance between the global exploration and local exploitation abilities of the original algorithm, a chaotic PSO with nonlinear time-varying acceleration coefficients is introduced. The proposed MPSO introduces the application of chaotic sequences to improve the original algorithms global seeking ability and prevent the early convergence to local minima. One of the dynamic systems showing a chaotic manner is a logistic map whose equation is described as follows:

$$\theta[k+1] = \mu\theta[k](1 - \theta[k]), \quad 0 \le \theta[1] \le 1 \tag{4}$$

where μ is a control parameter and has a real value in the range of $[0, 4]$, and k is the iteration number. The behavior of the system represented by Eq. (4) is greatly changed with the variation of μ. The value of μ determines whether stabilizes at a constant size, oscillates within limited bounds, or behaves chaotically in an unpredictable pattern. Equation (4) displays chaotic dynamics when $\mu=4.0$ and $\theta[1] \notin \{0, 0.25, 0.5, 0.75, 1\}$. The new equation for inertia weight obtained by multiplying Eq. (4) and Eq. (3) reads as follows:

$$w = \theta \times (w_{max} - \frac{w_{max} - w_{min}}{k_{max}}k) \tag{5}$$

Although the conventional inertia weight decreases monotonously from w_{max} to w_{min} , the new inertia weight decreases and oscillates simultaneously for total iteration when $\mu = 4.0$ and $\theta[1] = 0.55$. In addition, to further balance between global exploration and local exploitation abilities, nonlinear time-varying acceleration coefficients are introduced. In this approach the acceleration coefficients change according to the following equations:

$$c_1 = (c_{1i} - c_{1f}) \times exp[-(4 \times \frac{k}{k_{max}})^2] + c_{1f} \tag{6}$$

$$c_2 = (c_{2i} - c_{2f}) \times exp[-(4 \times \frac{k}{k_{max}})^2] + c_{2f} \tag{7}$$

where c_{1i} and c_{2i} are the initial values of the acceleration coefficient c_1 and c_2, and c_{1f} and c_{2f} are the final values of the acceleration coefficient c_1 and c_2, respectively. In the nonlinear time-varying acceleration coefficients strategy, the cognitive coefficient (c_1) is nonlinearly decreased during the course of run, however, the social coefficient (c_2) is nonlinear and it is inversely increased. The MPSO provides a larger value for the cognitive component and a smaller value for the social component at the beginning of the optimization procedure, which allow particles to move around the search space instead of moving toward the population best (*pbest*). In the later part of the optimization, the MPSO provides a smaller cognitive component and a larger social component, which allow the particles to converge to the global optimum. Consequently, the new velocity update equation for the proposed MPSO can be expressed as follows:

$$V_i[k+1] = \theta \times (w_{max} - \frac{w_{max} - w_{min}}{k_{max}} \times k) \times V_i[k]$$

$$+ ((c_{1i} - c_{1f}) \times exp[-(4 \times \frac{k}{k_{max}})^2] + c_{1f}) \times Rand(.) \times (pbest_i[k] - X_i[k])$$

$$+ ((c_{2i} - c_{2f}) \times exp[-(4 \times \frac{k}{k_{max}})^2] + c_{2f}) \times Rand(.) \times (gbest[k] - X_i[k])$$

$$(8)$$

The flowchart of the proposed MPSO algorithm is given in Fig. 1.

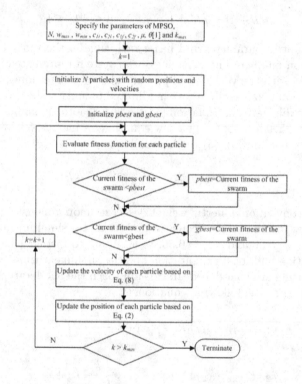

Fig. 1. Flowchart of MPSO algorithm

4   Models Verification

This section investigates the efficiency and robustness of the proposed MPSO algorithm for numerical optimization. To demonstrate, compare, and analyze the effectiveness and performance of the MPSO a set of five well-known standard benchmark functions are considered as follows:

$$F_1(x) = \sum_{i=1}^{n} x_i^2 \tag{9}$$

$$F_2(x) = \sum_{i=1}^{n} |x_i| + \prod_{i=1}^{n} |x_i| \tag{10}$$

$$F_3(x) = \sum_{i=1}^{n} ([x_i + 0.5])^2 \tag{11}$$

$$F_4(x) = -20exp\left(-0.2\sqrt{\frac{1}{n}\sum_{i=1}^{n} x_i^2}\right) -$$
$$exp\left(\frac{1}{n}\sum_{i=1}^{n} cos2\pi x_i\right) + 20 + e \tag{12}$$

$$F_5(x) = \frac{\pi}{n}\left\{10sin^2(\pi y_1) + \sum_{i=1}^{n}(y_i - 1)^2[1 + 10sin^2(3\pi y_{i+1})] + (y_n - 1)^2\right\}$$
$$+ \sum_{i=1}^{n} u(x_i, 10, 100, 4)$$
$$y_i = 1 + \frac{x_i + 1}{4} \tag{13}$$

These functions are chosen with regard to their particularities, from one simple function with a single minimum to one having a considerable number of local minima with similar fitness values. The dimension, the admissible range of the variable, and the optimum to be obtained for each function are given in Table 1.

To solve these optimization problems, the parameters of each algorithm are selected on the basis of several experimental studies that examine the effect of each parameter on the final solution and convergence of the algorithms. In order to get the optimum performance, N, k_{max}, w_{max}, w_{min}, c_1, c_2, c_{1i}, c_{2i}, c_{1f}, c_{2f}, $\theta[1]$ and μ are selected as 50, 2000, 0.9, 0.5, 2, 2, 2.5, 1.5, 1.5, 2.5, 0.55 and 4, respectively. The minimization results are summarized in Table 2. For each method, the worst, mean, median, best and standard deviations are calculated from the 30 simulated runs for comparison.

Table 1. Parameters of benchmark functions

Function	Name	Dimension(n)	Range	Optimum
F_1	Sphere	30	$[-100, 100]^n$	0
F_2	Schwefel P2.22	30	$[-10, 10]^n$	0
F_3	Step	30	$[-100, 100]^n$	0
F_4	Ackley	30	$[-32, 32]^n$	0
F_5	Generalized Penalized	30	$[-50, 50]^n$	0

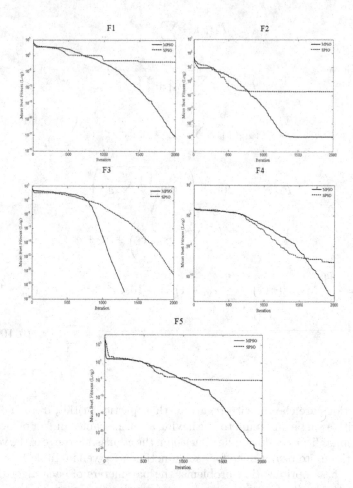

Fig. 2. Performance comparison of MPSO and SPSO for minimization of benchmark functions

Table 2. Result comparison on LN data

Function	Method	Worst	Mean	Median	Best	Standard Deviation
F_1	MPSO	$2.02e - 22$	$3.41e - 23$	$1.83e - 25$	$1.13e - 27$	$8.26e - 23$
	SPSO	0.0136	0.0073	0.0044	0.0038	0.0055
F_2	MPSO	$1.93e - 15$	$7.80e - 16$	$3.74e - 16$	$2.90e - 17$	$1.02e - 15$
	SPSO	2.4874	1.7497	1.5864	1.1825	0.5109
F_3	MPSO	0	0	0	0	0
	SPSO	$8.82e - 23$	$3.67e - 23$	$2.22e - 23$	$4.77e - 27$	$4.30e - 23$
F_4	MPSO	$7.32e - 13$	$2.96e - 13$	$2.03e - 13$	$6.48e - 14$	$2.57e - 13$
	SPSO	0.0029	$5.84e - 4$	$5.60e - 6$	$2.60e - 8$	0.0013
F_5	MPSO	$3.19e - 24$	$5.47e - 25$	$2.82e - 27$	$3.24e - 29$	$1.29e - 24$
	SPSO	0.1052	0.0217	$3.56e - 4$	$1.21e - 5$	0.0467

The results shown in Table 2 are very encouraging. In the numerical optimization problems, the mean and the best fitness values evaluated by the proposed MPSO for all functions are much lower than those computed by the SPSO. In addition, the standard deviation values of the MPSO method are less than those of the SPSO, illustrating the greater stability of the new method. Figure 2 demonstrates the convergence rate comparison between the proposed MPSO and SPSO algorithms on the five benchmark functions. In this figure, the vertical coordinates to indicate the mean best fitness value in the form of the logarithmic value.

As shown in Fig. 2, the varying curves of the fitness value using the MPSO descend much faster than those using the SPSO, and the fitness values descend to lower levels when the MPSO is used. The test results show that the convergence and the stability rates of the MPSO significantly surpass those of the SPSO on all the selected test functions.

5   Conclusion

In this paper, a modified optimization algorithm based on PSO has been developed. In the MPSO, the application of chaotic sequences combined with the nonlinear time-varying acceleration coefficient has been introduced. The performance comparison of the proposed algorithm has been examined on five well-known benchmark functions. Compared with the standard (original) algorithm, MPSO has been verified to possess excellent performance in terms of accuracy, convergence rate, stability, and robustness.

Acknowledgement. The authors are grateful to University Kebangsaan Malaysia (UKM) for supporting this study under grant UKM-DLP-2011-059.

References

1. Kennedy, J., Eberhart, R.: Particle swarm optimization. In: Proceeding of IEEE International Conference on Neural Networks, Perth, Australia, vol. 4, pp. 1942–1948 (1995)
2. Eslami, M., Shareef, H., Mohamed, A., Ghoshal, S.: Tuning of power system stabilizers using particle swarm optimization with passive congregation. Int. J. Phys. Sci. 5, 2574–2589 (2010)
3. Eslami, M., Shareef, H., Mohamed, A.: Power system stabilizer design using hybrid multi-objective particle swarm optimization with chaos. J. Cent. South Univ. Technol. 18, 1579–1588 (2011)
4. Eslami, M., Shareef, H., Mohamed, A.: Optimal Tuning of Power System Stabilizers Using Modified Particle Swarm Optimization. In: 14th International Middle East Power Systems Conference (MEPCON 2010), Cairo University, Egypt, pp. 386–391 (2010)
5. Eslami, M., Shareef, H., Mohamed, A., Khajehzadeh, M.: A hybrid PSO technique for damping electro-mechanical oscillations in large power system. In: IEEE Student Conference on Research and Development (SCOReD), Kuala Lumpur, Malaysia, pp. 442–447 (2010)

6. Eslami, M., Shareef, H., Mohamed, A., Khajehzadeh, M.: Coordinated design of PSS and SVC Damping Controller Using CPSO. In: 5th International Power Engineering and Optimization Conference (PEOCO), Malaysia, pp. 11–16 (2011)

7. Eslami, M., Shareef, H., Mohamed, A., Khajehzadeh, M.: Optimal location of PSS Using Improved PSO with chaotic sequence. In: 1st International Conference on Electrical, Control and Computer Engineering (InECCE), Malaysia, pp. 253–258 (2011)

8. Eslami, M., Shareef, H., Mohamed, A.: Optimization and coordination of damping controls for optimal oscillations damping in multi-machine power system. Int. Rev. Electr. Eng. 6, 1984–1993 (2011)

9. Eslami, M., Shareef, H., Mohamed, A.: Particle swarm optimization for simultaneous tuning of static var compensator and power system stabilizer. Prz. Elektrotechniczny (Electr. Rev.) 87, 343–347 (2011)

10. Eslami, M., Shareef, H., Mohamed, A., Khajehzadeh, M.: Improved particle swarm optimization with disturbance term for multi-machine power system stabilizer design. Aust. J. Basic Appl. Sci. 4, 5768–5779 (2010)

11. Eslami, M., Shareef, H., Mohamed, A., Khajehzadeh, M.: Design of UPFC Damping Controller using Modified Particle Swarm Optimization. Lecture Notes in Information Technology, vol. 13, pp. 441–447 (2012)

12. Khajehzadeh, M., Taha, M., El-Shafie, A.: Modified particle swarm optimization for probabilistic slope stability analysis. Int. J. Phys. Sci. 5, 2248–2258 (2010)

13. Khajehzadeh, M., Taha, M., El-Shafie, A., Eslami, M.: Modified particle swarm optimization for optimum design of spread footing and retaining wall. J. Zhejiang Univ-Sci. A (Appl. Phys. and Eng.) 12, 415–427 (2011)

14. Khajehzadeh, M., Taha, M., El-Shafie, A.: Reliability analysis of earth slopes using hybrid chaotic particle swarm optimization. J. Cent. South Univ. Technol. 18, 1626–1637 (2011)

15. Khajehzadeh, M., Taha, M., El-Shafie, A., Eslami, M.: Locating the general failure surface of earth slope using particle swarm optimization. Civil Eng. Environ. Syst. 29, 41–57 (2012)

16. Khajehzadeh, M., Taha, M., El-Shafie, A., Eslami, M.: Economic design of retaining wall using particle swarm optimization with passive congregation. Aust. J. Basic Appl. Sci. 4, 5500–5507 (2010)

17. Eslami, M., Shareef, H., Khajehzadeh, M., Mohamed, A.: A Survey of the State of the Art in Particle Swarm Optimization. Res. J. Appl. S. Eng. Technol. 4, 1181–1197 (2012)

18. Shi, Y., Eberhart, R.: A modified particle swarm optimizer. In: IEEE World Congress on Evolutionary Computation, pp. 69–73 (1998)

19. Angeline, P.J.: Evolutionary Optimization Versus Particle Swarm Optimization: Philosophy and Performance Differences. In: Porto, V.W., Waagen, D. (eds.) EP 1998. LNCS, vol. 1447, pp. 601–610. Springer, Heidelberg (1998)

Double Diffusive Natural Convection in Hydrothermal Systems: Numerical Simulation by Lattice Boltzmann Method

Wei Qiang[1,*] and Hui Cao[2]

[1] School of Computer Science
China University of Geosciences
Wuhan 430074, China
qw@cug.edu.cn
[2] Center of Information and Laboratory
China University of Geosciences
Wuhan 430074, China

Abstract. The double diffusive natural convection of ore-forming fluid in hydrothermal system is numerically simulated in two dimensions by lattice Boltzmann method (LBM). A buoyancy force proportional to the temperature and concentration is applied to the fluid in a rectangular enclosures according to the Boussinesq approximation. The non-equilibrium extrapolation boundary condition is employed in the horizontal direction, while the isothermal and isoconcentrations boundary conditions are imposed in the vertical direction. The temperature field and concentration field are passively advected by the fluid flow, which obey the same equations as those of the multicomponent mixtures within the lattice Boltzmann framework. The enrichment region of the ore-forming fluid is found to focus on the top of the uplift zone. The results are in agreement with those of the previous studies. The LBM provides an alternative to the numerical simulations of the enrichment of ore-forming elements in hydrothermal systems.

Keywords: Hydorthermal system, double diffusive, natural convection, lattice Boltzmann method.

1 Introduction

Recently the lattice Boltzmann method (LBM) has developed into a promising numerical scheme for simulating dynamical processes in fluids [1–3]. It originated from the lattice gas automata, a discrete particle kinetics utilizing a discrete lattice and discrete time. The fundamental idea of the LBM is to construct simplified microscopic models that simulate the motion of fluids by particles moving

* The work was supported in part by the National Natural Science Foundation of China under Grant No. 41172301,the Doctoral Program Foundation of Higher Education of China under Grant No. 20090145120007, and the Fundamental Research Funds for the Central Universities under Grant No. CUG090109.

Z. Li et al. (Eds.): ISICA 2012, CCIS 316, pp. 275–283, 2012.

and colliding on regular lattice, so that the macroscopic averaged properties, such as the density and velocity, satisfy the macroscopic Navier-Stokes equations [2, 3]. In comparison with the conventional numerical methods, which are based on discretization of macroscopic governing equations, the LBM provides clear physical pictures of molecular dynamics. The particle occupation is represented by the ensemble average distribution functions. The continuous distribution functions interact locally and propagate after collision to the next neighbor node. This eliminates statistical noise and leads to Galilei invariant macroscopic equations [4]. The collision operator can be linearized by assuming that the distribution is close to the local equilibrium state. Simplified collision model of Bhatnagar, Gross, and Krook (BGK), which make use of a single relaxation time towards the local equilibrium, is applied to the lattice Boltzmann equation to improve the computational efficiency [5]. The LBM was found numerically to be at least as stable, accurate, and computationally efficient as traditional computational fluid dynamics methods [2, 5].

The double diffusive natural convection has been the subject of intensive research due to its applications in various fields over the past decade [6]. It is caused by the temperature and concentration gradients in the fluid, in which the heat and mass transfer occur simultaneously under gravity. The system demonstrates a great variety of flow structures according to the orientation and magnitude of the gradients with respect to gravity. These phenomena motivated the applications on large-scale fluid transportation as well as the mineralization in the earth's crust [7]. Mathematical models are helpful for understanding the process of fluid transportation or the properties of heat conduction and convection during mineralization of ore deposit. Numerical solutions are available for double diffusive natural convection problems with certain boundary and initial conditions based on the discretization of the macroscopic continuum equation [7, 8]. In this paper, the temperature and concentration lattice BGK model is employed to model the double diffusive natural convection of ore-forming fluid in hydrothermal systems. The simulation is performed on a two-dimensional rectangular enclosure. The numerical results are found to be in agreement with those of the previous studies [8].

2   Model and Simulation

The schematic of the model under consideration is shown in Fig. 1, where a two-dimensional rectangular enclosure is considered as the hydrothermal systems filled with ore-forming fluid. The temperatures $(T_1 > T_2)$ and concentrations $(C_1 > C_2)$ are uniformly imposed along the top and bottom walls and the left and right walls are assumed to be adiabatic and impermeable to mass transfer. These temperatures and concentrations are kept constant thereafter and then the thermal and solutal gradients formed. The bottom wall is the source where the heat and ore-forming fluid diffuses to the top wall (sink). The non-equilibrium extrapolation boundary condition is employed in the horizontal direction, and the isothermal and isoconcentration boundary conditions are imposed in the vertical direction.

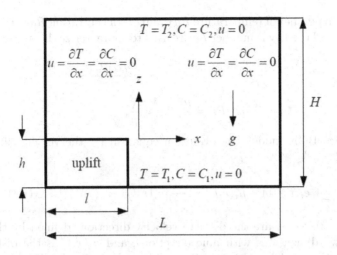

Fig. 1. Schematic of the model and coordinate for the system with boundary conditions

The fluid will be started to flow by the buoyancy forces induced by the density difference within the fluid, which can be described by a linear function of temperature T and concentration C with respect to a reference temperature T_∞ and concentration C_∞ [2]

$$\rho = \rho_\infty[1 + \beta_T(T - T_\infty) + \beta_C(C - C_\infty)], \tag{1}$$

where

$$\beta_T = -\frac{1}{\rho_\infty}\left(\frac{\partial\rho}{\partial T}\right)_C, \quad \beta_C = -\frac{1}{\rho_\infty}\left(\frac{\partial\rho}{\partial C}\right)_T, \tag{2}$$

with $\beta_T > 0$ and $\beta_C < 0$, $\rho_\infty(T_\infty, C_\infty)$ is the value of the fluid density at the reference point. If the viscous heat dissipation is negligible, their effect can be modeled by a body force term $\xi = g(\rho - \rho_\infty)$ in the momentum conservation equation according to the Boussinesq approximation [4]

$$\frac{\partial u}{\partial t} + (u \cdot \nabla)u = -\nabla p + \nu\nabla^2 u + \xi, \tag{3}$$

where u, p, g and ν are the velocity vector, press, gravitational acceleration and viscosity of the fluid. The equations governing the conservation of mass, energy and species concentration in nondimensional form can be written as [6]

$$\nabla \cdot u = 0, \tag{4}$$

$$\frac{\partial C}{\partial t} + (u \cdot \nabla)C = \eta\nabla^2 C, \tag{5}$$

$$\frac{\partial T}{\partial t} + (u \cdot \nabla)T = \chi\nabla^2 T, \tag{6}$$

where η and χ are the thermal and mass diffusivity. The characteristic velocity of the natural convection is determined by the system size $u_c = [g\beta_T(T_1 - T_2)H]^{\frac{1}{2}}$.

The Rayleigh number of temperature Ra_T and of concentration Ra_C, the Prandtl number Pr and the Lewis number Le are used to characterize the natural convection

$$Ra_T = \frac{g\beta_T(T_1 - T_2)H^3}{\nu\chi}, \quad Ra_C = \frac{g\beta_C(C_1 - C_2)H^3}{\nu\eta}, \tag{7}$$

$$Pr = \frac{\nu}{\chi}, \quad Le = \frac{\chi}{\eta}, \tag{8}$$

In the lattice BGK model, the evolution equation for the velocity field is expressed as [9, 10]

$$n_i(r + e_i, t + 1) - n_i(r, t) = -\frac{1}{\tau}\left[n_i(r, t) - n_i(r, t)^{eq}\right] + \xi_i(r, t), \tag{9}$$

where $e_i(i = 0, \cdots, 8)$ are the discrete velocity directions defined by the D2Q9 model of two dimensions with nine directions, and $n_i(x, t)$ is the distribution function for the particles with velocity e_i, whose deviation from the equilibrium distribution function n_i^{eq} gives the collision term if divided by the relaxation times τ [9]. The equilibrium distribution is given by

$$n_i^{eq} = w_i\left[1 + 3e_i \cdot u + \frac{9}{2}(e_i \cdot u)^2 - \frac{3}{2}u^2\right]\sum_{i=0}^{8} n_i. \tag{10}$$

In correspondence to equation (3), the body force term in the evolution equation (9) can be expressed as [10]

$$\xi_i = \left(1 - \frac{1}{2\tau}\right)\left[1 + 3e_i \cdot u + \frac{9}{2}(e_i \cdot u)^2\right] \cdot \xi, \tag{11}$$

where the momentum relaxation times τ is related to the viscosity of the fluid by $\nu = \frac{1}{3}(\tau - \frac{1}{2})$.

The temperature field and concentration field are passively advected by the fluid flow and obey the same equations as those of the multicomponent mixtures within the lattice Boltzmann framework. If the temperature and concentration are considered as passive scalars, the control equations of velocity, temperature and concentration can be treated individually [2, 10]. The evolution equations for the temperature and concentration field take the same form as equation (9) except for the omission of the body force term. The relaxation times are determined by $\chi = \frac{1}{3}(\tau_T - \frac{1}{2})$ and $\eta = \frac{1}{3}(\tau_C - \frac{1}{2})$, respectively. The equilibrium distribution functions for them take the form as equation (10), and the macroscopic quantities can be calculate as

$$u = \sum_{i=1}^{8} e_i n_i, \quad T = \sum_{i=1}^{8} T_i, \quad C = \sum_{i=1}^{8} C_i. \tag{12}$$

The solution in the hydrothermal system is assumed to be saturated in consideration of the dynamic equilibrium in the ore-forming fluids. The distributions of temperature and concentration are nonuniform and cannot be described by a simple linear relationship in the enclosure in respect to the double diffusive natural convection. In order to characterize the ore-forming fluids in the hydrothermal systems, the temperature and concentration are normalized to the same dimension ($0 \leq C \leq 1, 0 \leq T \leq 1$), hence, the degree of the element enrichment can be defined as [8]

$$q = \begin{cases} C - T, C > T \\ 0, \quad C \leq T \end{cases},$$ (13)

where T and C are dimensionless quantities that depend on their coordinate and time. If the normalized concentration C of the solution is higher than the normalized temperature T, the crystallization and precipitation occur. The assumption is reasonable in that the necessary factors are taken into consideration for the ore-forming element enrichment, including the physical and chemical properties of the hydrothermal fluid, the dimension of the enclosure, and the duration of the mineralization. The model provides a simple scenario in approaching the dynamics of ore-forming process in the hydrothermal system.

3   Results and Disscusion

The double diffusive natural convection of ore-forming fluid in hydrothermal system is simulated by LBM with the system size of $H \times L = 256 \times 256$, and the uplift zone size of $h \times l = 32 \times 64$. The basic characteristic dimensionless numbers include the Rayleigh number of temperature $Ra_T = 10^4$ and of concentration $Ra_C = 5.0 \times 10^4$, the Prandtl number $Pr = 0.7$, and Lewis number $Le = 10$. Temperature and concentration have opposing effect on the buoyancy force as indicated by different signs in equation (1). The convection of the ore-forming fluid is influenced by the buoyancy ratio. The fluid of higher temperature and lower concentration tends to rise to the top of the enclosure, and vice versa. The temperature and concentration of the fluid are initialized as the mean values of the top and bottom wall. The fluids near the bottom wall and the boundary of the uplift zone $T_1 = 1.0$, $C_1 = 1.0$ are heated firstly and dissolve more elements than those near the top wall $T_2 = 0$, $C_2 = 0$.

The unbalance of the temperature and concentration buoyancy fore drives the fluid and the convection starts. The isotherms and isoconcentrations of the ore-forming fluid in 2D simulation of $x - z$ plane at different time step are illustrated in Fig. 2 and Fig. 3. The stratification of temperature and concentration in the fluid can be clearly identified within the range of $t = 0 \sim 1.0 \times 10^4$, which indicate a steady convection in the system. The gradients of the temperature and concentration field are influenced by the geometry of the uplift zone. The steady convection is maintained until strong convection occurs during $t = 1.0 \times 10^4 \sim 1.44 \times 10^4$, herein the steady state are destroyed at $t = 10^4$ in Fig. 2 and Fig. 3. The hydrothermal system returns steady state henceforth, as shown $t = 10^5$ and $t = 1.7 \times 10^5$ ditto.

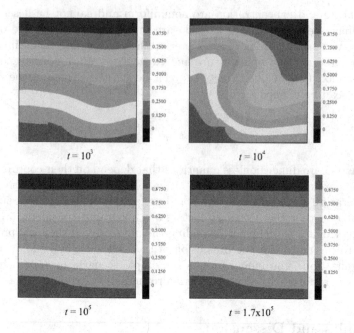

Fig. 2. Isotherms of the ore-forming fluids in 2D simulation

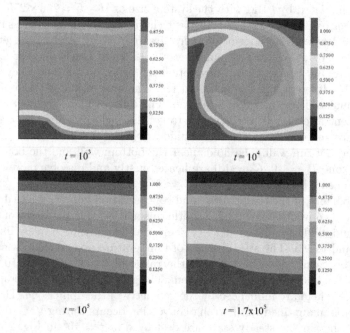

Fig. 3. Isoconcentrations of the ore-forming fluids in 2D simulation

The contours of the element enrichment of the ore-forming fluids in 2D simulation of $x-z$ plane are calculated by equation (13). The results at different time steps are illustrated in Fig. 4. The enrichment of ore-forming element initially occurs in the upper of the enclosure, as shown at $t = 10^3$. The hydrothermal system then experiences an intermediate state, wherein the enrichment region spread in the enclosure, as shown at $t = 10^4$. The enrichment region begins to focus on the top of the uplift zone after $t = 1.44 \times 10^4$, as shown at $t = 10^5$ and $t = 1.7 \times 10^5$ in Fig 4. The enrichment of ore-forming element will not occur elsewhere far from the uplift zone. The results are in agreement with those of the previous studies [8].

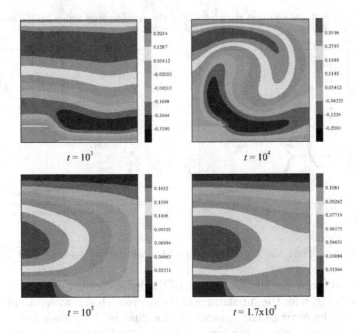

Fig. 4. Contour of the element enrichment of the ore-forming fluids in 2D simulation

The double diffusive natural convection of ore-forming fluid in enclosures without uplift zone is also simulated. In order to illustrate the effect of the uplift zone on the element enrichment of the ore-forming fluids, the contours in 2D simulation of $x - z$ plane at $t = 10^3$ and $t = 1.7 \times 10^5$ are presented for comparison in Fig. 5. The enrichment region migrates from the top to the bottom of the enclosure.

The velocities of the ore-forming fluids in the enclosure are recorded during the simulation. The convergence of the program are determined by the error of the velocity taken over the entire 2D fluid field in the $x - z$ plane

$$\varepsilon = \frac{\sum_{x,z}[u_{x,z}(t + \Delta t) - u_{x,z}(t)]^2}{\sum_{x,z}[u_{x,z}(t + \Delta t)]^2} : \tag{14}$$

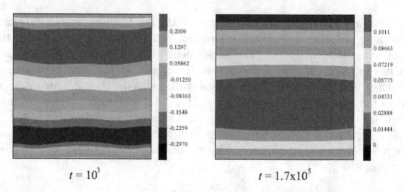

$t = 10^3$ $t = 1.7 \times 10^5$

Fig. 5. Contour of the element enrichment of the ore-forming fluids in 2D simulation

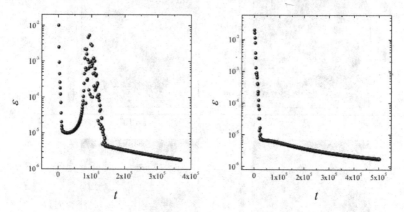

Fig. 6. Time history of the error of the velocity in the fluid field

ε is used to characterize the convection in the hydrothermal system. ε versus t are plotted in Fig. 6 for the simulations in enclosure with and without uplift zone, respectively. The error of the velocity in the fluid field decrease exponentially with the simulation time. Those data can be classified into two separate regimes, both of which take the form of $\varepsilon \sim e^{-t/t_0}$ in the left figure. In the first regime of $t = 0 \sim 1.0 \times 10^4$ for simulation in enclosure with uplift zone (see left figure), ε drops sharply with a very small t_0, and then the system enters an intermediate state with ε rising and drop, until the turbulence eventually settle down after $t = 1.44 \times 10^4$ and ε begins to decrease with a fairly large t_0. The simulation result in enclosure without uplift zone is shown in the right figure. There exist two regimes that can be clearly identified at the point $t = 2.7 \times 10^4$, however, the intermediate state is not observed.

4   Conclusion

The double diffusive natural convection of ore-forming fluid in rectangular enclosures is simulated by LBM. The hydrothermal system experiences a series of convection states until it eventually settle down. The enrichment region is found to focus on the top of the uplift zone. The results are in agreement with those of the previous studies. The LBM provide an alternative to numerical simulations of enrichment of ore-forming elements in hydrothermal system.

References

1. McNamara, G.R., Zanetti, G.: Use of the boltzmann equation to simulate lattice-gas automata. Phys. Rev. Lett. 61, 2332–2335 (1988)
2. Shan, X.: Simulation of rayleigh-bénard convection using a lattice boltzmann method. Phys. Rev. E 55, 2780–2788 (1997)
3. Chen, S., Doolen, G.D.: Lattice boltzmann method for fluid flows. Annu. Rev. Fluid Mech. 30, 329–364 (1998)
4. Wolf-Gladrow, D.A.: Lattice-Gas Cellular Automata and Lattice Boltzmann Models - An Introduction. Springer, Berlin (2005)
5. Bhatnagar, P.L., Gross, E.P., Krook, M.: A model for collision processes in gases. i. small amplitude processes in charged and neutral one-component system. Phys. Rev. 94, 511–525 (1954)
6. Young, Y., Rosner, R.: Numerical simulation of double-diffusive convection in a rectangular box. Phys. Rev. E 61, 2676–2694 (1988)
7. Yang, R., Ma, D., Bao, Z., Pan, J., Cao, S., Xia, F.: Geothermal and fluid flowing simulation of ore-forming antimony deposits in xikuangshan. Science in China D 49, 862–871 (2006)
8. Yang, R.: The Study of Metallogenesis and Concentration for Ore-forming in the Sedimentary Basins: A Case Study from the Xikuangshan Antimony Deposits. Press of China University of Geosciences, Wuhan (2005) (in Chinese)
9. Guo, Z., Shi, B., Zheng, C.: A coupled lattice bgk model for the boussinesq equations. Int. J. Numer. Meth. Fluids 39, 325–342 (2002)
10. Guo, Z., Zheng, C., Shi, B.: Discrete lattice effects on the forcing term in the lattice boltzmann method. Phys. Rev. E 65, 046308 (2002)

Research on Biogeography Differential Evolution Algorithm

Hongwei Mo, Zhenzhen Li, and Luolin Zhang

Automation College
Harbin Engineering University
Harbin 150001, P.R. China
Honwei2004@126.com
{zzli,lukeke}@163.com

Abstract. Biogeography-based optimization (BBO) is a population-based evolutionary algorithm (EA) that is based on the mathematics of biogeography. It mainly uses the biogeography-based migration operator to share the information among solutions. Differential Evolution (DE) is a fast and robust evolutionary algorithm for global optimization. In this paper, we propose a hybrid algorithm of BBO and DE, named BDE, for the global numerical optimization problem. To verify the performance of our proposed BDE, 12 benchmark functions with a wide range of dimensions and diverse complexities are employed. Experiment results indicate that our approach is effective and efficient.

Keywords: Biogeography-based Optimization, Differential Evolution, Hybrid Migration Over-cross Operator.

1 Introduction

Biogeography-based optimization (BBO) is a new developed algorithm for global optimization. Biogeography-based optimization was introduced by Simon in 2008 and demonstrated good convergence properties on various benchmark functions[1]. It is modeled after the immigration and emigration of species between islands. One characteristic of BBO is that the original population is not discarded after each generation but modified by migration. Various versions of biogeography-based optimization modals and theoretic research were proposed in[2-5].

The DE[6] algorithm was proposed by Storn and Price, and since then the DE algorithm has been used in different areas. The DE algorithm is a floating-point encoding evolutionary algorithm for global optimization over continuous spaces. DE has some good properties, such as efficiency and robustness, and it is a simple but powerful evolutionary algorithm. Many new versions of DE were proposed in[6-10]. Neri and Tirronen[11] gave a survey of recent advances in differential evolution.

BBO has a good exploitation ability for global optimization, while DE is good at exploring the search space and locating the region of global minimum.

Z. Li et al. (Eds.): ISICA 2012, CCIS 316, pp. 284–291, 2012.

However, it is not good at exploitation of solution. Based on these considerations, in order to balance the exploration and the exploitation of BBO and DE, we propose a hybrid approach, called BDE, which combines the exploitation of BBO with the exploration of DE effectively.

The remainder of this paper is organized as follows: Section 2 gives the problem definition. Section 3 describes the BBO algorithm and DE algorithm. Section 4 describes the proposed BDE algorithm. Analysis of BDE performance is presented in Section 5. And conclusions are given in Section 6.

2   BDE

2.1   BBO and DE

In geography, geographical areas(habitat,such as island) that are well suited as residences for biological species are said to have a high habitat suitability index (HSI), The variables that characterize habitability are called suitability index variables (SIVs)[14]. In BBO, each possible solution is an island and their features that characterize habitability are called suitability index variables (SIV). The goodness of each solution is called its habitat suitability index (HSI).For solving a engineering problem, a good solution is analogous to an island with a high HSI, and a poor solution represents an island with a low HSI. High HSI solutions resist change more than low HSI solutions. High HSI solutions tend to share their features with low HSI solutions. Poor solutions accept a lot of new features from good solutions. Suppose that we have a problem and a population of candidate solutions that can be represented as vectors of integers. Each integer in the solution vector is considered to be an SIV. With probability P_{mod}, each solution is modified based on other solutions. If a given solution is selected to be modified, then we use its immigration rate λ_s to probabilistically decide whether or not to modify each suitability index variable (SIV) in that solution. If a given SIV in a given solution S_i is selected to be modified, then the emigration rates μ_s of the other solutions are used to probabilistically decide which of the solutions should migrate a randomly selected SIV to solution . Based on the process of migration, the adaptive ability of habitat is improved by regulating immigration rate and emigration rate, migration topology, migration strategy. Thus, it can get optimal solution of problem.

Immigration rate λ_s and emigration μ_s can be evaluated by Eq. (1) and (2).

$$\lambda_s = I(1 - \frac{s}{s_{max}}) \tag{1}$$

$$\mu_s = \frac{Es}{s_{max}} \tag{2}$$

Where I and E are maximum immigration rate and maximum emigration rate.

Differential evolution (DE)[6] has been shown to be a simple yet efficient evolutionary algorithm for many optimization problems in real-world applications. It follows the general procedure of an evolutionary algorithm. After initialization,

DE enters a loop of evolutionary operations: mutation, crossover, and selection. There are several variant DE algorithms which are different in their different mutation strategies. Among them, the following mutation strategies are frequently used:

$$DE/best/1: \quad U_i = X_{best} + F(X_{r_1} - X_{r_2}) \tag{3}$$

$$DE/rand/1: \quad U_i = X_{r_1} + F(X_{r_2} - X_{r_3}) \tag{4}$$

Many schemes of creation of a candidate are possible in DE. We use both the DE/rand/1/bin and DE/best/1/bin[6] scheme in BDE.

2.2   BDE

In BDE, we design migration over-cross operator as follows:

Algorithm 1. Migration Over-cross Operator:

1: **for** $i = 1$ to NP **do**
2:     Select uniform randomly $r_1 \neq r_2 \neq r_3 \neg r_4 \neq r_5 \neq i$ and select the best individual;
3:     $j_{rand} = rndint(1, D)$;
4:     **for** $j = 1$ to D **do**
5:         **if** $rndreal(0, 1) < \lambda_i$ **then**
6:             **if** $rndreal_j[0, 1) > CR$ or $j == j_{rand}$ **then**
7:                 $U_i(j) = X_{r_1}(j) + F \times (X_{r_2}(j) - X_{r_3}(j))$
8:             **else**
9:                 Select X_{r_6} with high HSI;
10:                $U_i(j) = X_{r_6}(j)$
11:            **end if**
12:        **else**
13:            **if** $rndreal_j[0, 1) > CR$ or $j == j_{rand}$ **then**
14:                $U_i(j) = X_{r_4}(j)$
15:            **else**
16:                $U_i(j) = X_{best}(j) + F \times (X_{r_5} - X_{r_1}(j))$
17:            **end if**
18:        **end if**
19:    **end for**
20: **end for**

Based on migration over-cross operator, the main procedure of BDE is described as follows:

1: Generate the initial population P;
2: Evaluate the fitness for each individual in P;
3: **while** the halting criterion is not satisfied **do**

4:      For each individual, map the fitness to the number of species;
5:      Calculate the immigration rate λ_i and the emigration rate μ_i for each individual
        X_i according to (1) and (2);
6:      Modify the population with the migration over-cross operator;
7:      **for** $i = 1$ to NP **do**
8:          Evaluate the offspring U_i;
9:          **if** offspring U_i is better than X_{r_4} **then**
10:             $X_{r_4} = U_i$;
11:         **end if**
12:     **end for**
13:     **for** $i = 1$ to NP **do**
14:         Compute the probability P_i;
15:         Select a variable $X_i(j)$ with probability œP_i;
16:         **if** $rndreal(0,1) < m_i$ **then**
17:             Replace $X_i(j)$ with a randomly generated variable from its range;
18:         **end if**
19:     **end for**
20: **end while**

Where D is the number of dimension, $rndint$ a function producing an integer between 1 and D randomly, $rndreal$ is a uniformly distributed random real number in $(0,1)$. X_{best} is the best individual among the current population. F is scaling factor.

In step 16, we use $U_i(j) = X_{best}(j) + F \times (X_{r_1}(j) - X_{r_2}(j))$ to replace $U_i(j) = X_{r_1}(j) + F \times (X_{r_2}(j) - X_{r_3}(j))$, and the rest is the same with DE/rand/1/bin.

In the proposed BDE, it can be seen that the offspring U_i is possibly constituted by four components: the DE mutant (random and best), the migration of the other solutions (parents and others). The core idea of the proposed hybrid migration operator is based on two considerations. One hand, good solutions would be less destroyed, while poor solutions can accept a lot of new features from good solutions. In this case, the current population can be exploited sufficiently. On the other hand, the mutation operator of DE is able to explore the new search space.In literature[5], when $rndreal(0,1) < \lambda_i$, but when $rndreal(0,1) \geq \lambda_i$, $U_i(j) = X_i(j)$, while our approach allows more individuals to join the process of creation. when $rndreal(0,1) \geq \lambda_i$, we have two conditions to choose. If $rndreal_j[0,1) > CR$, we let $U_i(j) = X_{r_4}(j)$. $X_{r_4}(j)$ is selected randomly. Even if X_{r_4} has poor fitness HSI, it is different from literature[5] that the other individuals also have chance to migrate besides migrating ones. Secondly, we make use of the best individual, $U_i(k) = X_{best}(j) + F \times (X_{r_5}(j) - X_{r_1}(j))$ is added in order to speed up the convergence process.

3   Experimental Results

In Fig. 1, the relation between optimal value and generation of BDE, BBO and DE are shown.

Table 1. The Properties of Test function

Test Function	Property			
	Range	D	C	minimum
Sphere	$[-100, 100]$	30	US	0
Sumsquares	$[-10, 10]$	30	US	0
Step	$[-100, 100]$	30	US	0
Beale	$[-4.5, 4.5]$	2	UN	0
Matyas	$[-10, 10]$	30	UN	0
Schwefel1.2	$[-100, 100]$	30	UN	0
Ackley	$[-32, 32]$	30	MN	0
Griewank	$[-600, 600]$	30	MN	0
Penalty1	$[-50, 50]$	30	MN	0
Bohachevsky1	$[-100, 100]$	2	MS	0
Booth	$[-10, 10]$	2	MS	0
Rastrigin	$[-5.12, 5.12]$	30	MS	0

Table 2. The Formula of Test Functions

Test Function	Formulation
Sphere	$f(x) = \sum_{i=1}^{n} x_i^2$
Sumsquares	$f(x) = \sum_{i=1}^{n} i x_i^2$
Step	$f(x) = \sum_{i=1}^{n} (\lfloor x_i + 0.5 \rfloor)^2$
Beale	$f(x) = (1.5 - x_1 + x_1 x_2)^2 + (2.25 - x_1 + x_1 x_2^2)^2 + (2.625 - x_1 + x_1 x_2^3)^2$
Matyas	$f(x) = 0.26(x_1^2 + x_2^2) - 0.48 x_1 x_2$
Schwefel1.2	$f(x) = \sum_{i=1}^{n} (\sum_{j=1}^{i} x_j)^2$
Ackley	$f(x) = -20 exp(-0.2 \sqrt{\frac{1}{n} \sum_{i=1}^{n} x_i^2}) - exp(\frac{1}{n} \sum_{i=1}^{n} cos(2\pi x_i)) + 20 + e$
Ackley	$f(x) = \frac{1}{4000} \sum_{i=1}^{n} x_i^2 - \prod_{i=1}^{n} cos(\frac{x_i}{\sqrt{i}}) + 1$
Penalty1	$y_i = 1 + \frac{1}{4}(x_i + 1), u(x_i, a, k, m) = \begin{cases} k(x_i - a)^m, & x_i > a \\ 0, & -a \leq x_i \leq a \\ k(-x_i - a)^m, & x_i < a \end{cases}$
Bohachevsky1	$f(x) = x_1^2 + 2x_2^2 - 0.3 cos(3\pi x_1) - 0.4 cos(4\pi x_2) + 0.7$
Booth	$f(x) = (x_1 + 2x_2 - 7)^2 + (2x_1 + x2 - 5)^2$
Rastrigin	$f(x) = \sum_{i=1}^{n} [x_i^2 - 10 cos(2\pi x_i) + 10]$

In order to verify the performance of BDE, 12 benchmark functions are chosen. They are only briefly described in Table 3. The functions were divided into four categories: US, UN, MN, MS, where U means unimodal, M multimodal, S, separable, N, non-Swparable. They appear to be the most difficult class of problems for many optimization algorithms. For all experiments, we use the following parameters. Population $size = 50$; $D = 20$. Habitat modification probability = 1;Habitat mutation probability: 0.005. Scaling factor: $F = 0.5$; Crossover probability: $CR = 0.5$; DE mutation scheme: DE/rand/1/bin, DE/best/1/bin;Elitism parameter: 2.Each function is optimized over 30 independent runs. All the algorithms

Table 3. The test results of BDE on benchmarks

Test Function	Comparison of BDE,BBO and DE					
	BDE		BBO		DE	
	mean std	best worst	mean std	best worst	mean std	best worst
Sphere	9.8e-028 1.2e-027	6.1e-029 4.3e-027	2.3e-002 1.4e-002	0 5.0e-002	1.5e-010 6.8e-011	3.8e-011 3.1e-010
Sumsquares	2.2e-026 2.6e-026	6.7e-028 1.1e-025	7.8e-001 5.1e-001	3e-002 1.72	5.8e-009 3.5e-009	2.1e-009 1.5e-008
Step	0 0	0 0	7.4 3.9	3 18	0 0	0 0
Beale	8.1e-013 3.6e-012	0 1.6e-011	0.2 0.4	0 1.6	0 0	0 0
Matyas	4.3e-005 1.9e-004	0 8.6e-004	2.2e-002 2.4e-002	0 9e-002	0 0	0 0
Schwefel1.2	2.6 2.8	0.18 11.8	1022.4 519.8	303.1 2125.5	3973.0 886.3	2185.5 5570.5
Ackley	1.7e-013 8.2e-014	5.5e-014 3.3e-013	1.5 0.4	0.7 2.3	8.5e-005 1.8e-005	5.8e-005 1.3 e-004
Griewank	1 1.2e-016	1 1	1.0 0.0	1.0 1.1	1 2.6e-010	1 1
Penalty1	7.0e-026 3.1e-026	4.3e-028 2.1e-025	6.0e-002 5.0e-002	7.7e-003 2.2e-001	1.1e-008 7.6e-009	3.4e-009 2.8e-008
Bohachevsky1	0 0	0 0	6.4 9.9277	0.1 36.6789	0 0	0 0
Booth	0 0	0 0	0.251 0.36207	0 1.46	0 0	0 0
Rastrigin	4.2196 1.8381	0.8072 8.4165	1.112 1.032	0 3.0327	67.0003 5.8774	53.6351 80.3409

are implemented in standard MATLAB7.1. In order to show the superiority of our proposed BDE approach, we compare it with the original BBO algorithm and the DE algorithm. Table 1 and Table 2 show the characteristics and formulations of all test functions, respectively.

From Tables 3 and Fig. 1, it is obvious that BDE performs significantly better than BBO consistently for all test functions except Rastrigin. From Fig. 1, it be can seen that in the very beginning of the evolutionary process BBO converges faster than BDE, but BDE converges much faster in general and improve its solution steadily for a long run. From Table 3 it can be seen that BDE reach better values than DE in the same runs on 12 functions. The experimental results show that the BDE has the ability to escape from poor local optima and locate a better near global optimum.

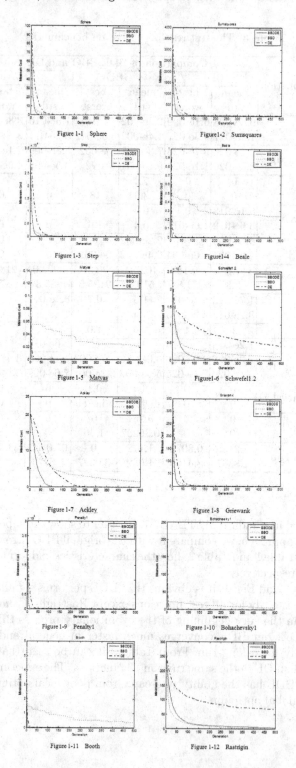

Fig. 1. Mean curves of BBODE, BBO, and DE for selected functions

4 Conclusion

In this paper, we propose a hybrid approach called BDE, which combines the exploitation of BBO with the exploration of DE and generates the promising solutions. Since the hybrid migration operator balances the exploration and the exploitation, it makes our proposed BDE approach be very effective and efficient. Compared with BBO, DE, the experimental results show that BDE is superior to BBO and DE on the test problems.

Acknowledgement. This work is partially supported by the National Natural Science Foundation of China under Grant No.61075113 and the Excellent Young Teacher Foundation of Heilongjiang Province of China under Grant No.1155G18. the Fundamental Research Funds for the Central Universities No.HEUCF110441. No.P043512010.

References

1. Simon, D.: Biogeography-based Optimization. IEEE Transactions on Evolutionary Computation 12(6), 702–713 (2008)
2. Ma, H.P.: An Analysis of the Behavior of Migration Models for Biogeography-Based Optimization. Information Sciences 180(18), 3444–3464 (2010)
3. Gong, W.Y., Cai, Z.H., Ling, C.X., Li, H.: A Real-Coded Biogeography-based Optimization with Neighborhood Search Operator. Applied Mathematics and Computation 216(9), 2749–2758 (2010)
4. Du, D.W., Simon, D., Ergezer, M.: Biogeography-based Optimization Combined with Evolutionary Strategy and Immigration Refusal. In: Proceedings of the IEEE Conference on Systems, Man, and Cybernetics, SanAntonio, Texas, pp. 1023–1028 (2009)
5. Simon, D., Ergezer, M., Du, D.: Population Distributions in Biogeography-based optimization algorithms with elitism. In: Proceedings of the IEEE Conference on Systems, Man, and Cybernetics, San Antonio, Texas, pp. 1017–1022 (October 2009)
6. Storn, R., Price, K.: Differential Evolution – A Simple and Efficient Heuristic for Global Optimization over Continuous Spaces. Journal of Global Optimization 11, 341–359 (1997)
7. Price, K.V., Storn, R.M., Lampinen, J.A.: Differential Evolution, A Practical Approach to Global Optimization. Springer, Heidelberg (2005)
8. Zhang, J.Q., Sanderson, A.C.: JADE: Adaptive Differential Evolution with Optional External Archive. IEEE Transactions on Evolutionary Computation 13(5), 945–958 (2009)
9. Das, S., Abraham, A., Chakraborty, U.K., Konar, A.: Differential Evolution Using a Neighborhood-based Mutation Operator. IEEE Transactions on Evolutionary Computation 13(3) (2009)
10. Rahnamayan, S., Tizhoosh, H.R., Salama, M.M.A.: Opposition-Based Differential Evolution. IEEE Transactions on Evolutionary Computation 12(1), 64–79 (2008)
11. Neri, F., Tirronen, V.: Recent Advances in Differential Evolution: A Survey and Experimental Analysis. Artificial Intelligence Review 33(1-2), 61–106 (2010)

Improving Multi Expression Programming Using Reuse-Based Evaluation

Wei Deng[1] and Pei He[1,2]

[1] School of Computer and Communication Engineering
Changsha University of Science and Technology
Changsha 410114, P.R. China
76995958@qq.com
[2] Key Laboratory of High Confidence Software Technologies(Peking University)
Ministry of Education, Beijing 100871, P.R. China
bk_he@126.com

Abstract. Multi expression programming is a linear genetic programming that dynamically determines its output from a series of genes of the chromosome. It works on a fixed-length individual, but gives rise to the complexity of the decoding process and fitness computations. To solve this problem, we proposed an improved algorithm that can speed up individual assessments through reuse analysis of evaluations. The experimental result shows that the present approach performs quite well on the considered problems.

Keywords: Multi Expression Programming, Linear Genetic Programming, Fitness Computations, Reuse Analysis of Evaluations.

1 Introduction

Multi Expression Programming (MEP) as a linear genetic programming approach is a Genetic Programming variant first proposed in 2002 by Oltean.M and Dumitrescu.D[1-7]. Compared to other variants of Genetic Programming, a unique feature of MEP is its ability of storing multiple solutions of problem in a single chromosome[9]. In MEP each chromosome contains a number of expressions which, called genes, consist of strings of variable length, and the number of genes per chromosome is constant. These features make it possible to greatly increase the probability of the problem solution [8-10]. So we can effectively solve complex problems. Now multi expression programming has been widely used in many fields, such as classification problems, stock market forecast, TSP, digital circuit design, and so on[11-13].

In this paper, we will present an improved algorithm for efficiently assessing the individuals of MEP. The idea lies in the fact that the fitness can be quickly computed based on encoded chromosome, and any genes inherited from parents dont be evaluated once again.

The paper is organized as follows. In section 2, the basic principle of the multi expression programming is presented. The improved algorithm in MEP is

Z. Li et al. (Eds.): ISICA 2012, CCIS 316, pp. 292–299, 2012.

described in section 3. In section 4, we will perform several numerical experiments. Section 5 concludes the paper.

2 Multi Expression Programming

2.1 MEP Algorithm

Standard MEP algorithm starts by creating a random population of individuals. The following steps are repeated until a problem is solved. Two parents are selected in a selection procedure[11]. Then the parents are recombined in order to generate new offspring by crossover. Finally the offspring are considered for mutation.

In MEP each gene encodes a terminal or a function whose arguments always have indices of lower values than the position of that function in the chromosome. In the selection process, the classical MEP will decode each gene in the population and assign it a fitness value according to how well it solves the problem. Usually the best solution is chosen for fitness assignment of the chromosome[9]. Create new individuals based on encoded genes by crossover and mutation. Standard MEP algorithm is described in detail as follows.

2.2 Encoding Principle

A chromosome of MEP consists of several genes. Its length is equal to the number of genes. Each gene represented by strings of a variable length is composed of function operator, terminal symbol and gene sequence number. MEP encoding rules are simple. The first symbol of the chromosome must be a terminal symbol, and function arguments of the gene must be smaller than the current genetic index.

For example, let $F = \{*, +, S\}$ be the set of function symbols where the symbol S represents the sin function, and the terminal set be $T = \{x, y\}$. Suppose a chromosome contains 6 genes, an example of chromosome using the sets F and T will take the form of Table 1.

Table 1. Gene encoding of a chromosome: the first row is gene indices, the second is the corresponding genes encoding

1	2	3	4	5	6
x	$*1,1$	y	$*1,3$	$+2,3$	$S5$

2.3 Decoding Principle and Fitness Assignment

In the evaluation process, the first step is to decode genes of the chromosome. For instance, in the previous example each gene would be decoded to a simple expression. These expressions are shown in Table 2.

Table 2. Gene encoding of a chromosome: the first row is gene indices, the second is the corresponding genes encoding

1	2	3	4	5	6
x	$x*x$	y	$x*y$	$x*x+y$	$sin(x*x+y)$

In general, phenotypic decoding process can be depicted based on Table 1 as follows. Genes 1 and 3 decode a simple expression by a single terminal symbol. Genes 2, 4, 5 and 6 include function symbol and gene indices, which, say gene 2, indicates the operation of multiplication on the operands gene1 and gene 1 of the chromosome. Therefore the translation of a gene into an expression involves complicate substitutions among sub expressions and gene indices. In this paper, we proposed an improved algorithm which does not need decoding, but evaluating genes by the encoded chromosome.

In MEP algorithm, the different fitness function has a different assessment effect for the gene. One of the most popular forms of fitness functions given below is the so called absolute error[11]:

$$fitness(E_i) = \sum_{j=1}^{N} |O_{j,i} - w_j| \qquad (1)$$

Where $O_{j,i}$ is the value of the expression E_i on the jth sample data and W_i corresponding target result. The fitness of the chromosome fitness(C) is equal to the lowest fitness of the genes in a single chromosome[9]. This fitness is widely used for solving regression problems.

$$fitness(C) = min f_i(E_i) \qquad (2)$$

2.4   Genetic Operators

The genetic operators used in the MEP algorithm are crossover and mutation. Since genetic operators preserve the chromosome structure, all the offspring are syntactically correct expressions[1].

Crossover: Two parents are selected randomly and recombined by crossover[1]. Cross point is also randomly generated. In this paper we consider one-point recombination.

Mutation: Terminal symbol, function symbol and gene index may be used to mutate the gene. In order to prevent the chromosome structure from damaging, the first gene must be encoded to a terminal symbol. If the current gene changes into a function symbol, the function arguments must be function pointers to the previous genes.

3   Improved MEP Algorithm

In order to evaluate chromosomes, we must translate each gene into an expression, and compute function and fitness value of the decoded genes. On the one hand, translation of expression consumes tremendous time and space resources; on the other hand the evaluation process will recalculate the same gene segments when the current gene contains gene indices. As a large number of the genes contains function pointer, this way will affect the evolution efficiency. So this paper presents an improved MEP algorithm which quickly computes the value of the expression based on the encoded genes, and don't need decoding genes. Some of the genes have not been modified by crossover and mutation, but their fitness will still be recalculated in a novel generation. So this paper proposed another improved algorithm which would preserve the primary fitness value of the unmodified gene segments, thus reducing the number of repeated computation. These two improved algorithm will be applied in MEP to solve practical problems.

The improved algorithm is described in MATLAB language as follows:

```
function chro_fit = evaluate(chro, flag,x_value)
%  Input arguments: a chromosome, modification flag of genes and
   sample data
%  Output arguments: fitness value of the chromosome
%  Global variables are initialized in the begin of evolutionary
   process, but not
%  in this m file. This is the m file of the evaluation.
%  y_value and gene_fit is initialized to null, and node_num
   is initialized to 0
global y_value;
%  output values of the parents
global gene_fit;
%  fitness values of the parents
global node_num;
%  the number of calculation nodes
for i = 1:length(chro)
%  ~ flag (i) = = 1 means that the current gene has not been
   modified.
if(gene_fit &~flag(i))
%  if gene_fit is empty, it represents the first generation
   continue;
%  ~flag(i) == 1 and gene_fit is not empty then continue;
end
node_num = node_num + 1;
%  the number of calculation nodes plus one
gene = chro(i);
if(length(deblank(gene))<2)
%  the current gene is a terminal symbol
     y_value(i) = x_value(argument_index)
else
%  the current gene contains function symbol
%  function symbol a binocular operator
```

```
    if(gene(1)is a binocular operator)
        y_value(i)=bin_operator(y_value((gene(function_pointer1)))),
        y_value((gene(function_pointer2));
else
%  function symbol a unary operator
        y_value(i) = unary_operator
                            (y_value(gene(function_pointer1)));
    end
end
    gene_fit = sum(abs(y_value (i)C x_value(sample_index)))
%  gene fitness value
end
chro_fit = min(gene_fit)
%  chromosome fitness value
```

The parameter flag is initialized to the value 0. Its value is modified only in genetic operation. Argument flag with the value 0 indicates that the gene has not been modified by the crossover and mutation, Otherwise the flag would be set to 1. If the gene contains function pointers and is not modified, we also need to consider whether the referred expression is modified. If the expressions that are pointed to by the function pointer have been modified, the gene flag should also be set to 1. In the process of evolution the number of calculation nodes is based on the flag. A node represents a gene.

The improved algorithm directly computes a function by scanning an encoded gene. If the gene is a terminal symbol, the function value is the argument sample data. If the gene contains function symbol, the result is calculated by the gene value of the function pointer. The standard MEP algorithm requires the translation of expressions to evaluate genes. But the improved MEP algorithm performs all arithmetic based on operators, and function pointer. In addition, unmodified genes dont participate in calculation except the first generation. The two improved ideas raise evolution efficiency.

4   Experiments and Analysis

In this section, several experiments with comparisons of standard MEP and the improved algorithm are performed on two well-known regression problems. The two functions to be examined are $y^4 + y^3 + y^2 + y$ and $sin(y^4 + y^2)$, respectively. We will compare the performance of these two methods in the time complexity and the number of calculation nodes by running them on the same problems based on the same 20 sample data for 50 times.

Fig. 1 makes a comparison between the standard and the improved MEP algorithm in the time complexity and the number of calculation nodes. Evolution parameters are shown in Table 3.

In the first diagram of Fig.1, abscissa axis represents the number of evolution, and the ordinate is the running time. It shows that the average evolution time of the improved MEP algorithm is about 9 times less than the standard algorithm.

Table 3. The experimental parameters of the two regression problems

Function Set	$\{+, *, S\}$
Terminal Set	$\{y\}$
Number of Chromosomes	20
Number of Genes	10
Number of Generations	1000
Crossover Probability	0.9
Mutation Probability	0.5

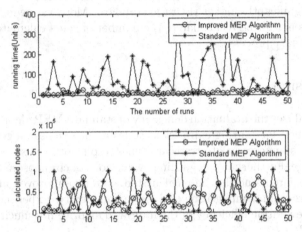

Fig. 1. Comparison of the two methods in solving $y^4 + y^3 + y^2 + y$

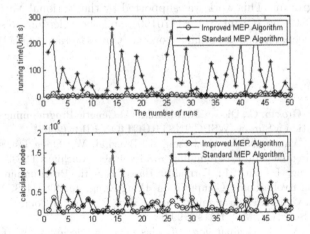

Fig. 2. Comparison of the two methods in solving $sin(y^4 + y^2)$

In the second diagram of Fig.1, abscissa axis is also the number of evolution, but the ordinate represents the calculated number of genes in the assessment process. The experimental results show that the number of calculation nodes by the improved algorithm was less than the traditional approach during the evaluation phase. The number of calculation nodes is based on the third part of the mentioned flag to decide whether to evaluate the gene.

In order to verify the validity of the improved algorithm, we conducted several experiments to test. The performance comparison of the other function $sin(y^4 + y^2)$ is given in the below Fig. 2.

The experimental result shows that the evolution efficiency of the improved algorithm is significantly higher than the standard. Moreover owing to that unmodified genes neednt be recalculated, the number of genes calculated decreases in the evolutionary process.

5   Conclusion

This paper analyzes the evaluation deficiency of standard MEP algorithmand put forward a method for sharing expression values among the same gene segment. Thus it becomes unnecessary to translate genes to phenotypic expressions in the evaluation stage. Furthermore, regarding evaluations of chromosomes, any genes inherited from parents neednt be recalculated. The two improved ideals greatly reduced the program running time and space complexity. Experimental results show that the evolution efficiency of the improved algorithm is much better than standard MEP.

Acknowledgment. This work was supported by the National Natural Science Foundation of China (Grant NO. 61170199), and the Scientific Research Fund of Education Department of Hunan Province, China (Grant NO. 11A004).

References

1. Oltean, M., Grosan, C., Diosan, L., Mihaila, C.: Genetic Programming With Linear Representation a Survey. WSPC/INSTRUCTION FILE (2008)
2. O'Nell, M., Vanneschi, L., Gustafson, S., Banzhaf, W.: Open Issues in Genetic Programming. Genetic Programming and Evolvable Machines 11, 339–363 (2010)
3. He, P., Kang, L., Fu, M.: Formality Based Genetic Programming. In: IEEE Congress on Evolutionary Computation, Hong Kong, pp. 4080–4087 (2008)
4. Koza, J.R.: Genetic Programming: On the Programming of Computers by Means of Natural Selection. MIT Press (1992)
5. Tsakonas, A.: A comparision of classification accuracy of four genetic programming-evolved intelligent structures. Informatin Sciences 176, 691–724 (2006)
6. Koza, J.R., Poli, R.: Genetic programming. In: Burke, E.K., Kendall, G. (eds.) Search Methodologies: Introductory Tutorials in Optimization and Decision Support Techniques, ch. 5. Springer (2005)

7. Oltean, M., Grosan, C.: A Comparison of Several Linear Genetic Programming Techniques. Complex Systems 14, 285–313 (2003)
8. He, P., Johnson, C.G., Wang, H.: Modeling grammatical evolution by automaton. Science China/Information Sciences 54(12), 2544–2553 (2011)
9. National Center for Biotechnology Information, http://www.ncbi.nlm.nih.gov; Oltean, M., Dumitrescu, D.: Multi expression programming, technical report, UBB-01-2002, Babes-Bolyai University, Cluj-Napoca, Romania, http://www.mep.cs.ubbcluj.ro
10. Chen, Y.H., Jia, G., Xiu, L.: Design of Flexible Neural Trees using Multi Expression Programming. In: Proceeding of Chinese Control and Decision Conference, vol. 1, pp. 1429–1434 (2008)
11. Oltean, M., Grosan, C.: Evolving Digital Circuits using Multi Expression Programming. In: 2004 NASA/DoD Conference on Evolvable Hardware, pp. 87–94. IEEE Computer Science, Washington (2004)
12. Wang, Y., Yang, B., Zhao, X.: Countour Registration Based on Multi-Expression Programming and the Improved ICP. IEEE (2009)
13. Cattani, P.T., Johnson, C.G.: ME-CGP: Multi Expression Cartesian Genetic Programming. In: IEEE Congress on Evolutionary Computation, pp. 1–6 (2010)

Improved Environmental Adaption Method
for Solving Optimization Problems

K.K. Mishra, Shailesh Tiwari, and A.K. Misra

MNNIT Allahabad
India-211004
kkm@mnnit.ac.in

Abstract. Recently a new optimization algorithm, Environmental Adaption Method (EAM) has been proposed to solve optimization problems.EAM target its search toward optimal solution using two operators adaption and mutation operator. Both of these operators perform random search of full search space until they got a good solution. Although EAM has a good convergence rate yet it can be further improved if instead of performing random search of overall search space, operators limit their search to a finite region that has a very high probability containing optimal solution. Proposed algorithm select this region by utilizing the information received from the known genomic structures of best solutions obtained in previous generations. A very similar idea was used in Particle Swarm Optimization algorithm however unlike PSO it does not require additional store. Updated version is very fast as compared to basic EAM algorithm. Different state of art algorithms are compared on benchmark functions to check its performance.

Keywords: Nature, Optimization, Evolutionary Algorithms.

1 Introduction

EAM was proposed by the authors K.K.Mishra, Shailesh Tiwari and A.K.Misra (2011) as an alternate of GA and PSO to solve optimization problems. Although it is comparable to other existing algorithms (Genetic Algorithm (GA), Particle Swarm optimization (PSO)) yet more efforts are required to improve the convergence rate of this algorithm. This paper proposes an updated version of EAM by incorporating some ideas which is very similar as of Particle Swarm optimization (PSO). The organization of paper is as follows Section 2 provides in brief the Particle Swam Optimization and Environmental Adaption Method, Section 3 presents technical details of proposed algorithm, Section 4 covers experimental analysis of proposed algorithm and finally in Section 5 conclusions are drawn and future direction of further work is explored.

Z. Li et al. (Eds.): ISICA 2012, CCIS 316, pp. 300–313, 2012.

2   Background Details

2.1   Environment Adaption Method [EAM]

EAM is computerized implementation of the process used by the species while they adapt themselves to survive in changing environment. According to this theory, all existing species has a tendency to update their genomic structure so that they can attain better genomic structure and contribute in creation of better environment(good as compared to previous). To measure which environment is better we use a term environmental fitness that can be taken as average fitness of all species residing in the current environment. Using its current fitness and environment fitness, each species refine its structure. Sometimes during refinement some mutation may occur due to environmental noise. Finally best species selected from newly created species and existing species form new environment. Those species which do not contribute in new environment will die. Since genomic changes require current environment, the possibility of getting good specie in next generation is very high.

Alike GA and PSO this algorithm is also a population based algorithm. This algorithm processes a randomly generated initial population using three operators. The first operator is named as an adaption operator and this causes every solution of current generation to update its genome structure taking its current fitness and environmental fitness in to account. Second operator simply performs mutation, which may result due to environmental noise. When both of these operators apply on the current generation, next generation is generated. Selection operator is used to select best solutions from the combination of next generation and previous generation form current generation with better environmental fitness. Those solutions which are not selected will be discarded from the current generation. This process is repeated until either we get the desired solution or maximum number of iterations has reached.

Adaption Operator: Adaption operator updates a solution P_i as follows:

$$P_{i+1} = (\alpha * (P_i)^{F(P_i)/F_{avg}} + \beta)/(2^l - 1) \tag{1}$$

Where $F(P_i)$ is the fitness value of P_i, α and β are the random numbers that are to be decided as per the requirement of problem. l represents the total number of bits in an individual. F_{avg} is the average fitness value of current population. F_{avg} represents the current environmental fitness.

Mutation Operator: Mutation operator generate new solution P_{i+1} by flipping one or more bit of P_i. Mutation is carried out according to the following pseudo-code:

- for each member in the population
- for each bit in the members genome
- Invert the bit with a probability of $Pmut$
- Each new member so generated is considered for selection for the next generation

Selection: Best solutions equal to size of initial population are to be selected from the combination of next generation and previous generation to form current generation.

2.2   Particle Swarm Optimization [PSO]

PSO is a population based optimization tool, where the system is initialized with a population of random particles and the algorithm searches for optima by updating generations. Suppose that the search space is n-dimensional, and then the particle i of the swarm can be represented by an n-dimensional vector $X_i = (x_{i1}, x_{i2}, ..., x_{in})$. The velocity of this particle can be represented by another n-dimensional vector $V_i = (v_{i1}, v_{i2}, ..., v_{in})$. The fitness of each particle can be evaluated according to the objective function of optimization problem. The best previously visited position of the particle i is noted as its individual best position $P_i = (p_{i1}, p_{i2}, ..., p_{in})$. The position of the best individual of the whole swarm is noted as the global best position $G = (g_1, g_2, ..., g_n)$. At each step, the velocity of particle and its new position will be assigned according to the following two equations:

$$V_i = \omega * V_i + c_1 * r_1 * (P_i - X_i) + c_2 * r_2 * (G - X_i) \tag{2}$$

$$X_i = X_i + V_i \tag{3}$$

where ω is called the inertia weight that controls the impact of previous velocity of particle on its current one. r_1, r_2 are independently uniformly distributed random variables with range $(0,1)$. c_1, c_2 are positive constant parameters called acceleration coefficients which control the maximum step size. In PSO, Eq. (3) is used to calculate the new velocity according to its previous velocity and to the distance of its current position from both its own best historical position and the best position of the entire population or its neighborhood. Generally, the value of each component in V can be clamped to the range $[-vmax, vmax]$ to control excessive roaming of particles outside the search space. Then the particle flies toward a new position according Eq.(2). This process is repeated until a user-defined stopping criterion is reached.

Based on the size of its neighbors, many variants of PSO have been developed. In canonical PSO size of its neighbors is equal to the size of population. In Local PSO model each particle update its position by considering some solutions as its neighbors. Interested reader may refer J. Kennedy and R. Eberhart(1995), R. Poli, J. Kennedy and T. Blackwell(2007),Yong Wang and Z.Cai (2009) for detailed information.

Although PSO has very good convergence rate but it may face problem of stagnation. Tuning of random parameters is very important issue and if properly done it can avoid premature convergence of PSO on local optimal solutions. Many papers   Yuhui Shi and Russell Eberhart(1998), Yuhui Shi and Russell Eberhart(2000) and M clerc and J kennedy(2002) suggested solution of this issue by giving various methods to tune random parameters(Like ω, c_1, c_2, r_1 and r_2).

A basic variant of the PSO algorithm works by having a population (called a swarm) of candidate solutions(called particles). These particles are moved around in the search-space according to a few simple formulae. The movements of the particles are guided by their own best known position in the search-space as well as the entire swarm's best known position. When improved positions are being discovered these will then come to guide the movements of the swarm. The process is repeated and by doing so it is hoped, but not guaranteed, that a satisfactory solution will eventually be discovered.

To remove basic limitations of PSO many new variants of PSO have been proposed. M. Senthil Arumugam, M. V. C. Rao, Aarthi Chandramohan (2008) proposed a new version of PSO (called as GLBestPSO) which approach global best solution in different manner. Further modification of same algorithm is done in 2009 by the same authors M. Senthil Arumugam, M. V. C. Rao, Aarthi Chandramohan (2009).

3   Proposed Approach

There are two main targets of any optimization algorithm. One is to search optimal solution in less iteration and other is to escape from local optimal solution. To improve the convergence rate of existing EAM and to prevent it to converge on local optimal solution, a slight change has been done in the adaption operator of basic EAM. In EAM each particle update its structure based on the direction received by changing environment there is no role of best particles. However genetic structures of best particles are also providing some guidelines to approaching optimal solution .This idea was implemented in PSO. In proposed version, same idea is used to modify the adaption operator of EAM. In proposed version, each solution will update its structure by receiving guidelines from the genomic structure of best solutions of previous and current generations. In updated EAM the adaption operator will work as follows.

$$X_{i+1} = (\alpha * (X_i)^{F(X_i)/F_{avg}} + \beta\{(P_{i-1} - X_i) + (G_i - X_i)\})/(2^l - 1) \quad (4)$$

Where G_i is the position vector of best particle of current generation and P_{i-1} is the best particle of previous generation (if it exists), X_i is the genomic structure of a particle that is updating its structure. In updated version there will be no change in mutation and selection operator.

Also in proposed version, setting of parameters α and β has done cleverly so that global optimal solution can be retrieved in late generations. In early generations the value of α is be randomly selected from set (0- 2L-1) and value of β is taken as $(F_{maxi} - F(X_i))$ where F_{maxi} is the fitness value of best solution of ith generation and $F(X_i)$ is the fitness value of solution X_i. This is done because in all optimization problems first requirement is to search optimal value as soon as possible, in initial generations proposed values will force algorithm to search from whole search space. In late generation, when all population is converging toward one structure (this is possible only if all solutions are targeting either local optimum or global optimum), we choose β as 0(all solutions will be equal) and the value of α as 1 so that when algorithm reaches optimal solution it will not update the value of optimal solutions. Once optimal solution is captured, A procedure is applied to check whether we are targeting global optimal solution or not. To perform this check, when solutions became stable, we again select from set (0-2L-1) .This is done to perturb solutions of current generations so that premature convergence can be avoided. The reason behind it can be explained as follows, if all solutions of previous generation were approaching toward global optimum, there will be no improvement in the structure of optimal solution. If the search was targeted toward local best then this stretching of will force new generation to capture new global/local solution and will again create an unstable population. We have to implement this stretching of α again and again until we get stable population. This algorithm will defiantly able to cover many optimal solutions in minimum number of iterations.

Proposed Algorithm:

```
Input:
          MAX_G (maximum number of generations)
          P_S (Population size)
Output:
          Q* (Final set)
Other variables:
          n (number of generation)
          O' (intermediate offspring)
          O (final offspring)
          P (temporary pool) in
             (ith individual of nth generation)
          Pi-1(Best individual of ith generation)
```

1: Set n=0
2: Generate initial population POP_0 by P_S random individuals
3: Apply Adaptation operator (calculation given below) to each member of population POP_n asexually and form O'_n
4: $P_{i+1} = (\alpha * (P_i)^{F(X_i)/F_{avg}} + \beta\{(P_{best\ i-1} - P_i) + (G_i - P_i)\}/(2^l - 1)$
5: Apply mutation operator as per the probability of mutation O'_n and form O_n
6: $P_n = \{POP_n \cup O_n\}$
7: $n = n + 1$
8: **if** $n < MAX\_G$ **then**
9:      go to 12
10: **elsego** to 9
11: **end if**
12: Set $Q^* = $ best solutions of P_n
13: Return Q^*
14: $POP_n = P\_S$ fittest individuals in P_{n-1}
15: go to 5

---

4   Result Analysis and Discussion

To check the efficiency of proposed algorithm, This algorithm is compared with other optimization algorithms EAM and ePSO . The following five benchmark functions were used for the analysis:

Spherical function:

$$f_1(x) = \sum_{i=1}^{7} x_i^2 \tag{5}$$

Rosenbrocks's function:

$$f_2(x) = \sum_{i=1}^{6} \left[100(x_{i+1} - x_i)^2 + (x_i - 1)^2\right] \tag{6}$$

Rastrigins's function:

$$f_3(x) = \sum_{i=1}^{7} (x_i^2 - 10cos2\pi x + 10) \tag{7}$$

Griewank function:

$$f_4(x) = \frac{1}{4000} \sum_{i=1}^{7} x_i^2 - \prod_{i=1}^{7} cos(\frac{x_i}{\sqrt{i}}) \tag{8}$$

Schwefel function:

$$f_5(x) = \sum_{i=1}^{7} x_i sin\sqrt{|x_i|} \tag{9}$$

The range of input parameters, the optimal fitness value and the input for which this best fitness comes is shown in the following table:

All algorithms were executed on P4 dual core processor for 40 rounds each of 100 iterations, In each table one block represents the average fitness of all best solutions generated in different generations. Size of initial population for all algorithms was 100 and to avoid any biasing it was generated randomly. The proposed algorithm Modified Environmental adaption method (MEAM) outperforms all other algorithms in all functions. It can be seen that performance of ePSO is good in comparison of EAM in functions which has only one optimal solution (Example Spherical function), in other functions the EAM perform better.

Table 1. Details of Benchmark Functions

Function	Range of x	Optimum	Optimal fitness
f_1	(-100, 100)	(0, 0, 0, ..., 0, 0)	0
f_2	(-30, 30)	(1, 1, 1, ..., 1, 1)	0
f_3	(-5.12, 5.12)	(0, 0, 0, ..., 0, 0)	0
f_4	(-600, 600)	(0, 0, 0, ..., 0, 0)	0
f_5	(-500, 500)	(420.96, ..., 420.96)	-2933

Table 2. Comparison of Results on Spherical's Function

Serial Number	ePSO	EAM	MEAM
1	6832.1	3124.9	3401.01
2	5163.6	3124.9	3401.01
3	4426.04	3124.77	3401
4	3073.24	3124.77	3401
5	3073.24	2948.79	2321.53
6	3073.24	2948.79	2321.53
7	3073.24	2948.79	2321.53
8	3070.49	2948.79	2267.84
9	2864.7	2872.31	2037.98
10	2832.38	1554.8	2037.98
11	2761.5	1554.8	1373.95
12	2638.18	1337.02	1373.95
13	2160.54	1239.4	1373.95
14	2095.17	1239.4	1373.95
15	1816.36	962.615	1065.41
16	1803.49	949.99	1048.81
17	1793.5	938.702	735.789
18	1793.5	938.508	357.8
19	1792.1	937.386	357.8
20	1591.27	937.386	173.502
21	1171.51	742.842	173.502
22	1171.51	649.4	173.502
23	1171.51	649.4	173.502
24	1171.51	649.4	96.3931
25	1147.35	649.073	96.3931
26	1040.17	638.179	47.0097
27	1040.17	448.879	39.834
28	999.062	448.879	39.834
29	999.062	448.879	39.834
30	914.876	441.652	39.5884
31	914.876	441.652	39.5884
32	914.876	349.343	39.0348
33	282.79	349.343	38.5596
34	282.776	349.232	38.5596
35	282.776	349.232	17.6911
36	282.751	195.889	17.6911
37	282.269	195.889	15.9318
38	227.311	195.883	15.9099
39	112.882	195.883	15.9099
40	112.882	142.952	15.9099

Table 3. Comparison of Results on Rosenbrocks Function

Serial Number	ePSO	EAM	MEAM
1	3.56E+06	2.09E+07	567664
2	3.56E+06	2.69E+06	567664
3	3.56E+06	2.69E+06	421737
4	2.85E+06	929560	140717
5	2.85E+06	929560	140717
6	2.85E+06	929560	140717
7	2.85E+06	925241	140715
8	894347	909016	140715
9	894347	812541	139381
10	828948	812541	87972.5
11	684184	780955	36245.6
12	684184	780955	10703
13	173209	574272	7933.58
14	101628	574272	7933.58
15	101628	468464	7933.58
16	101628	126551	7933.58
17	101628	88080.1	7259.03
18	67277.3	87194.7	7259.03
19	67198.6	86772.7	7258.63
20	67198.6	53475.1	6400.67
21	40374.4	51039.4	5696.12
22	30846.2	51039.4	5306.12
23	30846.2	51039.4	5306.12
24	30846.2	51039.4	5191.27
25	30776.2	50291.9	3645.98
26	26576.4	48253.6	3645.98
27	26576.4	47712.3	1319.8
28	26529.8	25927	1317.51
29	11930.7	25927	1319.8
30	11525.2	25927	282.149
31	11525.2	5793.61	282.149
32	2107.6	5793.61	282.149
33	2107.6	5793.61	282.149
34	2107.6	5783.19	282.054
35	2111.38	5797.76	282.435
36	2107.61	4110.41	282.082
37	1792.21	4055.99	257.482
38	1794.1	2612.63	257.526
39	1794.1	2612.63	253.459
40	486.359	933.32	202.21

Table 4. Comparison of Results on Rastrigins's function

Serial Number	ePSO	EAM	MEAM
1	54.4592	52.7785	49.0336
2	54.4592	52.7785	49.0336
3	50.4205	52.7731	49.0336
4	46.0709	38.7731	47.679
5	46.0709	38.6085	47.679
6	46.0709	35.0805	42.656
7	46.0709	35.0783	42.643
8	40.5637	35.0783	42.643
9	40.5637	34.2153	42.643
10	40.5637	34.1127	40.4798
11	31.7356	28.1269	40.4798
12	30.4034	17.35	40.4798
13	30.4034	17.35	20.9051
14	30.4034	17.35	20.9051
15	30.4034	17.35	20.9051
16	28.9629	17.3081	20.9051
17	23.9882	17.3058	20.9051
18	23.9871	17.2075	20.7704
19	23.9871	17.2075	20.7539
20	23.9871	16.3002	14.525
21	23.9744	16.3002	14.525
22	21.3299	16.3002	12.3339
23	21.3299	15.8745	12.3339
24	17.1344	14.8437	12.3339
25	16.227	14.8082	12.3241
26	16.2001	14.8082	12.3241
27	15.8518	14.4778	12.3191
28	14.5408	14.4544	11.2822
29	14.5408	14.4544	11.2822
30	13.0048	13.5376	11.2822
31	13.0048	13.5376	10.8337
32	12.7852	13.5376	10.5385
33	12.7852	13.5376	10.5385
34	12.7844	13.5329	9.21835
35	12.7844	13.5329	9.21835
36	12.7817	13.5329	9.21835
37	12.2923	13.5217	8.81107
38	12.2916	13.5196	8.80832
39	12.2916	13.5196	8.80832
40	12.2916	13.3078	8.80832

Table 5. Comparison of Results on Griewank Function

Serial Number	ePSO	EAM	MEAM
1	42.7324	46.9261	57.7216
2	42.7311	45.9261	56.3108
3	42.7311	45.9261	53.0596
4	14.9214	39.2958	49.1578
5	14.9214	39.2958	35.3437
6	14.9193	14.1983	35.3437
7	14.6399	14.1983	19.6726
8	12.5594	14.1527	19.6726
9	12.5594	14.1527	19.6726
10	12.113	14.1527	18.3314
11	12.113	13.1498	18.3314
12	12.113	13.1498	12.5213
13	12.113	5.87027	12.5213
14	9.49781	5.87027	12.5213
15	8.08872	5.87027	11.1594
16	7.46766	5.5224	10.3626
17	7.44002	4.93408	6.04311
18	7.43546	4.93408	6.04311
19	4.27107	4.93403	3.64772
20	4.27107	4.93403	3.64772
21	4.27107	4.93403	3.64772
22	4.27107	3.99827	3.64772
23	4.11751	3.99827	3.64772
24	4.03309	3.99827	3.33632
25	4.03309	2.72725	2.82536
26	2.69133	2.17733	2.4233
27	2.69133	2.13886	2.42198
28	2.41306	2.13886	2.41853
29	2.05465	2.12171	2.31442
30	2.05465	2.05042	2.31442
31	2.04982	2.05042	2.31442
32	1.87658	2.04304	1.58158
33	1.87658	2.04304	1.58158
34	1.73648	1.87351	1.58158
35	1.73648	1.52797	1.58158
36	1.72411	1.52797	1.57846
37	1.54847	1.49431	1.46583
38	1.54847	1.49431	1.34589
39	1.45414	1.49431	1.34589
40	1.37019	1.19723	1.31902

Table 6. Comparison of Results on Schwefel Function

Serial Number	ePSO	EAM	MEAM
1	-1259.01	-1204.85	-1573.13
2	-1259.01	-1420.42	-1573.13
3	-1259.01	-1536.19	-1573.13
4	-1624.69	-1556.32	-1603.25
5	-1624.69	-1556.32	-1603.25
6	-1755.66	-1556.32	-1978.22
7	-1755.84	-1611.1	-1978.34
8	-1876.12	-1611.36	-1978.34
9	-1876.12	-1838.58	-1978.34
10	-1876.12	-1848.3	-1994.54
11	-1939.66	-1848.3	-1994.54
12	-1939.66	-1848.3	-1994.54
13	-2062.18	-1944.22	-2054.87
14	-2062.18	-1944.22	-2090.67
15	-2062.18	-2018.56	-2093.39
16	-2078.7	-2018.56	-2141.02
17	-2291.33	-2018.56	-2200.2
18	-2291.45	-2053.92	-2200.2
19	-2293.37	-2053.92	-2250.06
20	-2339.46	-2057.52	-2250.06
21	-2339.46	-2155.63	-2285.18
22	-2349.97	-2155.63	-2285.18
23	-2349.97	-2155.63	-2285.18
24	-2370.47	-2182.35	-2285.18
25	-2370.67	-2182.6	-2285.18
26	-2370.67	-2222.14	-2285.59
27	-2370.67	-2231.03	-2306.24
28	-2428.71	-2246.67	-2306.24
29	-2428.71	-2246.67	-2306.24
30	-2428.71	-2300.42	-2312.97
31	-2428.71	-2300.42	-2375.66
32	-2468.73	-2309.22	-2384.13
33	-2468.73	-2309.22	-2384.13
34	-2474.53	-2309.44	-2384.13
35	-2474.53	-2309.44	-2384.13
36	-2474.56	-2316.5	-2579.21
37	-2474.56	-2316.5	-2589.54
38	-2476.05	-2319.21	-2589.55
39	-2476.05	-2319.21	-2589.55
40	-2476.74	-2334.82	-2625.59

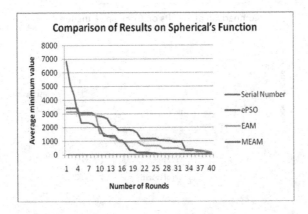

Fig. 1. Graph-1

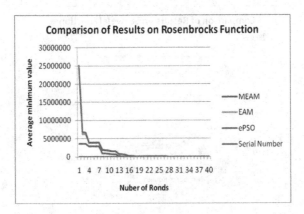

Fig. 2. Graph-2

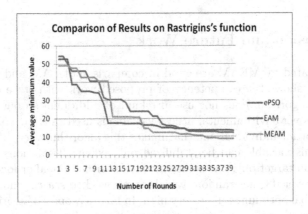

Fig. 3. Graph-3

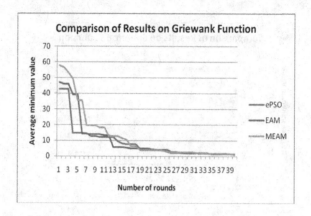

Fig. 4. Graph-4

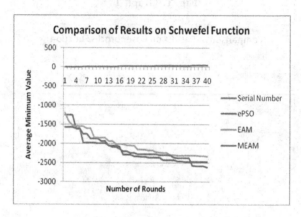

Fig. 5. Graph-5

5   Conclusion and Future Work

Results generated by MEAM are good in comparison of EAM and ePSO algorithm. Graph shows the competency of proposed work with state of art algorithms. This algorithm does not use any kind of additional storage to archive its goal. Moreover environmental fitness variable is used as a variable for whole population not for individual one that makes this algorithm very different from other one. This variable can be stabilized only when all solutions of the current population target only one solution which is either global or local optimum solution. However if other random variables are used to search whole space the chances of global optimum will be very high. In future proposed algorithm can be combined with other evolutionary algorithm to further improve its performance.

References

1. Mishra, K.K., Tiwari, S., Misra, A.K.: A bio inspired algorithm for solving optimization problems. In: 2011 2nd International Conference on Computer and Communication Technology (ICCCT), September 15-17, pp. 653–659 (2011)
2. Wang, Y., Cai, Z.: A hybrid multi-swarm particle swarm optimization to solve constrained optimization problems. Frontiers of Computer Science in China 3(1), 38–52 (2009)
3. Kennedy, J., Eberhart, R.: Particle swarm optimization. In: IEEE International Conference on Neural Networks, vol. 4, pp. 1942–1948 (November/December 1995)
4. Poli, R., Kennedy, J., Blackwell, T.: Particle swarm optimization: An Overview. In: Swarm Intelligence, pp. 33–57. Springer, New York (2007)
5. Shi, Y., Eberhart, R.C.: Parameter Selection in Particle Swarm Optimization. In: Porto, V.W., Waagen, D. (eds.) EP 1998. LNCS, vol. 1447, pp. 591–600. Springer, Heidelberg (1998)
6. Eberhart, R.C., Shi, Y.: Comparing inertia weights and constriction factors in particle swarm optimization. In: Proceedings of the 2000 Congress on Evolutionary Computation, vol. 1, pp. 84–88 (2000)
7. Clerc, M., Kennedy, J.: The particle swarm - explosion, stability, and convergence in a multidimensional complex space. IEEE Transactions on Evolutionary Computation 6(1), 58–73 (2002)
8. Senthil Arumugam, M., Rao, M.V.C.: Aarthi Chandramohan: A new and improved version of particle swarm optimization algorithm with global-local best parameters. Knowl. Inf. Syst. 16(3), 331–357 (2008)
9. Senthil Arumugam, M., Rao, M.V.C., Tan, A.W.C.: A novel and effective particle swarm optimization like algorithm with extrapolation technique. Appl. Soft Comput. 9(1), 308–320 (2009)

An Optimization Algorithm Based on Evolution Rules on Cellular System

Jieqing Xing[1] and Houqun Yang[2]

[1] Department of Information Technology
Qiongtai Teachers College
Haikou 571100, P.R. China
[2] College of Information Science & Technology
Hainan University
Haikou 570228, P.R. China

Abstract. As a new branch of natural computing, membrane computing has received increasing attention. The hierarchical and parallel structure of P system provides benefits for the resolving of optimization problems. In this paper, we combined membrane computing and evolutionary algorithms, and proposed an optimization algorithm to resolve the multi-variable optimization problems with constraints. The two standard testing functions were adopted to evaluate the proposed optimization algorithm. The results of the experiments showed the effectiveness of the proposed method.

Keywords: Evolution Rule, Optimization Algorithm, Testing Function.

1 Introduction

Biological system is a type of complicated computing system. Many intelligent techniques such as natural computing mimic the behaviors of biological system. As a new branch of natural computing, membrane computing has received much attention. The basic model of membrane computing is called as P system. Păun G. proposed the first variants of P systems in 1998[1]. P system is a distributed parallel computing model abstracted from the functionality and structure of tissues or organs of living cells. A P system consists of a hierarchical structure composed of several membranes, objects in each membrane, and its evolution rules applied to membranes and objects. The membrane structure delimits compartments in a hierarchical or network manner. Objects are arranged as multi-sets and dispersed across these compartments. Rules are provided for specified membranes to control the evolution of objects in a maximally parallel way[2].

P systems have been developed principally from a mathematical point of view, building a variety of computing models for different problems. In [3-5], a hybrid algorithm combining P systems and genetic algorithms was proposed to solve single-objective and multi-objective numerical optimization problems. In [6][7], a membrane algorithm with a nested membrane structure was introduced to

Z. Li et al. (Eds.): ISICA 2012, CCIS 316, pp. 314–320, 2012.

solve the TSP problem as well as the min storage problem[8]. In [9], a basic membrane system was applied to set up a mathematical model of the biological process triggered by an internal injury in the knee tissue. In [10], a tissue P system with cell division was presented to solve the 3-coloring problem. In [11], a class of tissue P systems with active membranes was proposed to generate iso-picture languages. In [12], a new method with discrete and stochastic membrane systems was employed to model metapopulations. In [13], a membrane algorithm combining one level membrane structure with quantum-inspired evolutionary algorithms (QIEA) was proposed to solve knapsack problems.

Evolutionary algorithms (EAs) are practical and robust optimization and search methods inspired by evolutionary processes occurring in natural selection and molecular genetics. The main features of EAs are the representation and evaluation of individuals, population dynamics, evolutionary operators such as selection, crossover and mutation[13][14]. As compared to conventional optimization methods, EAs are more suitable for solving complex optimization problems as they exhibit an intrinsic parallelism derived from dealing with multiple individuals, show remarkable adaptability and flexibility to various applications, and provide good search capability and robust results[15].

P system and EAs are different with respect to the objects and rules used, computational strategies employed, but both are nature-inspired models and are applied to solve complex problems. The possible interaction between P systems and EAs, also mentioned by the list of twenty-six open problems in membrane computing[16], represents a fertile research field. In this paper, we combine P system and EAs to propose an optimization algorithm based on evolution rules on cellular membrane. Here, evolutionary rules on cellular membrane are abstracted and integrated into the hierarchical structure.

2   P System

At present, there are three main types of P systems: cell-like P systems, tissue-like P systems and neural-like P systems[6]. A cell-like P system is made up of one membrane cell. A tissue-like P system consists of several one-membrane cells in a common environment. Neural-like P systems use neurons as objects. A cell-like P system, considered in this paper, is principally characterized by three ingredients: the membrane structure delimiting compartments, the multisets of abstract objects placed in compartments, and the evolution rules as-sociated to objects or membranes. The embrace structure of a cell-like P system, shown in Fig. 1, is hierarchical arrangement of membranes[4]. A skin membrane separates the system from its environment. Several membranes, each of which defines a region, are placed inside the skin membrane. An elementary membrane is the one without any membrane inside. Each region forms a different compartment of the membrane structure and contains a multiset of objects, membranes and a set of evolution rules.

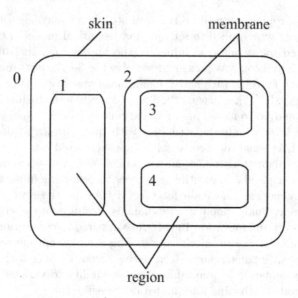

skin   membrane

0   1   2   3   4

region

Fig. 1. Membrane Structure

A cell-like P system with an output set of objects and using transformation and communication rules is formally defined as follows[4][5].

$$\prod = (V, T, \mu, w_1, ..., w_m, R_1, ..., R_m, i_0)$$

Where:

- V is an alphabet, its elements are called objects;
- $T \in V$, the output alphabet;
- μ is a membrane structure consisting of m membranes, with the membranes and the regions labeled in an one-to-one manner with elements of a given set Λ-usually the set $\{1, 2, ..., m\}$;
- m is called the degree of $\prod$;
- $w_i, 1 \leq i \leq m$, are strings which represent multisets over V associated with the regions $1, 2, ..., m$ of μ;
- $R_i, 1 \leq i \leq m$, are strings which represent multisets over V associated with the regions $1, 2, ..., m$ of μ;
- i_0 is a number between 1 and m which specifies the output membrane of $\prod$.

The rules of $R_i, 1 \leq i \leq m$, have the form $a \rightarrow v$, where $a \in V$ and $v \in (V \times \{here, out, in\})^*$. The multiset v consists of pairs $(b, t), b \in V$ and $t \in \{here, out, in\}$, where $here$ means that b will stay in the region where the rule is applied; out is used to show that b exists the region and in means that b will be communicated to one of the membranes contained in the current region which is chosen in a non-deterministic way.

A P system provides a suitable framework for distributed parallel computation that develops in steps. The multisets associated to regions form a configuration of a P system. The computation starts by processing the initial multisets,

$w_i, 1 \le i \le m$, and then the system will go from one configuration to a new one by applying the rules associated to regions in a non-deterministic and maximally parallel manner, i.e., all the objects that may be transformed or communicated must be processed. The system will halt when no more rules are available to be applied. A computation is a sequence of configurations obtained as it is described above, starting with the initial configuration and ending with the configuration when the system halts. The result of a computation, a multiset of objects, is obtained in the output region, . For more details about P systems definition see[1]. We notice that the rules presented above combine both transformation and communication, but these operations may be separated and then the transformation rules are responsible for evolving the objects and the communication rules will transfer objects among regions according to some targets. The initial multisets of objects may be replaced by strings or multisets of strings, the multiset rewriting by string rewriting and in the output region we obtain a set or multiset of strings.

3   Membrane Computing Based Optimization Algorithm

The main characteristics of P system are hierarchical structure and maximally parallel manner. In biological system, these characteristics are necessary for optimized evolution. Based on these characteristics, an algorithm for optimization problem with constraints, called membrane computing based optimization algorithm (MCOA), was proposed. This method is similar with the solution of random ensemble learning method.

3.1   Optimization Problem with Constraints

Optimization problems with constraints are popular practical problem in real engineering. It could be formularized as follows:

$$\begin{cases} min \ f(x) \\ s.t. \ C_i(x) \le 0, i = 1, ..., r \end{cases} \tag{1}$$

where, $x = (x_1, x_2, ..., x_n)$. $f(x)$ is the objective function. x is the optimization variable, which is represented as the set of objects. $C_i(x) \le 0$ is the constraint of the optimization problem, which is integrated into the communication rules.

3.2   Membrane System

The membrane system of MCOA is still the classic multi-membrane nesting hierarchical structure. It consists of two levels, surface membrane and inner membrane, which are illustrated as Fig. 2. The objects in each membrane are the possible solutions of the optimization problem. Each solution is coded as real numbers. In each inner membrane, the optimization problem is solved independently. Through several iterations, the rough solution is obtained in each

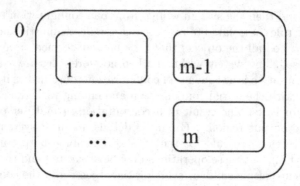

Fig. 2. The membrane system of MCOA

inner membrane. We do not consider the constraints in the solution process of each inner membrane. The inner membrane could communicate with the surface membrane. Here, the communication process is to transfer the computing results of the inner membrane to the surface membrane. The solutions which meet the constraints are exported. If there is no solution that meets the constraints in any inner membrane, the membrane has no exported solution.

3.3   Evolution Process of MCOA

Above all, the optimization problem without constraints is solved by each inner membrane independently. The evolution process is performed in each inner membrane, which is summarized as follows:

1. produce a set of random initial solutions;
2. cross and variation; the differential evolutionary operator based on the steepest descent method is adopted in the variation process;
3. the objective function is used as fitness function to assess the fitness of each possible solution;
4. delete the worse solutions; the left solutions are used as the new initial solutions to continue the evolution process;
5. terminate the evolution process of the inner membrane, and export all the solutions to the surface membrane;
6. perform cross and variation evolution in the surface membrane;
7. terminate the evolution process of the surface membrane, and export the final solution; otherwise, copy all the solutions and transfer each copy into each inner membrane;
8. the solutions obtained from the surface membrane are used as new initial solutions and jump to the step (2).

4   Experiment

We adopt two international standard test functions to evaluate the effectiveness of the algorithm MCOA. These test functions are multi-variable objective functions with multi-dimension constraints. The types of objective functions include linear and nonlinear ones. And the types of constraints include nonlinear equality, linear inequality and nonlinear inequality. The adopted test functions are as follows[17],

G1: Minimize:

$$f(\overline{x}) = (x_1 - 10)^2 + 5(x_2 - 12)^2 + x_3^4 + 3(x_4 - 11)^2$$
$$+ 10x_5^6 + 7x_6^4 + x_7^4 - 4x_6x_7 - 10x_6 - 8x_7$$

Subject to:

$$c_1(\overline{x}) = -127 + 2x_1^2 + 3x_2^4 + x_3^4x_4^2 + 5x_5 \leq 0$$
$$c_2(\overline{x}) = -282 + 7x_1 + 3x_2^4 + 10x_3^2 + x_4 - x_5 \leq 0$$
$$c_3(\overline{x}) = -196 + 23x_1 + x_2^2 + 6x_6^2 - 8x_7 \leq 0$$
$$c_4(\overline{x}) = 4x_1^2 + x_2^2 - 3x_1x_2 + 2x_3^2 + 5x_6 - 11x_7 \leq 0$$
$$- 10 \leq x_i \leq 10 \ (i = 1, ..., 7)$$

G2: Minimize:

$$f(\overline{x}) = x_2 - x_1^2 = 0$$

Subject to:

$$-1 \leq x_1 \leq 1, -1 \leq x_2 \leq 1$$

The parameters of experiments are set as follows: in each inner membrane, the size of the random initial solutions is 100. The number of iterations in each inner membrane is 50. The number of iterations in surface membrane is 50. The number of communications between inner membrane and surface membrane is 50. The experiment of each test function is repeated fifty times. The experimental results under the parameter configuration are summarized in Table 1. The best, mean and worst solutions are given in Table 1. From Table 1, it is noted that MCOA find the global optimal solution for G1, and find a closely approximate optimal solution for G2.

Table 1. The solutions for the two test functions

Function	Optimal	Best	Mean	Worst	St.dev
G1	680.630	680.631	680.633	680.651	6.6E-03
G2	0.7500	0.7500	0.7502	0.7508	1.1E-04

5 Conclusion

Inspired by the idea of membrane computing, we designed the optimization algorithm MCOA on P system. The method utilized the hierarchical and parallel structure of membrane system, and performed evolution rules of cellular membrane on its nesting structure. The experiments on standard testing functions evaluated the effectiveness of the proposed method. In particular, MCOA found a global optimal solution for G1 testing function.

Acknowledgement. This work is supported by the Natural Science Foundation of Hainan Province (No. 610216) , the scientific research project of Qiongtai Teachers College (qtky2010-20) and the scientific research project of colleges and universities in Hainan province (Hjkj2012-59).

References

1. Păun, G.: Computing with membranes. Journal of Computer and System Science 61(1), 108–143 (2000)
2. Păun, G.: Tracing some open problems in membrane computing. Romanian Journal of Information Science and Technology 10, 303–314 (2007)
3. Huang, L., Wang, N.: An Optimization Algorithm Inspired by Membrane Computing. In: Jiao, L., Wang, L., Gao, X.-b., Liu, J., Wu, F. (eds.) ICNC 2006. LNCS, vol. 4222, pp. 49–52. Springer, Heidelberg (2006)
4. Huang, L., He, X.X., Wang, N., Xie, Y.: P systems based multi-objective optimization algorithm. Progress in Natural Science 17, 458–465 (2007)
5. Huang, L., Wang, N., Zhao, J.H.: Multiobjective Optimization for Controller Design. Acta Automatica Sinica 34, 472–477 (2008)
6. Nishida, T.Y.: An approximate algorithm for NP-complete optimization problems exploiting P systems. In: Proc. Brainstorming Workshop on Uncertainty in Membrane Computing, pp. 185–192 (2004)
7. Nishida, T.Y.: Membrane Algorithms. In: Freund, R., Păun, G., Rozenberg, G., Salomaa, A. (eds.) WMC 2005. LNCS, vol. 3850, pp. 55–66. Springer, Heidelberg (2006)
8. Leporati, A., Pagani, D.: A Membrane Algorithm for the Min Storage Problem. In: Hoogeboom, H.J., Păun, G., Rozenberg, G., Salomaa, A. (eds.) WMC 2006. LNCS, vol. 4361, pp. 443–462. Springer, Heidelberg (2006)
9. Franco, G., Jonoska, N., Osborn, B., Plaas, A.: Knee joint injury and repair modeled by membrane systems. BioSystems 91, 473–488 (2008)
10. Daniel, D.P., Miguer, A., et al.: A Linear-time Tissue P System Based Solution for the 3-coloring Problem. Electronic Notes in Theoretical Computer Science 171, 81–93 (2007)
11. Annadurai, S., Kalyani, T., et al.: P systems generating ISO-picture languages. Process in Natural Science 18, 617–622 (2008)
12. Daniela, B., Paolo, C., et al.: Modelling metapopulations with stochastic membrane systems. Biosystem 91, 499–514 (2008)
13. Srinivas, M., Patnaik, L.M.: Genetic algorithms: a survey. IEEE Computer 27(6), 17–26 (1994)
14. Han, K.H., Kim, J.H.: Quantum-inspired evolutionary algorithm for a class of combinatorial optimization. IEEE Trans. Evolutionary Computation 6(6), 580–593 (2002)
15. Bäck, T., Hammel, U., Schwefel, H.-P.: Evolutionary computation: comments on the history and current state. IEEE Transactions on Evolutionary Computation 1(1), 3–17 (1997)
16. Păck, T., Hammel, U., Schwefel, H.-P.: Evolutionary computation: comments on the history and current state. IEEE Transactions on Evolutionary Computation 1(1), 3–17 (1997)
17. Mezura-Montes, E., Coello, C.A.: numerical comparison of some multi-objective-based techniques to handle constraints in genetic algorithms. Technical Report, EVOCINV-01-2003, México, pp. 1–34 (2003)

An Evolutionary Approach for Image Registration

Jing Zhang, Aimin Zhou, and Guixu Zhang

Department of Computer Science
East China Normal University
Shanghai 200235, China
cs.janezhang@gmail.com
{amzhou,gxzhang}@cs.ecnu.edu.cn

Abstract. Image registration plays an important role in many real-world applications such as remote sensing. A key issue of image registration is to find the hidden relationship between the input image and the reference image. In many cases, the hidden relationship is presented by a coordinate transformation matrix. Therefore, an image registration can be formulated as an optimization problem. In this paper, we propose to use evolutionary algorithms to optimize the transformation matrix. Instead of finding an optimal mapping between each pixel in the input and reference images, some local image features which are expressed as control points, are firstly extracted from the two images. An evolutionary algorithm is then applied to find the optimal mapping between the control points. Finally, the input image is registered by the optimal transformation. The proposed approach is applied to some remote sensing images and the statistical results show that our approach is promising for dealing with image registration.

Keywords: remote sensing, image registration, affine transformation, evolutionary optimization.

1 Introduction

Image processing plays a significant role in many real-world applications such as remote sensing, military and civilian uses. Registration is a fundamental issue among the typical operations of image processing. Image registration tries to match two or more images which are about the same objective but taken at different situations, such as different whether, cameras, or viewpoints [1]. In image registration, there usually exist two images: one is the reference image (or base image), and the other one is the input image. Let $I_1(x, y)$ be the intensity value at location (x, y) of the input image, and $I_2(x, y)$ be that of the reference image. The mapping between the input and the reference images can be expressed as

$$I_2(x, y) = g(I_1(f(x, y)))$$

where f is a 2D spatial coordinate transformation, and g is 1D intensity or radiometric transformation. An image registration problem could thus be converted

Z. Li et al. (Eds.): ISICA 2012, CCIS 316, pp. 321–330, 2012.

to find the optimal transformation f and g such that the images are matched mostly. In many cases, the g transformation is not necessary. Therefore, the image registration can be simply expressed as finding the optimal transformation f such that

$$I_2(x, y) = I_1(f(x, y))$$

In practice, the affine [2], projective, perspective, and polynomial transformations are widely used. Since the affine transformation is suitable for the two images which are taken form the same viewpoint but from the different positions, the work in this paper is based on the affine transformation.

Over the last few years, a number of techniques have been proposed for image registration. According to the registration process, these techniques can be classified into two categories, i.e., the area-based methods and the feature-based methods [3]. The area-based methods [4], such as correlation-like methods [5], Fourier methods [6] and mutual information (MI) methods [7, 8], directly use the statistical information of intensity values to process image. Since only the statistical information over the images is considered, these methods are sensitive to rotations. What's more, the statistical information is usually computational expensive. The feature-based methods firstly collect the feasters of an image, i.e., significant regions (forests, lakes, fields) [9], lines (region boundaries, coastlines, roads, rovers) [10] or points (region corners, line intersections) [11–13], and then use the features to do registration. These features can be represented by their point representatives, which are called control points (CPs). With all the CPs, the transformation model could be estimated. These methods can usually be implemented easily and converge fast. However, the finally quality highly depends on the captured features and the methods usually fall into local optimum.

An evolutionary algorithm (EA) is a generic population-based heuristic optimization method which simulates the biological evolution process [14]. By using reproduction, such as crossover, mutation, and selection, an EA can maintain a set of candidate solutions to tackle the optimization problem to be solved. Unlike traditional optimization methods, an EA does not make many assumptions about the optimization problems. Thus EAs are very suitable for real-world applications which may not have good mathematical properties. For this reason, we propose to use EAs to deal with image registration. The rest of this paper is organized as follows. Section 2 presents the details of our approach, including the chromosome, objective functions, and the algorithm framework. Section 3 briefly introduces the EA used in the paper. In section 4, the proposed method is applied to several widely used test images. Finally, the paper is concluded in Section 5.

2   Evolutionary Image Registration

Our basic idea is to find the optimal transformation between the input image and the reference image by using EAs. The affine transformation is used for this purpose. Therefore, an EA tries to find the optimal parameters of the affine transformation. The problem's dependent parts to apply an EA, i.e, the chromosome and objective functions are introduced in following with the algorithm framework.

2.1   Chromosome Definition

In affine, the translation matrix f can be illustrated as

$$f = \begin{pmatrix} z_1 & z_2 & z_3 \\ z_4 & z_5 & z_6 \\ 0 & 0 & 1 \end{pmatrix} \tag{1}$$

It can also be considered as the combination of a linear transformation and a translation. Assume (x, y) is the coordinate of the original pixel, and (x', y') is the new coordinate translated by the affine matrix from (x, y), then the relation between (x, y) and (x', y') is

$$\begin{pmatrix} x' \\ y' \end{pmatrix} = \begin{pmatrix} z_1 & z_2 \\ z_4 & z_5 \end{pmatrix} * \begin{pmatrix} x \\ y \end{pmatrix} + \begin{pmatrix} z_3 \\ z_6 \end{pmatrix}$$

or

$$\begin{pmatrix} x' \\ y' \\ 1 \end{pmatrix} = \begin{pmatrix} z_1 & z_2 & z_3 \\ z_4 & z_5 & z_6 \\ 0 & 0 & 1 \end{pmatrix} * \begin{pmatrix} x \\ y \\ 1 \end{pmatrix} \tag{2}$$

where $\begin{pmatrix} z_1 & z_2 \\ z_4 & z_5 \end{pmatrix}$ is the linear transformation and $\begin{pmatrix} z_3 \\ z_6 \end{pmatrix}$ is the translation. From (1) and (2), we can see in the affine matrix, there are six parameters, z_1, z_2, z_3, z_4, z_5, and z_6. Thus, we simply use the following vector to denote a chromosome

$$z = (z_1, z_2, z_3, z_4, z_5, z_6)$$

and let $a = (a_1, a_2, a_3, a_4, a_5, a_6)$ and $b = (b_1, b_2, b_3, b_4, b_5, b_6)$ be the lower and upper boundaries of the chromosome.

2.2   Fitness Definition

The control points of the features of the reference and input images are used to find the optimal affine transformation parameters. Let $M = \{A_1, A_2, ..., A_k\}$, $N = \{B_1, B_2, ..., B_k\}$ be the pixel sets of the control points from the reference image and input image respectively. A_i and B_i $(i = 1, 2, \cdots, k)$ are the corresponding points respectively. From the definition of chromosome, we know each chromosome z indicate a affine matrix f_z. By using f_z we can get a new set N' from N

$$B'_i = f_z(B_i)$$

where $N' = \{B'_1, B'_2, ..., B'_k\}$. If the f_z is optimal, the N' should be the same as M. So we can use the similarity between M and N' to measure the quality of the affine transformation. In this paper, we define the following three fitness functions.

$$f_1(x) = \sum_{i=1}^{k} \sqrt{(A_i(x) - B_i'(x))^2 + (A_i(y) - B_i'(y))^2} \tag{3}$$

$$f_2(x) = \sum_{i=1}^{k} (|A_i(x) - B_i'(x)| + |A_i(y) - B_i'(y)|) \tag{4}$$

$$f_3(x) = \max_{i=1}^{k} \sqrt{(A_i(x) - B_i'(x))^2 + (A_i(y) - B_i'(y))^2} \tag{5}$$

where $(A_i(x), A_i(y))$ is the coordinate of A_i, and $(B_i'(x), B_i'(y))$ is the coordinate of B_i'.

2.3 Framework of Our Approach

Our approach is based on feature points. Its major steps are illustrated in Fig.1. Firstly, the control point sets M and N are achieved by a preprocess. Secondly, an optimal transformation f_z is obtained by an EA. Finally, the optimal transformation f_z is applied to do the image registration.

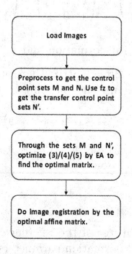

Fig. 1. The flowchart of our approach

3 Evolutionary Algorithm

It should be noted that in our approach, any evolutionary algorithm could be applied. In this paper, we use a recently published *Differential Evolution (DE)* algorithm, named as CoDE [15], to find the optimal affine transformation parameters. Recently, there are some DE reproduction strategies and algorithm parameters. Each strategy and each parameter may have different characteristics and may be suitable for different problems. The basic idea of CoDE is to combine these strategies and parameters to produce new solutions.

To generate a solution for a parent solution z^i, CoDE considers the following three mutation operators:

DE/rand/1 (Op_1):

$$u = z^{r1} + F(z^{r2} - z^{r3}) \tag{6}$$

DE/rand/2 (Op_2):

$$u = z^{r1} + F(z^{r2} - z^{r3}) + F(z^{r4} - z^{r5}) \tag{7}$$

DE/current-to-rand/1 (Op_3):

$$u = z^i + r(z^{r1} - z^i) + F(z^{r2} - z^{r3}) \tag{8}$$

where $z^{r1}, \cdots, z^{r5}$ are different parents which are randomly selected from the population, F is a parameter, and $r = rand()$ is a randomly generated from $[0, 1]$. CoDE also considers the following three parameter combinations:

$$Pa_1 = \{F = 1.0, CR = 0.1\}$$
$$Pa_2 = \{F = 1.0, CR = 0.9\}$$
$$Pa_3 = \{F = 0.8, CR = 0.2\}$$

The details of CoDE reproduction are shown in Algorithm 1. *Line 1* chooses five different parents randomly. For each mutation operator, a new trial solution is generated in *Line 4* with a parameter combination randomly selected in *Line 3*. *Lines 5-11* perform boundary checking and the trial solution is mutated in *Lines 12-20* to generate a new offspring solution. Finally, the best candidate solution is returned as the offspring in *Line 21*. More details about CoDE can be found in [15].

4   Experimental Results

In this section, the proposed evolutionary image registration method is applied to four remote sensing images. For each image, part is used as the reference image, and part is translated by an affine transformation to be the input image. We use the SIFT algorithm [16] to find the control points. The proposed method is implemented and executed in Matlab 7.13. The major parameters of CoDE are: population size $N = 10$, and maximum number of generations is 2000. CoDE is executed on each image for 10 times.

4.1   Statistical Results

Fig. 2 show the reference images, input images and the best approximations obtained by our approach according to the MI values over 10 runs. Firstly, by

---

Algorithm 1. CoDE Reproduction Procedure

---

1 For the parent z^i, randomly select other parents $z^{r1}, \cdots, z^{r5}$;

2 **foreach** *mutation operator $Op_k, k = 1, 2, 3$* **do**

3   Randomly select a parameter combination $Pa \in \{Pa_1, Pa_2, Pa_3\}$;

4   Generate a trial solution $u^k = Op_k(z^i, z^{r1}, \cdots, z^{r5}, Pa)$;

5   **foreach** *element $u_j^k, j = 1, \cdots, n$* **do**

6   **if** $u_j^k < a_j$ **then**

7              $u_j^k = 0.5(a_j + z_j^i)$;

8   **else if** $u_j^k > b_j$ **then**

9              $u_j^k = 0.5(b_j + z_j^i)$;

10   **end**

11   **end**

12   Randomly select $j_{rnd} \in \{1, \cdots, n\}$;

13   **foreach** *element $u_j^k, j = 1, \cdots, n$* **do**

14   **if** $rand() < CR || j == j_{rnd}$ **then**

15              $v_j^k = u_j^k$;

16   **else**

17              $v_j^k = z_j^k$;

18   **end**

19   **end**

20 **end**

21 **return** $z^* = \arg \min_{k=1,\cdots,3} f(v^k)$.

Table 1. Statistical results (mean±std.) of MI for different fitness definitions over 10 runs

Instance	f_1	f_2	f_3
image1	**1.60906**±0.00007	1.57356±0.001841	-0.88361±0.92646
image2	**12.91730**±0.00000	**12.91730**±0.00000	2.34450±0.62366
image3	2.36761±0.00002	**2.42851**±0.00087	0.28786±1.51231
image4	2.24770±0.00001	**2.26520**±0.00418	0.81222±0.07358

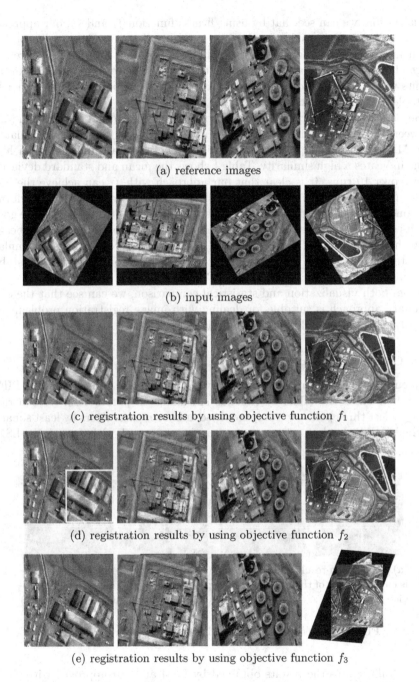

(a) reference images

(b) input images

(c) registration results by using objective function f_1

(d) registration results by using objective function f_2

(e) registration results by using objective function f_3

Fig. 2. The best approximations obtained by the evolutionary approach according to the MI values over 10 runs

visualization, we can see that by using fitness function f_1 and f_2, our approach can get very similar results on the four images and the results are acceptable. By using function f_3, however, our method fails on the 4th image. The reason might be that both f_1 and f_2 consider the average performance on all control points while f_3 focuses on the worst case of the control points. f_3 might not be suitable for image registration.

Secondly, we consider the performance metrics. The mutual information (MI) between the reference image and the transformed input image is calculated. The MI information represents the similarity between two images and a high value indicates a high similarity. Table 1 shows the mean and standard deviation values over 10 runs. It is clear that our approach with f_1 can achieve the best performance on the 1st and 2nd images; while the approach with f_2 performs best on the 2nd, 3rd and 4th images. The approach with f_3 fails to achieve good results on the four images. Although the method with f_2 performs slightly better than the method with f_1, the statistical results of the two methods are similar. The standard deviation values also indicate that the two approaches are stable in dealing with the four images.

From both visualization and statistical comparison, we can see that the evolutionary approach is suitable for dealing with image registration problems especially with the f_1 and f_2 fitness functions.

4.2 Comparison Study

As mentioned above, our approach firstly get the control points using the SIFT algorithm. Since the affine matrix has six unknown parameters, we can randomly select three points to compute these unknown parameters by least squares method (LSM). In this section, we compare our approach with f_2 with the LSM.

(a) registration result by LSM (Right is enlarge figure of the rectangular region in Left)

(b) registration results by our approach with f_2 (Right is enlarge figure of the rectangular region in Left)

Fig. 3. The results obtained by LSM and our approach with f_2

Fig. 3 illustrates the results obtained by LSM and our approach with f_2. It is clear from the zoomed figure that LSM does not work well. The reason might be that by using SIFT algorithm, not all control points are correctly matched. If the wrong points are used in LSM, the result will be wrong as well. This may also explain why our approach with f_3 does not work well.

5   Conclusion

In this paper, we proposed an evolutionary approach for image registration. The control points are firstly extracted from the reference and input images. The optimal transformation is then achieved by an evolutionary algorithm. Finally, the reference and input images are matched by the optimal transformation. The affine transformation, SIFT algorithm, and CoDE are used in our approach. The new approach is applied to four remote sensing images and the comparison studies show that our method with two fitness functions work well on the given images and it is also superior to the least squares method. In future, we will try other evolutionary algorithms as well as other fitness functions.

Acknowledgment. This work is supported by the Fundamental Research Funds for the Central Universities.

References

1. Brown, L.G.: A survey of image registration techniques. ACM Computing Surveys 24, 325–376 (1992)
2. Lin, H., Du, P., Zhao, W., Zhang, L., Sun, H.: Image registration based on corner detection and affine transformation. In: 2010 3rd International Congress on Image and Signal Processing (CISP), vol. 5, pp. 2184–2188 (October 2010)
3. Zitová, B., Flusser, J.: Image registration methods: a survey. Image and Vision Computing 21, 977–1000 (2003)
4. Fonseca, L.M., Manjunath, B.: Registration techniques for multisensor remotely sensed imagery. Photogrammetric Engineering and Remote Sensing 19, 1049–1056 (2001)
5. Pratt, W.K.: Digital image processing. Wiley (1991)
6. Bracewell, R.N.: The Fourier transoform and its applications. McGraw-Hill (1965)
7. Viola, P., Wells III, W.M.: Alignment by maximization of mutual information. International Journal of Computer Vision 24(2), 137–154 (1997)
8. Yamamura, Y., Kim, H., Yamamoto, A.: A method for image registration by maximization of mutual information. In: International Joint Conference on SICE-ICASE, pp. 1469–1472 (October 2006)
9. Pal, N.R., Pal, S.K.: A review on image segmentation techniques. Pattern Recognition 26(9), 1277–1294 (1993)
10. Canny, J.: A computational approach to edge detection. IEEE Transactions on Pattern Analysis and Machine Intelligence 8(6), 679–698 (1986)
11. Zheng, Z., Wang, H., Teoh, E.K.: Analysis of gray level corner detection. Pattern Recognition Letters 20(2), 149–162 (1999)
12. Pei, Y., Wu, H., Yu, J., Cai, G.: Effective image registration based on improved harris corner detection. In: 2010 International Conference on Information Networking and Automation (ICINA), vol. 1, pp. V1-93–V1-96 (October 2010)
13. Zhou, D., Gao, Y., Lu, L., Wang, H., Li, Y., Wang, P.: Hybrid corner detection algorithm for brain magnetic resonance image registration. In: 2011 4th International Conference on Biomedical Engineering and Informatics (BMEI), vol. 1, pp. 308–313 (October 2011)

14. Bäck, T., Fogel, D., Michalewicz, Z. (eds.): Handbook of evolutionary computation. Oxford University Press (1997)
15. Wang, Y., Cai, Z., Zhang, Q.: Differential evolution with composite trial vector generation strategies and control parameters. IEEE Transactions on Evolutionary Computation 15(1), 55–66 (2011)
16. Lowe, D.G.: Object recognition from local scale-invariant features. In: Proceedings of the International Conference on Computer Vision, pp. 1150–1157 (1999)

Non-negative Matrix Factorization: A Short Survey on Methods and Applications

Zhengyu Huang, Aimin Zhou, and Guixu Zhang

Department of Computer Science
East China Normal University
Shanghai 200235, China
`zyhuang@ecnu.cn`,
`{amzhou,gxzhang}@cs.ecnu.edu.cn`

Abstract. Non-negative matrix factorization (NMF) has been shown to be useful for a variety of practical applications. To meet the requirements of various applications, some extensions of NMF have been proposed as well. This paper presents a short survey on some recent developments of NMF on both the algorithms and applications. Some potential improvements of NMF are also suggested for future study.

Keywords: Non-negative Matrix Factorization, Dimensional Reduction.

1 Introduction

The amount of data has increased dramatically over the last 50 years [1]. Sufficient ways to handle the overwhelming amount of data automatically are urgently needed. The classical ways of data analysis such as locally-linear embedding (LLE) [2], principal component analysis (PCA) [3] can not afford to process these amounts of data efficiently. As a result, dimensional reduction is of great concern in the field of data analysis. The technology of dimensional reduction serves in two ways: (1) it properly represents data so that the inexactness is weakened and feasibility conditions are satisfied; and (2) it precisely defines variables while converting the original overlapping data caused by limitation of data gathering equipment to unambiguous data [4]. One important way of dimensional reduction is to use low-rank approximations [5]. Therefore they come into the forefront of many applications such as image processing and text mining. In these applications, the data to be analyzed is often non-negative. In order to have a precise physical meaning, the processed data is better to be non-negative as well [6]. Many classical tools do not guarantee to maintain the non-negative property. Naturally, a novel technology for low-rank approximation has come into our view- *Non-negative matrix factorization(NMF)*, which forces the processed data to be non-negative. In this paper, we present a brief summary to NMF. The basic concepts, the NMF algorithms and some applications are emphasized.

The rest of paper is organized as follows. Section 2 introduces the foundation of NMF with some properties of NMF. Section 3 discusses several algorithms

Z. Li et al. (Eds.): ISICA 2012, CCIS 316, pp. 331–340, 2012.

for implementing NMF framework in detail with their advantages and disadvantages. Section 4 illustrates some applications of NMF algorithms. Finally, Section 5 concludes the paper and outlines some directions to improve NMF algorithms.

2   NMF Foundation

It should be noted that there does exist the exact non-negative matrix factorization for the exact case and it may be a little misleading of the term *factorization*. In this paper, we focus on matrix approximation, i.e., approximating a non-negative matrix by a product of two non-negative matrix. In this section, we introduce the formulation of NMF problem and description of the solutions.

2.1   Problem Statement

NMF was first introduced by Paatero and Tapper in [7] using the name of *positive matrix factorization*. But it was only after Lee and Seung's work [6] that NMF can be easily understood and gain its popularity. NMF can be formally stated as follows [8].

Given a non-negative matrix $A \in \mathbb{R}^{m \times n}$ and a positive integer $k < \min(m,n)$, find non-negative matrices $W \in \mathbb{R}^{m \times k}$ and $H \in \mathbb{R}^{k \times n}$ to minimize

$$f(\mathbf{W}, \mathbf{H}) = \frac{1}{2} \|\mathbf{A} - \mathbf{WH}\|_F^2. \tag{1}$$

From the above NMF problem statement, it is clear that the aim of NMF is to find the approximation of $\mathbf{A}$ using the product of two matrices $\mathbf{W}$ and $\mathbf{H}$. It is also easy to deduce that the rank of both matrix $\mathbf{W}$ and $\mathbf{H}$ is at most k. The choose of k will influence the result a lot, however it is often hard to determine a proper value [9]. For the sake of dimensional reduction of NMF, we normally assume that k satisfies $k \leq \min(n,m)$ since k is the low rank representation of original matrix $\mathbf{A}$.

2.2   Properties of NMF

The foremost and straightforward property of NMF is that it allows only non-negative elements of the result matrices. This therefore forces the features in the basis matrix to allow only addition but no substraction. This property leads to the part learned property of NMF [6]. Indeed, the psychological [10] and physical [11][12] evidences have shown that the data presented in the brain is part-based.

Another important property of NMF is that the factorization result is not unique [13]. It can be easily seen by adding an invertible matrix $\mathbf{D}$ in the result matrix $\mathbf{W}$ and $\mathbf{H}$. In other words, if $\mathbf{W}$ and $\mathbf{H}$ are one potential solution of NMF, then $\mathbf{WD}$ and $\mathbf{D}^{-1}\mathbf{H}$ are also a potential solution of the same problem.

Besides, it has been proven that there exists local minimum for an NMF problem because for both **W** and **H**, the problem framework is not a convex function [8]. Therefore, it is common for an NMF algorithm to fall into local minimum.

Although NMF has many disadvantages as we have mentioned above, it has still been widely applied in many application regions and it has been proven to be successful in these applications.

3   NMF Algorithms

This section emphasizes the NMF algorithms to do factorization. We first introduce the standard NMF algorithm proposed by Lee and Seung [6]. Then we discuss some other extensions of NMF which changes the criteria or function of NMF. In the introduction, the basic idea and the procedure of each algorithm is presented in detail as well as some analysis.

3.1   Basic NMF Algorithms

The standard NMF algorithm suggested by Lee and Seung [6] is the most popular algorithm for NMF problems. In order to solve the NMF problem, they choose to fix one of the factors, for example, **W** or **H**, then try to minimize the cost function with respect to the other factor, i.e., **H** or **W**. The detailed description and proof is referred to [6]. The standard NMF starts from non-negative matrices **W** and **H**, and apply some update rules. The detailed description of deduction can be referred to [6]. It can be clearly seen that the iteration maintains the non-negative property of both matrices **W** and **H**. Besides, the sum of columns of **W** is also constrained to be unity. The Matlab pseudo code of the detailed procedure is as follows.

Multiplicative Update Algorithm for **NMF**

```
W = rand(m,k);  %Rank is k.
H = rand(k,n);
for i=1:maxiter:
(MU)     W = W .* ((A ./ (W * H + (A==0))) * H');
(MU)     H = H .* (W' * (A ./ (W * H + (A==0))));
end.
```

The term **A==0** added in each iteration is meant to avoid division by zero. Since it is the first well-known NMF algorithm, it has become a baseline with which many other extended NMF algorithms used to be compared. Practice of this basic NMF algorithm has shown that the convergence is notorious slow and not stationary [14]. Besides, unlike what illustrated in [6], this basic NMF algorithm is not sure to converge to a local minimization [15][16]. In order to overcome these shortcomings, two classes of modifications to the standard NMF problem have been proposed. One is called basic gradient decent method [17] and the other is called basic *alternating least squares method (ALS)* [7]. The Matlab pseudo codes of both basic gradient decent method and ALS are as follows.

Basic Gradient Descent Algorithm for **NMF**

```
W = rand(m,k);
H = rand(k,n);
for i=1:maxiter:
        H=H - epsilonH*diff(f,H);
        W=W - epsilonW*diff(f,W);
end.
```

Basic ALS Algorithm for **NMF**

```
W=rand(m,k);
for i=1:maxiter
(LS)                Solve for H in matrix equation
                    (W')WH=(W')A.
(NONNEG)            Set all negative elements in H to 0.
(LS)                Solve for W in matrix equation
                    H(H')(W')=H(A').
(NONNEG)            Set all negative elements in W to 0.
end.
```

For the basic gradient descent algorithm for NMF, the parameters *epsilonH* and *epsilonW* denote the step sizes which vary depending on the usage of this algorithm. Different algorithms use different ways to change step size. For example, Hoyer's algorithm [18] initially sets these step sizes to be one, then after each iteration reduces them. It is simple but not ideal for non-negative constraints. Other papers concerning gradient descent algorithms for NMF include [19]. The idea behind this modified algorithm is to find the best fitting for matrices **W** and **H**. Although it can not ensure that matrices **W** and **H** are non-negative after some iterations, some steps can be taken to let those entries who are negative to be zero after each iteration. However, choice of these parameters in this algorithm makes it hard to determine how fast can convergence be.

For the basic ALS algorithm for NMF, although the NMF problem is not convex for both matrices **W** and **H**, it is convex for **W** and **H** each respectively. As a result, when fixing matrix **W** or **H**, we can use *least squares method (LSM)* to solve **H** and **W** alternately. After each LSM, in order to avoid the appearance of negative entries of matrix **H** and **W**, all negative entries will be set to zero. It can easily be seen that this algorithm is inherently tend to be sparse, i.e., having more entries of zero in both matrix **W** and **H**. According to the implementation, ALS algorithm can be very fast. How to properly perform the non-negative constraint in the iterations is still being under discussion. The first ALS algorithm for NMF was introduced by Paatero [7]. Then some modifications of ALS algorithms for NMF appeared in [20] and [21], which added some constraints as what we will discuss in the next part of this paper in order to improve ALS algorithms for NMF.

3.2   NMF with Constraints

In this section, we review some of the most commonly used auxiliary constraints of the standard NMF. We can call these auxiliary constraints penalty terms since their existence is to satisfy some specific demands of NMF.

Normally, we can use the following formula to illustrate the NMF problem with auxiliary constraints.

$$f(\mathbf{W}, \mathbf{H}) = \frac{1}{2}\|\mathbf{A} - \mathbf{WH}\|_F^2 + \alpha J_1(\mathbf{W}) + \beta J_2(\mathbf{H}). \tag{2}$$

In the formula, function J_1 and J_2 serve for some measurements of matrix $\mathbf{H}$ and $\mathbf{W}$ of NMF such as sparsity or orthogonality. Besides, α and β are small regulate parameters to balance the trade-off between the approximation error and constraints. What follows is the detailed discussion about constraint function J for either $\mathbf{W}$ or $\mathbf{H}$ and some properties resulted by the added constraints.

Sparse NMF. Sparse is the most popular constraint for the extended NMF for the notion that sometimes only a few features can represent the whole data [19]. As what we mentioned above, the idea of NMF is for the part based representation of original data. Sparse is one of most important results of no-substraction linear combination of NMF algorithms. In some applications, sparse NMF shows its efficiency in both clustering performance and computation time when compared with other existing NMF algorithms.

For example, to apply spareness constraints on either $\mathbf{H}$ or $\mathbf{W}$, one of the famous penalty terms for sparse constraint is described by Hoyer [18] as follows.

$$sparseness(x) = (\sqrt{n} - (\sum |x_i| / \sqrt{\sum (x_i)^2})) / (\sqrt{n} - 1) \tag{3}$$

where n is the dimensionality of x. There also exists other criteria for sparsity measurements such as l^p for $0 \le p < 1$ [22].

Orthogonal NMF. Although Lee and Seung claimed their NMF algorithm to be coherent, later research did not seem to support this interpretation of their NMF algorithm [18][23]. As a result, Ding et.al., proposed the orthogonal NMF [24]. They imposed both matrix $\mathbf{W}$ and $\mathbf{H}$ with orthogonality by adding such constraints of $\mathbf{W}^T\mathbf{W} = \mathbf{I}$ and $\mathbf{H}^T\mathbf{H} = \mathbf{I}$. They also proved that orthogonal constraints could limit number of the results of NMF to only one. Besides, both the convergence to local minimization property and correctness of their algorithm have also been proved. In summary, orthogonal NMF overcomes one of the shortcomings of traditional NMF, i.e., an unique result will be obtained in an orthogonal NMF.

Smooth NMF. Smooth constraints are usually introduced in the penalty function in NMF problem for the sake of regulating the result in the presence of noise [8]. Smooth constraints can apply equally to both matrices $\mathbf{H}$ and $\mathbf{W}$ which is application-dependent. Since spectra such as the spectrum of music instrument change slowly over time [25], in some specific NMF applications, the rows in $\mathbf{H}$ are smooth. And this was also mentioned in the very first NMF paper [7]. Take the method in [26] as an example, its penalty function for smooth was

$$C(\mathbf{H}) = \|(\mathbf{I} - \mathbf{T})\mathbf{H}\|_F^2, \tag{4}$$

where $\mathbf{T}$ is a matrix calculated weighted average. It is also worth noticing that there are other kinds of definition of smooth such as the L_2 norm [27][28].

3.3   Relaxed NMF

Another way to improve its performance is to relax the standard NMF framework. For example, Semi-NMF relaxes the traditional NMF framework in the aspect of non-negative constraints. It allows the original data matrix $\mathbf{A}$ to have negative entries [29][30][31] or allows basis matrix $\mathbf{W}$ to have negative entries [32][33]. The former condition broadens the scope of the usage of NMF for a wider variety of applications while the corresponding algorithm designed for it still preserves the convergence property. And the latter condition relaxes the basis matrix in order to allow less flexible basis matrix calculated from NMF algorithm.

4   Applications of NMF

NMF has been attracting many attention after Lee and Seung's work [6]. It has also been proven to be successful in some application fields such as image processing, text mining and analysis music instrument spectrum, to name a few. In this section, we will show the applications of NMF algorithms in the following three fields although in some other fields, NMF algorithms also have proven to be useful.

Image Processing. In image processing, data can be presented as an $n \times m$ matrix $\mathbf{A}$ in which n represents the total number of images to be processed and m represents the dimension of the image. When applied NMF algorithms in it, matrix $\mathbf{A}$ is factorized into two non-negative matrix $\mathbf{W}$ and $\mathbf{H}$ such that $\mathbf{A}=\mathbf{WH}$. The columns of $\mathbf{W}$ and the columns of $\mathbf{H}$ each represents the basis matrix, in other words, the new features of the original matrix $\mathbf{A}$ and the coefficients to be linear combined to reconstruct the original matrix $\mathbf{A}$.

For the sake of space limitation, we just take face image processing as an example and omit the detailed description. NMF algorithms have successfully applied in the field of face image processing include recognition [34], classification [6] and representation [35].

Text Mining. There is also a success application for NMF proposed by Lee and Seung in [6]. Here, we take email semantic recognition as an example. In email semantic recognition, the original data matrix **A** is made up of n columns which represent n different emails and m represents m different possible semantic words which can show some important meanings in an email. By applying NMF algorithms in semantic recognition, matrix **A** is factorized into a basis matrix **W** which represents the hidden meanings of an email and the coefficient matrix **H** which shows how important one of the hidden meanings is in an email. In practice, Lee and Seung found that using NMF to do semantic recognition had another advantage that it could recognize different meanings of the same word. Other papers discussing NMF applications in text mining include [36]. And it can also be understood as a clustering method in a broad sense [37].

Music Analysis. Over the last decade, there is a growing interest in music analysis. Starting with [38], plenty of papers described how to use NMF algorithms to solve the problem. When we apply NMF algorithms in music analysis, A_{ij} means the i'th frequency at j'th time. And the columns of the result matrix **W** means basis vectors of one instrument playing one note and matrix **H** shows when the notes are active. In this field of application, the more sparse the result matrix **W** and **H** are, the better approximation will be. Other papers concerning NMF algorithms in analysis of music include [39] and [40].

5   Conclusion and Future Work

In this paper, we gave an outline of some basic conceptions and properties related to NMF, whether extended or not, together with some successful applications of NMF. But there are still some open questions for NMF and we list some possible trends for future work.

1. How to choose or produce the initial factors **W** and **H**? A random initialization may slow down the algorithm convergence and may also lead to local optimum.
2. Although many modified NMF algorithms have been proposed to overcome the limitations of traditional NMF algorithm, much work need to be done to further enhance the uniqueness and convergence speed of existing NMF algorithms.
3. The existing algorithms for NMF generally produce local minimizers for result matrix **W** and **H** even when various constraints are imposed. As a result, it may be of great interest to apply global optimization method to emit this shortcoming of NMF algorithms.

Acknowledgment. This work is supported by the Fundamental Research Funds for the Central Universities.

References

1. Kuo, F.Y., Sloan, I.H.: Lifting the curse of dimensionality. Notices of the AMS 52, 1320–1329 (2005)
2. Roweis, S.T.: Nonlinear dimensionality reduction by locally linear embedding. Science 290, 2323–2326 (2000)
3. Moore, B.: Principal component analysis in linear system: Controllability, observability, and model reduction. IEEE Transactions on Automatic Control 26(1), 17–32 (1981)
4. Lee, J.A., Verleysen, M.: Nonlinear dimensionality reduction. Springer Science Business Media (2007)
5. Ye, J.: Generalized low rank approximations of matrices. Machine Learning 61(1-3), 167–191 (2005)
6. Lee, D.D.: Learning the parts of objects by non-negative matrix factorization. Nature 401(6755), 788–791 (1999)
7. Paatero, P.: Positive matrix factorization: a non-negative factor model with optimal utilization of error estimates of data values. Environmetrics 5, 111–126 (1994)
8. Berry, M.W., Browne, M., Langville, A.N., Pauca, V.P., Plemmons, R.J.: Algorithms and applications for approximate nonnegative matrix factorization. Computational Statistics and Data Analysis 52(1), 155–173 (2007)
9. Wild, S., Curry, J., Dougherty, A.: Improving non-negative matrix factorizations through structured initialization. Pattern Recognition 37(11), 2217–2232 (2004)
10. Palmer, S.E.: Hierarchical structure in perceptual representation. Cognitive Psychology 9(4), 441–474 (1977)
11. Wachsmuth, E., Oram, M.W., Perrett, D.I.: Recognition of objects and their component parts: Responses of single units in the temporal cortex of the macaque. Cerebral Cortex 4(5), 509–522 (1994)
12. Logothetis, N.K., Sheinberg, D.L.: Visual object recognition. Annual Review of Neuroscience 19(1), 577–621 (1996)
13. Laurberg, H.: Uniqueness of non-negative matrix factorization. In: IEEE/SP 14th Workshop on Statistical Signal Processing (SSP 2007), pp. 44–48 (2007)
14. Chih-Jen, L.: On the convergence of multiplicative update algorithms for nonnegative matrix factorization. IEEE Transactions on Neural Networks 18(6), 1589–1596 (2007)
15. Finesso, L., Spreij, P.: Approximate nonnegative matrix factorization via alternating minimization. In: Proceedings of the 16th International Symposium on Mathematical Theory of Networks and Systems, pp. 1–10 (2004)
16. Gonzalez, E.F., Zhang, Y.: Accelerating the lee-seung algorithm for nonnegative matrix factorization. Department of Computational Applied Mathematics Rice University(CAAM) Houston TX Technical Reports, no. TR05-02, pp. 1–13 (2005)
17. Lin, C.J.: Projected gradient methods for nonnegative matrix factorization. Neural Computation 19(10), 2756–2779 (2007)
18. Hoyer, P.O., Dayan, P.: Non-negative matrix factorization with sparseness constraints. Journal of Machine Learning Research 5, 1457–1469 (2004)
19. Hoyer, P.O.: Non-negative sparse coding. In: Proceedings of the 12th IEEE Workshop on Neural Networks for Signal Processing, pp. 557–565 (2002)
20. Paatero, P.: The multilinear enginea table-driven, least squares program for solving multilinear problems, including the n-way parallel factor analysis model. Journal of Computational and Graphical Statistics 8(4), 854–888 (1999)

21. Albright, R., Cox, J., Duling, D., Langville, A.N., Meyer, C.D.: Algorithms, initializations, and convergence for the nonnegative matrix factorization. Matrix (919), 1–18 (2006)
22. Karvanen, J., Cichocki, A.: Measuring sparseness of noisy signals. In: Fourth International Symposium on Independent Component Analysis and Blind Singal Separation (ICA 2003), pp. 125–130 (2003)
23. Li, S.Z., Xin Wen, H., Hong Jiang, Z., Qian Sheng, C.: Learning spatially localized, parts-based representation. In: Proceedings of the 2001 IEEE Computer Society Conference on Computer Vision and Pattern Recognition (CVPR 2001), vol. 1, pp. 207–212 (2001)
24. Ding, C.: Orthogonal nonnegative matrix tri-factorizations for clustering. In: Proceedings of the 12th ACM SIGKDD International Conference on Knowledge Discovery and Data Mining, pp. 126–135 (2006)
25. Sha, F., Saul, L.K.: Real-time pitch determination of one or more voices by nonnegative matrix factorization. In: Advances in Neural Information Processing Systems, pp. 1233–1240 (2005)
26. Chen, Z., Cichocki, A.: Nonnegative matrix factorization with temporal smoothness and/or spatial decorrelation constraints. In: Laboratory for Advanced Brain Signal Processing, RIKEN, Tech. Rep. (2005)
27. Piper, J., Pauca, V.P., Plemmons, R.J., Giffin, M.: Object characterization from spectral data using nonnegative factorization and information theory. In: Proceedings of 2004 AMOS Technical Conference (2004)
28. Pauca, V.P., Piper, J., Plemmons, R.J.: Nonnegative matrix factorization for spectral data analysis. Linear Algebra and its Applications 416(1), 29–47 (2006)
29. Sajda, P., Du, S., Parra, L.: Recovery of constituent spectra using non-negative matrix factorization. In: Proceedings of SPIE, vol. 5207, pp. 321–331 (2003)
30. Sajda, P., Du, S., Brown, T.R., Stoyanova, R., Shungu, D.C., Xiangling, M., Parra, L.C.: Nonnegative matrix factorization for rapid recovery of constituent spectra in magnetic resonance chemical shift imaging of the brain. IEEE Transactions on Medical Imaging 23(12), 1453–1465 (2004)
31. Schmidt, M.N., Laurberg, H.: Nonnegative matrix factorization with gaussian process priors. In: Computational Intelligence and Neuroscience, pp. 1–10 (2008)
32. Ding, C.H.Q., Tao, L., Jordan, M.I.: Convex and semi-nonnegative matrix factorizations. IEEE Transactions on Pattern Analysis and Machine Intelligence 32(1), 45–55 (2010)
33. Tao, L., Ding, C.: The relationships among various nonnegative matrix factorization methods for clustering. In: Sixth International Conference on Data Mining (ICDM 2006), pp. 362–371 (2006)
34. Buciu, I., Pitas, I.: Application of non-negative and local non negative matrix factorization to facial expression recognition. In: Proceedings of the 17th International Conference on Pattern Recognition (ICPR 2004), vol. 1, pp. 288–291 (2004)
35. Feng: Local non-negative matrix factorization as a visual representation. In: Proceedings 2nd International Conference on Development and Learning (ICDL 2002), pp. 178–183 (2002)
36. Pauca, V.P., Shahnaz, F., Berry, M.W., Plemmons, R.J.: Text mining using nonnegative matrix factorizations. In: Proceedings of the SIAM International Conference on Data Mining, vol. 54, pp. 452–456 (2004)

37. Shahnaz, F., Berry, M.W., Pauca, V.P., Plemmons, R.J.: Document clustering using nonnegative matrix factorization. Information Processing and Management 42(2), 373–386 (2006)
38. Smaragdis, P., Brown, J.C.: Non-negative matrix factorization for polyphonic music transcription. In: IEEE Workshop on Applications of Signal Processing to Audio and Acoustics, pp. 177–180 (2003)
39. Asari, H., Olsson, R., Pearlmutter, B., Zador, A.: Sparsification for monaural source separation. In: Blind Speech Separation, pp. 387–410 (2007)
40. Abdallah, S.A., Plumbley, M.D.: Polyphonic music transcription by non-negative sparse coding of power spectra. Audio 510, 10–14 (2004)

Application of Differential Evolution to the Parameter Optimization of the Unscented Kalman Filter

Yao Jin

School of Computer Science
China University of Geosciences
Wuhan, 430074, P.R. China
Jinyao_wh@163.com

Abstract. The paper discusses choice for scaling parameter of the unscented transformation. By analyzing and comparing general method, the scaling parameter is selected as an optimization objective. Differential evolution algorithm is applied to the Unscented Kalman filter in offline model and online adaptive model. Experiment shows that the accuracy of UKF has been improved significantly by the two models.

Keywords: Differential Evolution, Unscented Transformation, Unscented Kalman Filter, Scaling Parameter.

1 Introduction

In the problem of nonlinear recursive state estimation of discrete-time stochastic dynamic systems, the Unscented Kalman filter (UKF)[1] is an effective filter technology. In Unscented Kalman filter, the Unscented Transformation (UT) which a certainty sampling strategy is used to parameterize the mean and covariance of a probability of the nonlinear dynamic system by a set of samples. The method yields a filter overcomes the drawback of the Extended Kalman filter (EKF)[2] have to calculate the Jacobian matrix; and is more accurate and easier to implement than an EKF or a Gauss second-order filter[1]. In addition, without the problems that need to simulate a large number of particle calculations and particle degradation in Particle filter (PF)[3], UKF is better than PF sometimes. The study of UKF has been a subject of considerable research interest for the last several decades.

There are many improvements for the UKF, [4] proposed a plain sigma point solution which for an n-dimensional space requires only n + 2 sigma points, its improves the efficiency of the filter in the high-dimensional problems. In [5] authors proposed an augmented UKF which concatenates the state and noise components together, so the effect of process and measurement noises can be used to better capture the odd-order moment information.[6] proposed an adaptive unscented particle filter algorithm based on predicted residual.

The scaling capabilities of the sampling points is the most obvious feature in unscented transform, scaling parameter to control the distribution of sampling

Z. Li et al. (Eds.): ISICA 2012, CCIS 316, pp. 341–346, 2012.

points and affect the accuracy of the approximate sampling. Although [1] gives a recommended setting of the scaling parameter, but the method may not lead to the best results. The discussion to scaling parameter of UT is few, so this paper focuses on optimization of the scaling parameter by Differential Evolution (DE)[7] and proposes the two methods of offline model and online adaptive model to select scaling parameter. The structure of the paper is as follows. Section 2 introduces the course of Unscented Transform. Section 3 describes the two models based on DE. The experiment is given in section 4 and conclusion is in section 5.

2   Unscented Transformation

UKF approximates nonlinear system by the samples which are defined by unscented transformation. The samples are called sigma-points $\{x_i\}$ with corresponding weights $\{w_m^i\}$.

$$x_0 = \hat{x}, W_m^0 = \frac{\lambda}{n_x + \lambda} \tag{1}$$

$$x_i = \hat{x} + (\sqrt{(n_x + \lambda)P_x})_i, W_m^i = \frac{\lambda}{2(n_x + \lambda)'} \tag{2}$$

$$x_{n_x+i} = \hat{x} - (\sqrt{(n_x + \lambda)P_x})_i, W_m^{n_x+i} = \frac{\lambda}{2(n_x + \lambda)'} \tag{3}$$

$$\lambda = \alpha^2(n_x + \kappa) - n_x \tag{4}$$

Where $i = 1, ..., n_x$, $\hat{x}$ is the mean of these sigma points, P_x is the covariance matrix of x, n_x is the dimension of the state, α and κ are the scaling parameters which fine λ. Each sigma-point is transformed the nonlinear function $y_i = g(x_i)$ and the next calculate is:

$$\hat{y}^{UKF} = \sum_{i=0}^{2n_x} W_c^{(i)} y_i \tag{5}$$

$$P_y^{UKF} = \sum_{i=0}^{2n_x} W_c^{(i)} (y_i - \hat{y}^{UKF})(y_i - \hat{y}^{UKF})^T \tag{6}$$

$$P_{xy}^{UKF} = \sum_{i=0}^{2n_x} W_c^i (x_i - \hat{x})(y_i - \hat{y}^{UKF})^T \tag{7}$$

$$W_c^0 = \lambda/(n_x + \lambda) + (1 - \alpha^2 + \beta) \tag{8}$$

$$W_c^i = 1/2(n_x + \lambda) \tag{9}$$

Where P_y^{UKF} is the covariance matrix of y, $W_c^{(i)}$ is the weight of y, P_{xy}^{UKF} is the cross covariance, β is antoher scaling parameter. (1)-(9) describes the course of scaled Unscented Transformation improved from original UT. Scaled UT allows sigma points to be scaled to an arbitrary dimension and second order accuracy is maintained and the algorithm is guaranteed to give a positive semi-definite covariance if all the weights on the sigma points are nonnegative[8].

There are three import parameters α, β, κ need to define, the parameters affect the filtering significantly. Recommending the choice of the parameters was given in[8], $\alpha \in [0,1], \beta = 2, \kappa = 3 - n_x$, but in practise the parameters is still a user-defined based on system to get a higher accuracy. The choice of the scaling parameter was discussed in[9], and proposed a method for adaptive setting of the parameter for original UT based on the likelihood function which depends on measurement predictive statistics with the parameter. But the method can only deal with one parameter; it's hard to use for Scaled UT. As the discussion to scaling parameter of UT is very few, so we use the Differential Evolution algorithm to optimize parameters for Scaled UT.

3   Differential Evolution to Parameter Optimization

Differential Evolution (DE) is a high efficient intelligent optimization method proposed by Storn and Price in 1995.With the advantages of simple structure, ease of use, speed and robustness, it has been successfully used in solving various optimization problems and it has been considered as a potential multi-disciplinary optimization algorithm[10].

In this paper, we use DE to solve the problem of parameter selection in UKF with Scaled UT. Scaling parameters is optimization objective and accuracy of UKF is evaluation function. Two methods of offline model and online adaptive model were discussed respectively.

3.1   Offline Model

Offline model is a posterior method to the entire filtering. In this model, the whole filtering in time T is regarded as a function. Scaling parameters α, β, κ are optimization objective of DE; the mean square error (MSE) of UKF in T is evaluation function. The results of DE make MSE of UKF minimum in T; it's the optimal parameter selection for the global filtering. The optimized filter is the same as the general optimization problem by DE. Generate a set of candidate sets in the given range, after mutation and crossover then input the scaling parameters to filter, select the candidate solutions which lead to minimum MSE of UKF into the next-generation until the end of the evolution.

This model need a set of filtering data used to train the optimal parameters. The optimal Scaling parameters can get a good result in different set of data with the same state. T is also an important status. The optimal parameters may change result in accuracy of UKF reduce with the different T. So the offline model is just suitable for equal interval period, the scaling parameters need to re-optimize if T has changed. That's the drawback of this model.

3.2   Online Adaptive Model

Compared with the offline model, online model needs to deal with scaling parameters adaptively. In this model, the filtering split into discrete-time filter calculation; scaling parameters are adjusted based on the estimated value of the state in each moment by DE. The filter calculation is a separate optimization task in every moment. Through solving every moment scaling parameters which get the minimum MSE, make the minimum filter MSE in T. Without the constraints of equal interval period, the algorithm of the UKF with the adaptive setting of the parameter can be summarized as follows:

1. Set the time instant $k = 0$ and define the state initial condition and the range of scaling parameters;
2. Optimize the filter scaling parameter by DE at present moment; the filtered value of optimum parameter is the result of UKF;
3. Set $k = k + 1$, continues by Step 2, until the end of filter.

With the advantages of fast convergence and good adaptability to various nonlinear functions, DE can get the extreme of unimodal function quickly. Time consumption for evolution at each time is not much; its suitable for online adaptive processing.

4   Experiment

To test the performance of two models, Univariate No stationary Growth Model (UNGM)[11] is used in the experiment. Its previously used as benchmark, what makes this model particularly interesting in this case is that its highly nonlinear and bimodal. The model for UNGM can be written as:

$$x_n = \alpha x_{n-1} + \beta \frac{x_{n-1}}{1 + x_{n-1}^2} + \gamma cos(1.2(n - 1)) + \mu_n \tag{10}$$

$$y_n = \frac{x_n^2}{20} + v_n, \quad n = 1, ..., N \tag{11}$$

Where $\mu_n \sim N(0, \sigma_\mu^2)$ and $v_n \sim N(0, \sigma_v^2)$, in experiment we set $\sigma_\mu^2 = 1, \sigma_v^2 = 1, x_0 = 0.1, \alpha = 0.5, \beta = 25, \gamma = 8$, and $N = 50$. Parameters of DE set if $F = 0.6, CR = 0.9, NP = 20, G = 50$, DE/rand1/bin. The range of three scaling parameters is set to $1 - 10$. Contrast the MSE of UKF with constant scaling parameters and two optimization models based on DE. A result of 100 independent experments is described in the table.

From the result, it can be seen the UKF with scaling parameters are optimized by DE is more accurate than the UKF with scaling parameters are chosen by the general method, the MSE of UKF is significantly improved. The error in every time instant is illustrated by Fig. 1. The accuracy of the two models is similar, online adaptive model is slightly better. The strategy of the parameter-adaptive adjustment at every moment is superior to the global parameter optimization

Table 1. MSE and Computational Costs of the UKF

	UKF with constant parameters		UKF based on DE	
	0.5/2/	0.9/2/2	Offline model	Online model
MSE	46.1613	44.4866	7.2151	6.3411
Times[s]	0.0026	0.0026	1.0699	1.0234

strategy. We also tried different value of F and CR with different evolution strategy, but the results are similar to Table 1.

The last row in Table 1 illustrates average computational costs of filtering in seconds at one time instant. The computational cost of UKF based on DE is more than UKF with constant parameters. But for the sake of a higher accuracy, it is worth to spend extra time to optimize parameters. The time consumption of two models is almost the same. So the online adaptive model has better applicability can be applied to any form of UKF to get good results, the offline model is a cheaper and faster way.

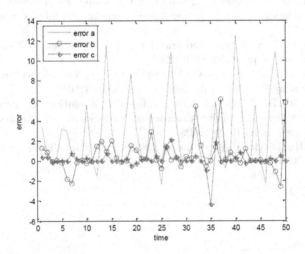

Fig. 1. The errors of UKF, a is the error of UKF with constant parameters; b is the error of offline model UKF; c is the error of online model UKF

5   Conclusion

In this paper, the choice of scaling parameter of the Unscented Kalman filter was discussed. Two models based on DE were proposed and analyzed. It was shown that the accuracy of UKF has been improved significantly by optimizing the scaling parameters. Online adaptive model get a better performance than offline model, it's more general and practical.

References

1. Julier, S.J., Uhlmann, J.K., Durrant-Whyte, H.F.: A new method for the nonlinear transformation of means and covariance in filters and estimators. Automatic Control 45(3), 477–482 (2000)
2. Gustafsson, F., Hendeby, G.: Some Relations Between Extended and Unscented Kalman filters. Signal Processing 60(2), 545–555 (2012)
3. Gordon, N.J., Salmond, D.J., Smith, A.F.M.: Novel approach to nonlinear none Gaussian Bayesian state estimation. Procom Radar and Signal Processing 140(2), 107–113 (1993)
4. Julier, S.J., Uhlmann, J.K.: Reduced sigma point filters for the propagation of means and co variances through nonlinear transformations. In: American Control Conference, Alaska, vol. 2, pp. 887–892 (2002)
5. Wu, Y., Hu, D., Wu, M., et al.: Unscented Kalman filtering for additive noise case: augmented versus no augmented. Signal Processing Letters 12(5), 257–360 (2005)
6. Wang, H.-J., Jing, Z.-R.: Adaptive unscented particle filter based on predicted residual. In: 6th IEEE Joint International Information Technology and Artificial Intelligence Conference, Chongqing, vol. 2, pp. 181–184 (2011)
7. Storn, R., Price, K.: Differential evolution CA simple and efficient heuristic for global optimization over continuous spaces. Journal of Global Optimization 11, 341–359 (1997)
8. Julier, S.J.: The scaled unscented transformation. In: American Control Conference, Alaska, vol. 6, pp. 4555–4559 (2002)
9. Dunik, J., Simandl, M., Straka, O.: Unscented Kalman filter: Aspects and Adaptive Setting of Scaling Parameter. Automatic Control 99 (2012)
10. Gong, W., Cai, Z.: A Multiobjective Differential Evolution Algorithm for Constrained Optimization. In: IEEE Congress on Evolutionary Computation, pp. 181–188. IEEE Press, Hong Kong (2008)
11. Kotecha, J.H., Djuric, P.A.: Gaussian particle filtering. Signal Process. 51(10), 2592–2601 (2003)

Multi-scale Segmentation Algorithm Parameters Optimization Based on Evolutionary Computation

Xin Zhang[1], Hengjian Tong[1], and Xiaowen Chen[2]

[1] School of Computer
China University of Geosciences
Wuhan 430074, P.R. China
[2] Peptrochina Pipeline Company(Pipeline Marketing Company)
Hainan University
Hebei 065000, P.R. China

Abstract. Multi-scale segmentation algorithm is the basis for classification and information extraction of object-oriented image analysis. Due to no obvious mathematical relationship between the scale parameter and the success of the segmentation, therefore, the selection of parameters highly depends on the user's experience. Users select parameters by trial and error method, which is iterative and time-consuming. The international famous object-oriented image analysis software eCognition also has not solved this problem. Aiming at the selection of multi-scale segmentation algorithm parameters (scale, shape, etc), this paper makes use of self-organizing, adaptive and self-learning characteristics of evolutionary computation to automatically optimize the parameters of the multi-scale segmentation algorithm according to the evaluation of segmentation results. This method eliminates blindness and subjectivity of parameter setting, makes the choice of the parameters not depend on the user's experience, and greatly improves the accuracy as well as efficiency of segmentation.

Keywords: Multi-scale Segmentation, Evolutionary Computation, Parameters Optimization.

1 Introduction

The high-resolution remote sensing images can provide a large number of surface features and rich spatial information. Inner detail information, size, shape of ground object and adjacent relationship between ground objects etc. have been well characterized. Due to the differences of spectral characteristics, texture characteristics, the size and shape among various ground objects, different ground object have different most suitable scale. If only use a single scale to make image segmentation, over segmentation and under segmentation are inevitable. If the ground object is relatively large, while the scale is too small, the ground object would be cut into too many pieces. If the size of ground object is small,

Z. Li et al. (Eds.): ISICA 2012, CCIS 316, pp. 347–358, 2012.

and the scale is too large, the segmentation may not be completed. According to this situationWe need to adopt different segmentation scales to extract different categories. Multi-scale segmentation method integrates the image information of different scales and combines the accuracy of fine scale with convenience of segmentation of rough scale. This theory originates from multi-scale systems theory and multi-scale estimation theory framework[1]. During segmentation process, multi-scale segmentation algorithm stores high-resolution pixel information in segmented image and polygon attributes for abstracting information from the target object. This algorithm segments image into meaningful polygon objects with minimum loss of detailed information. Fractal Net Evolution Approach (FNEA) is a potential multi-scale segmentation algorithm emerged in recent years. It is an object-oriented image segmentation method, which combines object-oriented method with image segmentation. FNEA is based on region merging algorithm of smallest internal heterogeneity. The method sets a specific threshold for target, called scale. It establishes segmentation criteria based on target object's spectral, texture, shape features to merging pixels whose spectral information is similar to the adjacent. Pixel merging starts at any pixel. At first, it merges single pixel into a smaller areathen merges smaller areas into polygon objects. Since Baatz and Schape presented object-based spatial information multi-scale cognitive model[2] and local best adaptive method[3]. FNEA gets wide applications. H. T. Li et al[4] presented a new high resolution QuickBird image multi-scale segmentation method on the basis of FNEA. The method used Statistical Region Merging (SRM) to do initial segmentation and used Minimum Heterogeneity Rule (MHR) to merge objects. Thomas Blaschke and Geoffrey J. Hay[5] model and evaluate for multi-scale landscape structure by using FNEA approach. Due to no obvious mathematical relationship between the scale parameter and the success of the segmentation. Therefore, the selection of parameters is highly depends on the user's experience. Users select parameters by trial and error method, which is iterative and time-consuming. The international famous object-oriented image analysis software eCognition also has not solved this problem. At presentthrough deep exploration and researchsome scholars have made some achievements in segmentation parameters automatic optimization. B. Bhanu et al[6] used GA to optimize the parameters of a segmentation method under various conditions of image acquisition. S. Chabrier et al[7] proposed a general scheme to segment images by a genetic algorithm. The developed method uses an evaluation criterion which quantifies the quality of an image segmentation result. The proposed segmentation method can integrate a local ground truth when it is available in order to set the desired level of precision of the final result. A genetic algorithm is then used in order to determine the best optimal parameters combination of information. Then, the article shows that this approach can either be applied for gray-levels or multicomponents images in a supervised context or in an unsupervised one. Gianluca Pignalberi et al[8] presented tuning range image segmentation by genetic algorithm .This method adopted a GA for tuning the set of parameters of a range segmentation algorithm. Its core theory is called GASE (Genetic Algorithm Segmentation Environment). The main objective of GASE is to provide the best optimal parameters combination for getting the optimal segmentation result. In this way, when this system finds a good segmentation for an image or for a particular surface, the authors can say that the same parameters will work correctly for the same class of images or for the same class of surfaces. Lucian Drăgut and Clemens Eisank presented local variance for multi-scale analysis in geomorphometry[9], object representations at multiple scales from digital elevation models[10], automated classification of topography from SRTM data using object-based image analysis[11], and developed software called Evaluation Scale Parameter (ESP)[12],which realizes automatically estimate scale parameter for multi-resolution image segmentation of remotely sensed data. In this paper, we respectively use genetic algorithm and differential evolution algorithm to implement the multi-scale segmentation algorithm parameters optimization by learning from the former experience.

2   Parameters Optimization Based on Genetic Algorithm

Genetic algorithm draws lessons from biological evolution process, simulates natural selection and genetic mechanisms, forms a search algorithm of self-generating and self-testing features. Fractal theory, formed in the late 1970s, is able to combines the spatial shape characteristics of the image at various scales [13]. FNEA combines fractal theory with object-oriented. In the range of pre-set scale, large-scale object and small-scale object coexist, which constitutes a multi-scale object network structure. Since Baatz and Schape presented object-based cognitive multi-scale spatial information model[2] and local best adaptive method[3], FNEA gets wide applications.

2.1   Design of Chromosome Structure

Segmentation parameters of FNEA include the segmentation scale (scale), the weight of shape heterogeneity (shape), and the weight of compactness (compactness). Among them, the range of interest scales is 0-127, the weight of shape heterogeneous, the weight of compactness is 0.0-1.0. The ranges of the three parameters are different. In order to make it easy to crossover and mutate between chromosomes, we adopt the binary coding method to map the range of three

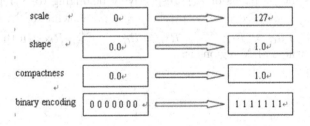

Fig. 1. The value range of the parameters and 7-bit binary string

parameters to the same binary space. The value range of the parameters and 7-bit binary strings show in Fig. 1.

As the mapping relation showed in Figure 1, we respectively convert the value of the segmentation scale, the weight of compactness, the weight of compactness into three 7 bit binary series.Chromosome is represented by a 21-bit binary string. In the string, from the high position to the low position, scale, shape, compactness are denoted. For example, while $scale = 70, shape = 0.164, compactness = 0.578$, the mapped binary string is 100011000101011001010.

2.2   Fitness Function

When we extract certain category objects from remote sensing image, there would be a kind of distance between the distribution of the objects in the ground truth image and in the machine-segmented image. Let G be the ground truth image, having N regions called R_{G_i} composed by pixels, $i = 1...N$, and let M be the machine-segmented image. As over segmentation and under segmentation are inevitable in the machine-segmented image, pixels in region R_{G_i} may belong to K regions. The K regions called R_{M_j} composed by P_{M_j} pixels. $j = 1, ..., K$. Then $P_{M_1} + P_{M_2} + ... + P_{M_K}$ is the minimum area in M which can cover region R_{G_i}. The minimum area in M covering region R_{G_i}. means the area sum of regions those include the pixels belonging to R_{G_i}.The fitness of single region R_{G_i} called F_i Thus:

$$F_i = \frac{P_{G_i}}{K \sum_{j=1}^{K} P_{M_j}} \tag{1}$$

The value of F_i is from 0 to 1.

1. When $K = 1$ and $P_{M_1} = P_{G_i}, F_i = 1$, expresses under segmentation. This means a certain region R_{M_1} in M is not only containing all the pixels of R_{G_i}, but also containing other pixels. $R_{M_1} \cap R_{G_i}, R_{M_1} \cup R_{G_i} = R_{M_1}$;
2. When $K > 1$ and $\sum_{j=1}^{K} P_{M_j} \geq P_{G_i}, 0 < F_i < 1$, expresses over segmentation. This means a certain region R_{M_1} in M containing all the pixels of R_{G_i}. Moreover, R_{M_1} doesn't have any other pixels. $R_{M_1} \cap = R_{G_i}, R_{M_1} \cup R_{G_i} = R_{G_i}$;
3. When $K > 1$ and $\sum_{j=1}^{K} P_{M_j} \geq P_{G_i}, 0 < F_i < 1$, expresses over segmentation. This means pixels of R_{G_i} respectively belonging to K regions in M, called R_{M_j} composed by P_{M_j} pixels, $j = 1, ..., K$, while $\sum_{j=1}^{K} P_{M_j} \geq P_{G_i}$, $\sum_{j=1}^{K} R_{M_j} \cap R_{G_i} = R_{G_i}, \sum_{j=1}^{K} R_{M_j} \cup R_{G_i} = \sum_{j=1}^{K} R_{M_j}$. In the same way, the total fitness of N regions is:

$$F = \sum_{i=1}^{N} F_i \tag{2}$$

Value of F is between 0 to N. When $F = N$, means the most ideal situation. When $0 < F < N$ means that over segmentation or under segmentation ocurs.

2.3 Parameters Optimization Process Based on GA

The basic process contains binary coding for first generation, roulette-wheel selection, multi-point crossover, multi-point mutation, calculation of the fitness of individuals. F_N represents the highest individual fitness among Nth generation. F_{N+1} represents the highest individual fitness among $(N+1)$th generation. The generation number is no more than the N_{MAX}. Therefore, terminal conditions are $N \geq N_{MAX}$ or $|F_N - F_{N+1}| < 0.01$. The specific work flow of parameters optimization is showed in Fig. 2.

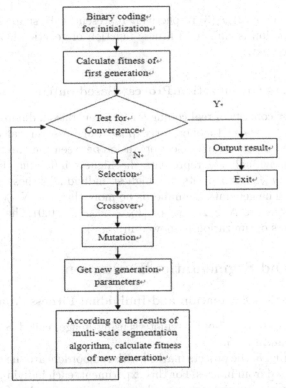

Fig. 2. Flow diagram of multi-scale segmentation algorithm parameters optimization based on GA

3   Parameters Optimization Based on Differential Evolution Algorithm

Differential Evolution algorithmDEwas proposed by Storn and Price[14] in 1995. DE mainly has two characteristics as follows:

1. Between the parent individuals and the child individual, DE adopt one to one select operation. Only if the child is superior to the parent, would the child replace the parent;

2. Mutation step and search direction of difference mutation operator are able to self-adaptive change according to actual situation[15].

3.1   Chromosome Structure and Fitness Function

Classical differential evolution algorithm adopts real coding method. In this paper, the multi-scale algorithm has three parameters, so the problem is three dimensions. The ith individual in population is expressed as :

$$X_i = \{x_i(1), x_i(2), x_i(3)\} \tag{3}$$

Among them, $x_i(1), x_i(2), x_i(3)$ respectively represent scale, shape and compactness. Fitness function is similar to fitness function of genetic algorithm, please refer to above context.

3.2   Parameters Optimization Process Based on DE

The basic process consists of real coding for first generation, difference mutation operation with DE/rand/1[15] operator, exponential crossover, calculation of the fitness of individuals, one to one select operation between the parent individuals and the child individual. F_N represents the highest individual fitness among Nth generation. F_{N+1} represents the highest individual fitness among $(N + 1)$th generation. The generation number is no more than the N_{MAX}. Therefore, terminal conditions are $N > N_{MAX}$ or $|F_N - F_{N+1}| < 0.01$. The specific work flow of parameters optimization is showed in Fig. 3.

4   Results and Segmentation Evaluation

4.1   Segmentation Evaluation and Individual Fitness Analysis

The test samples are two 5-band remote sensing image, called $Src1$ and $Src2$ shown as Fig. 4 and Fig. 5.

In Fig. 4 and Fig. 5, the objects marked with red borders are the targets which should be extracted from images. For this experiment, each individual consists of a segmentation scale, a weight of shape heterogeneity, a weight of compactness. We evaluate fitness of each individual like this:

Taking sample Src1 for example, the class which will be extracted from Src1, has three objects, called R_1, R_2, R_3. The sizes of R_1, R_2, R_3 are the number of pixels respectively belonging to them, named P_1, P_2, P_3. For each individual, we obtain a set of segmentation parameters according to individual gene sequence. We use the set of segmentation parameters as the input of multi-scale segmentation algorithm, getting a segmented image M. In M, we suppose that pixels of region R_1 respectively belong to the L regions, called $A_1, A_2, ..., A_L$, pixels of region R_2 respectively belong to the regions R_1, called $B_1, B_2, ..., B_M$, pixels of region R_3 respectively belong to the N regions, called $C_1, C_2, ..., C_N$. The size of $A_1, A_2, ..., A_L$ are $P_{A1}, P_{A2}, ..., P_{AL}$ in pixels, the size of $B_1, B_2, ..., B_M$, are $P_{B1}, P_{B2}, ..., P_{BM}$ in

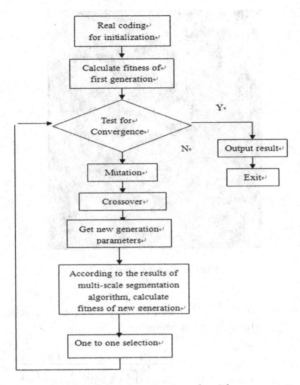

Fig. 3. Flow diagram of multi-scale segmentation algorithm parameters optimization based on DE

Fig. 4. Test sample Src1

Fig. 5. Test sample Src2

pixels and the size of $C_1, C_2, ..., C_N$ are $P_{C1}, P_{C2}, ..., P_{CN}$ in pixels. According to equation (1) and (2), the individual fitness is:

$$F = \frac{P_1}{L \sum_{i=1}^{L} P_{Ai}} + \frac{P_2}{M \sum_{j=1}^{M} P_{Bj}} + \frac{P_3}{N \sum_{k=1}^{N} P_{Ck}} \tag{4}$$

The range of F is from 0 to 3. The higher the value of F, the better the individual fitness.

4.2   Experimental Results and Analysis

To the two experimental samples, Src1 and Src2, we respectively repeat 10 times experiments by genetic algorithm and differential evolution algorithm. Before each experiment, we provide the same initial generation to genetic algorithm program and differential evolution algorithm program. The size of population is 30. In each experiment, program evolves 10 generation. The results of the experiments can be reproduced by software eCongnition.

The Result of Experimental Sample Src1. Segmentation result obtained by multi-scale segmentation algorithm parameters optimization based on genetic algorithm shows in Fig. 6. For this individual, the segmentation scale (scale) is 79, the weight of shape heterogeneity (shape) is 0.602, the weight of compactness (compactness) is 0.555, fitness is 2.95. Segmentation result obtained by multi-scale segmentation algorithm parameters optimization based on differential evolution algorithm is shown in Fig. 7. For this individual, the segmentation scale (scale) is 90, the weight of shape heterogeneity (shape) is 0.60, the weight of compactness (compactness) is 0.50, fitness is 2.94.

Fig. 6. The result of parameters optimization based on GA

Fig. 7. The result of parameters optimization based on DE

The Result of Experimental Sample Src2. Segmentation result obtained by multi-scale segmentation algorithm parameters optimization based on genetic algorithm shows in Figure 8.For this individual, the segmentation scale (scale) is 67, the weight of shape heterogeneity (shape) is 0.797, the weight of compactness (compactness) is 0.438, fitness is 2.88. Segmentation result obtained by multi-scale segmentation algorithm parameters optimization based on differential evolution algorithm shows in Figure 9.For this individual, the segmentation scale (scale) is 83, the weight of shape heterogeneity (shape) is 0.805, the weight of compactness (compactness) is 0.453, fitness is 2.91.

Fig. 8. The result of parameters optimization based on GA

Fig. 9. The result of parameters optimization based on DE

Analysis. According to experimental record, we make a statistic for the average individual fitness of the last generations in 10 times experiments. GA means Genetic Algorithm, representing the average individual fitness of the last generation evolved by genetic algorithm. DE means Differential Evolution, representing the average individual fitness of the last generation evolved by differential evolution algorithm.

In accordance with results showed in Fig. 6, Fig. 7, Fig. 8, Fig. 9, we conclude that evolution algorithm is a practical approach to multi-scale segmentation algorithm parameters optimization. The disadvantages are a lot of calculation and high time complexity. From Table 1, we can see that the last generation evolved by DE generally has a higher average individual fitness. Compared with former, the last generation evolved by GA generally has a lower average individual fitness. This indicates that the convergence rate of differential evolution algorithm

Table 1. The average individual fitness of the last generation

Experiment number	Sample Scr1		Sample Scr2	
	GA	DE	GA	DE
1	2.533	2.784	2.547	2.694
2	2.613	2.842	2.564	2.785
3	2.452	2.770	2.458	2.791
4	2.563	2.789	2.563	2.874
5	2.654	2.824	2.603	2.845
6	2.458	2.826	2.469	2.886
7	2.566	2.895	2.433	2.869
8	2.584	2.893	2.630	2.897
9	2.496	2.786	2.579	2.750
10	2.674	2.779	2.618	2.832
average	2.550	2.819	2.546	2.822

is higher than genetic algorithms. Therefore, DE is superior to GA in this case. We are more easily to obtain satisfactory result by multi-scale segmentation algorithm parameters optimization based on differential evolution algorithm in a relatively short time.

References

1. Benveniste, A., Nikoukhah, R., Willsky, A.S.: Multiscale system theory. In: Proc. 29th IEEE Conf. on Decision and Control, pp. 2484–2487. IEEE, Honolulu (1990)
2. Baatz, M., Schape, A.: Object-Oriented and Multi-Scale Image Analysis in Semantic Networks. In: 2nd International Symposium: Operationalization of Remote Sensing, August 16–20 (1999)
3. Baatz, M., Schape, A.: Multiresolution Segmentation: an optimization approach for high quality multi scale image segmentation, 'reviewed paper' angenommen (2000)
4. Li, H.T., Gu, H.Y., Han, Y.S., Yang, J.H., Han, S.S.: An Efficient Multi-Scale Segmentation for High Resolution Remote Sensing Imagery based on Statistical Region Merging and Minimum Heterogeneity Rule. The International Archives of the Photogrammetry (2008)
5. Blaschke, T., Hay, G.J.: Object-Oriented Image Analysis and Scale-Space: Theory and Methods for Modeling and Evaluating Multiscale Landscape Structure. International Archives of Photogrammetry and Remote Sensing 34(4), 22–29 (2001)
6. Bhanu, B., Lee, S., Ming, J.: Adaptive image segmentation using a genetic algorithm. IEEE Trans. Systems, Man, and Cybernetics 25(12), 1543–1567 (1995)
7. Pignalberi, G., Cucchiara, R., Cinque, L., Levialdi, S.: Tuning Range Image Segmentation by Genetic Algorithm. EURASIP Journal on Applied Signal Processing 2003 8, 780–790 (2003)
8. Chabrier, S., Rosenberger, C., Emile, B., Laurent, H.: Optimization-Based Image Segmentation by Genetic Algorithms. EURASIP Journal on Image and Video Processing (2008)

9. Drăgut, L., Eisank, C., Strasser, T.: Local variance for multi-scale analysis in geomorphometry. Geomorphology 130, 162–172 (2011)
10. Drăgut, L., Eisank, C.: Object representations at multiple scales from digital elevation models. Geomorphology 129, 183–189 (2011)
11. Drăgut, L., Eisank, C.: Automated classification of topography from SRTM data using object-based image analysis. Geomorphometry.org (2011)
12. Drăgut, L., Tiede, D., Levick, S.R.: ESP: a tool to estimate scale parameter for multiresolution image segmentation of remotely sensed data. International Journal of Geographical Information Science 24(6), 859–871 (2010)
13. Frontier, S.: Applications of Fractals Theory to Eeology. In: Legendre, P. (ed.) Developments in Numerical Ecology. Springer, Heidelberg (1987)
14. Storn, R., Price, K.: Differential evolution–A simple and efficient adaptive scheme for global optimization over continuous spaces. Technical Report: TR-95-012 (1995)
15. Price, K., Storn, R., Lapmpinen, J.: Differential evolution: A practical approach for global optimization. Springer, Berlin (2005)

An Architecture for Internet-Based Distributed Evolutionary Computation

Hui Li[1], Xiaoming Liu[1], Song Gao[2], and Dongdong Zhao[3]

[1] School of Computer Science
China University of Geosciences
Hubei, Wuhan 430074, P.R. China
huili@vip.sina.com
[2] Computer Science and Software Engineering
Auburn University, 3101 Shelby Center Auburn University
Auburn, AL 36849United States
song.gao@auburn.edu
[3] University of Science and Technology of China
Anhui, Hefei 230022, P.R. China
dongdongzhao.beta@gmail.com

Abstract. In this paper, an Internet-friendly architecture for distributed evolutionary computation is introduced. To make the architecture more flexible and fault-tolerable, it defines server (master) as a more important role. It serves not only as a simple communication center that transmits individuals during the migration operation, but also to maintain sub-populations. Also the progress of evolution is discredited into small tasks to assign to the clients (slaves). By working this way, it is also able to allocate computing resources dynamically, and is more efficient especially in heterogeneous environments. Browser/Server model is used in it. Native Client is used in Browser as the container to provide good performance for algorithms.

Keywords: Distributed Evolutionary Computation, Internet-based, Browser/Server.

1 Introduction

The number of personal computing devices is being enlarged fast, and the computing capability of them is also developing rapidly. However, for much of time, people's digital devices are just powered on, without taking more than half use of their computing resources (mainly CPU and GPU). This is a waste of computing resources as well as power resources. On the other side, universities and research institutes have to pay for high-performance computer clusters to support their scientists' research[1].

In this situation, we believe that establishing a connection over the Internet between research scientists and personal computing devices is in demand. To make the mechanism work, stable and flexible tools should be designed. In addition, psychology and social factors should be taken into consideration, to ensure

Z. Li et al. (Eds.): ISICA 2012, CCIS 316, pp. 359–367, 2012.

volunteers are willing to contribute their computing resources. There are some previous works implementing related things (such as BOINC[2] and FueySian Chong's work[3]), while they have limitations.

In the project it also uses some technologies which provided by google or other organizations and companies.

1.1   Native Client

Native Client[4], an open-source technology for running native code in web (browser) applications, is chosen as the container for the computation code in the browser. Employing this technology in Internet-based EC brings following advantages:

- High Performance. Native Client executes native code written by C/C++ natively on CPU. which makes algorithms running in the browser very fast.
- Security. Native Client runs native code in a sandbox, which keeps the host computer from any harm by the code. It ensures such safety by both hardware features and software fault isolation[5].
- Easily Porting Existing Code. Since code running in Native Client is written in standard C/C++(for security concerns, some of the features like physical _le access are disabled), and that it uses a GNU-based toolchain with customized versions of gcc, binutils and gdb, researchers are able to easily port their existing EC code to Native Client without much modification.
- Cross-Platform. Code running in Native Client directly runs on CPU, so Native Client code is classified by CPU architecture instead of operating system. At present, NaCl compiler supports x86 32, x86 64 and ARM (the bibliography[4,8] describes only the x86 architecture, however they have implemented the arm version), while theoretically speaking, any architecture available for LLVM is able to be supported.

1.2   Sandbox

Applications run in a secure environment that provides limited access to the underlying operating system. These limitations allow App Engine to distribute web requests for the application across multiple servers, and start and stop servers to meet traffic demands. The sandbox isolates your application in its own secure, reliable environment that is independent of the hardware, operating system and physical location of the web server[9].

1.3   Datastore

In this paper, it uses a distributed NoSQL data storage service to store data, this datastore is provided by App Engine provides and it is named app engine datastore. The App Engine datastore is not like a traditional relational database. Data objects, or "entities," have a kind and a set of properties. Queries can retrieve entities of a given kind filtered and sorted by the values of the properties. Property values can be of any of the supported property value types[10,11].

All the technologies used in this paper introduce an approach of building evolutionary computation which is more friendly to Internet-based computing environments.

2 Basic Architecture and Mainly Process

Browser/Server based architecture is extremely suitable for Internet-based distributed evolutionary computation. Browsers are perfect tools to establish communications between a user's computer and an external service. Compared to traditional desktop applications, browsers are able to provide almost equally excellent capabilities, in a more secure and more convenient way. Using a browser-based client in volunteer contributed evolutionary computation is more friendly to users. Volunteers are able to get rid of native software installation on his computer, as a result, they do not have to spend time on un-installation when they don't want it any more, and the secure mechanism of browsers, which has capability of keeping code running in it from doing harm to the users' computer, reduces volunteers' concern of security[8]. In addition, browser-based design makes cross-platform (Windows, Unix-like, mobile, etc.) work easier.

Internet-based EC is required to be much more fault-tolerable and adaptable than traditional distributed EC, since the system should tolerate volunteers to join or leave the computation whenever they like, meanwhile the whole computation shouldn't be affected too much. In addition, the difference of performance of computers should be taken into consideration.

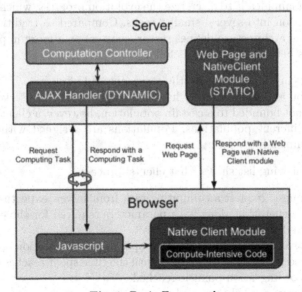

Fig. 1. Basic Framework

2.1 The Easy Situation

For problems requiring much computation on Evaluation, there's a simple solution: Distributing on Evaluation. This distributed modal is widely used for this kind of problem in traditional distributed EC. In this modal, client (browser) requests an individual, evaluates it, submits it, and repeats. Client is only working on evaluation, but do not handle the evolution process including breeding operations, etc. This modal meets the requirements: volunteers can join at any time and request individuals to evaluate, and if a client fails or leaves before it finishes the evaluation, the server is able to assign the individual to another client after timeout.

2.2 The Challenging Situation

For problems that evaluate fast but have large search-spaces, using Distributing on Evaluation may result in too frequent communications, which may seriously affect performance. In this situation, coarse-grained distribution is much more compatible: the client handles both evolution process and evaluation process including initialization, breeding, selection, etc., and exchanges individuals with other clients (migration) at some certain point. However, there's an issue: clients are better not allowed to leave or join at any time, because such actions destroy the original migration topology, thus producing a negative impact on the effect of distributed algorithms. Moreover, for synchronized migration, faster clients have to wait for slower clients to reach the migration point, which also have an impact on performance of the whole computation.

To solve this issue, a architecture particularly suitable for Internet-based EC is designed. The main idea is to discretize the evolution progress, which divides the whole computation into several smaller tasks. Compared to traditional coarse-grained EC, this architecture defines server (master) as a more important role. The differences between them are as follows:

- Populations are maintained on the server instead of clients
- Initialization and migration are handled completely by the server
- A client is not bounded to a certain population, however, a client may evolve on many different populations. Populations are assigned within tasks by server dynamically

Client Side. Following list shows what clients' process:

1. Step 1. Request: request a computing task from server, expecting a population at any generation along with parameters required for the evolutionary algorithm and the problem.
2. Step 2. Evolve: apply various breeding operators and selection operator onto the requested population iteratively, until the time spent reaches the timeout parameter, or migration point is reached.
3. Step 3. Submit: send the evolved population to the server.
4. Step 4. Goto Step 1

Notice that the stop/submit condition used by client is neither generation nor number of evaluations, however, timer is used. This seems unreasonable since computation time is unstable due to difference of performance of computers. However, in this architecture, timer is just the right condition to use. Since the server maintains the populations, synchronization among populations is also handled by the server. More importantly, Internet-based computation environment is extremely heterogeneous, in which generation synchronization or evaluation synchronization would be a nightmare, while time synchronization can not only reduce waiting time, but also enable server to assign tasks based on performance of clients, making use of available computing resources soundly.

Server Side On the server side, a lazy modal is used. The program behaves passively. It performs operations and pushes the evolution in response to clients call. When there's no client request, nothing is applied to the evolution.

Every time the server handles the request from a client, it choose an population for the client based on the client's performance record as well as the population's evolution progress (generation, etc.); Every time the server handles the submission from a client, it not only merge the submitted population into the population pool, but also check the whole population pool to see if every population reaches the migration point, determining whether to perform a migration operation. In addition, every submission triggers a check over all the joined clients, to withdraw the ones satisfying the timeout condition and re-mark its population assignable to other clients. To make the system more stable, safe, and fault-tolerant, populations, parameters, and evolution status are synchronized in the database.

On the whole, the browser requests computing task from the server, and transfers it to the Native Client module contained in the web page. When Native Client module finishes the computing task, browser pushes the result back to the server and starts another request. The overall architecture could be generalized with the Figure 1.

3   The Detailed Design and the Key Technologies

3.1   The Process of Server

1. Step 1. Initializing. To the algorithm, we should initialize the parameters, such as the generation, population, mutation and crossover probability and others. And after set down all parameters, the algorithm uses initialization function to get its first generation.
2. Step 2. Getting request. The server is starting to execute algorithm, and the algorithm will evaluate the individual of population, because this procedure may take long time and needs lots of resource of computer, and evolution algorithm has the characteristic of parallel, we choose clients to do this job. Because the architecture bases on B/S, the server must keep getting the request from clients, If there is no client, the server stays waiting and if server gets some clients' request, the server interrupts waiting, and puts the client into a queue.

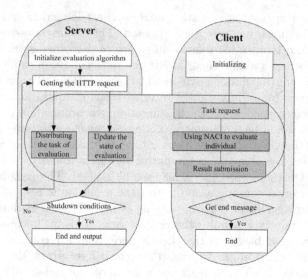

Fig. 2. The process of server and client

3. Step 3. Distributing task. Check current queue, if it is not empty, the server should produces a sub-population, marks it and send it to the top of queue's client, and delete the client in queue. When the queue is empty, the server should to keep waiting until the queue is not empty.
4. Step 4. Update. If the client returns the result of individuals which evaluated by itself to the server, the server will mark the current generation of the client and find whether there are better individuals than current individuals that has gotten. If the answer is yes, replace the bad result in server by the better individual.
5. Step 5. End conditions. Before the program start, it has setter down the shutdown conditions to end the program. Usually, the condition is that if there is an individual's fitness fitting the evolution algorithm, end the program on server and clients.
6. Step 6. End and output. Output the best individual. For multi-objective optimization problem, the best solution is pareto set, and for single objective optimization problem, the best solution is the maximal or minimal problem's solution.

3.2   The Process of Client

1. Step 1. Initializing. First we need open the specified HTTP page to connect the server. When the client just establishes a connection with the server, you need to initialize the NACI module first which can improve the assessment speed and reduce the assessment time and resource so that it can be called to assess the individual when getting the assigned task.

2. Step 2. Task request. After the initialization and task submit have been finished, the client will judge whether the computer resources are free, if so, the system will execute the request or submission , and send the information of task request or task submission to the sever. Otherwise, the system will wait for the task execution.

3. Step 3. Evaluation. In client, we use NACI to assess individual. NACI module can guarantee the security of client when it assesses individuals, besides, it has good portability which makes it transparent to operating system. The design of distributed framework can effectively reduce the burden of server, and NACI can reduce the burden of client, save the computing resources of client, and reduce the assessment time.

4. Step 4. Submission. When the client finishes the individual assessment through NACI module, it will send task submission request to the server which contains the gene information of the assessed individual, the assessed value and some other attribute information about the individual. When the server has accepted the task submission request, the client will write the submitted task information into the HTTP response stream and wait for the server to read. After the writing is completed, the client will turn to task request unless the server has sent program termination.

5. Step 5. End. If gets the ending message from the server, than end program.

3.3   The Entity Class of Architecture

In the App Engine distributed architecture, the Master node is responsible for the main process of evolutionary algorithm, the population evolution and migration, and data distribution to the slave node, task allocation, and the task control of slave node. The slave node mainly completes individual assessment, and returns the assessed individual to the master node. In this paper, the following entity classes are designed:

- Master entity class. This class contains The population migration options, child nodes information and population information;
- Slave entity class. This class contains the following three properties: the genetic manipulation generations, the sub-populations information distributed on the slave node, individual assessment in sub-populations.
- Population class. This class contains the following three properties: the information of the parent population, the information of the sub-population, the information of individual in population, the current generation and the current distribution of the slave node.
- Individual class. This class contains two properties: the individual's genetic value and individual's fitness information.
- Parameters class. Class contains two properties: the performance of slave nodes and the information of sub-population migration interval.

The following figure shows the relationship between each entity class.

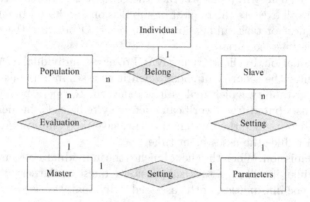

Fig. 3. The relationships of entities

4   Conclusion

This paper proposed an Internet-friendly architecture for distributed evolutionary computation. It can be use in internet, because of the using of native client. The program can compute as fast as supercomputer but cheaper. The architecture has test and find there are some problem.

- If the numbers of clients less than a number (we test the number is 4), the evaluation algorithm in the server often may can not quickly get result, because the server update population is not by the numbers of getting subpopulation, but depend on time.
- There are less people willing to contribute their computer to stay a long time on our web page. And the result is that we can not test it widely.

References

1. Alba, E., Tomassini, M.: Parallelism and Evolutionary Algorithms. IEEE Transactions on Evolutionary Computation 6(5), 443–462 (2002)
2. BOINC Project, http://boinc.berkeley.edu/
3. Chong, F.S.: Java based Distributed Genetic Programming on the Internet. In: Proceedings of the Genetic and Evolutionary Computation Conference (1999)
4. Yee, B., Sehr, D., et al.: Native Client: A Sandbox for Portable, Untrusted x86 Native Code. In: IEEE Symposium on Security and Privacy (May 2009)
5. Sehr, D., Muth, R., et al.: Adapting Software Fault Isolation to Contemporary CPU Architectures. In: The 19th USENIX Security Symposium (August 2010)
6. Castiglione, F., et al.: Optimization of HAART with genetic algorithms and agent based models of HIV infection. Bioinformatics Advance Access published (October 17, 2007)

7. Gong, L., Mueller, M., Prafullchandra, H., Schemers, R.: Going Beyond the Sandbox: An Overview of the New Security Architecture in the JavaTM Development Kit 1.2. In: Proceedings of the USENIX Symposium on Internet Technologies and Systems, Monterey, California, pp. 103–112 (December 1997)

8. Ames Jr., S.R., Gasser, M., Schell, R.G.: Security kernel design and implementation: An introduction. Computer, 14–22 (July 1983); Reprinted in Tutorial: Abrams, M.D., Podell, H.J. (eds.): Computer and Network Security, pp. 142–157. IEEE Computer Society Press (1987)

9. Google App Engine, The XMPP Java API

10. http://code.google.com/appengine/docs/java/xmpp/

11. Nurmi, D., Wolski, R., Grzegorczyk, C., Obertelli, G., Soman, S., Youseff, L., Zagorodnov, D., Eucalyptus, A.: Technical Report on an Elastic Utility Computing Architecture Linking Your Programs to Useful Systems, UCSB Computer Science Technical Report Number 2008-10 (August 2008)

An Improved GEP-GA Algorithm
and Its Application

Lei Yao and Hui Li

School of Computer Science
China University of Geosciences
Wuhan 430074, P.R. China
yaolei350124@gmail.com

Abstract. GEP is a powerful tool for automatically function model-ing. However, the classical GEP have some appearances such as lack of learning mechanism, search blindly, lack of diversity, prone to precocity when dealing with complicate problems. In light of these limitations the Improved GEP-GA Algorithm introduces the uniform initial population strategy, the adaptive mutation, the variation of population size strategy based on stagnant generations, and optimizes the coefficient of model by GA after the work of GEP. Then it proved that the Improved GEP-GA Algorithm is more effective than other similar algorithms in modeling and forecast though some experiments. It will make the algorithm hard to get into the local trap, and improve the fitting efficiency and forecast-ing accuracy of mining. The result of that applied the improved GEP-GA algorithm to the relay's parameter design shows that it will save time in calculating the electromagnetic suction force and improve the computa-tional efficiency in a large degree. It has a broad apace for development and application in relay parameters design.

Keywords: Gene Expression Programming, Generic Algorithms, Relay, Electromagnetic Suction Force Prediction.

1 Introduction

GEP is short for Gene Expression Programming[1]. It is a new adaptive type random search algorithm brought by Portuguese scientist Candida Ferreira in December, 2001 based on Genetic Algorithm (GA). GEP combine the advantages of the Genetic Algorithm (GA) and the Genetic Programming (GP). It encodes the individual with linear code to express complex computer program, which makes it simple, linear, compact, easy to genetic operation and fast to converge.

GEP has achieved good practical results in many fields such as function mining, classification, time series prediction with its strong feasibility and high precision[2-6] , and it has become hot issues in the research on the frontier of science worldwide. Usually, GEP uses random initialize strategy and fixed ge-netic operator when evaluating, which is simple and feasible, but make it lack of learning mechanism, search blindly, lack of diversity and easy to precocity

Z. Li et al. (Eds.): ISICA 2012, CCIS 316, pp. 368–380, 2012.

at the same time. The improved GEP-GA algorithm combined GEP with GA to ameliorate the limitations of the classical GEP. It uses uniform initialization strategy to ensure the diversity of population, which is conducive to global searching. Meanwhile, the adaptive mutation[7] is used to dynamically adjust the mutation probability which enables better local search ability. The introduction of the variation of population size strategy based on stagnant generations[8] is also employed to guide the algorithm out of local trap. The combination of these optimization strategies as well as optimizing the coefficient of model by GA after the work of GEP make the algorithm hard to get into the local trap, and improve the fitting efficiency and forecasting accuracy of mining. That will make the algorithm more fit for application.

2   The Improved GEP-GA Algorithm

2.1   The Uniform Initialization Strategy

The initial population of GEP describes the individuals in the solution space, and the operations to a population can indirectly mapping to the solution space. So if the initial population distributes uniformly in the solution space, the genetic operation will produce "combination effect", that is to say, individual modes combine with each other can produce new modes. That is helpful to improve the search efficiency, escape the local trap and converge to the global optimal.

The random initial strategy currently used produces population with limited diversity, and may even lost some gene segment. Only a large population size can make the distribution of the individual trend to uniform. Moreover, the random initial strategy may produce the population with negative highest fitness in some cases, and make it hard to evolve.

The Improved GEP-GA algorithm employs the uniform initialization strategy to increase population diversity of individuals. It distributes all symbols to every gene position averagely, so that the initial population can be rich in gene fragments and have a higher maximum fitness. The uniform initialization strategy can reduce the search process and leading to a faster search to the global optimal solution. The implementation steps are as follows:

1. Calculate the average probability of each symbol in each gene position. The gene is divided into three parts in the Improved GEP-GA algorithm, the first half of the head is only function operator, the other half can have function operator and terminator, and the tail can only contain the terminator. So the average probabilities of symbols in the three parts are as the formula (1), (2) and (3):

$$avgProHead1 = \frac{NumberOfGene * PopulationSize}{FSize} \quad (1)$$

$$avgProHead2 = \frac{NumberOfGene * PopulationSize}{FSize + TSize} \quad (2)$$

$$avgProTail = \frac{NumberOfGene * PopulationSize}{TSize} \qquad (3)$$

Here, $NumberOfGene$ is the number of the genes in every individual, $PopulationSize$ is the numbers of the individuals in every population, $FSize$ is the size of the function operator set, $TSize$, $FSize$ is the size of the terminator set.

2. Save times of each symbol appeared at each position in a two-dimensional array named counter, where counter $[i][j]$ means the jth symbol appeared counter $[i][j]$ times at the ith position.

3. Produce the ith gene position of $NumberOfGene*PopulationSize$. First, produce a symbol randomly, assuming it the jth symbol in the set, and then check the counter array. If counter $[i][j]$ is bigger than the average probability, change it into the symbol that appeared least at the ith position.

4. Divide the $NumberOfGene * PopulationSize$ genes into $PopulationSize$ individuals, and that is the initial population.

After the above steps, the function operators and the terminators tend to have a uniform distribution in each gene of the individuals, and leading to an even-distributed initial population.

2.2   The Adaptive Mutation Strategy

Usually, the mutation probability used in GEP is commonly fixed, that is to say, it will never change in any generation, and individuals with different quality will have common search space. However, the situation that the mutation probability changes dynamically during the evolution will be more reasonable.

Mutation is a favorable measure to keep the diversity of population. A small mutation probability is good for keeping genetic information, but against for generating new individual structure, while a big mutation probability can expand the search scope, but easy to lose direction. Moreover, the mutation is more effective in the early period of evolution than the later evolution.

The mutation probability in this paper is self-adaptive as the formula (4):

$$P = p * (1 - \frac{N_i}{N}) + p * (1 - \frac{f_{min}}{f_i}) \qquad (4)$$

Here, p is a weight that decides the change rate of the mutate probability, and it is commonly 0.05. N_i is the current generation and N is the max generation to go, while f_{min} is minimum fitness value in current generation and f_i is the fitness of the individual to be mutated.

As the result, the mutation probability will reduce as the evolution goes on, which is consistent with requirements to mutation operator of the algorithm. The adaptive mutation probability make the individuals adjust the search space to their value, so that individuals with low fitness value corresponds to a greater mutation probability and individuals with high fitness corresponds to a smaller mutation probability, which can not only maintain high-quality individual, but also conducive to improving the diversity of population, and lead to better convergence rate.

2.3   The Variation of Population Size Strategy Based on Stagnant Generations

It has good evolution performance in the early period of evolution, because it is easy to find better individuals with the good genetic diversity of population. But as the evolution goes on, the gene mode converges to the dominant ones due to the survival of the fittest. The reducing of the species diversity of the population makes the evolution slow down and search less efficiency.

In view of the above problems, the improved GEP-GA algorithm introduces the variation of population size strategy based on stagnant generations. Define the distance of the current population and the one that obtain the optimal individual as stagnant generations. And the steps of the strategy can be described as follows:

1. Using a small population size in the early period of the evolution.
2. When the stagnant generations reaches the pre-set threshold, which means the population can't produce a better gene mode and the evolution is in a stalemate, that may be two situations. If the population size has not reached the pre-set threshold, the population size will be enlarged by the best individuals in the parent population to increase the diversity and the search efficiency, which can improve the possibility and speed converging to a better solution. On the other hand, if the population size has reached the pre-set threshold, keep enlarge the population size will make it too long to evolve. So replace the worst 10% individuals with random individuals, which can introduce random gene mode and enrich the diversity will be a better plan.
3. A population with larger size will have a better search efficiency, but expend more time. So when the algorithm leave the stalemate (the stagnant generations return back to 0), the population size will be cut down to the initial size by random selection after a elitist selection.

The variation of population size strategy based on stagnant generations can adjust search range, control evolve speed and improve forecast accuracy.

2.4   Optimization with GA

When it comes to the later of the evolution, individuals in the population tends to be similar, leading to an inefficiency evolution. So the improved GEP-GA algorithm optimizes the factors of the GEP model with GA. It digs the function model from another side, which can find a more accuracy model in the same time than the classical GEP.

Take the function model dig by GEP as the test function of GA, and the original independent variable become the constant. Every gene of individuals in GA corresponding to a factor of the test function, and the last gene is the constant. It can fine adjust the function model without change the origin structure, and improve the accuracy of the model.

2.5   The Basic Flow

According to the above-mentioned improve strategies, the basic flow of the improved GEP-GA algorithm is showed in the Figure 1:

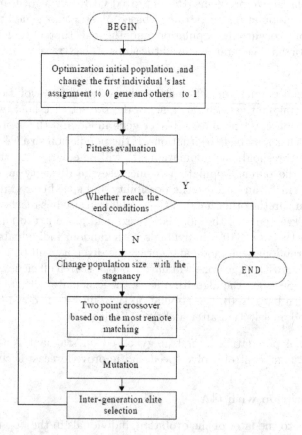

Fig. 1. The basic flow of the improved GEP-GA algorithm

3   Experiments and Analysis

3.1   Experiment One

According to the binary function $z = sin(2xy + 5)$, produce 30 sets of data meeting the condition that $x \in [-10, 10]$ at random, and then mine the function by classical GEP and the improved GEP-GA algorithm respectively. The max generation is 500, the initial population size is 50, the max population size is 200, and the function operator set is $\{+, -, *, /, S\}$, which represents the add, subtract, multiply , divide and sine in the arithmetic operation.

Running the classical GEP and the improved GEP-GA algorithm 20 times respectively, and compare the results in Table 1. The change trend of the best individual's fitness in the best result is showed in Figure 2, which shows that the improved GEP-GA algorithm has higher fitness than the classical GEP in the same generation, and the convergence accuracy is greatly improved.

Figure 3 is the change trend of the average fitness, which shows that the improved GEP-GA algorithm is more stable than the classical GEP and it has individuals with higher quality.

Table 1. The comparison between classic GEP and the improved GEP-GA algorithm

	Classical GEP	The improved GEP-GA Algorithm
Average Error	0.6000599867	0.2416684519
Unbiased Variance	0.5129536118	0.1497106927
Max Average Error	0.631819893333	0.36069943643
Max Unbiased Variance	0.66728038577	0.249949077307
Min Average Error	0.53869192969	1.58731570112e-007
Min Unbiased Variance	0.441380047675	4.66117973772e-014

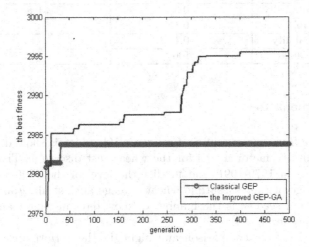

Fig. 2. The change trend of the best fitness

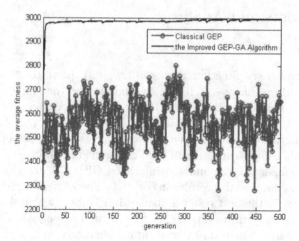

Fig. 3. The change trend of the average individual

Table 2. Parameter settings of the improved GEP-GA algorithm

Parameters	GEP	GA
Functions	+,-,*,/,Q,S,C,L,E,P	
Variables	as the number of variables	
Length of Head	8	
Number of Gene	5	as the number of parameters
Max Generation	1000	1000
Max Stay Generation	10	10
Initial Population Size	80	80
Max Population Size	400	400
Mutation Probability		0.05
Mating Probability	0.3	0.8
IS Probability	0.1	
RIS Probability	0.1	
Select Range	100	100

3.2 Experiment Tow

The data come from reference[9] and reference[10], which is showed in Table 3. Establish a suitable factor model for the wheat rust disease at Tianshui City, Gansu Province in 1979-1991, and predict the level of the 1996 pandemic. x is the percentage of the susceptible wheat acreage against all, y indicated the prevalence rate of seed, z is the number of snow days in winter and T is the actual prevalence level.

In order to facilitate comparison and highlight the performance of the improved GEP-GA algorithm, it uses the same parameter settings with the references, which is showed in Table 2.

Table 3. Parameter settings of the improved GEP-GA algorithm

	Physical Data					Fitted Value			Relative Error (%)		
Sample	Year	x	y	z	T	GP-GA	Improved GEP	Improved GEP-GA	GP-GA	Improved GEP	Improved GEP-GA
1	1979	61.0	0.405	12	3	2.9342	3.0045	2.8885	2.19	0.15	3.72
2	1980	64.8	0.397	16	3	2.9342	2.9955	3.0240	2.19	0.15	0.80
3	1991	50.8	0.002	10	1	0.9552	1.0078	1.0240	4.48	0.78	2.40
4	1982	52.9	0.317	18	2	2.0061	1.8469	1.9966	0.31	7.61	0.17
5	1983	61.2	0.111	18	3	3.0099	2.9931	2.9327	0.33	0.23	2.24
6	1984	76.0	0.521	22	4	3.9258	3.9370	4.0717	1.86	1.58	1.79
7	1985	80.4	0.887	39	5	5.0203	5.0001	4.9995	0.41	0	0
8	1986	50.9	0.391	6	1	0.9679	0.9997	1.0017	3.21	0.03	0.17
9	1987	50.4	0.082	10	1	1.0696	1.0344	1.0167	6.96	3.44	1.67
10	1988	41.2	0.081	9	1	0.9869	1.0000	0.9692	1.31	0	3.08
11	1989	60.0	0.900	25	3	3.0714	3.0000	3.0071	2.38	0	0.23
12	1990	80.0	1.910	27	5	4.8946	5.0264	5.0086	2.11	0.53	0.17
13	1991	70.0	0.750	9	3	3.0253	3.1473	2.9644	0.84	4.91	1.18
14	1996	80.0	1.020	36	5	4.9152	5.0650	4.9670	1.69	1.30	0.06
Average Error						0.0464	0.0345	0.0309		-	
Unbiased Variance						0.0030	0.0039	0.0019		-	
Standard Deviation						0.0548	0.0624	0.0446		-	
Forecasting Error						0.0848	0.0650	0.0330		-	

Running the improved GEP-GA algorithm 20 times, and the comparison between the best results of the improved GEP-GA algorithm and algorithms in the references is showed in Table 3.

Here,

$$AverageError = \frac{\sum_{i=1}^{n} |f(x_i, y_i, z_i) - T_i|}{n}$$

$$UnbiasedVariance = \frac{\sum_{i=1}^{n} (f(x_i, y_i, z_i) - T_i)^2}{n}$$

$$StandardDeviation = \sqrt{\frac{\sum_{i=1}^{n} [f(x_i, y_i, z_i) - T_i]^2}{n}}$$

$$ForecastingError = |f(x_{14}, y_{14}, z_{14}) - T_{14}|; n = 13$$

As we can see from Table 3, only results 1 and 10 of the improved GEP-GA algorithm have a larger relative error than others, in general, the fitting results of the improved GEP-GA algorithm are more stable, and the mining and forecast performance is better. The fitting results of the improved GEP-GA algorithm are 33.4% better than GP-GA in reference[9] and 10.4% better than the improved GEP in reference[10], and the improvement rate of forecast results are 61.1% and 49.2% respectively.

3.3   Experiment Three

The data come from reference[10], which is showed in table 4. Establish a gas
emission model in goal face, and select the coal seam burial depth, seam thickness, seam gas content, distance between coal seam and adjacent coal layer,
average daily progress, and average daily production as variables. In Table 5 a-f
are variables and T is the measured value. In the Table 4, sample 1-15 are used
for modeling, and 16-18 for forecasting.

Table 4. Samples and results

Sample	a	b	c	d	e	f	T	Improved GEP	Improved GEP-GA	Improved GEP	Improved GEP-GA
				Physical Data				Fitted Value		Relative Error (%)	
1	408	2.0	1.92	20	4.42	1825	3.34	3.38	3.3614	1.20	0.64
2	411	2.0	2.15	22	4.16	1527	2.97	3.19	2.9724	7.41	0.08
3	420	1.8	2.14	19	4.13	1751	3.56	3.44	3.6096	3.37	1.39
4	432	2.3	2.58	17	4.67	2078	3.62	3.619	3.6187	0.03	0.04
5	456	2.2	2.40	20	4.51	2104	4.17	3.97	4.1619	4.796	0.19
6	516	2.8	3.22	12	3.45	2242	4.60	4.67	4.6794	1.52	1.72
7	527	2.5	2.80	11	3.28	1979	4.92	4.921	4.8732	0.02	0.95
8	531	2.9	3.61	13	3.68	2288	4.78	4.82	4.7756	0.84	0.09
9	550	2.9	3.61	14	4.02	2325	5.23	5.18	5.0849	0.96	2.77
10	563	3.0	3.68	12	3.53	2410	5.56	5.63	5.5634	1.26	0.06
11	590	5.9	4.21	18	2.85	3139	7.24	7.38	7.2407	1.93	0.01
12	604	6.2	4.03	16	2.64	3354	7.80	7.71	7.887	1.15	1.11
13	607	6.1	4.34	17	2.77	3087	7.68	7.71	7.5788	0.39	1.31
14	634	6.5	4.80	15	2.92	3620	8.51	8.506	8.5137	0.047	0.65
15	640	6.3	4.67	15	2.75	3412	7.95	7.949	7.8915	0.001	0.73
16	450	2.2	2.43	16	4.32	1996	4.06	3.93	4.0597	3.20	0.01
17	544	2.7	3.16	13	3.81	2207	4.92	4.915	4.9292	0.1	0.187
18	629	6.4	4.62	19	2.80	3456	8.04	8.02	7.9817	0.25	0.73
Average Error								00.0781	0.0444	-	
Unbiased Variance								0.0098	0.0038	-	
Standard Deviation								0.0990	0.0616	-	
Forecasting Error								0.0517	0.0226	-	

The parameter settings are almost the same with Table 2, only the Length of
Head and Number of Gene change to 11 and 4 separately.

Running the improved GEP-GA in 20 times, and the comparison between the
best results of the improved GEP-GA algorithm and GEP in the reference[10]
is showed in Table 4.

$$AverageError = \frac{\sum_{i=1}^{n} |f(x_i, y_i, z_i) - T_i|}{n}$$

$$UnbiasedVariance = \frac{\sum_{i=1}^{n} (f(x_i, y_i, z_i) - T_i)^2}{n}$$

$$StandardDeviation = \sqrt{\frac{\sum_{i=1}^{n}[f(x_i, y_i, z_i) - T_i]^2}{n}}$$

$$ForecastingError = \sum_{j=16}^{18} |f(x_j, y_j, z_j) - T_j|; n = 15$$

As we can see from table 4, it can be seen from table 5 that only few results of the improved GEP algorithm have a larger relative error than the improved GEP in reference[10], and the improved GEP algorithm performs better at all other targets. The fitting results of the improved GEP-GA algorithm are 43.3% better than the GEP in reference[10], and the improvement rate of forecast results is 56.3%. That is to say, the improved GEP-GA algorithm has better performance in function mining and forecast than the GEP in reference[10].

4   Application of the Improved GEP-GA Algorithm

Relay is widely used in aerospace defense, communications technology, industrial automation, power system protection and other fields as a main electronic component. With the advancement of technology and manufacturing, all areas make strict limits on the relay's reliability, sensitivity, size and power consumption[11]. These restrictions require a more accurate calculation model and method on the parameter design which is one of the most important designs of relay. With development of technology for the calculation and analysis performance of the electromagnetic suction force, it has become a universal important basis for the parameter design of relay, which may also reflect the quality and reliability of productions[12].

How to determine the influence that key parameters to the magnetic system suction force in the case of knowing the mechanical reaction force is the basis of the optimization design for the polarization system in electromagnetic relay[13].As the rapid improvement of computer science, use some numerical value calculations such as finite element method and the difference time domain method in simulation software to analysis various electromagnetic problems have been a popular way. All these methods are iterative computation, and each of the calculations requires a huge calculation time spent. Actually, it needs only slightly changes in parameter values and constant state in calculation methods and models when calculating. If the GEP algorithm can be applied to set a prediction model base on the results of the previous calculations, and use the model to predict the output characteristic value of different combinations of the parameters in relay polarized magnetic system. That can save time in calculating the electromagnetic suction force and improve the computational efficiency in a large degree.

In view of the conclusions, the paper applied the improved GEP-GA algorithm to assist the optimal design of parameters in electromagnetic relay.

Take the polarized magnetic system in Figure 2 as example, and fitting and forecasting the electromagnetic suction force model with the improved GEP-GA

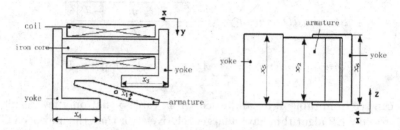

Fig. 4. Structure of the polarized magnetic system

algorithm. The parameters of the polarized magnetic system in Figure 4 are described below:

The electric current changes 16 times and the angle changes 21times;

- x1 is the thickness of armature (ArmThick), the range is [1.0,1.2], change times is 3;
- x2 is the height of armature (ArmHeigh), the range is [4.0,4.5], change times is 6;
- x3 and x4 is the length of the yoke(YokeLength), the range is [5.0,6.0], change times is 6;
- x5 and x6 is the height of the yoke(YokeHeight), the range is [4.5,5.5], change times is 6.

So there are 6 parameters in establishing the electromagnetic force model, including electric current, the angle, the thickness of armature, the height of armature, the length of the yoke and the height of the yoke. The data set should content 16*21*3*6*6*6=217728 samples, but there are only 16*21*18=6048 samples in the datum, the experiment uses 512 of which to modeling, and 129 samples to forecasting. The parameters settings are showed in table 4.

Fitting results are showed in figure 5, in which the average fitting error is 0.0009685, the unbiased variance is 2.18379450e-006, the average relative error is 2.81%, the max relative error is 12.1%, and the min relative error is 0.

Forecasting results are showed in figure 6, in which the average forecasting error is 0.00139663, the unbiased variance is 4.20543042e-006, the average relative error is 6.38%, the max relative error is 87.68%, and the min relative error is 0.67%.

These results indicate that the average fitting error of modeling with about 500 samples is about 3%, but in the practical engineering application the number of samples may reach 105 orders of magnitude, so the average error may be over 10application. Its mainly due to the limitation of the algorithm when applied in large scale data fitting, so it needs to be improved.

5   Conclusion

As a new member of evolutionary algorithm, GEP is a simple and effective tool for function mining, with high solving ability and excellent versatility. This paper

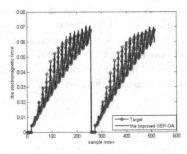

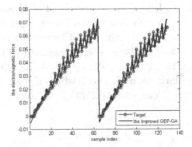

Fig. 5. Fitting results   **Fig. 6.** Forecasting results

introduced the basic principles of GEP, and then proposed an improved GEP with the adaptive mutation strategy and the inter-generational elitist selection strategy in allusion to the limitations of the classical GEP. The results of the experiments have testified that the improved strategies could speeding up the convergence rate and find better optimum solution. Applied in a gas emission model in goal face, the improved GEP algorithm has got a better performance than the GEP algorithm in the reference. That indicated that the improved GEP algorithm has vast development foreground and high value for application.

References

1. Candida, F.: Gene Expression Programming: A New Adaptive Algorithm for Solving Problems. Complex Systems 13(2), 87–129 (2001)
2. Tang, C., Zhang, T., Zuo, Y.: Knowledge discovery based on Gene Expression Programming evolution, achievement and development direction. Computer Application 10, 7–10 (2004)
3. Yuan, C., Tang, C., Zuo, J.: Function mining based on Gene Expression Programming convergence analysis and remnant-guided evolution algorithm. Journal of Sichuan University (Engineering Science Edition) 36(6), 100–105 (2004)
4. Tang, C., Chen, Y., Zhang, H.: Formula discovery based on Gene Expression Programming. Computer Application 27(10), 2358–2360 (2007)
5. Zuo, J., Tang, C., Li, C., Yuan, C.-a., Chen, A.-l.: Time Series Prediction Based on Gene Expression Programming. In: Li, Q., Wang, G., Feng, L., et al. (eds.) WAIM 2004. LNCS, vol. 3129, pp. 55–64. Springer, Heidelberg (2004)
6. Wang, R., Tang, C., Duan, L.: Polynomial function of GEP-based decomposition. Computer Research and Development 41, 442–448 (2004)
7. Fang, W., Zhang, K., Shao, L.: Improved Gene Expression Programming Based on a complex function modeling. Computer Engineering 32(21), 188–190 (2006)
8. Hu, J., Tang, C., Peng, J.: Quick jump out of local optimum VPS-GEP algorithm. Sichuan University (Engineering Science Edition) 39(1), 128–133 (2007)
9. Tang, L., Li, M., Zhang, J.: Mixed GP-GA model for the prediction of information systems. Computer Engineering and Applications 5, 44–48 (2004)
10. Lu, X., Cai, Z.: An improved GEP method and its evolutionary modeling prediction. Computer Applications 12(25), 2783–2786 (2005)

11. Zhai, G., Liang, H., Wang, H.: Polarization orthogonal design based on parameter optimization design method for magnetic studies. CSEE 10 (2003)
12. Zhou, X., Shi, J., Zhai, G.: Electromagnetic relay of Attraction force characteristics of test and analysis. Electromechanical Components 25(4), 6–9 (2005)
13. Liang, H., Ren, W., Wang, M.: Three-dimensional finite element method based on the polarization characteristics of magnetic systems of static suction. Electromechanical Components 25(3), 3–7 (2005)

Balancing Ensemble Learning between Known and Unknown Data

Yong Liu

The University of Aizu
Aizu-Wakamatsu, Fukushima 965-8580, Japan
yliu@u-aizu.ac.jp

Abstract. Without guidance on the unseen data, learning models could possibly approximate the known data by having different output on those unseen data. The results of such differences are the large variances in learning. Such large variances could lead to overfitting on many noisy data. This paper proposed one way of guidance by setting a middle value on the unknown data in balanced ensemble learning. Although balanced ensemble learning could learn faster and better than negative correlation learning, it also carried higher risk of overfitting in case of having limited number of training data points. Experimental results were conducted to show how such random learning could regulate the variances in balanced ensemble learning.

1 Introduction

It has been found that learning a given classification problem data set by neural networks could become hard or easy by setting different target values to the data points. Balanced ensemble learning [1–5] was developed from negative correlation learning [6] by shifting the learning targets differently to the learned and unlearned data. The idea of balanced ensemble learning is to weaken the error signals on the learned data points so that no further learning will be carried out to those data points. Therefore, the learning is turned to those not-yet-learned data points. Such idea has been tested on two structures of neural network ensembles between negative correlation learning and balanced ensemble learning. It was found that balanced ensemble learning could achieve much lower error rates than negative correlation learning on some hard training sets [3–5]. However, overfitting had appeared in both balanced ensemble learning and negative correlation learning especially when they were applied to the ensembles consisting of relatively large neural networks.

The larger neural networks are in the ensemble, the more the number of weights to be learned are. Generally speaking, the ensembles with large individual neural networks have the higher complexity, and could approximate the data points more flexible. Without the guidance on unseen data points, such flexibility could actually lead to the different output or even make different decisions on some unseen data. Such differences are the cause of possible large variances that might make overfitting happen. When there is no information on the unseen

Z. Li et al. (Eds.): ISICA 2012, CCIS 316, pp. 381–388, 2012.
© Springer-Verlag Berlin Heidelberg 2012

data, a safe learning is to set the target values in the middle of all possible values. It is to hope to reduce the variances by giving such constraints. The purposes of experiments conducted in this paper is to find out whether such learning on randomly selected data could control the complexity of the learned ensemble.

The rest of this paper is organized as follows: Section 2 describes ideas of balanced ensemble learning with learning on random data. Section 3 compares the learning behaviors of balanced ensemble learning with and without random learning on both ensemble level and individual level. Finally, Section 4 concludes with a summary of the paper.

2   Balanced Ensemble Learning with Random Learning

A balanced ensemble learning (BEL) [1] was developed by changing error functions in negative correlation learning (NCL) [6]. In NCL, the output y of a neural network ensemble is formed by a simple averaging of outputs F_i of a set of neural networks. Given the training data set $D = \{(\mathbf{x}(1), y(1)), \cdots, (\mathbf{x}(N), y(N))\}$, all the individual networks in the ensemble are trained on the same training data set D

$$F(n) = \frac{1}{M} \Sigma_{i=1}^{M} F_i(n) \tag{1}$$

where $F_i(n)$ is the output of individual network i on the nth training pattern $\mathbf{x}(n)$, $F(n)$ is the output of the neural network ensemble on the nth training pattern, and M is the number of individual networks in the neural network ensemble.

The idea of NCL [6] is to introduce a correlation penalty term into the error function of each individual network so that all the individual networks can be trained simultaneously and interactively. The error function E_i for individual i on the training data set D in negative correlation learning is defined by

$$E_i = \frac{1}{N} \Sigma_{n=1}^{N} E_i(n)$$
$$= \frac{1}{N} \Sigma_{n=1}^{N} \left[\frac{1}{2}(F_i(n) - y(n))^2 + \lambda p_i(n) \right] \tag{2}$$

where N is the number of training patterns, $E_i(n)$ is the value of the error function of network i at presentation of the nth training pattern, and $y(n)$ is the desired output of the nth training pattern. The first term in the right side of Eq.(2) is the mean-squared error of individual network i. The second term is a correlation penalty function. The purpose of minimizing is to negatively correlate each individual's error with errors for the rest of the ensemble. The parameter λ is used to adjust the strength of the penalty.

The partial derivative of E_i with respect to the output of individual i on the nth training pattern is

$$\frac{\partial E_i(n)}{\partial F_i(n)} = F_i(n) - y(n) - \lambda(F_i(n) - F(n))$$
$$= (1 - \lambda)(F_i(n) - y(n)) + \lambda(F(n) - y(n)) \tag{3}$$

In the case of $0 < \lambda < 1$, both $F(n)$ and $F_i(n)$ are trained to go closer to the target output $y(n)$ by NCL. $\lambda = 0$ and $\lambda = 1$ are the two special cases. At $\lambda = 0$, there is no correlation penalty function, and each individual network is just trained independently based on

$$\frac{\partial E_i(n)}{\partial F_i(n)} = F_i(n) - y(n) \tag{4}$$

At $\lambda = 1$, the derivative of error function is given by

$$\frac{\partial E_i(n)}{\partial F_i(n)} = F(n) - y(n) \tag{5}$$

where the error signal is decided by $F(n) - y(n)$, i.e. the difference between $F(n)$ and $y(n)$. For the classification problems, it is unnecessary to have the smallest difference between $F(n)$ and $y(n)$. For an example of a two-class problem, the target value y on a data point can be set up to 1.0 or 0.0 depend on which class the data point belongs to. As long as F is larger than 0.5 at $y = 1.0$ or smaller than 0.5 at $y = 1.0$, the data point will be correctly classified.

In BEL [1], the error function for each individual on each data point is defined based on whether the ensemble has learned the data point or not. If the ensemble had learned to classify the data point correctly, a shifting parameter β with values of $0 \leq \beta \leq 0.5$ could be introduced into the derivative of error function in Eq.(refrelation) for each individual

$$\frac{\partial E_i(n)}{\partial F_i(n)} = F(n) - |y(n) - \beta| \tag{6}$$

Otherwise, an enforcing parameter α with values $\alpha \geq 1$ would be added to the the derivative of error function for each individual

$$\frac{\partial E_i(n)}{\partial F_i(n)} = \alpha(F(n) - y(n)) \tag{7}$$

By shifting and enforcing the derivative of error function, the ensemble would not need to learn every data too well to prevent from learning hard data points too slowly.

In this paper, the ensemble was divided into two smaller sub-ensembles in which each sub-ensemble have one half of individual neural networks of the whole ensemble. BEL with $\beta = 0.25$ was applied on each sub-ensemble independently on the training data point. After learning a training data point, a random data point is generated. If the two sub-ensembles had the same classification on the random data point, 1.0 or 0.0 would be assigned as the guessed output depending on which classes the random data point belongs to. BEL with $\beta = 0.25$ was then carried out for the whole ensemble on the random data point. If the two sub-ensembles would not agree on the generated random data point, no learning would be performed on the random data point. Such an implementation of BEL is called BEL with random learning in this paper, which consists of learning both training data points and the randomly generated data points. In the next section, BEL with and without random learning has been compared in order to test whether such extra learning on random data point could help in preventing from overfitting.

3   Experimental Results

Three values were measured in BEL, including the average error rates of the learned ensembles, the average error rates of the individual neural networks, and the average overlapping rates of output between every two individual neural networks. The first two values represent the average performance of the ensemble and the individuals, while the third value shows how similar those individuals are. The overlapping rate with value 1 means that every two learners have the same classification on the measured data points, while the overlapping rate with value 0 implies that every two learners give the different classification on the measured data points.

3.1   Experimental Setup

The following four data sets were obtained from the UCI machine learning bench-mark repository. They were available by anonymous ftp at ics.uci.edu (128.195.1.1) in directory /pub/machine-learning-databases.

The diabetes data set is a two-class problem which has 500 examples of class 1 and 268 of class 2. There are 8 attributes for each example. The data set is rather difficult to classify. The so-called "class" value is really a binarised form of another attribute which is itself highly indicative of certain types of diabetes but does not have a one to one correspondence with the medical condition of being diabetic.

The Australian credit card assessment data set is to assess applications for credit cards based on a number of attributes. There are 690 cases in total. The output has two classes. The 14 attributes include 6 numeric values and 8 discrete ones, the latter having from 2 to 14 possible values.

The purpose of the heart disease data set is to predict the presence or absence of heart disease given the results of various medical tests carried out on a patient. This database contains 13 attributes, which have been extracted from a larger set of 75. The database originally contained 303 examples but 6 of these contained missing class values and so were discarded leaving 297. 27 of these were retained in case of dispute, leaving a final total of 270. There are two classes: presence and absence (of heart disease).

The breast cancer data set was originally obtained from W. H. Wolberg at the University of Wisconsin Hospitals, Madison. The purpose of the data set is to classify a tumour as either benign or malignant based on cell descriptions gathered by microscopic examination. The data set contains 9 attributes and 699 examples of which 458 are benign examples and 241 are malignant examples.

n-fold cross-validation were used in which n is set to 12 for the diabetes data, and 10 for the rest three data sets. 5 runs of n-fold cross-validation had been conducted to calculate the average results.

The ensemble architecture used in the experiments has 10 neural networks. Each individual neural network is a feedforward neural network with one hidden layer and 10 hidden nodes. $\beta = 0.25$ was used in BEL. The number of training epochs was set to 4000.

3.2 Results of Balanced Ensemble Learning without Random Learning

Table 1 presented the average results of error rates of the learned ensembles by BEL without learning on random data at different training epochs over 5 runs of n-fold cross-validation. BEL was indeed able to learn extremely well on the known data. The training error rate could fall below 0.04 on the diabetes data, while it could even reach 0.002 or lower on the rest of three data sets. However, the testing error gradually grew from the training epoch 50 to 4000, and became 0.02 bigger on the diabetes and the card data sets. In contrast, the testing error continuely dropped from 0.133 to 0.024 on the heart data, while it had little change on the cancer data from the training epoch 50 to 4000.

Table 2 presented the average results of error rates of the individual neural networks in the learned ensembles by BEL without learning on random data at different training epochs. On the training set, the individual error rates showed

Table 1. Average of error rates of the learned ensembles by BEL without learning on random data

No. of epochs	Diabetes		Card		Cancer		Heart	
	Train	Test	Train	Test	Train	Test	Train	Test
50	0.220	0.243	0.103	0.137	0.025	0.035	0.078	0.133
1000	0.091	0.259	0.013	0.146	0.007	0.039	0.001	0.030
2000	0.058	0.259	0.006	0.149	0.004	0.037	0.0004	0.029
3000	0.044	0.262	0.003	0.156	0.003	0.037	0.0002	0.030
4000	0.039	0.266	0.002	0.157	0.002	0.037	0	0.024

Table 2. Average of error rates of the individual neural networks in the learned ensembles by BEL without learning on random data

	Diabetes		Card		Cancer		Heart	
	Train	Test	Train	Test	Train	Test	Train	Test
50	0.355	0.367	0.283	0.298	0.197	0.202	0.259	0.291
1000	0.324	0.390	0.223	0.304	0.211	0.231	0.205	0.228
2000	0.302	0.385	0.214	0.308	0.209	0.232	0.200	0.222
3000	0.291	0.383	0.210	0.311	0.208	0.233	0.198	0.218
4000	0.285	0.384	0.208	0.311	0.206	0.232	0.197	0.217

Table 3. Average of overlapping rates of the individual neural networks in the learned ensembles by BEL without learning on random data

No. of epochs	Diabetes		Card		Cancer		Heart	
	Train	Test	Train	Test	Train	Test	Train	Test
50	0.583	0.582	0.601	0.601	0.672	0.672	0.624	0.613
1000	0.552	0.544	0.628	0.603	0.636	0.630	0.642	0.629
2000	0.565	0.549	0.634	0.602	0.636	0.628	0.648	0.635
3000	0.565	0.554	0.637	0.601	0.638	0.628	0.650	0.638
4000	0.577	0.555	0.639	0.602	0.639	0.629	0.650	0.639

a similar trend on the diabetes, the card, and the heart data sets by having relatively big drops from the training epoch 50 to 1000. In comparison, the individual error rates had the smaller changes on the cancer data. One of reasons causing such differences is that the ensemble could easily learn rather well just after 50 epoches on the cancer data. Different to what had been seen on the training data, the individual testing error rates only had the larger decrease on the heart data, while they had the much smaller increase on the rest of three data sets.

Table 3 presented the average results of overlapping rates of the individual neural networks in the learned ensembles by BEL without learning on random data at different training epochs. Although the big differences had been seen in the error rates on the training and testing sets, only small differences had been observed in the overlapping rates. It suggests that BEL could maintain the diversity among the learned ensemble in the learning process.

3.3   Results of Balanced Ensemble Learning with Random Learning

Table 4 presented the average results of error rates of the learned ensembles by BEL with learning on random data at different training epochs over 5 runs of n-fold cross-validation. When the ensemble was forced to take the same output to the unknown data in the same class being decided by the learned ensemble, it was prevented from learning too well on the known data. The training error rates of BEL with random learning became worse than those of BEL without random

Table 4. Average of error rates of the learned ensembles by BEL with learning on random data

No. of epochs	Diabetes		Card		Cancer		Heart	
	Train	Test	Train	Test	Train	Test	Train	Test
50	0.230	0.246	0.118	0.145	0.031	0.039	0.120	0.167
1000	0.150	0.246	0.053	0.138	0.024	0.035	0.034	0.087
2000	0.118	0.248	0.040	0.139	0.023	0.033	0.020	0.072
3000	0.104	0.244	0.033	0.144	0.022	0.034	0.015	0.062
4000	0.096	0.248	0.028	0.143	0.021	0.034	0.012	0.060

learning. However, learning a little less on the known data could actually helped on having stable performance in practical. Such a phenomenon is not new when it has been also observed in the early stopping developed for the neural network learning. However, the learning behaviors are different. In the early stopping, the ensemble stopped learning, and was impossible to make any changes further. With learning on the random data, BEL can be carried out in any lenght with little risk of overfitting without worrying about when to stop the learning.

Table 5. Average of error rates of the individual neural networks in the learned ensembles by BEL with learning on random data

No. of epochs	Diabetes		Card		Cancer		Heart	
	Train	Test	Train	Test	Train	Test	Train	Test
50	0.344	0.356	0.273	0.288	0.216	0.220	0.267	0.294
1000	0.308	0.361	0.236	0.295	0.213	0.225	0.219	0.247
2000	0.288	0.361	0.229	0.297	0.211	0.223	0.215	0.240
3000	0.279	0.360	0.226	0.297	0.212	0.225	0.213	0.236
4000	0.274	0.360	0.224	0.298	0.211	0.225	0.211	0.235

Table 5 presented the average results of error rates of the individual neural networks in the learned ensembles by BEL with random learning at different training epochs. With learning on the random data points, the rate difference between the training and testing set had been reduced on the diabetes, card, and cancer data except the heart data as been shown in Tables 2 and Table 5. In the testing results, the less the differences of the error rates between the training set and the testing set are, the better the performance on the testing set is.

Table 6. Average of overlapping rates of the individual neural networks in the learned ensembles by BEL with learning on random data

No. of epochs	Diabetes		Card		Cancer		Heart	
	Train	Test	Train	Test	Train	Test	Train	Test
50	0.599	0.599	0.628	0.628	0.649	0.650	0.644	0.642
1000	0.581	0.584	0.621	0.617	0.639	0.640	0.634	0.625
2000	0.587	0.586	0.624	0.616	0.640	0.641	0.634	0.627
3000	0.591	0.589	0.625	0.616	0.638	0.640	0.635	0.628
4000	0.593	0.590	0.626	0.615	0.638	0.639	0.636	0.629

Table 6 presented the average results of overlapping rates of the individual neural networks in the learned ensembles by BEL with random learning at different training epochs. It can been seen that BEL could still generated quite weak learners with the added learning on random data. It is one of uniqu capabilities for BEL to generate different weak learners without sampling data.

4   Conclusions

Without guidance on the unseen data, BEL could possibly approximate the known data by having different output on those unseen data. The results of such differences are the large variances in learning. Such large variances could lead to overfitting on many noisy data. This paper proposed one way of guidance by setting a middle value on the unknown data. In this paper, the number of the randomly selected data point is the same as the size of the training set. It should be interesting to see how the number of the randomly selected data point affect the learning behaviors.

References

1. Liu, Y.: A Balanced Ensemble Learning with Adaptive Error Functions. In: Kang, L., Cai, Z., Yan, X., Liu, Y. (eds.) ISICA 2008. LNCS, vol. 5370, pp. 1–8. Springer, Heidelberg (2008)
2. Liu, Y.: Balanced Learning for Ensembles with Small Neural Networks. In: Cai, Z., Li, Z., Kang, Z., Liu, Y. (eds.) ISICA 2009. LNCS, vol. 5821, pp. 163–170. Springer, Heidelberg (2009)
3. Liu, Y.: Create weak learners with small neural networks by balanced ensemble learning. In: Proceedings of the 2011 IEEE International Conference on Signal Processing, Communications and Computing (2011)
4. Liu, Y.: Target shift awareness in balanced ensemble learning. In: Proceedings of the 3rd International Conference on Awareness Science and Technology
5. Liu, Y.: Balancing ensemble learning through error shift. In: Proceedings of the Fourth International Workshop on Advanced Computational Intelligence
6. Liu, Y., Yao, X.: Simultaneous training of negatively correlated neural networks in an ensemble. IEEE Trans. on Systems, Man, and Cybernetics, Part B: Cybernetics 29(6), 716–725 (1999)

A Simulation Research on a Biased Estimator in Logistic Regression Model

Jiewu Huang

College of Science
Guizhou Minzu University
Guiyang, 550025, P.R.China

Abstract. In this article, a biased estimator is proposed to combat multi-collinearity in the logistic regression model. The proposed estimator is a general estimator which includes other biased estimators, such as the ridge estimator and the Liu estimator as special cases. Necessary and sufficient conditions for the superiority of the new biased estimator over the maximum likelihood estimator, the ridge estimator are obtained and some properties in the mean squared error sense are discussed. Furthermore, a Monte Carlo simulation study is given to illustrate some of the theoretical results.

Keywords: Logistic Regression, Maximum Likelihood Estimator, Ridge Estimator, Biased Estimator, Mean Squared Error Matrix, Mean Squared Error.

1 Introduction

In this paper we study estimation of Euclidean parameters $\beta \in R^p$ in logistic regression model based on the ith value of the dependent variable y_i which is $Be(\pi_i)$ distributed .The Bernoulli parameters π_i depends on β and regressors $x_1, x_2, ..., x_n$ from R_p with the following value:

$$\pi_i = \frac{exp(x_i\beta)}{1 + exp(x_i\beta)}, \quad i = 1, 2, ..., n \tag{1}$$

The most common method of estimating β is to apply the maximum likelihood method which is to maximize the following log-likelihood:

$$l = \sum_{i=1}^{n} y_i log(\pi_i) + \sum_{i=1}^{n}(1 - y_i)log(\pi_i) \tag{2}$$

The maximum likelihood estimator (ML) can be obtained by setting the first derivative of (2) to zero. Hence, the ML can be found by solving the following equation:

$$\frac{\partial l}{\partial \beta} = X'(y - \pi) \tag{3}$$

Z. Li et al. (Eds.): ISICA 2012, CCIS 316, pp. 389–395, 2012.

Where $X = (x_1', x_2', ..., x_n')', y = (y_1, y_2, ..., y_n)', \pi = (\pi_1, \pi_2, ..., \pi_n)'$.

The solution to the equation (3) is found by applying the following iterative weighted least square(IWLS) algorithm:

$$\hat{\beta}_{ML} = \left(X'\hat{W}X\right)^{-1} X'\hat{W}\hat{Z} \tag{4}$$

Where $\hat{Z}$ is a vector with the ith element equals $log(\hat{\pi}_i) + \frac{y_i - \hat{\pi}_i}{\hat{\pi}_i(1 - \hat{\pi}_i)}$ and $\hat{W} = diag\left(\frac{\hat{\pi}_i}{1 - \hat{\pi}_i}\right)$. The ML estimator is asymptotically normally distributed with a covariance matrix that equals the inverse of the Hessian matrix $\left(X'\hat{W}X\right)^{-1}$. Hence the mean squared error (MSE) based on the asymptotic covariance matrix equals:

$$MSE(\hat{\beta}_{ML}) = tr\left(X'\hat{W}X\right)^{-1} = \sum_{j=1}^{p} \frac{1}{\lambda_j} \tag{5}$$

Where λ_j is the jth eigenvalue of $(X'\hat{W}X)^{-1}$ are small. One of the drawbacks of the ML estimator is that the MSE becomes inflated when some eigenvalues of $(X'\hat{W}X)^{-1}$ are small. As a remedy to the problem caused by multi-collinearity, applying biased estimator is an effective way. A lot of shrinkage estimators are proposed to solve the multi-collinearity problem in linear model, such as the ordinary ridge estimator (RE) by Hoerl and Kennard(1970a, b), Liu estimator(LE) by Liu(1993).

Motivated by Nyquist(1991), Schaefer ea al.(1984), Albert and Anderson(1984), Mansson and Shukur(2011), and Mansson(2012), this paper is to introduce a biased estimator in the logistic regression model. A Monte Carlo simulation is conducted to study the properties of the proposed estimator.

The rest of the paper is organized as follows. In Section 2 the new shrinkage estimator in the logistic regression model is proposed and some properties are studied. A Monte Carlo simulation study is provided to illustrate some of the theoretical results in Section 3 .Finally; some conclusions are given in Section 4.

2   Properties of the Proposed Estimator $\hat{\beta}(k, d)$

The ridge estimator in the logistic regression model(LRE) proposed by Schaefer et al.(1984) is defined as:

$$\hat{\beta}_k = \left(X'\hat{W}X + kI\right)^{-1} X'\hat{W}X\hat{\beta}_{ML}, k > 0 \tag{6}$$

We can also define the Liu estimator in the logistic regression(LLE) model as:

$$\hat{\beta}_d = \left(X'\hat{W}X + I\right)^{-1} \left(X'\hat{W}X + dI\right)\hat{\beta}_{ML}, 0 < d < 1 \tag{7}$$

To hope that the combination of two different estimators might inherit the advantages of both estimators, we propose a new biased estimator(LBE) as:

$$\hat{\beta}(k, d) = \left(X'\hat{W}X + kI\right)^{-1} \left(X'\hat{W}X + kdI\right)\hat{\beta}_{ML} \tag{8}$$

Where $k > 0, 0 < d < 1$.

For the definition, it's easy to see that $\lim_{d \to 1} \hat{\beta}(k, d) = \hat{\beta}_{ML}$, $\lim_{k \to 0} \hat{\beta}(k, d) = \hat{\beta}_{ML}$,

$$\lim_{d \to 0}(k, d) = \left(X'\hat{W}X + kI\right)^{-1} X'\hat{W}X\hat{\beta}_{ML}$$

and

$$\hat{\beta}(k, 1) = \left(X'\hat{W}X + I\right)^{-1} \left(X'\hat{W}X + dI\right) \hat{\beta}_{ML}$$

.

For the convenience of the following discussions, we denote $\alpha = Q'\beta$, $\Lambda = diag(\lambda_1, ..., \lambda_p) = Q'\left(X'\hat{W}X\right)Q$, where $\lambda_1 \geq \lambda_2 \geq ... \geq \lambda_p > 0$ are the ordered eigenvalues of $X'\hat{W}X$, Q is the orthogonal matrix whose columns constitute the eigenvectors of $X'\hat{W}X$ and the ith element of $Q'\beta$ is denoted as $\alpha_j, j = 1, 2, ..., p$.

The mean squared error and the mean squared error matrix(MSEM) of an estimator $\hat{\beta} = Z\hat{\beta}_{ML}$, where Z is a matrix with proper order, are defined as:

$$MSE(\hat{\beta}) = E(\hat{\beta} - \beta')(\hat{\beta} - \beta) \tag{9}$$

$$MSEM(\hat{\beta}) = E(\hat{\beta} - \beta)(\hat{\beta} - \beta)' \tag{10}$$

Using (6), (7), (8) and (9), it is easy to compute that:

$$MSE(\hat{\beta}_k) = \sum_{j=1}^{p} \frac{\lambda_j}{(\lambda_j + k)^2} + \sum_{j=1}^{p} \frac{k^2 \alpha_j^2}{(\lambda_j + k)^2} \tag{11}$$

$$MSE(\hat{\beta}_d) = \sum_{j=1}^{p} \frac{(\lambda_j + d)^2}{\lambda_j(\lambda_j + 1)^2} + \sum_{j=1}^{p} \frac{(d - 1)^2 \alpha_j^2}{(\lambda_j + 1)^2} \tag{12}$$

$$MSE(\hat{\beta}(k, d)) = \sum_{j=1}^{p} \frac{(\lambda_j + kd)^2}{\lambda_j(\lambda_j + k)^2} + \sum_{j=1}^{p} \frac{k^2(1 - d)^2 \alpha_j^2}{(\lambda_j + k)^2} = \gamma_1(k, d) + \gamma_2(k, d) \tag{13}$$

where the first term is the asymptotic variance and the second term is the squared bias. The objective of the biased estimator is to choose appropriate values of k and d such that the reduction in the variance term is greater than the increase of the squared bias.

In Hoerl and Kennard(1970a) three theorems regarding the MSE properties of the ridge estimator in linear regression were derived. Here, we will show that the proposed biased estimator has some similar properties.

Theorem 2.1. The asymptotic variance $\gamma_1(k, d)$ and the squared bias $\gamma_2(k, d)$ are two continuous functions of k and d; for a fixed d^* between zero and one, $\gamma_1(k, d^*)$ and $\gamma_2(k, d^*)$ are monotonically decreasing and increasing functions of k respectively; for a fixed $k^* > 0$, $\gamma_1(k^*, d)$ and $\gamma_2(k^*, d)$ are monotonically increasing and decreasing functions of d respectively.

Proof. It is obvious that $\gamma_1(k,d)$ and $\gamma_2(k,d)$ are two continuous functions of k and d. Differentiating $MSE(\hat{\beta}(k,d))$ with respect to k and d respectively, we obtain:

$$\frac{\partial MSE(\hat{\beta}(k,d))}{\partial k} = -2\sum_{j=1}^{p} \frac{(1-d)(\lambda_j + kd)}{(\lambda_j + k)^3} + 2\sum_{j=1}^{p} \frac{k\lambda_j(1-d)^2\alpha_j^2}{(\lambda_j + k)^3} \tag{14}$$

$$\frac{\partial MSE(\hat{\beta}(k,d))}{\partial d} = 2\sum_{j=1}^{p} \frac{k(\lambda_j + kd)}{\lambda_j(\lambda_j + k)^2} - 2\sum_{j=1}^{p} \frac{k^2(1-d)\alpha_j^2}{(\lambda_j + k)^2} \tag{15}$$

Since $k > 0$, $\frac{\partial\gamma_1(k,d)}{\partial k} = -2\sum_{j=1}^{p} \frac{(1-d)(\lambda_j+kd)}{(\lambda_j+k)^3}$, for a fixed d^* between zero and one, it's easy to see that $\frac{\partial\gamma_1(k,d^*)}{\partial k} < 0$, i.e. $\gamma_1(k,d^*)$ is a monotonically decreasing function of k. In the same way, one get other results presented in Theorem 2.1. Proof is completed.

Theorem 2.2. For a fixed d^* between zero and one, there always exists a $k > 0$, such that $MSE(\hat{\beta}(k,d^*)) < MSE(\hat{\beta}_{ML})$.

Proof. In Theorem 2.1 it was proved for a fixed d^* between zero and one, $\gamma_1(k,d^*)$ and $\gamma_2(k,d^*)$ are monotonically decreasing and increasing functions of k respectively. Their first derivatives are always non-positive and non-negative respectively. Thus, to prove the theorem, it is only necessary to show that there always exists a $k > 0$, such that $\frac{\partial MSE(\hat{\beta}(k,d^*))}{\partial k} < 0$. A sufficient condition for this is:

$$k < \frac{1}{\alpha_{max}^2} \tag{16}$$

Proof is completed. Furthermore, we have the following general conclusions.

Theorem 2.3. If $k > 0$ and $0 < d < 1$, then $MSEM(\hat{\beta}(k,d)) < MSEM(\hat{\beta}_{ML})$, iff $k(1-d)\alpha'\left(k(1+d)\Lambda^{-1} + 2I\right)^{-1}, \alpha < 1$.

Proof. Denote $A = (\Lambda + kI)^{-1}(\Lambda + kdI)$, using (5) and (13), we obtain

$$MSEM(\hat{\beta}_{ML}) - MSEM(\hat{\beta}(k,d)) = Q\left(\Lambda^{-1} - A\Lambda^{-1}A - (A - I)\alpha\alpha'(A - I)'\right)Q'$$

For $\frac{1}{\lambda_j} - \frac{(\lambda_j+kdI)^2}{(\lambda_j+kI)^2\lambda_j} > 0$, we can see that $\Lambda^{-1} - A\Lambda^{-1}A > 0$. Thus $\Lambda^{-1} - A\Lambda^{-1}A - (A - I)\alpha\alpha'(A - I)' > 0$ if and only if

$$\alpha'(A - I)'(\Lambda^{-1} - A\Lambda^{-1}A)^{-1}(A - I), \alpha < 1 \tag{17}$$

Simplifying expression (17), we have $k(1-d)\alpha'(k(1+d)\Lambda^{-1} + 2I)^{-1}, \alpha < 1$. Proof is completed.

Corollary 2.3.1. if $k(1-d)\alpha'\left(k(1+d)\Lambda^{-1} + 2I\right)^{-1}, \alpha < 1$, then $MSE(\hat{\beta}(k,d)) < MSE(\hat{\beta}_{ML})$.

Theorem 2.4. if $k > 0$ and $0 < d < 1$, then $MSEM(\hat{\beta}(k,d)) < MSEM(\hat{\beta}_k)$ iff $\lambda_{min}(A_1B_1^{-1}) < 1$, where $B_1 = k^2(2-d)d(\Lambda+kdI)^2\Lambda^{-1}$ and $A_1 = (\Lambda+kI)^{-2}\Lambda - (\Lambda + kI)^{-2}(\Lambda + kdI)^2\Lambda^{-1}$.

Proof: It's easy to compute that

$MSEM(\hat{\beta}_k) - MSEM(\hat{\beta}(k,d)) = Q(A_1 + B_1)Q'$, where $A_1 = (\Lambda + kI)^{-2}\Lambda - (\Lambda + kI)^{-2}(\Lambda + kdI)^2\Lambda^{-1}$, $B_1 = k^2(2 - d)d(\Lambda + kI)^{-1}aa'(\Lambda + kdI)^{-1}$, since $A_1 < 0$, $B_1 > 0$, hence $A_1 + B_1 > 0$ if and only if $\lambda_{min}(A_1B_1^{-1}) < 1$.

Corollary 2.4.1. if $\lambda_{min}(A_1B_1^{-1}) < 1$, then $MSE(\hat{\beta}(k,d)) < MSE(\hat{\beta}_k)$, where:

$$A_1 = (\Lambda + kI)^{-2}\Lambda - (\Lambda + kI)^{-2}(\Lambda + kdI)^2\Lambda^{-1}$$
$$B_1 = k^2(2 - d)d(\Lambda + kI)^{-1}aa'(\Lambda + kI)^{-1}$$

3   Monte Carlo Simulation

The design of the experiment is briefly described as follows:

The dependent variable of the logistic regression is generated using pseudo-random numbers from the $Be(\pi_i)$ distribution and five explanatory variables are generated using the following Monte Carlo equations: $x_{ij} = (1 - \rho^2)^{1/2}z_{ij} + \rho z_{i5}$ where z_{ij} are independent standard normal pseudo-random numbers and ρ is specified so that correlation between any two explanatory variables is given by ρ^2.

The parameter values of ρ are chosen so that $\sum_{j=1}^{p}\beta_j^2 = 1$ which are common restrictions in many simulation studies (See Kibria, 2003 and Alkhamisi and Shukur, 2008).

Four different values of ρ corresponding to 0.85, 0.9, 0.95, 0.99 are considered and the sample size is equal to 100.

Table 1. Estimated MSE values of the ML, LRE, LLE, LBE when $\rho = 0.85$

d	0.2			0.4			0.6		
k	0.1	0.3	0.5	0.1	0.3	0.5	0.1	0.3	0.5
ML	0.8041	0.8041	0.8041	0.8041	0.8041	0.8041	0.8041	0.8041	0.8041
LRE	0.7731	0.7184	0.6719	0.7731	0.7184	0.6719	0.7731	0.7184	0.6917
LLE	0.6199	0.6199	0.6199	0.6606	0.6606	0.6606	0.7084	0.7408	0.7408
LBE	0.7792	0.7347	0.6961	0.7854	0.7514	0.7215	0.7916	0.7685	0.7479

Table 2. Estimated MSE values of the ML, LRE, LLE, LBE when $\rho = 0.9$

d	0.2			0.4			0.6		
k	0.1	0.3	0.5	0.1	0.3	0.5	0.1	0.3	0.5
ML	1.1703	1.1703	1.1703	1.1703	1.1703	1.1703	1.1703	1.1703	1.1703
LRE	1.1046	0.9929	0.9022	1.1046	0.9929	0.9022	1.1046	0.9929	0.9022
LLE	0.8101	0.8101	0.8101	0.8882	0.8882	0.8882	0.7943	0.9743	0.9743
LBE	1.1175	1.0263	0.9507	1.1305	1.0607	1.0017	1.1437	1.0962	1.0553

All simulation results are summarized in Tables 1-6.

From Table1-Table4, we can see that the new biased estimator is superior to the ML in the MSE sense, but the new biased estimator is not superior to the LRE and LLE in this experiment.

Table 3. Estimated MSE values of the ML, LRE, LLE, LBE when $\rho = 0.95$

d	0.2			0.4			0.6		
k	0.1	0.3	0.5	0.1	0.3	0.5	0.1	0.3	0.5
ML	2.2769	2.2769	2.2769	2.2769	2.2769	2.2769	2.2769	2.2769	2.2769
LRE	2.0346	1.6668	1.4054	2.0346	1.6668	1.4054	2.0346	1.6686	1.4054
LLE	1.2006	1.2006	1.2006	1.4222	1.4222	1.4222	1.6754	1.6754	1.6754
LBE	2.0816	1.7781	1.5554	2.1293	1.8948	1.7176	2.1778	2.0168	1.8919

Table 4. Estimated MSE values of the ML, LRE, LLE, LBE when $\rho = 0.99$

d	0.2			0.4			0.6		
k	0.1	0.3	0.5	0.1	0.3	0.5	0.1	0.3	0.5
ML	11.182	11.182	11.182	11.182	11.182	11.182	11.182	11.182	11.182
LRE	6.7743	3.3980	2.1419	6.7743	3.3980	2.1419	6.7743	3.3980	2.1419
LLE	2.0902	2.0902	2.0902	3.5936	3.5936	3.5936	5.1606	5.1601	5.6101
LBE	7.5589	4.5481	3.2968	8.3921	5.9017	4.7783	9.2738	7.4586	6.5865

As be shown in Table1-Table6, when $k(1-d)\alpha'\left(k(1+d)\Lambda^{-1}+2I\right)^{-1}, \alpha < 1$, the estimated values of the LBE are smaller than those of the ML which agree with our theoretical findings. We can also see fromTable1-Table4 and Table 6 that when $k(1-d)\alpha'\left(k(1+d)\Lambda^{-1}+2I\right)^{-1}, \alpha < 1$ is not satisfied, we may also have the estimated values of the LBE are smaller than those of the ML which tell us $k(1-d)\alpha'\left(k(1+d)\Lambda^{-1}+2I\right)^{-1}, \alpha < 1$ is a sufficient condition not a necessary and sufficient condition for $MSE(\hat{\beta}(k,d)) < MSE(\hat{\beta}_{ML})$.

For fixed value of d, the mean square error of the LBE decreases with the increase of k; for fixed value of k, the mean square error of the LBE increases with the increase of d.

The mean square error of the estimators increases with the increase of ρ; and the difference between the mean square errors of the ML and the LBE increases with the increase of ρ.

Table 5. Estimated values of $k(1-d)\alpha'\left(k(1+d)\Lambda^{-1}+2I\right)^{-1}, \alpha < 1$

(k,d)	$\rho = 0.85$		$\rho = 0.9$		$\rho = 0.95$		$\rho = 0.99$	
$(0.1, 0.2)$	0.0396	< 1	0.0395	< 1	0.0392	< 1	0.0367	< 1
$(0.3, 0.2)$	0.1169	< 1	0.1159	< 1	0.1130	< 1	0.0973	< 1
$(0.5, 0.2)$	0.1916	< 1	0.1890	< 1	0.1818	< 1	0.1492	< 1
$(0.1, 0.4)$	0.0297	< 1	0.0296	< 1	0.0293	< 1	0.0272	< 1
$(0.3, 0.4)$	0.0873	< 1	0.0864	< 1	0.0840	< 1	0.0713	< 1
$(0.5, 0.4)$	0.1428	< 1	0.1405	< 1	0.1345	< 1	0.1089	< 1
$(0.1, 0.6)$	0.0198	< 1	0.0197	< 1	0.0194	< 1	0.0179	< 1
$(0.3, 0.6)$	0.0580	< 1	0.0573	< 1	0.0555	< 1	0.0465	< 1
$(0.5, 0.6)$	0.0945	< 1	0.0929	< 1	0.0885	< 1	0.0708	< 1

Table 6. The maximum eigenvalue of $\Delta = MSEM(\hat{\beta}(k,d)) - MSEM(\hat{\beta}_{ML})$

(k,d)	$\rho = 0.85$		$\rho = 0.9$		$\rho = 0.95$		$\rho = 0.99$	
$(0.1, 0.2)$	-0.0240	<0	-0.0233	<0	-0.0227	<0	46.5812	>0
$(0.3, 0.2)$	-0.0238	<0	-0.0231	<0	-0.0225	<0	16.0437	>0
$(0.5, 0.2)$	-0.0235	<0	-0.0229	<0	-0.0223	<0	7.1509	>0
$(0.1, 0.4)$	-0.0240	<0	-0.0233	<0	-0.0227	<0	40.9964	>0
$(0.3, 0.4)$	-0.0238	<0	-0.0231	<0	-0.0225	<0	11.4996	>0
$(0.5, 0.4)$	-0.0236	<0	-0.0229	<0	-0.0223	<0	4.3275	>0
$(0.1, 0.6)$	-0.0240	<0	-0.0233	<0	-0.0227	<0	36.3237	>0
$(0.3, 0.6)$	-0.0238	<0	-0.0231	<0	-0.0225	<0	8.5567	>0
$(0.5, 0.6)$	-0.0236	<0	-0.0229	<0	-0.0223	<0	2.7924	>0

4 Conclusion

To combat multi-collinearity, we propose a new biased estimator in logistic model and discuss some properties about this estimator. All the theoretical and simulation results show that the proposed two-parameter estimator is meaningful in practice.

Acknowledgement. This work is supported by Guizhou Science and technology department and Guizhou Minzu University (Grant No. Qian Science co-J word LKM [2011]09).

References

1. Albert, A., Anderson, J.A.: On the existence of maximum likelihood estimates in logistic regression models. Biometrika 71, 1–10 (1984)
2. Alkhamisi, M., Shukur, G.: Developing ridge parameters for SUR model. Communications in Statistics Theory and Methods 37, 544–564 (2008)
3. Hoerl, A.E., Kennard, R.W.: Ridge regression: biased estimation for non-orthogonal Problems. Technometrics 12, 55–67 (1970a)
4. Hoerl, A.E., Kennard, R.W.: Ridge Regression: Application to Non-orthogonal Problems. Technometrics 12, 69–82 (1970b)
5. Khalaf, G., Shukur, G.: Choosing ridge parameters for regression problems. Communications in Statistics- Theory and Methods 34, 1177–1182 (2005)
6. Kibria, B.M.G.: Performance of some new ridge regression estimators. Communications in Statistics- Theory and Methods 32, 419–435 (2003)
7. Liu, K.: A new class of biased estimate in linear regression. Communications in Statistics-Theory and Methods 22, 393–402 (1993)
8. Mansson, K., Shukur, G.: On Ridge Parameters in Logistic Regression. Communications in Statistics, Theory and Methods 40, 3366–3381 (2011)
9. Mansson, K.: On Ridge Estimators for the Negative Binomial Regression Model. Economic Modelling 29, 178–184 (2012)
10. Nyquist, H.: Restricted estimation of generalized linear models. Applied Statistics 40, 133–141 (1991)
11. Schaefer, R.L., Roi, L.D., Wolfe, R.A.: A ridge logistic estimator. Communications inStatistics Theory and Methods 13, 99–113 (1984)

Multi-model Combination Techniques for Flood Forecasting from the Distributed Hydrological Model EasyDHM

Weihong Liao* and Xiaohui Lei

China Institute of Water Resource and Hydropower Research
Beijing, 100038, P.R. China
behellen@gmail.com

Abstract. The independently-developed distributed hydrological model EasyDHM, supplies four different runoff-generation simulation methods and three different conflux simulation methods. Each has its own advantages and disadvantages. In order to increase the accuracy of flood forecast of EasyDHM, this paper studied on several multi-model combination techniques for these different simulation methods, including the simple model average method, weighted average method, adaptive simple average method and adaptive weighted average method. These combination techniques were applied to the Taoer River in Northeast China, and evaluated from the intercom parison of results by single-model simulations and results by other combination methods in this paper. The study revealed that the multi-model simulations are generally better than any single-member model simulations. Furthermore, different multi-model combination techniques also showed different affection to the accuracy levels of flood forecasting.

Keywords: Flood Forecasting, Multi-model Combination, Distributed Hydrological Model, EasyDHM.

1   Introduction

During the past several decades, considerable achievements have been made in the conceptual understanding and mathematical description of water cycle in both the earth and atmosphere. A large number of models of different degrees of complexity and dimensionality are now available for predicting the hydrological cycle. In this situation, research on distributed hydrological models originated in 1970s in all over the world. The concept and framework of distributed hydrological physical models was firstly proposed by Freeze and Halranin *'Blueprint for a Physically-based, Digitally-simulated Hydrologic Response Model'*[1]. Since then, many distributed hydrological models have been developed around the world, such as MIKE SHE, SWAT, HSPF, GBHM and WEP, which can successfully be applied in both practical water resource and flood management[1-3]. As for

---

* Corresponding author.

Z. Li et al. (Eds.): ISICA 2012, CCIS 316, pp. 396–402, 2012.

flood forecasting, the most important objective is the accuracy of the predicted results. Thus, besides improving old models and developing new models, some other research to increase the reliability of flood forecast, like real-time correction and multi-model forecasting, have been carried on by researchers[4-5].

Multi-model forecasting is based on the technique of the multi-model combination of results of several similar models. It has been commonly used in other fields, such as statistics, management, economics and meteorology[6]. Shamseldin et al. (1997) first applied this technique to rainfallCrunoff models[7]. Multi-model techniques utilize the outputs of different rainfall-runoff models to provide an overall combined estimated output that can be more accurate than any one of the individual models. Hence, such a combined output can sensibly be used as an alternative to that of the best single individual model. Later Shamseldin and OConnor (1999) developed a real-time model output combination method (RTMOCM)[8], and tested it using three rainfallCrunoff models on five watersheds. The results indicated that the combined streamflow forecasts were much superior to those from the individual rainfallCrunoff models.

EasyDHMEasy Distributed Hydrological Modelis a distributed hydrological model developed independently for the practical projects about water resources[9]. A flood module of this hydrological model is also developed for flood management project. In EasyDHM, there are four different runoff-generation simulation methods and three different conflux simulation methods for usage. These methods are developed by ourselves or introduced from other models. In its flood module, users can choose any runoff-generation or conflux method to simulate and forecast the flood case. Thus, users can obtain different forecasting results from this very model just through different simulation methods. Based on this, multi-model combination techniques can be introduced to this model to improve the accuracy of flood forecast by any simulation method. This improvement is successfully applied to the Taoer river in China, which will be descript in the following contents.

2   The Distributed Hydrological Model EasyDHM

The distributed hydrological model EasyDHM was developed independently by our research group[9]. And its modeling system named MWEasyDHM was also developed for the pre-processing and post-processing of applications[10]. As mentioned above, this new hydrological model is that there are four different runoff-generation methods and three different conflux methods for simulation an integral hydrological process. These methods are chosen by users arbitrarily.

2.1   Runoff-Generation Simulation

The four methods in this module are: EasyDHM method, improved WetSpa method, Xinanjiang method and Hymod method[9]. EasyDHM method, whose runoff-generation process is shown as Fig. 1, is developed by our research group independently. Its rain-runoff mechanism includes 3 main steps that are surface process (interception, evapotranspiration, depression storage, infiltration, snow process, surface runoff), soil process (interflow, recharge and freeze thawing) and ground process. The other three methods are all introduced from those mature open-source models, rewritten in Fortran language with some improvement by our research group.

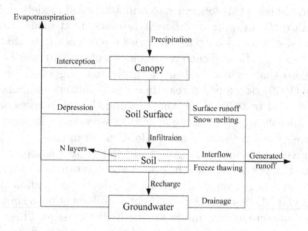

Fig. 1. The runoff-generation process of the EasyDHM method

2.2   Conflux Simulation in Rivers

In EasyDHM model, digital networks can be generated and classified automatically, and conflux process is simulated reach by reach. While sub-basins in Easy-DHM model are divided according to the topology of river network, thus, each sub-basin contains one and only one reach. Therefore, conflux simulation is based on these reaches. Like runoff-generation simulation, there also are three methods for conflux simulation in reaches, which are Muskingum, Variable Storage and Manning methods[9].

However, the difference of the results from different conflux methods is rather small. Thus, only the simulation results from different runoff-generated methods are compared and combined for better results in this paper.

3   Multi-model Combination Techniques

Since no single method can perfectly simulate the real flood process, the simulated results of any method differ from the observed ones in various respects. For the flood module of the model EasyDHM, we introduced the following multi-model combination techniques. And these different can be chosen by users accordingly.

3.1   Simple Average

The Simple average method is the most direct multi-model combination technique. It has been proved that multi-model ensemble averages perform better than any single-model simulations, including the best-calibrated single-model simulations[11]. Hence, the multi-model ensemble averages are more skillful and reliable than the single-model simulations. The simple average model can be expressed by the following equation:

$$Q(t) = \frac{\sum_{i=1}^{N} Q_i(t)}{N} \tag{1}$$

where, $Q_i(t)$ is the simulated or forecasted discharge at the time t by the ith runoff-generation method; N is the number of runoff-generation methods; and $Q(t)$ is the final result. This technique is quite simple and need none historical simulated and observed data.

3.2   Weighted Average Technique

Based on the simple average technique, weighted average technique is specifically raised for flood forecast models by Shamseldin et al. in 1997[7]. In this technique, different models have different model weight. The model weights are always positive and sum to 1. This technique considers the different action of different runoff-generation algorithms. It can be expressed as:

$$Q(t) = \sum_{i=1}^{N} \omega_i Q_i(t) \tag{2}$$

Where ω_i is the weight of the ith runoff-generation algorithm. The weights should be assigned first in multi-model simulation. And these weights are assigned according to simulation effect of different model. In EasyDHM, weights are calculated according to the errors between simulated and observed discharges of all runoff-generation methods. That is:

$$\omega_i = \frac{\left(\sum_{t=1}^{T} (Q_i(t) - Q_0(t))^2 \right)^{-1}}{\sum_{i=t}^{N} \left(\sum_{t=1}^{T} (Q_i(t) - Q_0(t))^2 \right)^{-1}} \tag{3}$$

Where, $Q_0(t)$ is the observed discharge of time t, T is the time steps for calculating weights. Weights can be decided only through historical simulated and observed discharges, or together with those have been forecasted and observed. And weights can be keep the same with time or change all the time as well.

3.3   Adaptive Simple Averag/Weighted Average Technique

Adaptive simple average/weighted average technique is also based on the simple average/weighted average technique, with only those effective models participating in calculation. Effective models are picked out automatically at first according to the Nash-Sutcliffe coefficient of each model[12]. The Nash-Sutcliffe coefficient describes the relationship between the residual variance of each model and the variation amplitude of the observed time series. This coefficient can be described as

$$Ns(i) = 1 - \frac{\sum_{t=1}^{t} (Q_i(t) - Q_0(t))^2}{\sum_{t=1}^{t} \left(Q_0(t) - \overline{Q_0(t)}\right)^2} \qquad (4)$$

Those models with their Nash-Sutcliffe coefficients above 0.9, can be seen as picked effective models. Accordingly, in EasyDHM model, those runoff-generation methods with their Nash-Sutcliffe coefficients above 0.9, can be seen as picked effective methods and will participate in the multi-model combinations.

4   Application

Tao River originates from Gaoyue mountain in East of Daxinganling in Northeast China, with its length as $466km$, and watershed area as 25966 km^2. Average annual rainfall of its watershed is 400 450mm, with rainfall during June August accounted for 84% -89% of the whole year.By using the flood module of the hydrological model EasyDHM, the flood of Taonan station in 1998 is simulated through four different runoff-generation model. The results of each single-model are compared with the observed ones in Fig. 2. As shown in Fig. 2, different models obtain different results and all models acts quite well but still need improvement.

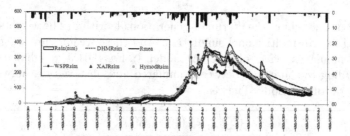

Fig. 2. The simulated results of each single model

Thus, multi-model combination techniques are introduced in this flood module for the flood forecasting in Taonan Station. As mentioned above, these multi-model combination techniques include the simple average method, weighted average method, adaptive simple average method and adaptive weighted average method. As shown in Fig. 3 and 4, the multi-model combination results from single-model simulations are generally better than any single-member model simulations. The simulated results of simple average method and weighted average method are much better than those of any single model as shown in Fig. 3. And by using adaptive methods, effective models can be picked out firstly, thus the forecasting results can be improved further. In this situation, the two different average techniques can obtain almost the same results, as the weights of all the effective models are almost the same.

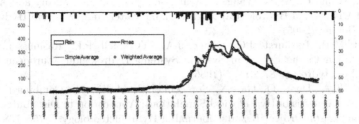

Fig. 3. The simulated results of multi-model combination

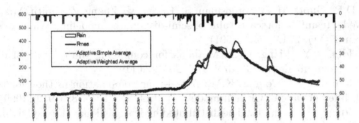

Fig. 4. The simulated results of adaptive multi-model combination

5   Conclusion

In this paper, the distributed hydrological model EasyDHM, and several multi-model combination techniques for its flood module are described first. These combination techniques include the simple model average method, weighted average method, adaptive simple average method and adaptive weighted average method. The model EasyDHM and these combination techniques were applied to the Taoer River in Northeast China, and evaluated from the intercom parison of results by single-model simulations and results by different multi-model combination simulations. The results revealed that the multi-model combination results are generally better than any single-member model. Furthermore, different multi-model combination techniques also showed different affection to the accuracy levels of flood forecasting. By using adaptive techniques, effective models can be picked out at first. Thus, the simulated results can be improved further.

Acknowledgement. This paper was supported by funds from Hydrological Simulation & Regulation of WaterSheds (NO. 50721006) of the Funds for Creative Research Groups of China.

References

1. Zhao, R.: Watershed Hydroloigcal Simulation. Hydraulic and Electric Power Press, Beijing (1984)
2. Jia, Y., Wang, H., Ni, G., Yang, D., Wang, J., Qing, D.: The Theory and Application of Distributed Hydrological Model. China Water Resources and Hydropower Press, Beijing (2005)
3. Abbott, M.B., Bathurst, J.C., Cunge, J.A., O'Connell, P.E., Rasmussen, J.: An Introduction to the European System: Systeme Hydrologique Europeen (SHE). Journal of Hydrology 87, 61–77 (1986)
4. Newsha, K.A., Duan, Q., Gao, X., Sorooshian, S.: Multimodel Combination Techniques for Analysis of Hydrological Simulations: Application to Distributed Model Intercomparison Project Results. Journal of Hydrometeorology 7, 755–768 (2006)
5. Asaad, Y.S., O'Connor, K.M.: A real-time combination method for the outputs of different rainfall-runoff models. Hydrological Sciences Journal 44(6), 895–912 (1999)
6. Boyle, D.P., Gupta, H.V., Sorooshian, S., Koren, V., Zhang, Z., Smith, M.: Toward improved streamflow forecast: value of semi-distributed modeling. Water Resources Research 37(11), 2749–2759 (2001)
7. Shamseldin, A.Y., Liang, G.C.: Methods for combining the outputs of different rainfall Crunoff models. J. Hydrol. 197, 203–229 (1997)
8. Shamseldin, A.Y., O'Connor, K.M.: A real-time combination method for the outputs of different rainfall Crunoff models. Hydrol. Sci. J. 44, 895–912 (1999)
9. Lei, X., Jiang, Y., Wang, H.: The distributed hydrological model: EasyDHM. China Water Power Press, Beijing (2010)
10. Lei, X., Wang, Y., Liao, W., Jiang, Y., Tian, Y., Wang, H.: Development of efficient and cost-effective distributed hydrological modeling tool MWEasyDHM based on open-source MapWindow GIS. Computers and Geosciences 37(9), 1476–1489 (2011)
11. Georgakakos, K.P., Seo, D.J., Gupta, H., Schake, J., Butts, M.B.: Characterizing streamflow simulation uncertainty through multimodel ensembles. J. Hydrol. 298, 222–241 (2004)
12. Kachroo, R.K., Natale, L.: Non-linear modelling of rainfall Crunoff transformation. J. Hydrol. 135, 341–369 (1992)

The Study of Item Selection Method in CAT

Peng Lu[1,3], Dongdai Zhou[1,2,3], Shanshan Qin[1],
Xiao Cong[3], and Shaochun Zhong[1,2,3]

[1] Ideal Institute of Information and Technology
Northeast Normal University
Changchun 130017, P.R. China
[2] Software School
Northeast Normal University
Changchun 130017, P.R. China
[3] Engineering & Research Center of E-learning
Northeast Normal University
Changchun 130017, P.R. China
{lup595,ddzhou}@nenu.edu.cn

Abstract. Item selection method is one of the most important parts of computerized adaptive testing. Traditional method is based on the item information function to select the item which has maximum information, with the maximum information test to achieve the purpose of accurate estimates the examinee's ability level. However, this method has high item exposure rate and the test content imbalance problem, etc. To solve these problems, this article introduces a new heuristic item selection method. The results of the study show that compared with the maximum information method, with the assurance of the condition of the same measurement accuracy, the new method can better control item exposure rate and achieve content balancing.

Keywords: Computerized Adaptive Testing, Item Selection Method, Exposure Control, Content balancing.

1 Introduction

In recent years, with the development of personalized learning, computerized adaptive testing as a new type of way to assess candidates ability or knowledge level was paid more and more attention[1]. Computerized Adaptive Testing (CAT) is one test form which manages the test through computer, and according to the examinee's response, the computer is dynamically select the questions and decided to terminate the test[2]. It has been used in educational assessment, psychological testing, personnel selection, and medical diagnostics and other fields etc..

According to the study of the measurement expert Lord[3], only choosing the items that close to the examinee's ability level, then the measurement is the most efficient. The core idea of CAT is: the system according to the examinee's answer, to calculate the value of the ability level constantly, and adjust the

Z. Li et al. (Eds.): ISICA 2012, CCIS 316, pp. 403–415, 2012.
© Springer-Verlag Berlin Heidelberg 2012

strategies based on these parameters in real-time, select test item that adapt to examinee's ability, and finally give the examinee a proper evaluation[4]. Therefore, compared with the traditional test form (paper-and-pencil, P&P), CAT implements tailor-made test for every examinee, examinees who during the test not need to face many too simple or too difficult items, therefore, it can provide more accurate estimate of the candidates ability with less items[5].It has a lot of potential advantages, such as the ability to estimate more efficient, high precision, more flexible implementation of the test and so on[6]. In the past few decades, some large-scale examinations (e.g., GRE, GMAT and TOEFL) have begun part or completely using CAT form. In CAT, item response model describes the assessment process of the examinee's knowledge level. At present, the most commonly used item response model is Three-Parameters Logistic model (3PL) which based on Item Response Theory (IRT), as shown in the following formula (1).

$$P(\theta) = c + (1 - c)\frac{1}{1 + e^{-Da(\theta - b)}} \tag{1}$$

Where θ is the examinee's ability level; $P(\theta)$ is the probability of correct response; $D = 1.702$, a constant; a is the discrimination of item; b is the difficulty of item; c is the guessing factor of item.

Based on the IRT, the main parts in the process of CAT implementation includes item pool, the set of examinee initial ability, item selection method, the method of estimate ability level, termination conditions of test, etc. Item selection method as one of the important parts of CAT, and it has great influence on the effectiveness of estimate examinees ability level and the contrast among several examinees performance, but also affects the examinee experience. In CAT, currently, the most widely used item selection method is based on the item maximum information. The core idea of this method is, in the testing process, according to the candidates current ability , through the item information function(IIF) select the item with has maximum information as a testing item[7], as shown in the following formula (2).

$$I_i(\theta') = \frac{[P_i'(\theta')]^2}{P_i(\theta')Q_i(\theta')} = \frac{[P_i'(\theta')]^2}{P_i(\theta')\left[1 - P_i(\theta')\right]} \tag{2}$$

MI method, in theory, has proven that it cans accurate estimate the examinee's ability. However, in item selection process in addition to consider item information, it still should consider other constraints, such as the using frequency of every test item (exposure rate), the proportion of the selected questions in every content domain (content balancing), the distribution of the correct answer in the options (answer balance), and allow a certain percentage of the 'special' test item in the test and so on. A detailed study of the item selection method of CAT will be carried out by this article.

2   Overview of the Item Selection Approach in CAT

Item selection method is the most important parts of CAT. Traditional item selection algorithm based on the candidate ability level and to select item of maximum information for testing[8], this method is called the maximum information method (MI), and the theoretical basis for doing so can improve the measurement accuracy.

However, this method often leads to imbalance use of test item in the item pool[10]. On one hand, the use of the item which has high-discrimination is more frequent, resulting in excessive exposure of these items; On the other hand, the low-discrimination items with little or not be used[11][12].For testing, this imbalance of item use can caused a serious impact on the test safety and development costs[13]. Ideally, all items in the item pool should have a similar exposure to meet the test safety and the efficiency of the use of item[14]. In order to solve these problems, scholars have propose many kinds of methods, mainly includes the following categories.

1. Random Method
   It uses a simple random strategy in the item selection process. For example, in the first, select several (e.g. 5) items, and then use the maximum information method to choose the most appropriate item [15]. Philp E. Cheng and Michelle Liou propose the Nearest-Neighbors Criterion method(NNCM)[16]. The basic idea of this method is, first search for the test item which difficulty parameter b is very close to the current ability estimate of candidates, and then calculate the maximum information value of these items, from which choose the test item with maximum information.

2. Combination Method
   The item selection strategy in adaptive testing system Flip (PeWePro part of the project)[17] is based on the way of the combination of the course structure, item information and test history. In the end, choose the best item which meet candidates. And stratified methods[18] based on the parameters of the item divide the test into several strata of fixed number. The corresponding test will choose test item from different strata.

3. Heuristic Algorithm
   Huang,Y.M. and others[19] develop a CAT system for mobile learning, and Particle Swarm Optimization(PSO) was introduced into the item selection method to improve the speed of searching questions from a large item pool; Stocking and Swanson propose Weighted Deviation Model(WDM),it can deal with many constraints at the same time, and provide an acceptable solution[18].This method does not require all the constraints should be strictly meet, allow to break some constraints; Cheng, Y. and Chang, H.[20] introduced the Maximum Priority Index method(MPI) to select the test item. It can accommodate a variety of non-statistical constraints, in CAT, is very easy to implement, and do not need to adjust the weight between the constraint and information.

In this paper, in order to solve the problems in CAT, such as imbalance test content, item exposure rate is too high, the use of item is uneven and other issues, we propose a new heuristic method, which test item selection method based on the simulated anneal algorithm (TIS-SA).The method ensure test accuracy, controls item exposure rate and the content balancing effectively. In addition, in the implementation of the CAT also very easy, don't like WDM needs to adjust the relative weights between the various constraints and item information before choosing item, time complexity is low. The next part will introduce the TIS-SA method in detail.

3   The Item Selection Method Base on TIS-SA

For the various constraints faced by the CAT in the implementation process, as well as the existing shortcomings of the major item selection method to deal with constraints, this paper propose TIS-SA item selection method. During the test, the method continue to record the test information (use times, the content domain it belongs to, etc.) of the use item to the database, continue to select item by the full consideration of the history records in the item pool to achieve the control of item exposure and content balancing, to improve the quality of selecting item, and meet the actual testing requirements.

3.1   The Design of Method

Specifically, in the item selection process, considering the item information, item exposure rate and content balancing three evaluation indicators. TIS-SA method design objective function based on these three indicators, as shown in the following formula (3).

$$Max \ F(I_i) = \alpha \cdot I + \beta \cdot B + \gamma \cdot E \tag{3}$$

Where I defined as the information indicator of item i; B defined as the indicator of content balancing; E defined as the indicator of exposure rate; α, β and γ are the weight coefficients for every indicators, the experience values is 10, 2 and 3, respectively.

According to the item selection objective function, during the performance retrieving the item which has the maximum value and as the next test item. In this way, the selected item not only has high information (to ensure the validity of the test), and lower exposure rate (make sure the test safety). In addition, after the test, the using items distribute evenly in the respective content domain(to content balancing),so the more comprehensive evaluation of the examinees can be taken, the reasonable comparison of performances among all examinees can also be taken. Next, three indicators of item information, the control of item exposure and content balancing will be introduced.

Item Information. Item information is still one of the most important indicators to select the test item; it can ensure that the validity of the test and test accuracy.3PL item information function as shown in the following formula (4).

$$I_i(\theta') = \frac{[P_i'(\theta')]^2}{P_i(\theta') \left(1 - P_i(\theta')\right)} = \frac{D^2 a_i^2 (1 - c_i)}{\left[c + e^{Da_i(\theta - b_i)}\right] \times \left[1 + e^{-Da_i(\theta - b_i)}\right]^2} \tag{4}$$

Scanning the item pool to calculate information of item by this function, and then show the proper item for the examinee to test, and providing candidates with the accurate estimates of ability.

The Control of Item Exposure. According to the problem of uneven item exposure, the item exposure should be controlled to ensure test security. However, both the S-H method and stratified methods control the item exposure rate from the whole. A test of computerized adaptive testing does not refer to the frequency of the use of each item. Therefore, no matter what kind of item exposure control algorithm may make the same item repeat in one test. In this study, through calculating the exposure frequency of every item, and according to the similar normal distribution to proceed mapping to control item exposure rate. The method is shown in the following formula (5).

$$E = e^{-\frac{n^2}{\sigma}} \tag{5}$$

Where n is the frequency of item exposure(include 0); σ is a constant, and equal 100 in case(Practice, it need to set this value according to the maximum item exposure rate).

Through the mapping, the frequency of every item exposure is converted to the range $(0, 1]$. If the more frequency of item exposure is closer converted value to 0, so the selected probability in the test is higher; On the contrary, if item has never been used, then the converted value is closer to 1, the probability of being selected is higher. In this way, through the control of item exposure, on the one hand, by reducing exposure rate of the item which has high item exposure rate, to ensure the safety of the test; On the other hand, by improving the utilizing rate of the item which has low exposure rate to achieve the purpose of improving the utilizing rate of item pool.

Content Balancing. In need of having a comprehensive, integrated evaluation for candidates, therefore, in the actual test should according to the degree of importance of each content domain (chapter), in which making a relatively uniform choice of test items to achieve content balancing. This can avoid the difference of test content causes the ability level estimates are not comprehensive and the unfair among candidates. And similar to the item exposure control method, the following equation denotes the selected test item and the degree of correlation of each content domain.

$$B = e^{-\frac{w_k m_k^2}{\sigma}} \tag{6}$$

Where w_k is the weight of chapter k, m^k is the number of selected item in chapter k. Other parameters refer to the parameters of item exposure control.

In the same way, through this equation, converting the frequency of selected item of each content domain into the range $(0, 1]$. In a content domain, the more using frequency of item, the closer converted value to 0, so the selected probability in the test is lower, contrariwise; So improving the using frequency of item with low utilizing rate and reducing the using frequency of item with high utilizing rate, the results of the test achieve balance, proceeding a comprehensive estimate for examinee, to meet the actual requirements.

Through the above three indexes, constraining the test item, ensure that the selected item has high item information, low exposure rate and content balancing. By scanning the item pool, solving the objective function $Max\ F(I_i)$, to select the optimal item for testing. However, the solving of objective function is combinatorial optimization problem; and the speed of selecting item will be an important issue.

3.2   TIS-SA Method

In order to meet practical needs, and improve the system response speed, this study uses Simulated Anneal Algorithm (TIS-SA) to solve the objective function to quickly select the appropriate test items. Simulated annealing algorithm is efficient, robust, general, flexible and other characteristics[21], it first starts from an initial solution(random generation), after the transformation of a large number of solutions through performing Metropolis algorithm, we can obtain relative optimal solution of the combinatorial optimization problem when obtain a given control parameter value; And then reduce the value of temperature control parameter t, performing Metropolis method repeatedly, finally obtaining optimal solution of the combinatorial optimization problem when value of temperature control parameter t tends to zero.

Using simulated anneal algorithm for solving the objective function $Max\ F(I_i)$, to improve the speed of selecting item. Although the final solution obtained by this method is not necessarily the optimal solution, but it can still test examinee and give an accurate estimate by extending the length of the test .The algorithm flow chart as shown in Fig. 1, including the following steps.

1. Cooling Schedule Initialization
   The initial temperature T is equal 100; decay parameter is 0.95; the length of Markov chain is 25; and termination condition is that the difference between the new optimal solution the previous is less than the threshold e (0.0001);
2. Randomly select the first item, set the Id of the item as previously optimal item and the optimal item, that is Id = PreId = PreBestId = BestId. And according to the item Id, calculating the value of objective function;
3. According to this item Id, randomly select next item NextId from adjacent domain, calculating the difference of objective function according to the Id, if the value of new item's objective function is greater than or equal to the value of provisionally optimal item's objective function, then update the optimal solution, and keep the Id of optimal solution as BestId;
4. If the difference between the value of objective function of the newly created item NextId and the value of objective function of previous item NextId is greater than 0,then accept NextId for new PredId, the next iteration point begin with the new accepted point;
5. Otherwise, if the difference between NextId and PreId is less than 0, then comparing the absolute value of difference with random number generated by the random number generator. If the value is greater than random number, then accept that update the new acceptable point NextId as the next iteration point PreId; otherwise, discard the Id, still use PreId as the point of the next iteration;
6. Repeat steps (3) (4) (5) according to the Markov chain, when achieve the maximum Markov chain, and finish the internal circulation;

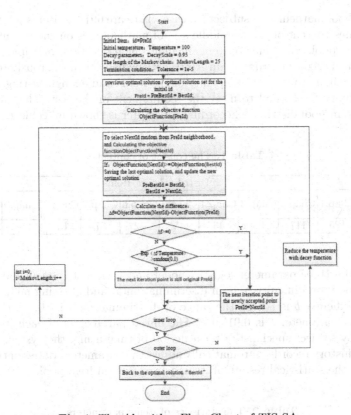

Fig. 1. The Algorithm Flow Chart of TIS-SA

7. 'Slowly' anneal, reduce the temperature, and repeat steps (3) (4) (5) (6), until the encounter the standard of termination or attenuation function to a minimum, at this time BestId is the optimal solution to find;

8. Finally, according to the BestId, selected items are show to the candidates for testing.

Through the objective function $Max\ F(I_i)$, achieve item exposure control and content balancing; solve the problems of the MI method in the item selection process. In order to improve the speed of selecting item, use simulated anneal algorithm (SA) to solve objective function. Although the solution obtained by this method is not necessarily the optimal solution, by extending the length of the test, still accurately estimate the candidates.

4   Experimental Studies

4.1   Data

In the experiment, based on the method of TIS-SA and has developed computerized adaptive testing prototype system - CAT-GD. This system chooses junior middle school mathematics subject for case, there are 691 test items relate to the two volumes content of junior middle school mathematics on the seventh grade in the item pool. The kinds of items include single selection, multiple selection and judgment right or wrong item. All test items have been rigorously tested by subject teachers, to ensure the quality of the items and enough testing items in various ability levels, and to meet the requirement of the test. The number of items of item pool distribute according to chapter is shown in Table 1.

Table 1. The consist of content

Content Domain	Distribution of Items									
	Chap.1	Chap.2	Chap.3	Chap.4	Chap.5	Chap.6	Chap.7	Chap.8	Chap.9	Chap.10
Num	125	117	95	43	45	42	87	61	18	58

Using the three-parameter logistic model and each test item includes three parameters: item difficulty b, item discrimination a, and guessing factor c. The range of difficulty b is [-3, 3], the range of discrimination a is (0, 2], the range of guessing parameter c is 0.01, 0.25.The initial parameters of each item were assigned by subject specialist, after a period of time using, the system analyze the items history records, automatically adjust the parameters values of the item. Table 2 is the statistical results of items parameters in item pool.

Table 2. The statistics of items parameters

	discrimination a	difficulty b	guessing factor c
Mean	0.74	0.55	0.24
Range	0 2	-3-3	{0.01,0.25}
Maximum	2.0	3	0.25
Minimum	0.1	-3	0.01

Through experiment studies is comparison the effect of different item selection methods, such as TIS-SA item selection method, MI item selection method, FB(function based) item selection method and the random item selection method. There are 50 students in the experiment, and four tests are carried out according to the different item selection methods.

4.2   The Analysis of Experimental Results

The following analysis include three aspects, measurement accuracy, item exposure rate and content balancing of the TIS-SA, MI and other two item selection methods, respectively.

Measurement Accuracy. The measurement accuracy is an important indicator to measure the ability of candidates, and estimate the accuracy and effectiveness use Bias and MES to measure the accuracy of estimating the ability of candidates in different methods.

$$Bias = \frac{\sum_{i=1}^{N}\left(\theta_i - \overline{\theta_i}\right)}{N} \tag{7}$$

$$\sigma = \sqrt{\frac{1}{N}\sum_{i=1}^{N}\left(\theta_i - \overline{\theta}\right)^2} \tag{8}$$

Table 3 introduces the overall measurement accuracy indicators. It can be seen from the table, the highest measurement accuracy when using the maximum information method (MI), and the lowest measurement accuracy when using random item selection method; Compared with MI method, TIS-SA method has some slight loss in measurement accuracy, resulting in Bias and MES.

This is understandable, because of item exposure rate and content balancing are considered when the TIS-SA method selects items, and the selected item information is not necessarily optimal. But, due to the small deviation, improving the length of test appropriately, still can offer the accurate assessment for candidates.

Table 3. The indicators of measurement accuracy

	MI method	Random method	FB method	TIS-SA method
Bias	4.56	5.83	4.71	4.78
MES	4.52	5.80	4.69	4.73

Item Exposure Rate. Item exposure rate is an important indicator in CAT, if the item exposure rate is not well controlled, and easily to appear the problem of over exposure for some items, and it final affects the safety test. The following explains the indicator firstly, and then four item selection methods are compared and analyzed.

Item exposure rate is shown in the following formula (9).

$$r_i = \frac{T_i}{m} \tag{9}$$

Where T_i is the frequency of the use of item i; m is the number of candidates.

Statistical results are shown in below table. It can be seen from the table, relative to the MI method, the TIS-SA item selection method better controls the item exposure rate.

The following Fig. 2 shows the comparison results of the four item selection methods' item exposure rate. It can be seen from the figure, part of the item exposure rate based on MI method is relatively high, even up to 0.9. Using random item selection method makes the existing problem small in item exposure

Table 4. The indicators of measurement accuracy

	MI method	Random method	FB method	TIS-SA method
Overall average	0.13	0.02	0.07	0.08
The overall maximum value	0.9	0.2	0.5	0.6
Overall the minimum value	0	0	0	0
Overall overexposed(%)	48	6	21	24
Overall has never been exposed(%)	48.3	10.1	34.5	30.2
Maximum exposure times in the single test	19	2	4	3

rate, the exposure of all the items is even average. Using the method of FB is generally better to the item exposure rate, the exposure rate of most of the items are less than 0.2, and the exposure rate of some parts of the items reach 0.5. The TIS-SA item selection method, the exposure rate of most of the items is less than 0.3, but the exposure rate of a few parts of items are high. Overall, relative to the MI method, TIS-SA item selection method greatly reduces the item exposure rate. Another problem needs to pay attention to is the utilization rate of item pool, 334 items based on MI method without exposure, the utilization rate of item pool is 52%, this will waste a lot of test items; 70 items of random item selection method have not been used, the utilization rate of item pool at 89%; 238 items of FB item selection method have not been used, the utilization rate of item pool is 66%; 207 items of TIS-SA item selection method have not been used, the utilization rate of item pool is 69%, so, relative to MI method, TIS-SA improve the utilization rate of item pool. Through the contrast and analysis can conclude that TIS-SA method compared with MI method had a better control in item exposure rate, final improve the test safety.

Content Balancing. Fig. 3 is the comparison results of MI, random item selection, FB and TIS-SA methods in the content balancing. It can be seen from the figure, the items selected by the MI method mainly concentrated in chapter four, chapter six and chapter seven, the maximum frequency of use is 520, however, the use of the items in other chapters is less, even never select test items from chapter three, chapter eight and chapter nine, lead to a serious issue of content balancing. Random item selection method selected the items from the item pool randomly each time, has not causes the problem of content imbalance, selects test items from every chapter; FB item selection method uses content balancing method, and generally achieves content balancing. Relative to random item selection method, it selects more items in chapter four and chapter eight; The TIS-SA item selection method selects test items in each chapter, but the number of selected items in each chapter and are not equal, selecting more

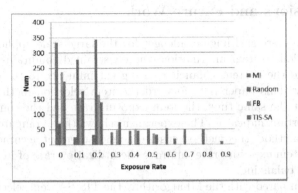

Fig. 2. The comparison results of item exposure rate

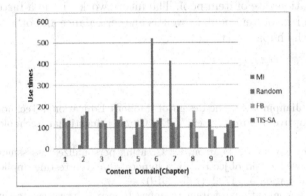

Fig. 3. The comparison results of content balancing

items in the second and seventh chapters, for a maximum of 201, and less of the selected items in Chapters eight and Chapters nine, which is mainly caused by the uneven number of items of each chapter. In addition, item information and the frequency item exposure affect the content balancing with a certain extent. Through comparison and analysis we can see, the TIS-SA method compared to the MI method is more balanced in content balancing, comprehensive evaluation of the candidates is more reasonable.

In the experiment, contrast and analyze the maximum information method, random selection method, FB item selection method and TIS-SA item selection methods in the aspects of measurement accuracy, item exposure rate, the content balancing and so on. Overall, under the condition, the TIS-SA item selection method presented by this article ensure a certain degree of measurement accuracy, well control the item exposure rate, and achieve good result of content balancing.

5 Conclusions and Future Work

At present, there are still huge challenges for the large-scale application of the CAT. In addition to item information, the constrained conditions of item exposure rate and the content balancing have great influence to the effectiveness and test safety. This paper puts forward the item selection method based on TIS-SA, and at the same time, the item exposure rate and the content balancing as an important indicator. The experiment shows that, compared to the MI item selection method, this method ensure the measurement accuracy, while the relatively low item exposure rate, improve the utilization rate of item pool, and ensure content balancing.

Although compared with the MI algorithm, the TIS-SA item selection method has made progress on several key issues, but still has some shortcomings, such as more test items are required in the testing process, improve the test length, and low utilization rate of item pool. The future work will do a further study of the utilization rate of item pool, as well as answer balance and other constrained conditions of the impact test.

References

1. Shipin, C., Jianping, Z.: The design of 'Modern Educational Technology' Master Course Adaptive Testing System. CET China Educational Technology 9, 93–97 (2008)
2. Lazarinis, F., Green, S., Pearson, E.: Creating personalized assessments based on learner knowledge and objectives in a hypermedia Web testing application. Computers & Education 55, 1732–1743 (2010)
3. Lord, F.M.: Applications of item response theory to practical testing problems. Krlbaum, NJ (1980)
4. Chuanhai, Z., Min, W., Yan, Y.: College English Vocabulary Online Adaptive Testing System Design based on IRT. Modern Educational Technology 12, 87–90 (2008)
5. Jinzhong, Y., Weimin, W.: Design and Implementation of the Online Adaptive Testing System based on Grails. Open Education Research 6, 96–103 (2009)
6. López-Cuadrado, J., Pérez, T.A., Vadillo, J.A., Gutiérrez, J.: Calibration of an item bank for the assessment of Basque language knowledge. Computers & Education 55, 1044–1055 (2010)
7. Thissen, D., Mislevy, R.J.: Testing algorithms. In: Wainer, H. (ed.) Computerized Adaptive Testing: A Primer Krlbaum, 2nd edn., NJ, pp. 101–113 (2000)
8. Lord, F.M.: A broad-range tailored test of verbal ability. Applied Psychological Measurement 1, 95–100 (1977)
9. Kozierkiewicz-Hetmańska, A., Nguyen, N.T.: A Computer Adaptive Testing Method for Intelligent Tutoring Systems. In: Setchi, R., Jordanov, I., Howlett, R.J., Jain, L.C. (eds.) KES 2010, Part I. LNCS, vol. 6276, pp. 281–289. Springer, Heidelberg (2010)
10. Huang, Y.M., Lin, Y.T., Cheng, S.C.: An Adaptive Testing System for Supporting Versatile Educational Assessment. Computers & Education 52, 53–67 (2009)
11. Yi, Q., Chang, H.: a-stratified CAT design with content blocking. British Journal of Mathematical and Statistical Psychology 56, 359–378 (2003)

12. Barla, M., Bielikova, M., Ezzeddinne, A.B., Kramar, T., Simko, M., Vozar, O.: On the impact of adaptive test question selection for learning efficiency. Computers & Educations 55, 846–857 (2010)
13. Weiss, D.J.: Adaptive testing by computer. Journal of Consulting and Clinical Psychology 53, 774–789 (1985)
14. Mills, C.N., Stocking, M.L.: Practical issues in large-scale computerized adaptive testing. Applied Measurement in Education 9, 287–304 (1996)
15. Tzai-hsing, K., Hsien-tang, L., Shyan-ming, Y.: Web-Based Adaptive Testing System. WuHan University Journal of Natural Sciences 1, 313–322 (2006)
16. Cheng, P.E., Liou, M.: Computerized Adaptive Testing Using the Nearest-Neighbors Criterion. Applied Psychological Measurement 3, 204–216 (2003)
17. Barla, M., Bielikova, M., Ezzeddinne, A.B., Kramar, T., Simko, M., Vozar, O.: On the impact of adaptive test question selection for learning efficiency. Computers & Educations 2, 846–857 (2010)
18. Leung, C.K., Chang, H.H., Hau, K.T.: Computerized adaptive testing: A mixture item selection approach for constrained situations. British Journal of Mathematical & Statistical Psychology 3, 239–257 (2005)
19. Huang, Y.M., Lin, Y.T., Cheng, S.C.: An Adaptive Testing System for Supporting Versatile Educational Assessment. Computers & Education 52, 53–67 (2009)
20. Cheng, Y., Chang, H.: The maximum priority index method for severely constrained item selection in computerized adaptive testing. British Journal of Mathematical and Statistical Psychology 62, 369–383 (2009)
21. Lishan, K., Yun, X., Shiyong, Y., Zhuha, L.: Non-numerical parallel algorithms -simulated annealing algorithm. Science Press, Beijing (2003)

Weighted Splicing Systems

S. Turaev[1], Y.S. Gan[2], M. Othman[1], N.H. Sarmin[2,3], and W.H. Fong[3]

[1] Faculty of Computer Science and Information Technology
Universiti Putra Malaysia
43400 UPM Serdang, Selangor, Malaysia
{sherzod,mothman}@fsktm.upm.edu.my
[2] Department of Mathematical Sciences, Faculty of Science
Universiti Teknologi Malaysia
81310 UTM Johor Bahru, Johor, Malaysia
ysgn88@gmail.com, nhs@utm.my
[3] Ibnu Sina Institute for Fundamental Science Studies
Universiti Teknologi Malaysia
81310 UTM Johor Bahru, Johor, Malaysia
fwh@ibnusina.utm.my

Abstract. In this paper we introduce a new variant of splicing systems, called *weighted splicing systems,* and establish some basic properties of language families generated by this type of splicing systems. We show that a simple extension of splicing systems with weights can increase the computational power of splicing systems with finite components.

1 Introduction

DNA computing is one of the most exciting new developments of computer science from both theoretical and practical points of view. DNA computing models use *Watson-Crick complementary* of DNA molecules (sequences), which are double stranded structures composed of four nucleotides A (*adenine*), C (*cytosine*), G (*guanine*), and T (*thymine*). According to Watson-Crick complementary, adenine always bonds with thymine, and guanine with cytosine. This feature of DNA molecules makes possible only to check the information encoded on a single strand and makes far-reaching conclusions since the information on the other strand can be decoded according to the complementary. Another feature of DNA molecules is the *massive parallelism* of DNA strands, which allows constructing many copies of DNA strands and carrying out operations on the encoded information simultaneously. The use of these two fundamental features of DNA molecules has already illustrated that DNA based computers can solve many computationally intractable problems: Hamiltonian path problem (Adleman, [1]), the satisfiability problem for arbitrary contact networks (Lipton, [2]), the satisfiability problem for Boolean circuits (Boneh et al., [3]), etc.

One of the early theoretical proposals for DNA based computation was made by Head [4] who used the *splicing operation* – a formal model of the cutting and recombination of DNA molecules under the influence of restriction enzymes. This

Z. Li et al. (Eds.): ISICA 2012, CCIS 316, pp. 416–424, 2012.

process works as follows: two DNA molecules are cut at specific subsequences and the first part of one molecule is connected to the second part of the other molecule, and vice versa. This process can be formalized as an operation on *strings*, described by a so-called *splicing rule*, which are the basis of a computational model called a *splicing system* (or *H system*). A system starts from a given set of strings (*axioms*) and produces a *language* by iterated splicing according to a given set of splicing rules.

Since splicing systems with finite sets of axioms and rules generate only regular languages (see [5]), several restrictions in the use of rules have been considered (see [6]), which increase the computational power up to the recursively enumerable languages. This is important from the point of view of DNA computing: splicing systems with restrictions can be considered as theoretical models of *universal programmable DNA based computers*.

In this paper we define another restriction of splicing systems, called *weighted splicing systems*, associating weights with the axioms, and calculating the weight $w(z)$ of the string z generated from two strings x and y from their weights $w(x)$ and $w(y)$ according to the operation $\odot$ defined on weights, i.e., $w(z) = w(x) \odot w(y)$. Then we consider several types of threshold languages generated by weighted splicing systems considering different weighting spaces and cut-points. We show that the selection of weighting spaces and cut-points effect to the generative power of splicing systems.

We should mention that weighted grammars and automata have been widely investigated in formal language theory since they were introduced in different forms in the 1960's (for instance, see[7-12]). On the one hand, the study of weighted grammars and automata shows that weights can increase the generative power of usual grammars and automata, and on the other hand, they help to construct more accurate models for stochastic phenomena and processes appearing in many applications of formal language theory. For instance, the use of weights makes possible to develop more efficient parsing and tagging algorithms for the natural and programming language processing.

This paper is organized as follows. Section 2 contains some necessary definitions and notations from the theories of formal languages and splicing systems. The concepts of weighted splicing systems and threshold languages generated by weighted splicing systems are introduced in Section 3. Section 4 illustrates the power of weighted splicing systems: it shows that some weighted splicing systems of finite components can generate even non-context-free languages. Section 5 discusses some open problems and possible topics for future research in this direction.

2   Preliminaries

In this section we recall some prerequisites, by giving basic notions and notations of the theories of formal languages and splicing systems which are used in sequel. The reader is referred to [6,13,14] for further information.

Throughout the paper we use the following general notations. The symbol $\in$ denotes the membership of an element to a set while the negation of set membership is denoted by $\notin$. The inclusion is denoted by $\subseteq$ and the strict (proper) inclusion is denoted by $\subset$. $\emptyset$ denotes the empty set. The sets of integers and positive rational numbers are denoted by $\mathbb{Z}$ and $\mathbb{Q}_+$, respectively. The cardinality of a set X is denoted by $|X|$. The families of recursively enumerable, context-sensitive, context-free, linear, regular and finite languages are denoted by **RE**, **CS**, **CF**, **LIN**, **REG** and **FIN**, respectively. For these language families, the next strict inclusions, named *Chomsky hierarchy*, hold:

Theorem 1 ([??]). **FIN** $\subset$ **REG** $\subset$ **LIN** $\subset$ **CF** $\subset$ **CS** $\subset$ **RE**.

Further, we briefly cite some basic definitions and results of iterative splicing systems which are needed in the next section.

Let V be an alphabet, and $\#, \$ \notin V$ be two special symbols. A *splicing rule* over V is a string of the form

$$r = u_1 \# u_2 \$ u_3 \# u_4, \text{ where } u_1, u_2, u_3, u_4 \in V^*.$$

For such a rule $r \in R$ and strings $x, y, z \in V^*$, we write

$$(x, y) \vdash_r z \text{ if and only if } x = x_1 u_1 u_2 x_2, \ y = y_1 u_3 u_4 y_2, \text{ and } z = x_1 u_1 u_4 y_2,$$

for some $x_1, x_2, y_1, y_2 \in V^*$.

z is said to be obtained by splicing x, y, as indicated by the rule r; $u_1 u_2$ and $u_3 u_4$ are called the *sites* of the splicing. We call x the *first term* and y the *second term* of the splicing operation.

An *H scheme* is a pair $\sigma = (V, R)$, where V is an alphabet and $R \subseteq V^* \# V^* \$ V^* \# V^*$ is a set of splicing rules. For a given H scheme $\sigma = (V, R)$ and a language $L \subseteq V^*$, we write

$$\sigma(L) = \{z \in V^* \mid (x, y) \vdash_r z, \text{ for some } x, y \in L, r \in R\},$$

and we define

$$\sigma^*(L) = \bigcup_{i \geq 0} \sigma^i(L)$$

by

$$\sigma^0(L) = L,$$
$$\sigma^{i+1}(L) = \sigma^i(L) \cup \sigma(\sigma^i(L)), i \geq 0.$$

An *extended H system* is a construct $\gamma = (V, T, A, R)$, where V is an alphabet, $T \subseteq V$ is the *terminal* alphabet, $A \subseteq V^*$ is the set of *axioms*, and $R \subseteq V^* \# V^* \$ V^* \# V^*$ is the set of *splicing rules*. When $T = V$, the system is said to be *non-extended*. The language generated by γ is defined by

$$L(\gamma) = \sigma^*(A) \cap T^*.$$

EH(F_1, F_2) denotes the family of languages generated by extended H systems $\gamma = (V, T, A, R)$ with $A \in F_1$ and $R \in F_2$ where

$$F_1, F_2 \in \{\mathbf{FIN}, \mathbf{REG}, \mathbf{CF}, \mathbf{LIN}, \mathbf{CS}, \mathbf{RE}\}.$$

Theorem 2 ([6]). *The relations in the following table hold, where at the intersection of the row marked with F_1 with the column marked with F_2 there appear either the family $\mathbf{EH}(F_1, F_2)$ or two families F_3, F_4 such that $F_3 \subset \mathbf{EH}(F_1, F_2) \subseteq F_4$.*

F1＼F2	FIN	REG	LIN	CF	CS	RE
FIN	REG	RE	RE	RE	RE	RE
REG	REG	RE	RE	RE	RE	RE
LIN	LIN, CF	RE	RE	RE	RE	RE
CF	CF	RE	RE	RE	RE	RE
CS	RE	RE	RE	RE	RE	RE
RE	RE	RE	RE	RE	RE	RE

3   Definitions

In this section we introduce the notion of weighted splicing systems which is specified with a weighting space and operations over weights closed in the weighting space.

Definition 1. *A weighted splicing system is a 7-tuple $\gamma = (V, T, A, R, \omega, M, \odot)$ where V, T, R are defined as for a usual extended H system, M is a weighting space, $\omega : V^* \to M$ is a weight function, $\odot$ is the operation over the weights $\omega(x)$, $x \in V^*$, and A is a subset of $V^* \times M$.*

Further, we define a weighted splicing operation and the language generated by a weighted splicing system.

Definition 2. *For $(x, \omega(x)), (y, \omega(y)), (z, \omega(z)) \in V^* \times M$ and $r \in R$,*

$$[(x, \omega(x)), (y, \omega(y))] \vdash_r (z, \omega(z))$$

iff $(x, y) \vdash_r z$ and $\omega(z) = \omega(x) \odot \omega(y)$. Then the language generated by the weighted splicing system γ is defined as

$$L_\omega(\gamma) = \{z \in T^* \mid (z, \omega(z)) \in \sigma^*(A)\}.$$

Remark 1. As a weighting space one can consider different sets and (algebraic) structures, for instance, the sets of integers, rational numbers, real numbers, the sets of Cartesian products of the sets of numbers, the set of matrices with integer entries, the set of lattices, groups, etc. Then, the operations over weights of strings are defined with respect to the chosen weighting space. In this paper we consider as weighting spaces the sets of integers, positive rational numbers, the set of Cartesian products of integers and the set of 2×2 matrices with integer entries.

Definition 3. *Let $L_\omega(\gamma)$ be the language generated by a weighted splicing system $\gamma = (V, T, A, R, \omega, M, \odot)$. A threshold language $L_\omega(\gamma, *\tau)$ with respect to a threshold (cut-point) $\tau \in M$ is a subset of $L_\omega(\gamma)$ defined by*

$$L_\omega(\gamma, *\tau) = \{z \in T^* \mid (z, \omega(z)) \in \sigma^*(A) \text{ and } \omega(z) * \tau\}$$

where $* \in \{=, >, <\}$ is called the mode of $L_\omega(\gamma, *\tau)$.

Remark 2. We can also consider as a threshold a subset of M. Then, the mode for such a threshold is defined as a membership to the threshold set, i.e., for a threshold set $A \subseteq M$, the modes are $\in$ and $\notin$.

The family of threshold languages generated by weighted splicing systems of type (F_1, F_2) is denoted by $w\mathbf{EH}(F_1, F_2)$ where

$$F_1, F_2 \in \{\mathbf{FIN}, \mathbf{REG}, \mathbf{CF}, \mathbf{LIN}, \mathbf{CS}, \mathbf{RE}\}.$$

From the definition, the next lemma follows immediately.

Lemma 1. *For all families $F_1, F_2 \in \{\mathbf{FIN}, \mathbf{REG}, \mathbf{CF}, \mathbf{LIN}, \mathbf{CS}, \mathbf{RE}\}$,*

$$\mathbf{EH}(F_1, F_2) \subseteq w\mathbf{EH}(F_1, F_2).$$

Proof. For any splicing system γ, we define the weighted splicing system by choosing $\{0\}$ as the weighting space, the usual addition as the weighting operation, and associating 0 with each axiom. Then $L_\omega(\gamma', = 0) = L(\gamma)$.   □

4   Examples and Results

In this section we consider examples of weighted splicing systems with different weighting spaces, and show that for the same finite sets of axioms and splicing rules, the selection of weighting spaces effect to the generative power of the weighted splicing system, i.e., the same splicing system with different weighting spaces can generated regular, context-free and context-sensitive languages.

Example 1. Let us consider the weighted splicing system

$$\gamma_1 = (\{a, b, x, y\}, \{a, b, x\}, \{(xay, \tau_1), (ybx, \tau_2)\},$$
$$\{r_1 = a\#y\$x\#ay, r_2 = yb\#x\$y\#b, r_3 = a\#y\$y\#b\}, \omega, M, \odot).$$

It is not difficult to see that the application of the rule r_1 iteratively to the axiom xay and the generated strings results in the strings $xa^k y$ for all $k \geq 1$, and the application of the rule r_2 iteratively to the axiom yax and the generated strings results in the strings $yb^m x$ for all $m \geq 1$. Further, applying the rule r_3 to the strings $xa^k y$, $k \geq 1$, and $yb^m x$, $m \geq 1$, we obtain the strings $xa^k b^m x$. Thus,

$$L_\omega(\gamma_1) = \{xa^k b^m x \mid (xa^k b^m x, \omega(xa^k b^m x)) \in \sigma^*(A), k, m \geq 1\},$$

where $A = \{(xay, \tau_1), (ybx, \tau_2)\}$.

First, we choose the set $\mathbb{Z}$ of all integers as the weighting space M, the addition $+$ of integers as the operation $\odot$, and $\tau_1 = 1, \tau_2 = -1$. Then, it is clear that

$$L_\omega(\gamma_1) = \{xa^k b^m x \mid (xa^k b^m x, k - m) \in \sigma^*(A), \, k, m \geq 1\}$$

and

$$L_\omega(\gamma_1, = 0) = \{xa^n b^n x \mid n \geq 1\},$$
$$L_\omega(\gamma_1, > 0) = \{xa^k b^m x \mid k > m \geq 1\},$$
$$L_\omega(\gamma_1, < 0) = \{xa^k b^m x \mid m > k \geq 1\}.$$

These three languages are context-free and not regular.

Second, we choose the set $\mathbb{Z} \times \mathbb{Z}$ as the weighting space M. We define the operation $\odot$ as the componentwise addition of pairs from $\mathbb{Z} \times \mathbb{Z}$, i.e., for any two $(x_1, x_2), (y_1, y_2) \in \mathbb{Z} \times \mathbb{Z}$, the sum is defined by $(x_1 + y_1, x_2 + y_2)$ and the ordering relation is also defined by componentwise, i.e., $(x_1, x_2) > (y_1, y_2)$ if and only if $x_1 > y_1$ and $x_2 > y_2$. Let $\tau_1 = (1, 0), \tau_2 = (-1, 0)$. Then with the cut-point $(0, 0)$, we generate the same languages above.

Third, the same languages above are also generated, when we consider the set of all 2×2 matrices with integer entries as the weighting space M, the componentwise addition as the operation $\odot$, the componentwise ordering as the ordering relation, $\tau_1 = \begin{pmatrix} 1 & 0 \\ 0 & 1 \end{pmatrix}$ and $\tau_2 = \begin{pmatrix} -1 & 0 \\ 0 & -1 \end{pmatrix}$ as the weights, and $\tau = \begin{pmatrix} 0 & 0 \\ 0 & 0 \end{pmatrix}$ as the cut-point.

Example 2. Consider a weighted splicing system

$$\gamma_2 = (\{a, b, c, w, x, y\}, \{a, b, w\}, \{(wax, \tau_1), (xby, \tau_2), (ycw, \tau_3)\},$$
$$\{r_1 = a\#x\$w\#ax, r_2 = b\#y\$x\#by, r_3 = c\#w\$y\#cw,$$
$$r_4 = a\#x\$x\#b, r_5 = b\#y\$y\#c\}, \omega, M, \odot).$$

One can see that for all $k, m, n \geq 1$,

$$(wa^k x, wax) \vdash_{r_1} wa^{k+1} x, \quad (xb^m y, xby) \vdash_{r_2} xb^{m+1} y, \quad (yc^k w, ycw) \vdash_{r_3} yc^{n+1} w.$$

Further,

$$(wa^k x, xb^m y) \vdash_{r_4} wa^k b^m y, k, m \geq 1,$$

and

$$(wa^k b^m y, yc^k w) \vdash_{r_5} wa^k b^m c^n w, k, m, n \geq 1.$$

Then, the language generated by the weighted splicing system γ_2 is

$$L_\omega(\gamma_2) = \{wa^k b^m c^n w \mid (wa^k b^m c^n w, \omega(wa^k b^m c^n w)) \in \sigma^*(A), \, k, m, n \geq 1\},$$

where $A = \{(wax, \tau_1), (xby, \tau_2), (ycw, \tau_3)\}$.

Next, we define different threshold languages with different weighting spaces and operations.

First, let $M = \mathbb{Q}_+$, the operation $\odot$ be the usual multiplication, and $\tau_1 = 3^{-1}$, $\tau_2 = 5^{-1}$, $\tau_3 = 15$. Then,

$$L_\omega(\gamma_2) = \{wa^k b^m c^n w \mid (wa^k b^m c^n w, 3^{n-k} 5^{n-m}) \in \sigma^*(A),\, k, m, n \geq 1\}.$$

We choose $\tau = 1$ as a cut-point, and define the following threshold languages:

$$L_\omega(\gamma_2, = 1) = \{wa^n b^n c^n w \mid n \geq 1\} \in \mathbf{CS} - \mathbf{CF},$$
$$L_\omega(\gamma_2, > 1) = \{wa^k b^m c^n w \mid n > k, m \geq 1\} \in \mathbf{CF} - \mathbf{REG},$$
$$L_\omega(\gamma_2, < 1) = \{wa^k b^m c^n w \mid k, m > n \geq 1\} \in \mathbf{CF} - \mathbf{REG}.$$

Second, let $M = \mathbb{Z} \times \mathbb{Z}$, the operation $\odot$ is defined as in Example 1, and $\tau_1 = (1, 0)$, $\tau_2 = (-1, 1)$, $\tau_3 = (0, -1)$. Then,

$$L_\omega(\gamma_2) = \{wa^k b^m c^n w \mid (wa^k b^m c^n w, (k - m, m - n)) \in \sigma^*(A),\, k, m, n \geq 1\}.$$

Consequently,

$$L_\omega(\gamma_2, = (0,0)) = \{wa^n b^n c^n w \mid n \geq 1\} \in \mathbf{CS} - \mathbf{CF},$$
$$L_\omega(\gamma_2, > (0,0)) = \{wa^k b^m c^n w \mid k > m > n \geq 1\} \in \mathbf{CS} - \mathbf{CF},$$
$$L_\omega(\gamma_2, < (0,0)) = \{wa^k b^m c^n w \mid n > m > k \geq 1\} \in \mathbf{CS} - \mathbf{CF}.$$

Third, the same languages above can be generated by γ_2 if we choose the set of all 2×2 matrices with integer entries as the weighting space, the componentwise addition as the operation $\odot$,

$$\tau_1 = \begin{pmatrix} 1 & 0 \\ 0 & 1 \end{pmatrix}, \quad \tau_2 = \begin{pmatrix} -1 & 1 \\ 1 & -1 \end{pmatrix} \text{ and } \tau_3 = \begin{pmatrix} 0 & -1 \\ -1 & 0 \end{pmatrix},$$

and $\tau = \begin{pmatrix} 0 & 0 \\ 0 & 0 \end{pmatrix}$ as the cut-point for the threshold languages.

Remark 3. Example 2 shows that the use of a Cartesian product of integers and matrices as weighting spaces make possible for splicing systems with finite components to generate non-context-free languages. For instance, one can easily construct a weighted splicing system which generates the language

$$\{wa_1^n a_2^n \cdots a_k^n w \mid n \geq 1\}$$

for any $k \geq 2$, if the Cartesian product $\underbrace{\mathbb{Z} \times \mathbb{Z} \times \cdots \times \mathbb{Z}}_{k}$ is used as the weighting space.

Combining the results of Theorem 2, Lemma 1 and Examples 1 and 2 above, we obtain the following results:

Theorem 3. *For $F_1 \in \{\mathbf{LIN}, \mathbf{CF}\}$,*

$$w\mathbf{EN}(\mathbf{FIN}, \mathbf{FIN}) - \mathbf{EN}(F_1, \mathbf{FIN}) \neq \emptyset.$$

Theorem 4.
$$\mathbf{REG} \subset w\mathbf{EN}(\mathbf{FIN}, \mathbf{FIN}) \subseteq \mathbf{RE}.$$

5 Conclusion

This paper introduces a new definition of weighted splicing systems and establishes some new facts. We have shown that even a simple extension of splicing systems with weights increases the generative power of splicing systems with finite components: in some cases they can generate non-context-free languages. The problem of the incomparability of the family of linear and context-free languages with the family of threshold languages generated by weighted splicing systems with finite components (the inverse inequality of that in Theorem 3) and the strictness of the second inclusion in Theorem 4 remain open. Most probably, linear languages as well as simple matrix languages (see [14]) can be generated by weighted splicing systems if matrices are used as weighting spaces but recursively enumerable languages may not be generated since the simulation of the context-sensitivity property of phrase-structure grammars is impossible with weights.

Acknowledgment. The first and third authors would like to acknowledge MOHE and UPM for the research grants with Vote No FRGS/1/11/SG/ UPM/01/1 and RUGS 05-01-10-0896RU/F1 respectively. The second author is indebted to MOHE for his MyBrain15 Scholarship, and the fourth and fifth authors have been supported by UTM Research University Grant Vote No. 02J65.

References

1. Adleman, L.: Molecular computation of solutions to combinatorial problems. Science (266), 1021–1024 (1994)
2. Lipton, R.: Using DNA to solve NP Ccomplete problems. Science (268), 542–545 (1995)
3. Boneh, D., Dunworth, C., Lipton, R., Sgall, J.: On the computational power of DNA. Special Issue on Computational Molecular Biology (71), 79–94 (1996)
4. Head, T.: Formal language theory and DNA: An analysis of the generative capacity of specific recombination behaviors. Bull. Math. Biology (49) (1987)
5. Pixton, D.: Regularity of splicing languages. Discrete Applied Mathematics (69), 101–124 (1996)
6. Păaun, G., Rozenberg, G., Salomaa, A.: DNA computing. In: New Computing Paradigms. Springer (1998)
7. Salomaa, A.: Probabilistic and weighted grammars. Information and Control 15, 529–544 (1969)

8. Fu, K.S., Li, T.: On stochastic automata and languages. International Journal of Information Science (1969)
9. Mizumoto, M., Toyoda, J., Tanaka, K.: Examples of formal grammars with weights. Information Processing Letters 2, 74–78 (1973)
10. Mizumoto, M., Toyoda, J., Tanaka, K.: Various kinds of automata with weights. Journal of Computer and System Sciences 10, 219–236 (1975)
11. Alexandrakis, A., Bozapalidis, S.: Weighted grammars and kleenes theorem. Information Processing Letters 24, 1–4 (1987)
12. Droste, M., Kuich, W., Vogler, H. (eds.): Handbook of Weighted Automata. Springer (2009)
13. Rozenberg, G., Salomaa, A. (eds.): Handbook of formal languages, vol. 1-3. Springer (1997)
14. Dassow, J., Páun, G.: Regulated rewriting in formal language theory. Springer, Berlin (1989)

Individual Paths in Self-evaluation Processes

Sylvia Encheva

Stord/Haugesund University College
Bjønsonsg. 45, 5528, Haugesund, Norway
sbe@hsh.no

Abstract. A large number of approaches for evaluating students' knowledge are made available by the research community. Much less is known when it comes to discussing qualities of assessment materials. In this work we propose use of graphical representations of individual paths in self-evaluation processes for unveiling dependences between tests' content, alternative answers and learning.

Keywords: Individual Paths, Self-evaluation, Graphical Representations.

1 Introduction

The effect of technology on students learning is well explored area of research. Questions however about content qualities are seldom asked. Usually automated tests for self-evaluation of knowledge and/or concept understanding are composed on the fly by random selection of items from a pull of questions. If the test is taken repeatedly the same procedure is applied where some more advanced systems replace all of the old questions with new ones regardless the outcome of the previous tests.

In this work we propose use of graphical representations of individual paths in self-evaluation processes for unveiling dependences between tests' content, alternative answers and learning. Automated tests with number of questions varying from three to seven are suggested to students. Their responses are recorded and included in tables. It is not a big surprise to see that responses from a single individual are not homogeneous. Response representations in a graph form can help to choose which questions or explanations ought to be reconsidered.

The rest of the paper is organized as follows. Related work and supporting theory may be found in Section 2. The model of the proposed system is presented in Section 3. The paper ends with a conclusion in Section 4.

2 Related Work

Some of the methods designed to handle impresize data are fuzzy logic [2], [6], [8], rough sets theory [12], [13], [14], [15], [16], grey theory, [5], [9] [17], many valued logics, [10] and formal concept analysis, [11].

Z. Li et al. (Eds.): ISICA 2012, CCIS 316, pp. 425–431, 2012.

With no interactivity, the ICT resource merely presents standard information, such as a prepared sequence of slides or pages on an IWB which are presented to the class, [1]. A personalized learning path generator based on metadata standards is presented in [3]. The effect of micro-level tutorial decisions is discussed in [4]. The study described in [7] investigates the advantages and disadvantages of the use of Smart-Board and offers suggestions for effective use of Smart-Boards basing on teacher opinions. Use of interactive whiteboards for studing mathematics was discussed in [18].

We interested to find out which questions and help functions are most fruitful for learning.

3   Application

Students' understanding, knowledge and skills have been a subject of research for decades. Various methods, approaches, reasonings, and decision support systems have been developed for that. What seems to be somewhat less discussed is ways to evaluate qualities of tests' questions, hints and help functions.

Lets first consider a test for self-assertion that a new term or concept is correctly understood. Suppose the test is of multiple-choice type, contains one question and the options are correct answer, incorrect answer and nor answer is selected. It can be taken three times since the purpose is to provide reassurance to students only.

Every time a student chooses a correct answer the system acknowledges it and provides a new question. In case an incorrect answer or missing answer options are selected the system first provides help in a form of short explanation followed by a new question. A correctly answered question is followed by a slightly more difficult question. Wrongly answered or missing answer options are followed by a question of similar level of difficulty.

Table 1. Sequences for three questions

Questions	Sequences							
	1st	2nd	3rd	4th	5th	6th	7th	8th
1	0	0	0	0	1	1	1	1
2	0	1	0	1	0	1	0	1
3	0	0	1	1	0	0	1	1

Table 1 summarizes all possible triplets of responses where correct answer is denoted by 1 and a wrong or missing answer by 0. Responses concerning less than three questions are not included. Fig. 1 illustrates all paths and where they cross. Obviously there are two junctions to be considered, one related to the second question and another one related to the third question. Majority voting is often employed while drawing conclusions from experience with web-based evaluations. We focus on the content of provided questions and help functions. Following individual paths in Fig. 1 we believe might give better answers to questions like is the content of questions to be re-examined or the content of help functions or both.

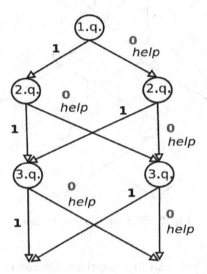

Fig. 1. Three questions outcomes

We next consider a test with four questions. Outcomes are listed in Table 2 and individual paths are graphically presented in Fig. 2. The two junctions related to the second and the third questions are to be considered, but the one related to the forth question is of even bigger interest.

Table 2. Sequences obtained from four questions test

Questions	Sequences							
	1st	2nd	3rd	4th	5th	6th	7th	8th
1	0	0	0	0	1	1	1	1
2	0	0	1	1	0	1	0	1
3	0	1	0	1	1	0	0	1
4	0	0	1	1	0	1	0	1

To facilitate better readability we omit writing 'help' after every '0' entrance to the rest of the paper.

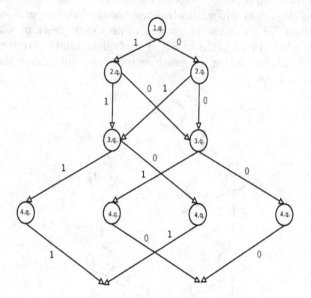

Fig. 2. Four questions outcomes

We next consider a test with five questions. Outcomes are listed in Table 3 and individual paths are graphically presented in Fig. 3. The junction related to the second question is to be considered while the ones related to the third, forth and fifth questions deserve additional attention.

Table 3. Sequences obtained from five questions test

Questions	Sequences							
	1st	2nd	3rd	4th	5th	6th	7th	8th
1	0	0	0	0	1	1	1	1
2	0	0	0	1	0	1	0	1
3	0	1	1	1	0	0	1	1
4	0	0	1	0	1	1	1	1
5	0	0	0	1	0	1	0	1

We next consider a test with six questions. Outcomes are listed in Table 4 and individual paths are graphically presented in Fig. 4. Junctions related to the second, third, sixth and seventh questions are to be considered while the ones related to the forth and fifth questions deserve additional attention.

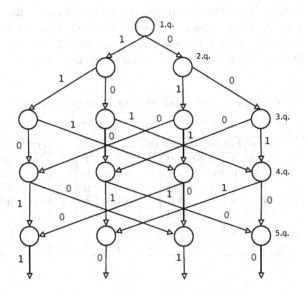

Fig. 3. Five questions outcomes

Table 4. Sequences obtained from six questions test

Questions	Sequences															
	1st	2nd	3rd	4th	5th	6th	7th	8th	9th	10th	11th	12th	13th	14th	15th	16th
1	0	0	0	0	0	0	0	0	1	1	1	1	1	1	1	1
2	0	0	0	0	1	1	1	1	1	0	1	1	1	0	0	0
3	1	1	0	0	1	1	0	0	0	0	1	1	0	0	1	1
4	1	1	1	1	0	0	1	1	1	1	1	1	1	1	0	0
5	1	0	0	1	0	1	1	0	1	0	1	0	1	1	0	1
6	1	0	0	1	1	0	0	1	0	1	1	0	1	0	1	0

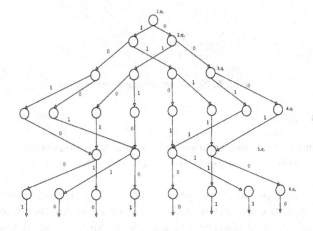

Fig. 4. Six questions outcomes

We next consider a test with seven questions. Outcomes are listed in Table 5 and individual paths are graphically presented in Fig. 5. Junctions related to the second, third and fifth questions are to be considered while the ones related to forth and sixth questions deserve additional attention.

Table 5. Sequences obtained from seven questions test

Questions	Sequences											
	1st	2nd	3rd	4th	5th	6th	7th	8th	9th	10th	11th	12th
1	0	0	0	0	0	1	1	1	1	1	1	0
2	0	0	0	1	1	0	0	0	0	1	1	0
3	0	0	1	0	0	1	1	0	0	0	0	1
4	0	1	1	0	0	1	0	1	0	1	1	0
5	0	1	0	1	0	0	1	1	0	1	0	1
6	0	1	1	1	1	1	0	1	0	0	0	0
7	0	0	0	0	0	1	1	1	1	1	1	0

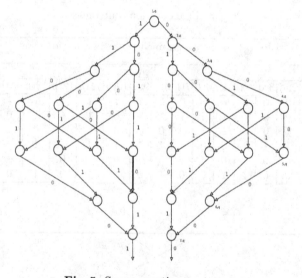

Fig. 5. Seven questions outcomes

4   Conclusion

We believe that similar investigations and graphical representations can be used in many other web-based systems providing automated evaluations and recommendations.

The included graphical representation provides better overview of existing interrelationships in comparison to a three structure. Another interesting advantage is that one can observe and analyse various developments due to an increasing number of questions in a test.

References

1. Beauchamp, G., Kennewell, S.: Interactivity in the classroom and its impact on learning. Computers & Education 54(3), 759–766 (2010)
2. Carlsson, C., Fuller, R.: Optimization under the fuzzy if-then rules. Fuzzy Sets and Systems 119(1) (2001)
3. Colace, F., De Santo, M., Vento, M.: A Personalized Learning Path Generator Based on Metadata Standards. International Journal on E-Learning, Association for the Advancement of Computing in Education 4(3), 317–335 (2005)
4. Chi, M., VanLehn, K., Litman, D.J.: Do Micro-Level Tutorial Decisions Matter: Applying Reinforcement Learning to Induce Pedagogical Tutorial Tactics. Intelligent Tutoring Systems (1), 224–234 (2010)
5. Deng, J.L.: Introduction to grey system theory. Journal of Grey Systems 1, 1–24 (1989)
6. Fuller, R., Zimmermann, H.-J.: Fuzzy reasoning for solving fuzzy mathematical programming problems. Fuzzy Sets and Systems 60, 121–133 (1993)
7. Gursula, F., Tozmaza, G.B.: Which one is smarter? Teacher or Board. Procedia Social and Behavioral Sciences 2, 5731–5737 (2010)
8. Herrmann, C.S.: Fuzzy Logic as Inferencing Techniques in Hybrid AI-Systems. In: Martin, T., L. Ralescu, A. (eds.) IJCAI-WS 1995. LNCS, vol. 1188, pp. 69–80. Springer, Heidelberg (1997)
9. Hu, Y.C.: Grey relational analysis and radical basis function network for determining costs in learning sequences. Applied Mathematics and Computation 184, 291–299 (2007)
10. Lei, Y., Wang, Y., Cao, B., Yu, J.: Concept Interconnection Based on Many-Valued Context Analysis. In: Zhou, Z.-H., Li, H., Yang, Q. (eds.) PAKDD 2007. LNCS (LNAI), vol. 4426, pp. 623–630. Springer, Heidelberg (2007)
11. Liu, J., Yao, X.: Formal concept analysis of incomplete information system. In: Seventh International Conference on Fuzzy Systems and Knowledge Discovery (FSKD), pp. 2016–2020 (2010)
12. Pawlak, Z.: Rough Sets. International Journal of Computer and Information Sciences 11, 341–356 (1982)
13. Pawlak, Z.: Rough Sets: Theoretical Aspects of Reasoning about Data, vol. 9. Kluwer Academic Publishers, Dordrecht (1991)
14. Polkowski, L., Skowron, A.: Rough mereological approach to knowledge-based distributed AI. In: Lee, J.K., Liebowitz, J., Chae, J.M. (eds.) Third World Congress on Expert Systems, Soeul, Korea, February 5-9, pp. 774–781 (1996)
15. Polkowski, L., Skowron, A.: Rough mereology: A new paradigm for approximate reasoning. International Journal of Approximate Reasoning 15(4), 333–365 (1996)
16. Polkowski, L., Skowron, A.: Rough mereology in information systems. A case study: Qualitative spatial reasoning. In: Polkowski, L., Lin, T.Y., Tsumoto, S. (eds.) Rough Set Methods and Applications: New Developments in Knowledge Discovery in Information Systems. STUDFUZZ, vol. 56, ch. 3, pp. 89–135. Springer, Heidelberg (2000)
17. Rahimnia, F., Moghadasian, M., Mashreghi, E.: Application of grey theory organizational approach to evaluation of organizational vision. Grey Systems: Theory and Application 1(1), 33–46 (2011)
18. Torff, B., Rose Tirotta, R.: Interactive whiteboards produce small gains in elementary students' self-reported motivation in mathematics. Computers & Education 54(2), 379–383 (2010)

Learning Sequential Investment Strategy in High-Frequency Environment

Hao Cui, Yunyan Zhang, and Hanming Chen

Beijing Institute of Technology
Beijing 100081, P.R. China
bitcuihao@gmail.com

Abstract. The paper proposes a new approach to build stock trading decision support system. This system, aiming at maximizing the sequential investment outcome, regards the sequential investment strategy for decision-making in high-frequency environment. Moreover, TD(λ) method is applied in learning the strategy, and substantial results are obtained in practical environment.

Keywords: Stock, Investment, Strategy, TD(λ).

1 Introduction

In the research of stock trading decision support system, there is a conventional method, namely to analyze historical stock price through data mining technology, and try to forecast the future price. By this means, advice for trading decision-making is provided for investors. (Tsang and Martinez-Jaramillo, 2001).

However, some controversy over this method still exists(Hilgers, 2001). Whether we can predict the stock price depends on the key to three problems. First, does some kind of pattern do exist in the stock price? Second, if the pattern surely exists, will it be possible for learning system to obtain? Third, supposed that the pattern is undoubtedly acquired; will the price certainly follow this pattern in the future? Different opinions on those questions are held, especially on the first and the last. Besides, in practical investment, the stock price pattern is not intrinsic, but influenced by external environment. For example, good news of certain stock will bring a sharp rise to its price. Whereas it is hard for predictive model to reflect these impacts, thus an obvious limitation of practicability emerges in price prediction approach.

This paper puts forward a new approach, which is to investigate the sequential investment strategy in a relatively short term. We argue that in a short term, price change mainly depends on the market itself, rather than outside world. Thus, this approach differs from conventional in many aspects, and has better feasibility. Furthermore, a learning system is built to learn the strategy, and the experiment ran in real stock market proves that this approach is feasible.

Z. Li et al. (Eds.): ISICA 2012, CCIS 316, pp. 432–439, 2012.

2 SICL Strategy

2.1 Sequential Investment with Certain Length

This paper mainly discusses the strategy of sequential investment in high-frequency environment. The sequential investment is namely a combination of actions in a time series within certain length. Sequential investment treats the action combination as a whole, and it maximizes the final outcome of the action series through strategic combinational operation. In the T+0 mode of market, it is feasible for trading in a very short time. Thus by means of intraday trading, some investors are available to do high speed trade and gain profit. The purpose of this paper is to explore how to maximize the final outcome of sequential investment within certain length, in such a high-frequency trading environment. The strategy to solve the above-mentioned problem is named as SICL(Sequential Investment within Certain Length) strategy hereinafter.

Compared with conventional price predication method, this research has a better maneuverability and practicability. First, action series within a certain length constructs a finite state set. Repeated searching in such set guarantees the convergence of the algorithm. This also means that if a good strategy does exist in sequential investment, it probably can be learned. Second, in short term, more obvious regularity can be seen in stock market, and learning that regularity is more achievable. For instance, in practical, invertors always tend to chase the winner and cut the loser, resulting in the herding of stock market. Therefore, if the shares persistently fall in a span, the possibility of continue dropping is increased (although it does not necessarily happen). In contrast with the new approach, the conventional predicts the shares by observing the historical daily price, which seems influenced mainly by outside world news, rather than the market itself. Thus it can be demonstrated that the regularity embodied in the fluctuation of price in shorter time is intuitively stronger than that in the longer. Third, theoretically, by way of sequentially investing, better profit expectation and risk management can be achieved; hence research of such strategy has greater practical value.

2.2 Comparison with Conventional Method

The approach in this paper is essentially different with the conventional one. First, target result is different. The target of conventional method is to predict specific numeric value, while the new approach aims at acquiring a strategy to make decisions. Actually, this strategy also implicitly predicts the price; moreover, based on the prediction, it acquires the ability of reasoning and decision-making. It is worth mentioning that the prediction is not for a specific value, but the trend of the future price change. Second, target environment is distinct. The conventional approach concentrates on observing data in days or longer time units, while the new one observes the quick change of price in high-frequency environment. Third, target form is disparate. The conventional concerns the price in the discrete time, however, the new pay close attention to the overall trend of the stock in a certain length of time series.

3   Learning Algorithm

3.1   TD(λ) Method

Sutton(1988) discussed predicting problems, and SICL strategy, discussed in this paper, belongs to problems of learning to predict, that is, of using past experience with an incompletely known system to predict its future behavior. At any time step t_k, within a time series $t_1, t_2, t_3, ..., t_n$, the investment decision only consider n-k steps later when the final outcome of the sequential investment comes out.

TD(λ) method is put into use to learn SICL strategy in this paper. TD(λ) pertains to reinforcement learning, and is specialized for predicting problem. Conventional prediction-learning methods are driven by the error between predicted and actual outcomes, whereas TD(λ) methods are similarly driven by the error or difference between temporally successive predictions; with them, learning occurs whenever there is a change in prediction over time. Sutton puts forward that in the learning of prediction problem, TD(λ) has two advantages over the conventional. First, it is more incremental and easier to compute. Second, it tends to make more efficient use of their experience. More analysis can be seen in Sutton(1988).

3.2   The Reason for Choosing TD(λ)

Admittedly, conventional methods also work when learning SICL strategy is applied. At any time step k, the system preserves the prediction pk, pairs the final rewards and each preserved pk at the end of the investment sequence, and then learns all the two tuples.

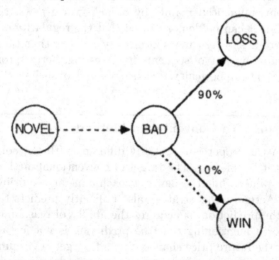

Fig. 1. Comparison Between TD(λ) and Conventional Methods

Nevertheless, inherent detriments are emerged in conventional methods when learning the SICL strategy. Sutton(1988) discussed the advantage of TD(λ) over conventional methods by using a specific position in game, as it has shown in figure 1. He argues that in that case, it is more reasonable and easier for algorithm to convergence and gains a better result when pairing NOVEL and BAD, rather than associating NOVEL with the terminal states. He thus proposed that the main advantage of TD(λ) is it shields the noises in the experience, by pairing state with its neighborhood, rather than the terminal.

It is noteworthy that using TD(λ) to learn the SICL strategy is dominant. The total amount of the assets an investor owns is limited, hence the possible decisions in the following state depends on the decision made in the previous state to a great extent. For instance, buying shares using all the cash in state A and entering state B, results in the only choice of hold or sell in B. We can come to the conclusion that in sequential investment, the transformation between adjacent states is a key point. Considering that there are numerous unstable factors in the stock market, linking a certain state with the final state will bring noises into the learning experience, and may lead to unsatisfying learning result.

4   Learning System Design

4.1   Design of Neural Network

BP neural network comes into use in constructing the learning system. There is a good combination of various neural networks and TD(λ), as it has been discussed by Sutton in 1998. In all the neural networks, at present, BP neural network is widely used and famous for its stability.

The system contains a three-tier topology, in which each unit is a sigmoid unit. The input structure is a 33-dimentional vector. In this vector, the previous 30 data are comprehensive analysis of recent stock price, first buying price and first selling price, such as whether the stock price is continuously rising in the last three states. Each of the 30 data is either 0 or 1. The 31st and 32nd data are the proportion of cash and stock in all the assets of the system. The 33rd data is the ratio of current step and length of the sequential investment. The 34th data is the ratio of the assets system owns currently and that at the very beginning of the sequence. The hidden layer contains thirty units. The output of the system is designed to be a real number, denoting the current prediction of the final outcome of the sequential investment. All the weights are randomly initialized in the range of $[-0.1, 0.1]$. The assessment of the assets at any time step has a good guidance for the final profit, hence the weights between the last input unit and all the hidden units are initialized to be 0.05, in order to speed up the learning.

All legal actions in the system are presented as a two tuple $< N, P >$. N reflects the proportion of the cash used to buy stock, or the proportion of stock to sell. P indicates the difference between the price in the system decision and the possible transaction price. Possible values of N and P are listed in Table [1-2]. By combining N and P, the system can obtain all the possible decisions. After that, system tries all the candidates, and then selects the one which maximizes

Table 1. Possible Values of N

NO.	1	2	3	4	5	6	7	8	9	10
N	0.1	0.2	0.3	0.4	0.5	0.6	0.7	0.8	0.9	1.0

Table 2. Possible Values of P

NO.	1	2	3	4	5	6	7
P	-0.03	-0.02	-0.01	0	0.01	0.02	0.03

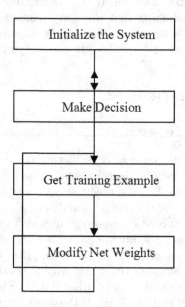

Fig. 2. Work Flow Of the Learning System

the system output as the final decision. Finally, the decision is executed, and the state is transformed.

At the new state, the system evaluates the state, and denotes the evaluation by Yt+1. Along with evaluation of the previous state Yt, the system gets a training example, which presented as two tuple $< Yt, Yt + 1 >$.

After that, using the training example, system modifies the weights of the neural network based on (1).

$$w_{t+1} - w_t = \alpha(Y_{t+1} - Y_t) \sum_{k=1}^{t} \lambda^{t-k} \nabla_w Y_k \qquad (1)$$

More specific analysis of (1) can be seen in Sutton(1988).

After the training, system enters to the next time step. The overall operating flow of the system is shown in Figure 2.

4.2 Other Important Issues

The system takes precautions against the possible over training. During running, its performance is calculated and recorded periodically. The length of the period is presented by T. In actual operation, if a T is found, making the system performance in the following 10 Ts are all worse than that in T, that is to say over training may occur. If that happens, the system stops training and performance in that T will be considered as the final performance of the system. Details about over training can been seen in M.Mitchell(1997).

Besides, with the learning progressing, the system dynamically adjusts the learning rate alpha. At the very beginning, alpha is somewhat greater, and set to 0.4. The closer the systems performance getting to the expected, the smaller the alpha turns until alpha reaches to its smallest value 0.01. This design aims at speeding up the convergence at the beginning, and carefully modifies the weights in the later stage for the purpose of avoiding missing the optimal solution. Related topics can been seen in M.Mitchell(1997).

Finally, for the sake of speeding up learning, multi-agent cooperation is applied in the system. 20 homogeneous agents, with same initialization, parallel to learn the strategy independently. All the agents average their network weights at certain frequency. This method brings obvious benefit. On one hand, a larger assumption space is searched; on the other hand, the convergence of the algorithm is easier. More specific discussion about multi-agent cooperation can be seen in Tan(1993).

5 Running Learning System in the Market

The learning system has run in the real stock market. The chosen stock for investigation is SH600312 in China, and all the market data, acquired by real time query, is actual information of it. In order to obtain more useful training example, the 20 agents began learning one by one with one second interval, and each agent queried the data and executed learning every 20 seconds. Those steps are for the purpose of ensuring the market data of every second had been utilized, and that for each agent, a long enough interval, which means adequate change, existed. For simplifying the problem, the matching system is only a simulation, and no real trade is executed in the market.

More than 14000 examples were obtained each day. After about 500,000 times of learning, the system performance obviously improved. The learning result, measured by performance in the market, using the difference between system procession change of the system, namely P, and price change of the market, namely R, was shown in figure 3.

It was found that in real market, the learning system made some achievements. Firstly, owing to about 200,000 times of learning, the system performance,

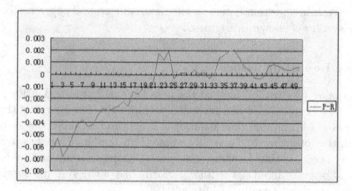

Fig. 3. Learning Performance in Real Market

measured by P-R, had become greater than 0 steadily. It suggests that since 200,000 times, the performance of the system was better than that of the real market.

Moreover, two peaks were found in the performance curve. After analyzing the data, we discovered that the time spans corresponded to the peaks happened to be the spans when SH600312 were going down. These two peaks pointed out that it was probably because the system had gained the ability to stabilize during the learning in real market, and this was quite encouraging.

Finally, continuous rise of SH600312 had been discovered, and it occurred when the learning ran about 310,000 to 320,000 times. The learning records showed that during that time, P-R was less than 0. Maybe it was because the volatility of real market was much stronger than the ideal. Although the system grasped the rise, the experience gained in the real market made it more deliberate. By analyzing the decision records, we found that in early stage of the continuous rise, system use most, instead of all, of his cash to buy shares. In the later stages of the sequence, system made such decision: a sell immediately followed by a buy. We could demonstrate that although it did not gain the potential profit, it took substantial actions to manage risks.

6 Summary

The paper proposes that, it is feasible to build stock trading decision support system, by learning SICL strategy. Furthermore, TD(λ) method is applied, in order to learn that strategy. In ideal market, the learning result shows if fluctuation of the market in short term does contain some pattern, that pattern can be learned by the system. While in the real market, the learning result indicates that even if in a complicated real environment, it is still feasible to learn SICL strategy through this system.

However, the existence and repeatability of the regularity of stock price, is still not confirmed. The experiment in real market does makes some achievements, nevertheless, it is not adequate. By running learning system on more than one stock for longer time, it is possible for us to do more analysis.

References

1. Hilger, M.G.: Computational Finance Models. IEEE Potentials 19, 8–10 (2001)
2. Tsang, E., Martinez-Jaramillo, S.: Computational Finance. In: IEEE Computational Intelligence Society Newsletter, pp. 8–13 (2004)
3. Yingsaeree, C., Treleaven, P.: Computational Finance. Computer 43, 36–43 (2010)
4. Sutton, R.S.: Learning to predict by the methods of temporal differences. Machine Learning 3(1), 9–44 (1988)
5. Tan, M.: Multi-Agent reinforcement learning: Independent VS. Cooperative Agents. In: 10th International Conference on Machine Learning, pp. 330–337 (1993)
6. Mitchell, M.: Machine Learning, p. 109. McGraw-Hill Companies, Inc. (1997)
7. John, T.J., Samuel, R.: Herd immunity and herd effect: new insights and definitions. European Journal of Epidemiology 16, 601–606 (2000)
8. Dayhoff, J.: Neural Network Architectures. Van Nostrand Reinhold, 115 Fifth Avenue, New York, NY 10003 (USA), 220 (1989)

An Efficient ID-Based Directed
Signature Scheme from Optimal Eta Pairing

Junhua Ku[1,2], Dawei Yun[1], Bing Zheng[1], and She Wei[1]

[1] Department of Information Engineering
Hainan Institute of Science and Technology
Hainan Haikou 571126, P.R.China
kujunhua@163.com
[2] School of Computer Science and Technology
China University of Geosciences
Hubei Wuhan 430074, P.R.China
cogemm@163.com

Abstract. A directed signature scheme allows a designated verifier to directly verify a signature issued to him, and a third party to check the signature validity with the help of the signer or the designated verifier as well. In this paper, starting from the Vercauteren's work on optimal pairings, we describe how to exploit the action of the $2^{3m}th$ power Verschiebung in order to reduce the loop length of Miller's algorithm even further brief than the $genus - 2\eta_T$ approach. At the same time, we propose an efficient identity-based directed signature scheme from Optimal Eta Pairing on Supersingular Genus-2 Binary Hyperelliptic Curves.

Keywords: Directed Signature, ID-based Cryptography, Optimal Eta Pairing.

1 Introduction

Digital signature is one of the most important techniques in the modern information security system for its functionality of providing data integrity and authentication. In a directed signature scheme, the signature is generated for a designated verifier, who can directly verify the signature while others know nothing about its validity. In addition, at the time of trouble or if necessary, both the signer and the designated verifier can prove to a third party that the signature is valid. However, many directed signature schemes work in the Public-Key Infrastructure (PKI) based settings. To eliminate the burden of certificate management, Shamir introduced the notion of identity-based cryptography[1]. ID-based cryptography is supposed to provide a more convenient alternative to the conventional public key infrastructure. In 2008, Xun Sun et al. [2] proposed an ID-based directed signature scheme from bilinear pairings. The scheme is based on modified SOK identity-based signature scheme due to Bellare et al.[3]. Also, Xun Sun et al. based on the work of Libert and Quisquarter[4], they proved that their scheme is existentially unforgeable and invisible in the random oracle model based on the Computational Diffie-Hellman (CDH) and the

Z. Li et al. (Eds.): ISICA 2012, CCIS 316, pp. 440–448, 2012.
© Springer-Verlag Berlin Heidelberg 2012

Decisional Bilinear Diffie-Hellman (DB-DH)assumption respectively. In 2009, Jianhong Zhang et al. [5] proposed an ID-based directed signature scheme using Waters signature scheme[6]. They proved that their scheme is secure against existential unforgeability and invisibility in the standard model.

Miller described the first iterative algorithm to compute the Weil and Tate pairings back in 1986[7]. The Tate pairing seems to be more suited to efficient implementations, and has therefore attracted a lot of interest in the research community. A large number of articles, culminating in thepairing algorithm[9], focused on shortening the loop of Miller's algorithm in the case of supersingular abelian varieties. The Ate pairing, introduced by Hess et al. [10] for elliptic curves and by Granger et al. [11] for the hyperelliptic curves, is generalized to ordinary curves. Eventually, several variants of the Ate pairing aiming at reducing the loop length of Miller's algorithm have been proposed in 2008[12,17]. Supersingular curves over or are better suited to hardware implementation, and offer more efficient point doubling and tripling formulae than BN-curves. Moreover, supersingularity allows the use of a distortion map and thus provides Type-1 (or symmetric) pairings [13], which cannot be obtained withh ordinary curves. However, the embedding degree of a supersingular elliptic curve is always less than or equal to 6[14]. As a consequence, the security on the curve is too high with respect to the security of the group of $l - th$ roots of unity, and one has to consider curves defined over very large finite fields.

To alleviate the effect of the bounded embedded degree, Estibals proposed to consider supersingular elliptic curves over field extensions of moderately composite degree[15]. Curves are vulnerable to Weil descent attacks[16], but a careful analysis allowed him to maintain the security above the 128-bit threshold. Yet another way to reduce the size of the base field of the Tate pairing in the supersingular case is to consider a genus-2 binary hyperelliptic curve with embedding degree $k = 12$[8], which is the solution investigated in this work. We indeed show that, thanks to a novel pairing algorithm, these curves can be actually made very effective in the implementations of ID-based directed signature scheme from optimal Eta pairing.

In this paper, we propose an efficient ID-based directed signature from Optimal Eta pairings on supersingular genus-2 curves. This scheme uses the Hess signature scheme[8] as the basal scheme. The proposed scheme is more efficient than the X.Sun et al. Scheme[2] and Jianhong Zhang et al. scheme[5]. The rest of the paper is organized as follows: Section 2 briefly describes the necessary background material and notations; Section 3 presents an optimal Eta pairing on and describes an algorithm that further reduces the loop length of Miller's algorithm compared to the approach.; Our ID-based directed signature scheme and its correctness proof are proposed in Section 4; Finally, we conclude our work in Section 5.

2 Background Material and Notations

In this section, we briefly recall a few definitions and results about hyperelliptic curves, and Eta Pairing on Supersingular Genus-2 Binary Curves. At the same time, we briefly review the basic concepts on some computational problems and Hess ID-based signature scheme.

2.1 Reminder on Hyperelliptic Curves

Let C be an imaginary nonsingular hyperelliptic curve of genus g defined over the finite field F_q, where $q = p^m$ and p is a prime, and whose affine part is given by the equation $y^2 + h(x)y = f(x)$, where $f, g \in F_q[x]$, deg $f = 2g + 1$, and deg $h \le g$. For any algebraic extension F_{q^d} of F_q, we define the set of F_{q^d}-rational points of C as:

$$C(F_{q^d}) = \{(x, y) \in F_{q^d} \times F_{q^d} | y^2 = h(x)y = f(x)\} \cup \{P_\infty\} \qquad (1)$$

Where P_∞ is the point at infinty of the curve. For simplicity, we also write $C = C(F_{q^d})$. Additionally, denoting by ϕ_q the q-th power Frobenius morphism: $\phi_q : C \to C, (x, y) \to (x^q, y^q)$ and $P_\infty \to P_\infty$ note that point $P \in C$ is F_{q^d}-rational if and only if $\phi_q^d(P) = P$.

We also denote by $JacC$ the Jacobian of C, which is an an abelian variety of dimension g defined over F_q, and whose elements are represented by the divisor class group of degree-0 divisors $Pic_C^0 = Div_C^0/Princ_C$. In other words, two degree-0 divisors D and D' belong to the same equivalence class $\overline{D} \in \boldsymbol{Jac_C}$ if and only if there exists a non-zero rational function $z \in \overline{F_q}(C)^*$ such that $D' = D + div(z)$. Naturally extending the Frobenius map to divisors as:

$$\sum_{P \in C} n_p(P) \to \sum_{P \in C} n_p(\phi_p(P)) \qquad (2)$$

We say that D is F_{q^d}-rational if and only if $\phi_q^d(D) = D$.

It can also be shown that any divisor class $\overline{D} \in Jac_C(F_{q^d})$ can be uniquely represented by an F_{q^d}-rational reduced divisor $\rho(\overline{D}) = \sum_{i=1}^r (P_i) - r(P_\infty)$, with $r \le g$, $P_i \ne P_\infty$ and $P_i \ne P_\infty$ for $i \ne j$, where the negative of a point $P = (x, y)$ is given via the hyperelliptic involution by $P = (x, y - h(x))$.

Using the Mumford representation, any non-zero F_{q^d}-rational reduced divisor:

$$D = \rho(\overline{D}) = \sum_{i=1}^r (P_i) - r(P_\infty) \qquad (3)$$

can be associated with a unique pair of polynomials $[u(x); v(x)]$, with $u, v \in F_{q^d}[x]$ and such that u is monic, $deg(v) < deg(u) = r \le g$, and $u|(v^2 - vh - f)$. Furthermore, given two reduced divisors D_1 and D_2 in Mumford representation, Cantor's algorithm[19] can be used to compute the Mumford representation of $(D_1 + D_2)$, the reduced divisor corresponding to their sum on the Jacobian.

Let l be a prime dividing $\#JacC(F_q)$ and coprime to q, Let also k be the corresponding embeding degree, i.e, the smallest integer such that $l|(q^k - 1)$ We denote by $Jac_C(F_{q^k})[l]$ the F_{q^k}-rational l-torsion subgroup of $\boldsymbol{Jac_C}$. The Tate pairing on C is then the well-defined, non-degenerate, and bilinear map:

$$<... >_l: Jac_C(F_{q^k})[l] \times Jac_C(F_{q^k})[l]/lJac_C(F_{q^k}) \to F_{q^k}^*/(F_{q^k}^*)^l \qquad (4)$$

$$< \overline{D_1}, \overline{D_2} >_l \equiv f_{l,D_1}(D_2) \tag{5}$$

Where D_1 and D_2 represent the divisor classes $\overline{D_1}$ and $\overline{D_2}$ respectively.

2.2 Eta Pairing on Supersingular Genus-2 Binary Curves

We consider the family of supersingular genus-2 hyperelliptic curves defined over F2 by the equation $C_d : y^2 + y = x^5 + x^3 + d$, where $d = 0$ or 1. Because of their supersingulrity, which provides them with a very efficient arithmetic, along with their embedding degree of 12, which is the highest among all supersingular genus-2 curves, these curves are a target of choice for implementing pairing-based cryptography. They have therefore already been studied in this context in several articles[9,17].

For m positive integer coprime to 6, the cardinlity L of the Jacobian of C_d over F_{2^m} is $L = Jac_{C_d}(F_{2^m}) = 2^{2m} + \delta \cdot 2^{(3m+1)/2} + 2^m + \delta \cdot 2^{(m+1)/2} + 1$, where the value of δ is:

$$\delta = \begin{cases} (-1)^d & when\ m \equiv 1,7,17\ or\ 23(mod\ 24) \\ -(-1)^d & when\ m \equiv 5,11,13\ or\ 19(mod\ 24) \end{cases} \tag{6}$$

The embedding degree of C_d is $k = 12$, and $\#JacC_d(F_{2^m})|(2^{12m} - 1)$. The Tate pairing and its variants will then map into the degree-12 extension $F_{2^{12m}}$, which we represent as the tower field $F_{2^{12m}} \cong F_{2^m}[\tau, s_{\tau,0}]$, where $\tau \in F_{2^6}$ is such that $\tau^6 + \tau^5 + \tau^4 + \tau^3 + \tau^2 + \tau^0 = 0$, and $s_{\tau,0} \in F_{2^{12}}$ is such that $s_{\tau,0}^2 + s_{\tau,0} + \tau^5 + \tau^3 = 0$.

As $l|N$ we can then take the Miller function $f_{N,D_1} = f_{N+1,D_1} = f_{T^4,D_1} = \prod_{i=0}^{3} f_{T,[T^i],D_1}^{T^{3-i}}$. Furthermore, since D_1 and D_2 are F_{2^m}-rational reduced divisors, we also have that $[2^{i \cdot 3m}]D1 = \phi_{2^{3m}}^{-2}(D_1)$ and $(\varepsilon(\psi(D_2))) = \phi_{2^{3m}}^{-2}(\varepsilon(\psi(D_2)))$ for all i. Putting the preceding all together, we finally obtain.

$$\tau(\overline{D_1}, \overline{D_2})^M = f_{2^{3m}, D_1}(\varepsilon(\psi(D_2)))^{4 \cdot 2^{3 \cdot 3m} \cdot (2^{12m} - 1)/L} \tag{7}$$

From the bilinearity and the non-degeneracy of the Tate pairing, we can then conclude that the η_T pairing defined as follows is also bilinear and non-degenerate[9]:

$$\eta_T : Jac_{C_d}(F_{2^m})[l] \times Jac_{C_d}(F_{2^m})[l] \to u_l \subseteq F_{2^{12m}}^* \tag{8}$$

$$(\overline{D_1}, \overline{D_2}) \to f_{2^{3m}, D_1}(\varepsilon(\psi(D_2)))^{4 \cdot 2^{3 \cdot 3m} \cdot (2^{12m} - 1)/L} \tag{9}$$

With μ_l the subgroup of l-th roots of unity.

2.3 Computational Problems

Now, we give some computational problems, which will form the basis of security for our scheme.

Discrete Logarithm Problem (DLP): Given two elements P and Q, find an in-teger n such that $Q = nP$ whenever such an integer exists.

Decisional Diffie-Hellman Problem (DDHP): For $a, b, c \in_R Z_q$, given P, aP, bP, cP decide whether $c \equiv ab\ mod\ q$.

Computational Diffie-Hellman Problem (CDHP): For $a, b, c \in_R Z_q$, given P, aP, bP, Compute abP.

2.4   Hess-ID- Based Signature Scheme

To prepare for the proposed scheme, we first give a review of the ID-based signature scheme [8] given by Hess as follows.

Setup: The Private Key Generator (PKG) chooses $s \in Z_q^*$ as his master secret key and computers the global public key pub $P.$, sP. The PKG also selects a map-to-point hash function $H : \{0,1\} \to G_1^*$ and another cryptographic hash function $h : \{0,1\}^* \times G_2 \to Z_q^*$.

PKG publishes system parameters params $< G_1, G_2, e, P, P_{pub}, H, h >$. and the master key s is kept secret.

Extract: Given the public identity information on ID compute the secret key for the identity as $d_{ID} = sQ_{ID}$. The component $Q_{ID} = H(ID)$ plays the role of the corresponding public-key.

Signature: To sign a message $M \in \{0,1\}^*$, using the secret key d_{ID}, the signer chooses an arbitrary $P \in G_1$ picks a ramdom integer $k \in Z_q^*$. Then signer respectively computes:

$$R = e(P_1, P)^k, V = h(M, R), U = V d_{ID} + k P_1 \tag{10}$$

The signature on message M is $\sigma = (U, V) \in G_1 \times Z_q^*$.

Verification: To verify the signature $\sigma = (U, V)$ of an identity ID on a message M, the verifier computes $R = e(U, P)e(Q_{ID}, -P_{pub})^V$. He accpets the signature if and only if $V = H(M, R)$.

3   Optimal Eta Pairing on C_d

In order to further decrease the loop length in Miller's algorithm, we adapt in this work the optimal pairing technique as introduced by Vercauteren[18] to the case of the action of the 2^{3m}-th power Verschiebung $\phi_{2^{3m}}^{-2}$ and the Eta pairing mentioned in the section 2. Letbe a large prime dividing $L = \#JacC_d(F_{2^m})$. After ensuring that η_T paring with $T = 2^{3m}$ and $l|L|N$ for $N = 2^{12m}-1 = T^4-1$, we can then define the optimal Eta pairing as the non-degenerate bilinear map:

$$\bar{\eta}_T : Jac_{C_d}(F_{2^m})[l] \times Jac_{C_d}(F_{2^m})[l] \to u_l \subseteq F_{2^{12m}}^* \tag{11}$$

$$(\overline{D_1}, \overline{D_2}) \to (f_{c_1,D_1}^{3^m} \cdot f_{c_0,D_1})(\varepsilon(\psi(D_2)))^{(2^{12m}-1)/L} \tag{12}$$

We can reach a conclusion that:

$$\tau(\overline{D_1}, \overline{D_2})^{M'} = \bar{\eta}_T(\overline{D_1}, \overline{D_2}) \cdot \overline{\eta_T}(\overline{D_1}, \overline{D_2})^{c_1} \tag{13}$$

$$\tau(\overline{D_1}, \overline{D_2})^W = \bar{\eta}_T(\overline{D_1}, \overline{D_2}). \tag{14}$$

Where $[c_1, c_0] = [\delta 2^{(m-1)/2} + 1, 2^m + \delta 2^{(m-1)/2}]$, $N' = c_1 2^{3m} + c_0 = M'L$ and $W = 2^{2m} + \delta 2^{(3m-1)/2} + 2^m + \delta 2^{(m-1)/2} + 1$.

Note that the optimal Eta pairing $\bar{\eta}_T$ is also bilinear and non-degenerate.

The computation of the optimal Eta pairing relies on the evaluation of the two Miller functions f_{c_0,D_1} and f_{c_1,D_1} at $\varepsilon(\psi(D_2))$. We can take the following algorithm.

Input: $\overline{D_1}, \overline{D_2} \in Jac_{C_d}(F_{2^m})[l]$ represented by the reduced divisors D_1 and D_2.
Output: $\overline{\eta}_T(D_1, D_2)$ or $\overline{\eta}_T(D_1, D_2)^2 \in u_l \subseteq F_{2^{12m}}^*$ depending on whether $m = 1$ or $5(mode\ 6)$, respectively.

Algorithm 1. Computation of the optimal Eta pairing

1: **if** $m \equiv 1 \bmod 6$ **then**
2:     $m' \leftarrow m - 1$ else $m' \leftarrow m + 1$
3: **end if**
4: $G_1 \leftarrow 1; R_1 \leftarrow [\delta]D_1; E_1 \leftarrow \varepsilon(\psi(D_2));$
5: **for** i=1 to $m'/6$ **do**
6:     $G_1 \leftarrow G_1^8 \cdot f_{8,R_1}(E_1)$
7:     $R_1 \leftarrow [8]R_1$
8: **end for**
9: $G_1 \leftarrow G_1^\delta; R_2 \leftarrow [\delta]R_1;$
10: **for** i=1 to $m'/6$ **do**
11:     $G_2 \leftarrow G_2^8 \cdot f_{8,R_2}(E_1)$
12:     $R_2 \leftarrow [8]R_2$ //$G_2 = f_{2^{m'},R_2}(E_1)$ and $R_2 = [2^{m'}]D_1$
13: **end for**
14: **if** m=1 mod 6 **then**
15:     $G_2 \leftarrow G_2^2 \cdot f_{2,R_2}(E_1)$ //$G_2 = f_{2^{m'},D_1}(E_1)$
16:     $H_1 \leftarrow G_1 \cdot G_2 \cdot f_{2,R_2}(E_1)$ //$H_1 = f_{c_1,D_1}(E_1)$
17:     $H_2 \leftarrow G_1 \cdot G_2 \cdot f_{2^m,D_1,R_1}(E_1); H_2 \leftarrow G_1 \cdot G_2 \cdot f_{2^m,D_1,R_1}(E_1)$
18: **end if**
19: Return $(H_1^{2^{3m}} \cdot H_2)^{(2^{12m}-1)/L}$ //$L = \#Jac_{C_d}(F_{2^m})$

---

4   Proposed ID-Based Directed Signature Scheme

In this section, we propose our ID-based directed signature scheme from optimal Eta pairings on supersingular genus-2 curves. Our scheme is based on the Hess ID-based signature scheme. The proposed scheme is described as follows:

Setup: The PKG chooses $x \in Jac_{C_d}(F_{2^m})$ as his master secret key and computes the global public-key P_{pub} as $x\overline{D} \in Jac_{C_d}(F_{2^m})[l]$. The PKG also selects hash functions:

$$Hash_1, Hash_2 : \{0,1\}^* \rightarrow Jac_{C_d}(F_{2^m})[l] \tag{15}$$

And another cryptographic hash function:

$$Hash_3 : \{0,1\}^* \times \mu_l \rightarrow Jac_{C_d}(F_{2^m}) \tag{16}$$

$$\overline{\eta}_T : Jac_{C_d}(F_{2^m})[l] \times Jac_{C_d}(F_{2^m})[l] \rightarrow \mu_l \subseteq F_{2^{12m}}^* \tag{17}$$

$Params < Jac_{C_d}(F_{2^m})[l], \mu_l, \overline{\eta}_T, \overline{D}, P_{pub}, Hash_1, Hash_2, Hash_3 >$ Master key x.

Extract: Given an identity $ID \in \{0,1\}^*$, the PKG computes $Q_{ID} = Hash_1(ID) \in Jac_{C_d}(F_{2^m})[l]$, then computes the user private key $d_{ID} = xQ_{ID} \in Jac_{C_d}(F_{2^m})[l]$.

Signature: To sign a message $M \in \{0,1\}^*$ to a designated verifier IDv, the signature with identity IDs and private key d_{IDs} perform the signature.

1. choose an arbitrary $\overline{D_1} \in Jac_{C_d}(F_{2^m})[l]$, and pick a random integer $r, k \in Jac_{C_d}(F_{2^m})$.
2. compute $U = \overline{\eta}_T(\overline{D_1}, \overline{D})^k$, $R = rD_{ID}$;
3. compute $V = Hash_3(H, U)$,
   where $H = Hash_2(IDs, IDv, M, \dot{U}, \overline{\eta}_T(d_{IDs}, rQ_{IDv}))$.
4. compute $W = Vd_{IDs} + k\overline{D_1}$.

The signature is $\sigma = (V, W, R)$.

DVerify: Given a purported signature $\sigma = (V, W, R)$, on signature IDs, verifier IDv and message M, IDv verifies it with his key d_{IDv} as follows:

1. Compute $U = \overline{\eta}_T(W, \overline{D}) \cdot \overline{\eta}_T(Q_{IDs}, -P_{pub})^V$;
2. Compute $H = Hash_2(IDs, IDv, M, U, \overline{\eta}_T(d_{IDv}, R))$.

Accept the signature if $V = Hash_3(H, U)$. Reject it otherwise.

PVerify: Given a purported signature $\sigma = (V, W, R)$, on signature IDs, verifier IDv and message M, to enable third party T to verify it either IDs or IDv computes $A_{third} = \overline{\eta}_T(d_{IDv}, R)\overline{\eta}_T(d_{IDs}, rQ_{IDv})$ and send it to T.
Now T computes:

$U = \overline{\eta}_T(W, \overline{D}) \cdot \overline{\eta}_T(Q_{IDs}, -P_{pub})^V$
and $H = Hash_2(IDs, IDv, M, U, A_{third})$.
Accept the signature if $V = Hash_3(H, U)$, reject it otherwise.

The security of identity based directed signature scheme from Optimal Eta Pairing consists of two properties: unforgeability and invisibility. Here we leave out the security of identity based directed signature scheme .Our proofs are same as given in the[19]. Now, we provide the Proof of correctness of my scheme. To show correctness of my scheme, we show that the DVerify algorithm is consistent with the signature algorithm because:

$$
\begin{aligned}
U &= \overline{\eta}_T(W, \overline{D}) \cdot \overline{\eta}_T(Q_{IDs}, -P_{pub})^V \\
&= \overline{\eta}_T(Vd_{IDs} + k\overline{D_1}, \overline{D}) \cdot \overline{\eta}_T(Q_{IDs}, -P_{pub})^V \\
&= \overline{\eta}_T(Vd_{IDs}, \overline{D}) \cdot \overline{\eta}_T(k\overline{D_1}, \overline{D}) \cdot \overline{\eta}_T(Q_{IDs}, -P_{pub})^V \\
&= \overline{\eta}_T(d_{IDs}, \overline{D})^V \cdot \overline{\eta}_T(\overline{D_1}, \overline{D})^k \cdot \overline{\eta}_T(Q_{IDs}, -P_{pub})^V \\
&= \overline{\eta}_T(xQ_{IDs}, \overline{D})^V \cdot \overline{\eta}_T(Q_{IDs}, -x\overline{D})^V \cdot \overline{\eta}_T(\overline{D_1}, \overline{D})^k \\
&= \overline{\eta}_T(\overline{D_1}, \overline{D})^k = U
\end{aligned}
\tag{18}
$$

Correctness of PVerify algorithm is straightforward.

5   Conclusion

The directed signature is very useful in some practical applications, where a signed message is personally or commercially sensitive. In this paper, starting from the Vercauteren's work on optimal pairings, we describe how to exploit the action of the 23m-th power Verschiebung in order to reduce the loop length of Miller's algorithm even further than the genus-2 η_T approach.

We propose an efficient identity based directed signature scheme from optimal pairings. Our scheme is less than the previous identity-based directed signature schemes in the loop length of Miller's algorithm. So our scheme is more efficient than the existing directed signature schemes. At the same time, our scheme is unforgeable under the Computational Diffie-Hellman assumption, and invisible under the Decisional Bilinear Diffie-Hellman assumption.

Acknowledgment. This work was supported by the Department of Education Research Projects of Hainan Province under grant no.Hjkj2012-61. Our deepest thanks to You Lin who advised us to get an attempt at genus-2 pairings. He shall receive here our utmost gratitude.

References

1. Shamir, A.: Identity-Based Cryptosystems and Signature Schemes. In: Blakely, G.R., Chaum, D. (eds.) CRYPTO 1984. LNCS, vol. 196, pp. 47–53. Springer, Heidelberg (1985)
2. Sun, X., Li, J.-H., Chen, G.-L., Yung, S.-T.: Identity-Based Directed Signature Scheme from Bilinear Pairings, http://eprint.iacr.org/2008/305.pdf
3. Bellare, M., Namprempre, C., Neven, G.: Security Proofs for Identity-Based Identification and Signature Schemes. In: Cachin, C., Camenisch, J.L. (eds.) EURO-CRYPT 2004. LNCS, vol. 3027, pp. 268–286. Springer, Heidelberg (2004)
4. Libert, B., Quisquater, J.J.: The exact security of an Identity based signature and its applications, http://eprint.iacr.org/2004/102
5. Zhang, J., Yang, Y., Niu, X.: Efficient Provable Secure ID-Based Directed Signature Scheme without Random Oracle. In: Yu, W., He, H., Zhang, N. (eds.) ISNN 2009, Part III. LNCS, vol. 5553, pp. 318–327. Springer, Heidelberg (2009)
6. Waters, B.: Efficient Identity-Based Encryption Without Random Oracles. In: Cramer, R. (ed.) EUROCRYPT 2005. LNCS, vol. 3494, pp. 114–127. Springer, Heidelberg (2005)
7. Miller, V.: The Weil pairing, and its efficient calculation. J. Cryptol. 17(4), 235–261 (2004)
8. Hess, F.: Efficient identity based signature schemes based on pairings. In: Nyberg, K., Heys, H.M. (eds.) SAC 2002. LNCS, vol. 2595, pp. 310–324. Springer, Heidelberg (2003)
9. Barreto, M., et al.: Efficient pairing computation on supersingular Abelian varieties. Des. Codes Crypt. 42, 239–271 (2007)
10. Hess, F., Smart, N., Vercauteren, F.: The Eta pairing revisited. IEEE Trans. Inf. Theory 52(10), 4595–4602 (2006)
11. Granger, R., Hess, F., Oyono, R., Thériault, N., Vercauteren, F.: Ate Pairing on Hyperelliptic Curves. In: Naor, M. (ed.) EUROCRYPT 2007. LNCS, vol. 4515, pp. 430–447. Springer, Heidelberg (2007)
12. Hess, F.: Pairing Lattices. In: Galbraith, S.D., Paterson, K.G. (eds.) Pairing 2008. LNCS, vol. 5209, pp. 18–38. Springer, Heidelberg (2008)

13. Galbraith, S., Paterson, K., Smart, N.: Pairings for cryptographers. Discrete Applied Mathematics 156, 3113–3121 (2008)
14. Menezes, A., Okamoto, T., Vanstone, S.: Reducing elliptic curves logarithms to logarithms in a finite field. IEEE Trans. Inf. Theory 39(5), 1639–1646 (1993)
15. Estibals, N.: Compact Hardware for Computing the Tate Pairing over 128-Bit-Security Supersingular Curves. In: Joye, M., Miyaji, A., Otsuka, A. (eds.) Pairing 2010. LNCS, vol. 6487, pp. 397–416. Springer, Heidelberg (2010)
16. Aranha, D.F., López, J., Hankerson, D.: High-Speed Parallel Software Implementation of the η_T Pairing. In: Pieprzyk, J. (ed.) CT-RSA 2010. LNCS, vol. 5985, pp. 89–105. Springer, Heidelberg (2010)
17. Galbraith, S.D.: Supersingular Curves in Cryptography. In: Boyd, C. (ed.) ASIACRYPT 2001. LNCS, vol. 2248, pp. 495–513. Springer, Heidelberg (2001)
18. Vercauteren, F.: Optimal pairings. IEEE Trans. Inf. Theory 56(1), 455–461 (2010)
19. Cantor, D.: Computing in the Jacobian of a hyperelliptic curve. Math. Comput. 48(177), 95–101 (1987)

WCD-New Approach Combining Words, Concepts and Documents Based on Ontology

Haoming Wang, Ye Guo, and Xibing Shi

School of Information
Xi'an University of Finance and Economics
Xi'an Shaanxi 710100, P.R. China
hmwang@mail.xaufe.edu.cn,
{guoyexinxi,xbshine}@126.com

Abstract. In traditional Information Retrieval (IR) system, the document is represented by the set of words or terms. If the words or terms are regarded as the components of a vector, the model is called the vector space model (VSM). VSM has been widely used in IR systems in recently decades. As the the new words appear dramatically in the Internet era, the amount of computation is very large and it draws back the IR system's performance. This paper puts forward a new approach according to the relations among the words, concepts and the document by using the concept of the ontology. The new approach has two levels, the Word-Concept (WC) level and the Concept-Document (CD) level. In the WC level, the transition probability matrix is constructed by using the word-word pairs appeared in the same paragraph, and the biggest eigenvector of matrix is computed. The eigenvector reflects the importance of the word to the concept. In the CD level, the distance matrix is constructed by using the distance between words in the concept, and the average variance values of elements is computed. The value determines the relevance of the document to the concept. In order to expand the query sentence, the Personal Information Profile (PIP) of the user is defined by using the query history of the user. It is proofed to be more effective than previous one.

Keywords: Ontology, Word-Concept level, Concept-Document level, Personal Information Profile.

1 Introduction

In Internet era, search engine (SE) is used to help the user to find the information in Internet. The user always expects to find the most relevant information to his query. The recall and precision are used to test the effects of retrieval results. Normally, the SE cannot always feed back an ideal result list as the various reasons. The user has to spend a considerable time to find the useful information in the feedback list. The main reason of causing the poor results is that the SE does not really know what the user wants to get. The SE splits the query sentence

Z. Li et al. (Eds.): ISICA 2012, CCIS 316, pp. 449–458, 2012.

into terms, and computing the page value by using the contents or the page-links of the pages or the both.

The first kind of SE works on the contents of pages or documents. The typical example is vector space model (VSM). This model regards the document as a vector which is consisted of many component, and each component is a term. In the most time, the term is the word. So, the more the different terms in a document, the more the dimensions of the vector. This method works well for traditional documents, but the performance drops significant when applied to the web pages. The main reason is that the web pages are not organized as strict as the documents do. Web pages contain much irrelevant information, and the useful information is submerged in. On the other hand, the VSM model neglects the relevant pages that do not contain the index terms which are specified in the user's queries.

The second one takes the hyperlink structures of web pages into account in order to improve the performance. The examples are Pagerank and HITS. They are applied to Google and the CLEVER project respectively. The Google search engine is based on the popular Pagerank algorithm first introduced by Brin and Page in[1]. Considering the pages and the links as a graph $G = P(Page, Link)$, we can describe the graph by using the adjacency matrix. Computing the eigenvector which represents the value of the pages of the matrix, the Google feedbacks the top $-n$ pages to the user. The [2,3] introduced the algorithm, but they did not introduce the deep technologies of the SE, which they are the trade secrets. The [4] shows the result that Google cannot get the high level of precision and recall if they compute the page value just by using the links between the pages. They must have other technologies to improve the results of retrieval.

Normally, in order to improve the effect of retrieval, the specific domain knowledge should be added to the queries, which is called query expansion. Ontology is one of the knowledge which can be used to expand the connotation of the query.

Ontology is a conceptualization of a domain knowledge. It is a concept set with the machine readable, human understandable format. It consists entities, attributes, relationship, and axioms. Those elements construct the network in order to present the knowledge of a special domain. The network is checked by the experts of the domain in order to guarantee the concepts represent the knowledge of the domain indeed.

For an ontology-based information retrieval system, it tries to insert the ontology knowledge to the query expression in order to enhance the ability of representing. In the conceptual level, documents having very different vocabularies could be similar in subject and, similarly, documents having similar vocabularies may be topically very different.

The paper is organized as follows: Section 2 introduces the related concept of ontology and query expansion. Section 3 discusses the relevance computing of words to concepts and documents to concepts. Section 4 puts forward the method of query expansion. In section 5, a summary of the paper and directions for future work are discussed.

2   Related Work

In this section, the basic concepts, such as ontology, conceptual representing, document representing, query expansion, and WordNet are introduced.

2.1   Conceptual Representation

If a document is considered as a set of words, the relation between words is neglected. Due to the ambiguity and the limitation of the ability of expressing of the single word, it is difficult to decide which word is more important than others for the document. It is obviously that the importance of the word is decided by the document or set of documents which the word appeared. There is no experimental result shows that a word is always important that the others. So, we try to discuss the importance of words in an ontology or a concept environment.

One way of deciding which word is more important than others is Term Frequency-Inverse Document Frequency (TFIDF). The main idea of TFIDF is the more vocabulary entry in document set, the lower separate ability of document property, and then the weight value is small. On the other hand, the higher frequency for a certain vocabulary entry in a document, the higher separate ability, and then the weight value is big. This method is widely used in selecting text feature[5]. But it has many disadvantages. First, the method undervalues that this term can represent the characteristic of the documents of this class if it only frequently appears in the documents belongs to the same class while infrequently in the documents of the other class. Second TFIDF neglects the relations between the feature and the class or the terms[6].

The another way is Latent Semantic Indexing (LSI). The most improvement is mapping the documents from the original set of words to a concept space. Unfortunately, LSI maps the data into a domain in which it is not possible to provide effective indexing techniques. Instead, conceptual indexing permits to describe documents by using concepts that are unique and abstract human understandable notions. After that, several approaches, based on different techniques, have been proposed for conceptual indexing.

One of the well-known mechanism for conceptual representation is conceptual graph (CG). In [7], two ontologies are implemented based on CGs: the Tendered Structure and the abstract domain ontology. And, the authors first survey the indexing and retrieving techniques in CG literatures by using these ontologies.

2.2   Ontology

According to [8,9], an ontology is considered as a set of definitions of concepts and relations between these concepts with typically structure. It provides the semantic context by adding semantic information to models. It is machine-processable, and it can be used in natural language processing, reasoning capabilities, domain enrichment, domain validation, etc.

Ontology is explicit representations of a shared conceptualization, i.e., an abstract, simplified view of a shared domain of discourse. More formally, an ontology defines the vocabulary of a problem domain, and a set of constraints (axioms and rules) on how terms can be combined to model specific domains.

In the traditional IR system based on VSM, documents and queries are simply represented as a vector with the term weight, and the similarity is computed by the cosine distance between the vectors. This approach does not require any extraction or annotation phases. Therefore, it is easy to implement, however, the precision values are relatively low. Compared with the traditional approach, the new one expands the query by using the knowledge of relative domains. It is hoped to improve the precision and the recall.

2.3   Query Expansion

Normally, the query expression is consisted of 3 to 7 words. The information including in those words is not clear enough to let the IR system know what the user wants really. The IR system has to feed back the much more results to the user. This causes the lower precision and recall. Query expansion is one of the ways to solve this problem[10].

Query expansion technology was brought forward in [11]. It consists of expanding a query with the addition of terms that are semantically correlated with the original terms of the query. Several works demonstrated the performance of IR system was improved by using it. As the terms, which are added to the query, play a decision rule in the query process, they should be selected carefully. Experimental results show that the incorrect choice of terms might harm the retrieval process by drifting it away from the optimal correct answer[12].

There are many ways to select the words to add to the query sentence. One way is to add the synonyms of that words appeared in the query sentence. The task of defining the relation between the words is urgent. Further more, it is better to consider that in semantic level. In this paper, we select the synonyms in conceptual level of an ontology.

3   Document Representing

The document is an objective reality, and it can be dealt with by many kinds of ways[13]. The VSM discussed above is one of the methods. VSM regards the document as a vector combined with the words. Each word is regarded as a component of the vector, and the relations between the words are neglected.

In the following, we construct a new approach with two-level structures. The first level, called Word-Concept (W-C) level, is used to reflect the relevance between the words and the concepts. The second level, called Concept-Document (C-D) level, is used to reflect the relevance between concepts and documents.

We discuss the W-C level first. For a document, there are three tasks in representing it based on the concept:

1. Marking the words in the document. The document consists of words, most of them should be belonged to one or more concepts. Some of the words are not so closed to the concept, and they are omitted in this step. The document is represented by the remain words.

2. Computing the relevance of word-document and document-concept. For a given concept, no experimental result shows that some words are always more important than others. The importance of the word is different in different documents. We want to get a word list by the importance decrease order to the concepts.

3. Deciding the attribution of the document to the concept. For a given document, it may have relevance with two or more concepts. In the other side, there are many documents have relevance to a given concept. Sometime, we need to answer the question that if two documents had same words set but different words order list, which one is more important for a query or a concept?

3.1 Marking Words

In our discussion, the first task is marking the words in a document. For the concept and the document, we can assume the facts:

1. An ontology is a very large set of concepts, and there are several hundreds of concepts in it. Each concept is consisted of many words, meanwhile each word may belong to more than one concept.

2. For a document, which is a instance for a given topic, it is impossible to include all of the words in a special concept or a ontology. In other words, it is impossible that all of the words in a document are belonged to one concept or one ontology.

3. Assuming the word d is one of the words of a document $D(d \in D)$, d can be labeled to concept C_1 or C_2 or both according to the term-list of the concepts.

We select the concepts in WordNet as the working level. The Word-Concept level of the new approach can be described as:

1. Constructing the matrix UC for each concept C, it is:

$$UC = \begin{pmatrix} uc_{11} & uc_{12} & \cdots & uc_{1n} \\ uc_{21} & uc_{22} & \cdots & uc_{2n} \\ \cdots & \cdots & \cdots & \cdots \\ uc_{n1} & uc_{n2} & \cdots & uc_{nn} \end{pmatrix}$$

Where the element uc_{ij} is the times which word d_i and d_j appear synchronously in a paragraph, and $u_s c_{ii}$ is the times which word d_i appears in a paragraph by itself. In the beginning, all of the elements is 0.

2. Scanning the document D from the first word to the end, we mark the words to the different concepts. If a word is belong to two or more concepts, marking it to each of the concepts. After the scanning, we sum the times, of which word d_i and d_j appear synchronously in the same paragraph. The value of element uc_{ij} in matrix UC is updated by it.

3. Dealing with the matrix UC. If the column i is all zero, it means the word d_i never appear in document D. The column i and row i of this matrix should be deleted.

The matrix UC is symmetric matrix. In order to decrease the amount of computation, we set a threshold for the value of elements. Deleted the rows i and columns i synchronously, the matrix keeps the characters of symmetric.

The document D may have relevance with concepts $C_1, C_2, ..., C_k$, $k \in [1, n]$. We denote the relevance by matrix UC_p, $p \in [1, n]$. In the following section, we indicate the matrix UC_p, $p \in [1, n]$ with Q for convenience.

3.2   Computing Relevance

The elements $q_{ij}(i, j \in [1, n])$ of matrix Q responds to the times of word pair $d_i - d_j$ appeared in the same paragraph in a document D. The $row(i)$ means the probability of word d_i and the word d_j, $j \in [1, n]$ appear at the same time in this document. Normalizing the matrix Q, we explain it as:

We have a set of words $D = \{d_1, d_2, ..., d_n\}$, and we name each word with the state. The process starts in one of these states and moves successively from one state to another. Each move is called a step. If the chain is currently in state d_i, then it moves to state d_j at the next step with a probability denoted by q_{ij}, and this probability does not depend upon which states the chain was in before. The word set $D = \{d_1, d_2, ..., d_n\}$ can be regarded as Markov Chain. The matrix Q is row-stochastic matrix, and the elements q_{ij} is transition probabilities.

According to the Chapman-Kolmogorov equation[14],

$$q_{i_1,...,i_{n-1}}(f_1, ..., f_{n-1}) = \int_{-\infty}^{+infty} q_{i_1,...,i_n}(f_1, ..., f_n) df_n \tag{1}$$

For the Markov chains, we can get[15],

$$q_{ij}^{n+m} = \sum_{k=0}^{\infty} q_{ik}^n q_{kj}^m (n, m \geq 0, \forall i, \forall j) \tag{2}$$

If we let $Q^{(n)}$ denote the matrix of $n - step$ transition probabilities q_{ij}^n, then we can asserts that:

$$Q^{(n+m)} = Q^{(n)} \cdot Q^{(m)}$$
$$Q^{(2)} = Q^{(1)} \cdot Q^{(1)} = Q \cdot Q = Q^2 \tag{3}$$
$$Q^{(n)} = Q^{(n-1+1)} = Q^{(n-1)} \cdot Q^{(1)} = Q^{n-1} \cdot Q = Q^n$$

That is, the $n - step$ transition matrix can be obtained by multiplying the matrix Q by itself n times.

The elements of the matrix Q are connected to others, and the matrix cannot be divided into two parts. So the Q is irreducible. Meanwhile the Q is aperiodic too. The Perron-Frobenius theorem guarantees the equation $x^{(k+1)} = Q^T x^{(k)}$

(for the eigensystem $Q^T x = x$) converges to the principal eigenvector with eigenvalue 1, and there is a real, positive, and the biggest eigenvector[16].

Because Q corresponds to the stochastic transition matrix over the graph G, the stationary probability distribution over all words induced by a random selection of words on document D can be defined as a limiting solution of the iterative process:

$$x_j^{(k+1)} = \sum_i Q'_{ij} x_i^{(k)} = \sum_{i \to j} x_i^{(k)} / deg(i) \qquad (4)$$

The biggest eigenvector means the importance of word d_i to the concept C_S.

3.3 Deciding Affiliation

In this section, we will discuss the C-D level, which is the relevance of the concept to the document.

Assuming the relevance of two sets, which come from the documents D_1 and D_2 respectively, to the concept C have been computed, we should decide which document has much importance to the concept?

According to the definition of ontology, there are four kinds of relation between the words, $part - of$, $kind - of$, $instancd - of$, and $attribute - of$. We define the distance between the words as,

Define 1. *Assuming w_i are the nodes of graph G. If w_i does not connect to w_j directly, there is a path from w_i to w_j. The distance between them is the minimum of the steps from w_i to w_j.*

$$distance(w_i, w_j) = Min(n|w_i \to w_1 \to w_2 \to ... \to w_n \to w_j) \qquad (5)$$

Define 2. *If w_i connected to w_j directly, the distance between them is,*

$$distance(w_i, w_j) = \begin{cases} 1 \ (if \ \ Rela(w_i, w_j) \in \{part - of, attribute - of\}) \\ 2 \ (if \ \ Rela(w_i, w_j) \in \{instance - of, kind - of\}) \end{cases} \qquad (6)$$

Here $Rela(w_i, w_j)$ is the one of the four relations between words in a ontology.

According to the $Define1$ and $Define2$, we construct the distance matrix $Dis(C, D)$, which represents the distance of the concept C to de document D.

$$Dis(C, D) = \begin{pmatrix} d_{11} & d_{12} & ... & d_{1n} \\ 0 & d_{22} & ... & d_{2n} \\ ... & ... & ... & ... \\ 0 & 0 & ... & d_{nn} \end{pmatrix}$$

In $Dis(C, D)$, $d_{ij}(i, j \in [1, n])$ is the distance between word w_i and w_j. The word $w_k(k \in [1, n])$ is one of the words in set D, which is discussed in section 3.2. In order to reduce the amount of computation, we sort the words in decrease order by the importance and set the thresholds. We deal with the Top-n words only.

The column's order of $Dis(C, D)$ can be exchange in order to keep the column i is the hypernym of column j when $i < j$.

The sum of the row i, named it with $Dis(i) = \sum_j dis_{ij}, j \in [1, n]$, means the degree of words representing the concept or ontology. In general, the more the sum, the much irrelevance of the words to the concept or ontology.

So, assuming the matrix $Dis(C, D_1)$ and $Dis(C, D_2)$ represent the relevance of document D_1 and D_2 to the concept C respectively, we compute the distance respectively just as follows:

1. Computing the Average Variance of each rows $AV_i, i \in [1, n]$;
2. Sum the Average Variance value $V = \sum AV_i, i \in [1, n]$.

Hence, we get two Average Variance values V_1 and V_2 for the document D_1 and D_2 to the concept C respectively. We consider that the document with the less Average Variance of V_1 and V_2 has much relevance with the C.

Section 3.1, Section 3.2 and Section 3.3 introduce the our new approach, combining the word, concept, and document based on ontology.

4   Query Expansion

Search Engine (SE) plays the important role in finding information in Internet. The user inputs the query sentence to SE, and he hope the SE can feed back the results what he want to get exactly. In normal, the query sentence is not clear enough to let the SE know what the question is. Query expansion is used to solve this problem.

As we know, it is difficult to expand the user's query sentence without any other help, such as the domain information, surfing history or log records. In this paper, we require the user to register if he want to get the personalized service. The personal information of the user is used to construct the Personal Information Profile (PIP). After the IR system feeds back the results, he checks the results and estimates them. The IR system re
ne the PIP according to the estimation. The PIP works as a filter between the user and the feedback results.

By using the PIP, the method for query expansion can be described as,

1. Splitting the query to words and marking them in the domain words pool. The weight of the word plus 1 for each time appeared in the query. It is obviously that the more times the word appear in the query, the more weight it is in the domain words pool.
2. Selecting the concept and the words involved in the concept according to the user's PIP, we order the words belong to the concept just as following steps,
   (a) Ordering two word-lists. The first one is that the words order by the relevance, which are computed in W-C level. We named it as,

$$M(wi) = M(w_{i1}, w_{i2}, ..., w_{im}) \tag{7}$$

The second one is that the words order by the appearance in the domain words pool in a given period. We named it as,

$$N(wj) = N(w_{j1}, w_{j2}, ..., w_{jn}) \tag{8}$$

(b) Setting the final word list according to the Eq.(7) and Eq.(8) as,

$$P(M, N) = \alpha M(wi) + (1 - \alpha)N(wj), \alpha \in (0, 1) \tag{9}$$

(c) Setting the thresholds, and selecting the $Top - R$ words. The R words have much relevance with the words appeared in the query sentence.
3. The R words selected in the last step will be submitted to the SE, and SE feedbacks the results to user according to the these words. The user reviews the results, and he presents his owner opinion for the retrieval results. The opinion will be used to refine the parameter $\alpha \in (0, 1)$ in the formula.

In the step 2 of query expansion, the amount of computation is very huge because we have to consider the words in whole concept level. With the time going, if we could focus on the some of the words in the concept, and those words have more relevance than the others, we can reduce the amount of computation. So, it can be imaged that the effect of this way is not ideal in the beginning as the limited of the words in query sentence. With the times of query input increased, the accuracy will be better.

5   Conclusion

The paper introduces the concepts of ontology, query expansion, and representing the document by using the ontology. We construct a new approach with two levels, the Word-Concept level and Concept-Document level, which reflects the relevance among the words, concepts, documents and queries. By computing the biggest eigenvector of words matrix to determine the relevance of words to the concepts, and computing the average variance to determine the distance of document to the concept. In the last paragraph, we introduce the way to expand the query sentence by constructing the Personal Information Profile (PIP) of user. According to the forecast, the feedback results will be fine than before.

Acknowledgement. This work was supported by Scientific Research Program Funded by Shanxi Provincial Education Department, P.R.China (Program No.09JK440), and Natural Science Foundation of Shaanxi Province of China (Program No.2012JM8034).

References

1. Brin, S., Page, L.: The anatomy of a large-scale hypertextual web search engine. In: Proceedings of the 7th International Conference on World Wide Web, pp. 107–117 (1998)

2. Bianchini, M., Gori, M., Scarselli, F.: Inside pagerank. ACM Transactions on Internet Technology 5(1), 92–128 (2005)
3. Altman, A., Tennenholtz, M.: Ranking systems: the pagerank axioms. In: Proceedings of the 6th ACM Conference on Electronic Commerce, EC 2005, pp. 1–8. ACM, New York (2005)
4. Wang, H.-m., Rajman, M., Guo, Y., Feng, B.-q.: NewPR-Combining TFIDF with Pagerank. In: Kollias, S.D., Stafylopatis, A., Duch, W., Oja, E. (eds.) ICANN 2006. LNCS, vol. 4132, pp. 932–942. Springer, Heidelberg (2006)
5. Jones, K.S.: A statistical interpretation of term specificity and its application in retrieval. Journal of Documentation 28, 11–21 (1972)
6. Qu, S., Wang, S., Zou, Y.: Improvement of text feature selection method based on tfidf. In: International Seminar on Future Information Technology and Management Engineering, FITME 2008, pp. 79–81 (2008)
7. Kayed, A., Colomb, R.M.: Using ontologies to index conceptual structures for tendering automation. In: Proceedings of the 13th Australasian Database Conference, ADC 2002, vol. 5, pp. 95–101. Australian Computer Society, Inc., Darlinghurst (2002)
8. Kara, S., Alan, O., Sabuncu, O., Akpnar, S., Cicekli, N.K., Alpaslan, F.N.: An ontology-based retrieval system using semantic indexing. Information Systems 37(4), 294–305 (2012)
9. Kang, X., Li, D., Wang, S.: Research on domain ontology in different granulations based on concept lattice. Knowledge-Based Systems 27, 152–161 (2012)
10. Myoung-Cheol Kima, K.S.C.: A comparison of collocation-based similarity measures in query expansion. Information Processing and Management 35(1), 19–30 (1999)
11. Efthimiadis, E.N.: Query expansion. Annual Review of Information Science and Technology 31, 121–187 (1996)
12. Cronen-townsend, S., Zhou, Y., Croft, W.B.: A framework for selective query expansion. In: Proceedings of Thirteenth International Conference on Information and Knowledge Management, pp. 236–237. Press (2004)
13. Wu, C.C., Chou, C.H., Chang, F.: A machine-learning approach for analyzing document layout structures with two reading orders. Pattern Recognition 41(10), 3200–3213 (2008)
14. Gardiner, C.: Stochastic Methods: A Handbook for the Natural and Social Sciences. Springer Series in Synergetics. Springer (2009)
15. Mian, R., Khan, S.: Markov Chain. VDM Verlag Dr Muller (2010)
16. Serre, D.: Matrices: theory and applications. Graduate texts in mathematics. Springer (2010)

A Knowledgeable Decision Tree Classification Model for Multivariate Heart Disease Data-A Boon to Healthcare

G. NaliniPriya[1,*], A. Kannan[1], and P. Anandhakumar[2]

[1] Department of Information Science and Technology
Anna University, Chennai
nalini.anbu@gmail.com
[2] Department of IT
Anna University, Chennai

Abstract. In a smart hospital, effective decision supports are useful for medical diagnosis. Recent advances in the field of data mining, pervasive computing and other computing methods are ready to meet this kind of challenges. However, few techniques can be gracefully adopted for generating accurate and reliable as well as biologically interpretable rules. The objective of this paper is to introduce a novel method for classifying coronary artery disease dataset based on the principle of decision trees. We extend classical decision tree building algorithms to handle data sets with Multivariate in nature. Extensive experiments have been conducted which shows that the resulting classifiers are more accurate than the existing classifiers. The performance of the algorithm is evaluated with the coronary artery disease (CAD) data sets taken from University California Irvine (UCI).

Keywords: Knowledge Discovery, Data Mining, CAD, Heart Disease, Classification, Multivariate Data, Decision Tree.

1 Introduction

Modern hospital practices depend upon different modern hardware and software based technologies. Classification and other data mining[27] algorithms play major role in designing computing environments in smart hospitals. For instance, classification algorithms are often useful in patient activity classification[2][3] and the diagnosis of a disease using a multivariate clinical data, which were acquired from the hospital environment using different technologies. This data may be the combination of different types. Development of computer methods for the diagnosis of heart disease attracts many researchers. In the past time, the use of computers was to build knowledge based decision support system which uses knowledge from medical experts and transfer this knowledge into computer algorithms manually. This process is time consuming and really depends on

---

* Corresponding author.

Z. Li et al. (Eds.): ISICA 2012, CCIS 316, pp. 459–467, 2012.

medical expert's opinion which may be subjective. To handle this problem, new classifiers[15] [16] have been developed in this work to gain knowledge automatically from examples or raw data.

1.1   Motivation of the Research

The clinical data which will be used to diagnose a disease will be a mixed type of data, which contains different types of attributes. Classification of a data can be solved by using a lot of methods from simple methods to complex methods such as decision trees, neural networks and genetic[14] algorithms. However, it is known that classification of multivariate data are a difficult problem because of several reasons. Generally, this kind of medical data or multivariate data[11][12] will contain an error and missing values and will not always be pure. Further, the mutual dependence of attributes or variables causes distortion of the space. Due to an effect called 'boundary effect', the nearest points seem to be rather far and farther points near and this causes considerable error in distance calculation during the classification process. So most of the algorithms which are used for classification cannot be applied on multivariate data.

Motivated by the need of such an expert system, this paper, we propose a decision tree for classifying multivariate dataset.

1.2   Coronary Artery Disease (CAD)

Heart disease, which is usually called coronary artery disease (CAD), is a broad term that can refer to any condition that affects the heart. CAD is a chronic disease[12][13] in which the coronary arteries gradually hardens and narrow. It is the most common form of cardiovascular disease and the major cause of heart attacks in all countries. Many factors to analyze and diagnose the heart diseases, physicians generally make decisions by evaluating the current test results of the patients. The previous decisions made on other patients with the same condition are also examined by the physicians. These complex procedures are not easy when considering the number of factors that the physician has to evaluate. So, diagnosing the [15][16] heart disease of a patient involves highly skilled physicians and experience.

Recent advances in the field of artificial intelligence and data mining have led to the emergence of expert systems for medical applications. Moreover, in the last few decades computational tools have been designed to improve the experiences and abilities of physicians for making decisions about their patients. In this paper, a new classification model for [9][10] multivariate representation of CAD data is proposed for effective storage, retrieval and analysis.

2   Proposed Work

2.1   Decision Tree

A Decision Tree[1] is a tree-structured plan of a set of attributes to test in order to predict the output. Decision tree learning algorithm has been successfully used in expert systems in capturing knowledge and determines appropriate classification according to decision tree rules. The Fig. 1 mentioned the proposed method stages. It involves IDTC (Intelligent Decision Tree construction) algorithm, Multivariate decision tree classifier[23][24] and classification model. Traditional data analysis techniques cannot support huge and complex medical data set[9]. New data analysis technique such as data mining can be helpful in analysis large and complex medical data set[11].

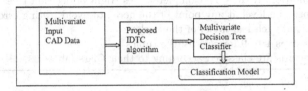

Fig. 1. Stages involved in the proposed multivariate decision tree methodology

Researchers may need data and knowledge which was discovered by other researchers for their research[10][11] that is distributes multidimensional data format. New systems are needed to analyse large and complex medical data for accurate diagnosis. It helps the doctors to diagnose the disease in a better way .Here pruning is done up to 7 levels. Pruning is heuristically trying to find the simplest tree structure for which all within-leaf node disagreements can be explained by chance. In our method we are utilizing common pruning[24][26] techniques. Then we are reducing the pruning levels automatically the branches of decision tree also get reduced. Our proposed systems contribute that need to assists in decision making process and help researchers to precede their research work.

2.2   The Proposed IDTC Algorithm

A decision tree based IDTC algorithm for classifying Multivariate heart disease is detailed is below. The value of Spn is empirically chosen to grow the decision tree for classification from the set of attributes Ajn.In this algorithm the search direction can be chosen with the condition of split point values (Spn). Justification of this strategy is that the value is tangent to the stable of the multivariate decision system[1].However computing split point values is computationally challenging for multivariate data. The root node has been selected in an arbitrary way. Hence, exploitation of the structure of multivariate decision tree under study will prove fruitful in classification of multivariate data set. In the case of traditional decision tree approach based on C4.5 algorithm[23][24][25][26] utilizing statistical values.

To illustrate this IDTC algorithm using the example records. This set consists of sample ten records of two class label '1' and '2'. Each record has set of 13 attributes 'Ajn'. First step is to find the point value of each attributes 'ps'.

1: Input: Coronary Artery Disease Data
2: Assign set of attributes as SA, and point values as ps, $SA = \{A_{j1}, A_{j2}, ..., A_{jn}\}$
3: Select the attribute A_{jn} from SA as a root node and convert its value into point value(ps)
4: Check for $C_i \in C$, where C Set of class labels and calculate point values for each attributes A_{jn}, where $ps = \{p_1, p_2, ..., p_n\}$
5: Assign Input values as Iv, Where Iv is set of input values from the tuples/records
6: Place the tuples into two subset 'left' and 'right' based on input value (Iv)
7: Test the condition, If $Iv \leq Sv$ Completely lies Left of decision tree
8: Test the condition, If $Sv \leq Iv$ Completely lies Right of decision tree
9: Check for pruning level at any point of the node, to avoid testing error
10: Check for C_i at each leaf node, of the tuples
11: Repeat the steps 1 to 9 for classify the tuples
12: Output: Records are classified according to their Class labels

---

Compare each Iv value from record with ps, where Iv is input value from record . Based on the split value spn the attributes are placed left or right of the decision tree. Here class labels '1' for normal and label '2' for Sick. To get good results we fix the pruning level in the higher side. In this paper the pruning level is 7. At each child node we are checking for class labels. In our example data it is 1 and 2.

3   Experimental Setup

The simulation of the Decision tree is constructed by using MATLAB (Release 2009a) Statistical Tool Box. To evaluate the algorithm under consideration, a suitable and standard multivariate data set is needed. A suitable UCI data set called 'cleveland.data'[21][22], concerning heart disease diagnosis is used for the evaluation of the algorithms under consideration. This data[22][13] was originally provided by Cleveland Clinic Foundation. This database contains 303 records with 13 attributes which have been originally extracted from a larger set of 75 attributes and a class attribute [21][22]. Among the 303 records, 164 belong to healthy and remaining are from diseased.

We have successfully implemented the decision tree algorithm in Mat lab statistical Tool box and repeated the experiments with different set of parameters to obtain optimum performance[9][10] in that algorithm. The following Section explains the performance of the algorithms.

The classification[2][3] was done using a algorithm for several times and the best results were tabulated and the average value is used to evaluate the Performance.

In the decision tree figure the records are classified according to their point value. The labels $X_1, X_2, ..., X_{13}$ are represented as attributes. In this dataset we are making use of 13 attributes. Each leaf node represent the class labels. Since it is a two classification problem it is classified as 1 for normal and 2 as sick. Here we are taking pruning level is 7 to avoid test error.

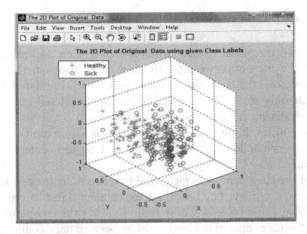

Fig. 2. The 2D plot of cleavant heart disease data set

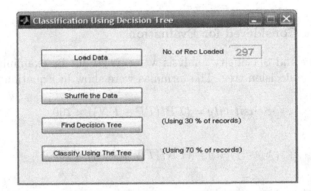

Fig. 3. Screen Shots of Multivariate Decision Tree construction

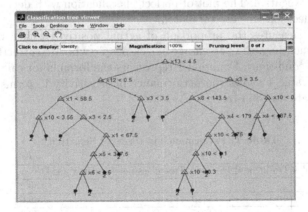

Fig. 4. Screen Shots of Multivariate Decision Tree

4   Result and Discussion

In this paper two diagnosis classes are considered such as healthy and sick. From the related work section we came to know that various methods have been proposed for diagnosis the heart disease. The accuracy is tabulated. In the Resul Das[9][10] paper only statically similar data set is considered and evaluated. If there is no dissimilarity in data then the classification is very easy and we can get the accuracy in the higher side. Since we are handling multivariate dataset classification is very challenging. Here we are making use of decision tree with IDTC algorithm.

The result of the classification is measured in terms of metrics such as sensitivity, specificity and accuracy. The best three experimental results are tabulated in the Table 1, Table 2 and table 3. These table showing the comparative result of classification. Another topic of discussion is the performance comparison of multivariate decision tree. The average best result of the data set is been achieved and tabulated after conducting more experiments.

4.1   Metrics Considered for Evaluation

The sensitivity and Specificity analysis was carried out by examining the performance of the decision tree. The formulas were show in Equation 1, 2 and 3.

$$Sensitivity = (TP/(TP + FN)) \times 100 \qquad (1)$$

$$Specificity = (TN/(TN + FP)) \times 100 \qquad (2)$$

$$Efficiency = ((TP + TN)/(TP + TN + FP + FN)) \times 100 \qquad (3)$$

Where $TP = Ture$, Positive, $TN = True$, Negative, $FP = False$, Positive and $FN = False$, Negative. The plot of perfect classification with decision tree and multivariate decision tree are tabulated.

Sensitivity measures the proportion of actual positives which are correctly identified as such (e.g. the percentage of sick people who are correctly identified as having the condition). Equation 1 represents the formula for calculating sensitivity. The Table 1 shows the performance of sensitivity in terms of decision and multivariate decision tree.

Table 1. Performance in terms of Sensitivity

Training Samples	Traditional Decision Tree classifier	Multivariate Decision Tree classifier
1	79.10	80.27
2	59.10	69.02
3	69.27	70.21
Avg	69.1	73.1

Specificity measures the proportion of negatives which are correctly identified as such (e.g. the percentage of healthy people who are correctly identified as not having the condition). Equation 2 represents the formula for calculating specificity. The Table 2 shows the performance of specificity in terms of decision and multivariate decision tree.

Table 2. Performance in Terms of Specificity

Training Samples	Traditional Decision Tree classifier	Multivariate Decision Tree classifier
1	78.60	79.63
2	78.30	79.59
3	86.42	87.76
Avg	81.10	82.3

Accuracy of measurement system is the degree of closeness of measurements of a quantity to its actual (true) value. Equation represents the formula for calculating accuracy. The Table 3 shows the performance of accuracy in terms of decision and multivariate decision tree.

Table 3. Performance in Terms of Accuracy

Training Samples	Traditional Decision Tree classifier	Multivariate Decision Tree classifier
1	78.78	79.78
2	69.79	70.79
3	78.78	79.78
Avg	75.78	76.78

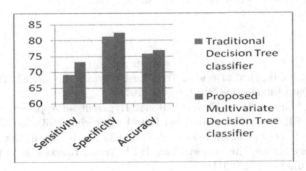

Fig. 5. Performance Comparisons of Multivariate decision tree and Traditional Decision Tree classifier

In the graph shown below, the two well known methods have been compared in terms the performance metrics such as Sensitivity, Specificity and Accuracy. According to the graph the sensitivity of multivariate decision tree is performed well compare to existing decision tree method. In terms of specificity the performance of multivariate decision tree is seems to be good.

In the graph shown in Fig. 5 represents the performance comparisons of Multivariate decision tree and Traditional decision tree .In the X-axis represents metrics such as sensitivity, specificity and accuracy and Y-axis represents the percentage. The accuracy is important collective method which is directly showing the overall classification performance of the algorithm.

The performance of multivariate decision tree is seems to be good. The proposed multivariate Decision tree algorithm giving satisfied results compared to existing traditional based decision tree methods.

5   Conclusion

In this paper, a new algorithm called IDTC has been proposed to classify the multivariate data. We have implemented this algorithm on decision and multivariate decision trees. We have evaluated decision tree classification algorithms for the diagnosis of coronary artery disease dataset and obtained encouraging results .The results obviously shows the complex nature of data set restricts these algorithms from achieving better accuracy, while testing the same algorithms with a normal synthetic data set with normal distribution, they provided ideal performance and better accuracy. Though the algorithm was giving satisfied classification results while applying this on Coronary artery disease data, it is having certain limitation when were calculating split values.

In fact, the goal we wanted to achieve with multivariate decision tree is powerful classifier which performs well for complex data. To improve tree construction efficiency suitable pruning techniques may be adopted. In future we plan to implement a classification model for the same data set in Machine learning based algorithm and hope to produce better results.

References

1. Tsang, S., Kao, B., et al.: Decision Trees for Uncertain Data. IEEE Trans. Knowledge and Data Engineering 23(1), 64–78 (2011)
2. Wei, J.-M., Shu-Qin, et al.: Ensemble Rough Hypercuboid Approach for classifying Cancers. IEEE Trans. Knowledge and Data Engineering 23(3), 381–389 (2010)
3. Masud, M.M., et al.: Classification and Novel Class Detection in Concept-Drifting Data Streams under Time Constraints. IEEE Trans. Knowledge and Data Engineering 23(6), 859–874 (2011)
4. Cheng, H., et al.: Efficient Algorithm for Localized Support Vector Machine. IEEE Trans. Knowledge and Data Engineering 22(4), 381–389 (2010)
5. Chen, H., et al.: Multiobjective Neural Network Ensembles Based on Regularized Negative Correlation Learning. IEEE Trans. Knowledge and Data Engineering 22(12), 1738–1743 (2010)
6. Wang, B.: ELITE: Ensemble of Optimal Input-Pruned Neural Networks Using TRUST-TECH. IEEE Trans. Neural Networks 22(4), 96–107 (2011)

7. Sahbi, H.: Context-Dependent Kernels for Object Classification. IEEE Trans. Pattern Analysis and Machine Intelligence 33(4), 699–708 (2011)
8. Antonio, P.: Logistic Regression by Means of Evolutionary Radial Basis Function Neural Networks. IEEE Trans. Neural Networks 22(2), 246–278 (2011)
9. Ince, T., Kiranyaz, S., et al.: Evaluation of global and local training techniques over feed-forward neural network architecture spaces for computer-aided medical diagnosis. Elsevier–Expert Systems with Applications 37(12) (December 2010)
10. Das, R., Turkoglu, I., et al.: Effective diagnosis of heart disease through neural networks ensembles. Elsevier–Expert Systems with Applications 36(4) (May 2009)
11. Setiawan, N.A., Venkatachalam, P.A., et al.: Rule Selection for Coronary Artery Disease Diagnosis Based on Rough Set. International Journal of Recent Trends in Engineering 5(2) (November 2009) ISSN: 1797-9617
12. Setiawan, N.A., Venkatachalam, P.A., et al.: Diagnosis of Coronary Artery Disease Using Artificial Intelligence Based Decision Support System. In: Proceedings of the International Conference on Man-Machine Systems (ICoMMS), Malaysia (October 2009)
13. Bhatia, S., Prakash, P., et al.: SVM Based Decision Support System for Heart Disease Classification with Integer-Coded Genetic Algorithm to Select Critical Features. In: Proceedings of the World Congress on Engineering and Computer Science, WCECS 2008 (2008) ISBN: 978-988-98671-0-2
14. Khan, S.S., Kant, S.: Computation of Initial Modes for K-modes Clustering Algorithm using Evidence Accumulation. In: IJCAI (2007)
15. Yang, J.-B.: Determination of Global Minima of Some Common Validation Functions in Support Vector Machine. IEEE Trans. Neural Networks 22(4), 656–678 (2011)
16. Windeatt, T.: Embedded Feature Ranking for Ensemble MLP Classifiers. IEEE Trans. Neural Networks 22(6), 907–968 (2011)
17. Deng, Z.: Robust Relief-Feature Weighting, Margin Maximization, and Fuzzy Optimization. IEEE Trans. Fuzzy Systems 18(4) (August 2010)
18. Alvarez, G., Coiera, E.: Interruptive communication patterns in the Intensive Care Unit Ward Round. Journal of Medical Informatics 74(10), 779–781 (2005)
19. Miller, M.J., Ferrin, D.M., et al.: Using RFID Technologies to Capture Simulation Data in a Hospital Emergency Department, pp. 1365–1371
20. NaliniPriya, G., AnandhaKumar, P.: Neural Network Based Efficient Knowledge Discovery in Hospital Databases Using RFID Technology. In: Proceeding of International IEEE Conference Tencon (November 2008)
21. Frank, A., Asuncion, A.: UCI Machine Learning Respository (2010) (online)
22. Robert Detrano, M.D.: The Cleveland Data, V.A. Medical Centre, Long Beach and Cleveland Clinic Foundation
23. Quinlan, J.R.: C4.5: Program for machine learning. Morgan Kaufmann (1993)
24. Quinlan, J.R.: Induction of decision trees. Machine Learning 1(1), 81–106 (1986)
25. Petre, E.G.: Elia Georgiana Petre, A Decision Tree for weather Prediction LXI(1), 77–82 (2009)
26. http://www.americanheart.org
27. Gang, K., Yong, P., Yong, S., et al.: Privacy-preserving data mining of medical data using data separation-based techniques. Data Science Journal (2007)

The Research of Abnormal Target Detection Algorithm in Intelligent Surveillance System

Junkai Yang and Mengyu Hua

International School
Beijing University of Posts and Telecommunications
West Tocheng Road. 10, 100876 Beijing, China
{yangjunkai91,dreamrain91}@gmail.com

Abstract. To implement real-time surveillance of the home security, an intelligent surveillance system based on embedded system platform and wireless communication technology was designed. To overcome interference coming from small target movement or background light changes, moving target detection algorithm was proposed. Abnormal matters at the monitor area can be detected exactly by analyzing the real time image information, and effective security actions will be taken with corresponding patterns. The experimental results show that the system possesses the advantages of real-time standard, high credibility and practicability.

Keywords: Intelligent Surveillance System, Abnormal Target Detection, Frame Subtract, Adaptive Threshold Value.

1 Introduction

Video monitoring system has already became one of the most important components of the security system. With the development of Computer technology, wireless communication and the image processing technology, Intelligently video security monitoring system is displacing traditional manual work monitoring pattern step by step, with its advantages such as all-weather and high reliability.

In the intelligent video monitoring system, abnormal target detection algorithm (mainly refers to the detection of moving item) is the core content of it. The merits of the algorithm determines the win or lost of the system. Current moving item monitoring detection including optical flow [1], inter-frame difference method [2], background differencing method [3], and statistical background model [4] method. However, the above four methods all have limitations: optical flow method do not need background information, but having complex computation, low instaneity, could not meet the requirement of real-time detection, especially not useful in the micro-processor; background differencing method, having easier implementation, but having low anti-influence performance, also need to revise the background instantly; inter-frame difference method has strong adaptivity, however, the requirement of the chance to choose is high, for the detection base on the moving velocity of the item, having low reliability.

Z. Li et al. (Eds.): ISICA 2012, CCIS 316, pp. 468–478, 2012.

The paper put forward an algorithm based on the abnormal detection according to background subtraction. The algorithm combines the static threshold as well as dynamic adaptive threshold [5] and doing the differential operation, also introducing the detecting and tracking for small moving targets to dealing the detected image sequence. The method efficiently solved the illumination variant of background and the moving of small targets, having high reliability and practical applicability.

2 System Structure

The hardware platform researched by the writer me is the Intelligent Image Security Monitoring System, the system is based on the design of mobile phones and embedded system, made up of CMOS camera, S3C44BOX Microprocessor, TR800 MMS Module, ZigBee Module and sensors. The system chart is as follow:

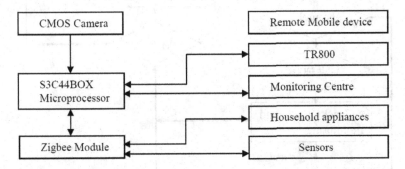

Fig. 1. The System Chart of the Intelligent Image Monitoring system

When monitoring the key areas via the CMOS camera, sending the JPEG format image data through the serial port to microprocessor timely. The latter deal with and analyses the image sequence by abnormal target detecting algorithm. Although, perform the relative actions exactly, then sending alarm MMS to the user, sending alarming signal to warning center, store the current image and video, all this are completed by the support of GPRS/ 3G. At the time, the alarming information made by the alarming sensors in the system is able to go through the embedded controller and control the household appliances, users also are able to set the working patterns via mobile phone messages, getting information such as the monitored images.

3 Abnormal Target Detection Algorithm

The abnormal target detection algorithm discussed by us, the entrance of the system is the monitored image sequence transferred by the camera, however the exit of the system is the control signal that in the condition waiting to be dealt with.

The algorithm based on the background differencing method, using the dot-based dynamic threshold combining method, the main steps are mainly resolved into three steps:

Image pre-processing→Quantification detection→Abnormal state processing
The detection algorithm flow chart is as follows:

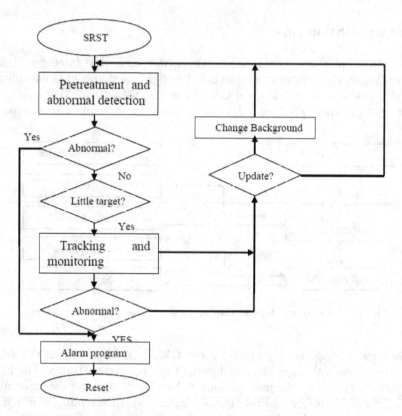

Fig. 2. The Flow chart for the Abnormal Detection Algorithm

When the system received the new image transmitted by the camera, it would first do the pretreatment operation, in order to get the abnormal two-value image from the monitored area. Go further, the system analysis and judge the monitored area using the Quantitative function, in order to get the abnormal quantitative index. In the end, the system would do the subsequent security operation according to the Quantitative function, such as abnormal alarm, tracking detection operation, etc.

3.1 Image Pretreatment

In order to achieve more efficient information about abnormal difference, the system would put the achieved monitoring image in the ARM processor and do the pretreatment. The main sections are differential operation, Threshold value generation, binaryzation, denoising and morphological anti-aliasing[7], the flow chart is as follows:

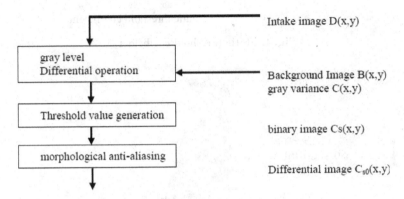

Fig. 3. The process of image pretreatment

The reasonable or not for the threshold has huge influence on the reliability for the algorithm. The algorithm decides to use the combination of Static threshold and dynamic threshold. The choice avoids not only the dead angle that the static threshold has when meeting the background light, but also avoids the defect than generated by the over-sensitivity when pure-dynamic background meets the light.

In the research we define S is the binary value threshold, the equation represent this is as follow:

$$\mu K \leq \lambda \frac{1}{N_A} \sum_{A(x,y)\in A} |D(x,y,t) - B(x,y,t)| \tag{1}$$

- K-gray level of the image
- A-Detection area image spot collection
- N-pixel
- μ-static threshold index

Static threshold μK is mainly used to avoid the tiny change brightness of the background, it are able to choose static threshold index by experiment. The statistic coefficient of room environment is as follow:

We can see from the picture that when μ is bigger than 8%, the order of magnitude of Misjudgment proportion is 10^{-4}, thus the influence of this to the Quantitative function can be ignored.

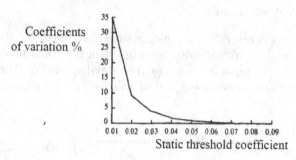

Fig. 4. Static threshold coefficient

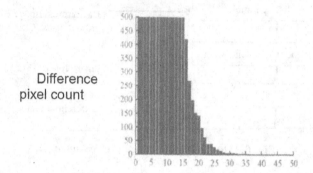

Fig. 5. Background gray variance

When in the same background, the Stat-Histogram of Differences pixels is follow: We can see the majority of the gray variance is among the range of 25(gray variance 256), that is, $\mu = 25/256 = 9.6\%$.

So based on the result of experiment, we use $\mu = 10\%$ at the static threshold coefficient.

For $\lambda \frac{1}{N} \sum_{A(x,y) \subset A} |D(x,y,t) - B(x,y,t)|$, the dynamic threshold is used to overcome the influence caused by the slow changing of background light. Also provide judgment for changing of background. The advantage comparing with other methods are: Simple in operation, high applicability, especially suitable in embedded system.

Another main problem of image pretreatment is about the elimination of noise, this also the key step for the detection of abnormal area accurate. The research use erosion and dilation algorithm in the morphological anti-aliasing, do further research on the Image corrosion expansion.

3.2   Detection of Abnormal Target

In order to judge the detection area, we need to do the Abnormal rate computation for the Binary image, also define the abnormal rate P (The image differences between the total area of area and cumulative) is the abnormal of target index, thus to complete the judgment of abnormal state. The usual expression is:

$$P = \frac{\sum C_{s0}(x,y,t)}{N}, C_{s0}(x,y,t) \in A \tag{2}$$

where

$$C_S(x,y,t) = \begin{cases} 1 \to C(x,y,t) \geq S \\ 0 \to C(x,y,t) < S \end{cases} \tag{3}$$

State threshold is influenced by monitoring the parameter selection regional environment, foreign bodies, location and the influence of the layout of the camera, in order to increase the accuracy of the system, the paper put out the subsection Threshold parameters: p1 and p2, below is the detail explanation.

when $P \geq P1$, which means the abnormal rate is very high, we judge it's in the abnormal mode, the system send MMS to users' mobile phones, send alarm signal to alarming center, store the abnormal image. When necessary, control the electrical equipment in case of safety.

when $P1 > P \geq P2$, there exists small area abnormal, but the harmfulness is not large enough to judge, thus the system require for the continuous tracking, doing the small moving target tracking.

when $P < P2$, the detecting area is normal type.

For the choice of State threshold, the research via theoretical derivation and experimental test, get the equation:

$$\begin{cases} P_1 = \frac{K}{L^2}\% \\ P_2 = 0.6 * P_1 \end{cases} \tag{4}$$

The derivation of equation (4) as follows:

The problem is solved in the frequency domain using the Fourier transforms $U(f)$, $H(f)$ and $R(f)$ of $u(t)$, $h(t)$, $v(t)$ and $r(t)$, respectively. The receiver output $q(t)$ consists of $q_s(t)$, the signal component, and $q_n(t)$, the noise component. The output SNR at time $0\ t$ is defined as:

$$SNR_{t_0} \doteq \frac{|q_s(t_0)|^2}{E|q_n(t_0)|^2} \tag{5}$$

Which is the energy ratio between the signal component and the noise component. The energies of the signal and noise components are given by:

$$|q_s(t_0)|^2 = |\int_{-B/2}^{B/2} H(f+f_c)U(f)R(f)e^{j2\pi f t_0}df|^2$$

$$E|q_n(t_0)|^2 = \int_{-B/2}^{B/2} |R(f)|^2 P_{nn}(f)df$$

Therefore, the output SNR is:

$$SNR_{t_0} = \frac{|\int_{-B/2}^{B/2} H(f+f_c)U(f)R(f)e^{j2\pi f t_0}df|^2}{\int_{-B/2}^{B/2} |R(f)|^2 P_{nn}(f)df} \tag{6}$$

Applying the Cauchy-Schwarz inequality to Eq, gives:

$$SNR_{t_0} \leq \int_{-B/2}^{B/2} \frac{|H(f + f_c)U(f)|^2}{P_{nn}(f)} df$$

with equality if and only if:

$$R(f) = \frac{K\overline{U(f)}H(f + f_c)e^{-j2\pi f t_0}}{P_{nn}(f)}, f \in [-B/2, B/2] \qquad (7)$$

Which is a function of the transmit waveform. The objectives to find a $u(t)$ that maximizes for the following constraints: $u(t) = 0$ when $[-T/2, T/2]$ and $|u(t)| = E/T$, where E is the total energy of the waveform. The output SNR is then bounded as follows:

$$0 \leq SNR_{t_0} \leq max_{-B/2 \leq f \leq B/2}(|H(f + f_c)|^2/P_{nn}(f))E$$

$$SNR_{t_0} =$$

$$\int_{-B/2}^{B/2} \left[\int_{-T/2}^{T/2} u(t)e^{-j2\pi f t} dt \right] \left[\int_{-T/2}^{T/2} \overline{u(\tau)}e^{j2\pi f \tau} d\tau \right] (|H(f + f_c)|^2/P_{nn}(f)) df =$$

$$\int_{-T/2}^{T/2} u(t) \int_{-T/2}^{T/2} \overline{u(\tau)} \left[\int_{-B/2}^{B/2} (|H(f + f_c)|^2/P_{nn}(f))e^{j e \pi f(\tau - t)} df \right] d\tau dt =$$

$$\int_{-T/2}^{T/2} \int_{-T/2}^{T/2} u(t)\overline{u(\tau)}L(\tau - t)d\tau dt$$

$$(8)$$

At last, The target autocorrelation matrix L has the form:

$$L = \begin{bmatrix} L_0 & L_{-1} & ... & L_{-N+1} & L_{-N} \\ L_1 & L_0 & L_{-1} & ... & L_{-N+1} \\ ... & L_1 & L_0 & ... & ... \\ L_{N-1} & ... & ... & ... & L_{-1} \\ L_N & L_{N-1} & ... & L_1 & L_0 \end{bmatrix}$$

Which is a Hermitian Toeplitz matrix, where the elements are the samples of $L(t)$ given by:

$$\int_{-B/2}^{B/2} (|H(f + f_c)|^2/P_{nn}(f))e^{j2\pi f k T_s}), k \in [-N, N] \qquad (9)$$

The optimal waveform that maximizes the SNR for this assumption is proportional to the eigenvector of L corresponding to the largest eigenvalue. Unfortunately, the eigenvector does not usually have a constant modulus, therefore, is not the desired result. The phase vector φ is used to design the phase-modulated waveform. Moreover, the Hessian matrix of the objective function is given by:

$$\nabla_\varphi^2 SNR_{t_0} = \frac{\partial^2 SNR_{t_0}}{\partial\varphi\partial\varphi^T} = jT_s^2 \frac{\partial(diag(u)\overline{L}\overline{u} - diag(\overline{u})Lu)}{\partial\varphi^T}$$

$$= jT_s^2\left(diag(u)\frac{\partial\overline{L}\overline{u}}{\partial\varphi^T} + diag(\overline{L}\overline{u})\frac{\partial u}{\partial\varphi^T} - diag(u)\frac{\partial Lu}{\partial\varphi^T} - diag(Lu)\frac{\partial\overline{u}}{\partial\varphi^T}\right) =$$

$$T_s^2(diag(u)\overline{L}diag(\overline{u}) + diag(\overline{u}Ldiag(u)) - diag(diag(u)\overline{L}\overline{u}) - diag(diag(\overline{u}Lu)) \tag{10}$$

Here, the derivation is completed.

3.3 Small Target Motion Tracking Detection

For we use abnormal rate P as one of the index for abnormal detection, the space distribution in the monitoring area would influence the reliability of judging index. That's the reason why we introduce the Small target motion tracking detection. If the system cannot decide whether the abnormal rate is caused by interloper, the system would apply for the continuous taking of several pictures, accumulate the differential effect.

If the abnormal is introduced by moving target, the P would be increasing step by step, at the time when $P \geq P1$, the system would sure that it is caused by moving target.

The introduction of small target detection helps the system to avoid Blind alarm, offset the short come for the detection of small target, increasing the reliability of the system greatly.

3.4 The Change of Background

The moving target detection algorithm has another difficulty: the Live Update for the referenced background. Nowadays there exists several main Background Updating algorithms: Gaussian Mixture Model Method, novel nonparametric multimodal background model method [8], Background statistical technology method [9], etc.

By using the dynamic threshold and static threshold relation as the background of conditions change, under the condition that having no security problem, when the dynamic threshold achieves or over the static threshold, the changing of background begins, the condition is:

$$\mu K \leq \lambda \frac{1}{N_A} \sum_{A(x,y)\in A} |D(x,y,t) - B(x,y,t)| \tag{11}$$

When the type founded, indicates that it is no abnormal condition, is caused by dynamic threshold value increase at this time is the main reason of the background of their own changes (mainly is light Line of light and shade change). Therefore, in order to guarantee the real-time background, avoid background the accumulation of tiny change to the reliability of the system influence, need to change the background.

4   Analysis of Experimental Results

Using the above algorithm- abnormally detection algorithm, detecting the person into the background and the light transforming a large number of experimental, we achieve good results.

4.1   Pretreatment of Abnormal Target Extraction Effect

Using the pretreatment algorithm can effectively eliminate the pixels, shade, etc. interference, reached the extraction of accurate information is to eliminate the faint change, shadow background or camera dithering on the influence of the P value, making the anomaly target tracking more reliable.

4.2   Effect of Small Target Motion Tracking Detection

As can be seen from the graph, for small the target of the campaign, after tracking treatment abnormal ratio P will increase, and to enhance the small target P value, to avoid to small target motion (or distal target) omission. The link is good to improve the target tracking system reliability, expand the scope of the application of the system.

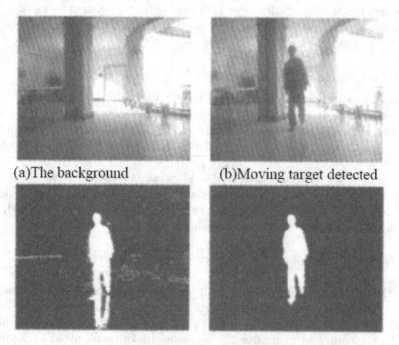

(a)The background   (b)Moving target detected

Fig. 6. Result of image pretreatment

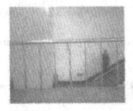

Fig. 7. Detected image

Fig. 8. Small target motion detection

5   Conclusion

The paper proposed the monitoring system based on embedded system, discussed the system structure and working principle of the abnormal detection system, including the design and realization of the algorithm. This system comprehensives on image processing, wireless transmission, embedded system and so on. Compare to real exiting algorithm and current cable video monitoring system, it has the simpler configuration, flexible installation, high reliability, good real-time characteristics. Having extensive application prospect in important background real-time detection.

References

1. Jaffer, A.J., Bar-Shalom, Y.: A.: On optimal tracking in multiple target environments. In: Proceedings of the Third Symposium on Non-Linear Estimation Theory and Its Applications
2. Jiang, M.: The arithmetic research of the selection recognition and tracking for medium infrared image. Department of Electronics Engineering Nanjing University of Aeronautics and Astronautics, 83–86 (2002)
3. Guo, W.: Infrared target detection and tracking in complex background. Xidian University, 46–48 (2008)
4. Micheloni, C., Foresti, G.: A new feature clustering method for object detection with an active camera. In: IEEE International Conference on Image Processing, Singapore, pp. 271–275 (2004)
5. Wang, J.A., Xiao, W.A.: Analysis of infrared radiation feature of targets based on double bands. Laser and Infrared 31(6), 351–354 (2001)

6. Kolomenkin, M.: Image matching using photometric information. IEEE Trans. on Computer Science 77(6), 654–658 (2006)
7. Lowe, D.G.: Distinctive image features from scale-invariant key points. International Journal of Computer Vision 60(2), 91–100 (2004)
8. Barton, D.K.: Radar System Analysis and Modeling. Artech House, Norwood (2005)
9. Leshem, A., Naparstek, O., Nehorai, A.: Information theoretic adaptive radar waveform design for multiple extended targets. IEEE J. Selected Topics in Sig. Process. 1(1), 42–55 (2007)

A Simulation Study of Modular Robot Self-replication

Lei Zhang, Zhenhua Li, Hao Zhang, and Huaming Zhong

School of Computer Science
China University of Geosciences
Hubei, Wuhan 430070, P.R. China
zhli@cug.edu.cn

Abstract. There are many research institutions dedicated to reconfigurable modular robot modeling and simulation. Various forms of reconfigurable modular robot that developed by hardware, some institutions design entity robot for self-reconfigurable experimental study. In this paper we analyze a couple of popular reconfigurable modular robot simulation platforms, and based on the analysis, we choose the Molecube simulation platform to fulfill a self-replication experiment, after design the self-replication and motion plan, The experiment proves the feasibility of proposed approach.

Keywords: Modular Robotics, Self-replication, Simulation Platform, Motion planning.

1 Introduction

Conceptually, the modular robot system is composed by a number of independent modules, such as animal cells. Modular robot system is a huge complex, this either independently but also collective action. It can form a variety of configurations, and be re-assembled into any desired shape or replicate them, it has broad application prospects.

The basic concept of the modular robot is introduced in the 1980s by Toshio Fukuda through CEBOT robot system[1]. So far, the module based on reconfigurable and replication has been a hot area of research of international robot[2–5]. With the deepening of the robot copy theory, micro-electro-mechanical and micro-electromagnetic technology continues to evolve, as well as the huge benefits behind the robot copy. At domestic and international many research institutions dedicated to reconfigurable modular robot modeling and simulation that have developed a various types of modular robot simulation platform. There are various forms of reconfigurable modular robot developed by hardware support. The more representative likes M-TRAN[6], Polybot[7], ATRON[8], CONRO[9], Superbot[10] and Molecubes[11]. These models can be simple to achieve self-reconfigurable or self-replicating simulation; there are some institutions entity robot self-reconfigurable experimental studies. With the development of computer technology, today's computer simulation has reached a higher level,

Z. Li et al. (Eds.): ISICA 2012, CCIS 316, pp. 479–489, 2012.
© Springer-Verlag Berlin Heidelberg 2012

through the physics engine and graphics engine to support, we can simulate the movement of various objects on the computer.

2   Related Works

In 2005, the evolution of robot research has entered a new field with Cornell University's self-replicating robot was born. Soon after Britain, Japan, Korea and other countries have also presented research results of the self-replicating and self-reconfigurable robots. In 2008, European scientists launched a project called "symbiotic robot" (Symbrion)[12] that focused on the development of the ability of evolution of symbiotic robot organisms. The goal is to design and develop the ultra-large-scale popularization of robot systems[13], and to enable them to adapt to the dynamic and open environment of a high degree of autonomy. In 2009, European EPFL launched roombots project based on Webots simulation platform. The robot system is composed of each individual roombots module. Each module of the degree of freedom for the Six Degrees of Separation, and composed of two semi-circular, there is inter-module through the shaft connection with mechanical latch lock. Through this new feature, so that a single roombots module whether flat or convex and concave surfaces are available through[14]. Modular robot system research is in the theory of the exploratory stage in domestic.

Cornell University's success and the excellent properties of modular robot-specific, modular robot self-replication, self-reconfigurable become a research hotspot. This paper attempts to analyze Webots and Molecubes simulation platform to understand the property of Webots simulation platform under Yamors module robot. We write the algorithm to control the various modules of the robot's movement with replication planning, And to achieve modular robot self-replication simulation in Molecubes simulation platform.

3   Webots Simulation Platform and Its Yamor Modular Robot

Webots simulation platform is developed by the Swiss Federal Institute of Technology Research Institute; it is a set of modeling, programming, simulation, and the transplant procedure as one of the multi-functional robot development software. We select Yamor modular robot system on Webots simulation platform. Each module has a connector and the receiver. The robot system move by two ways, one is rolling up and down, another is like circular. All modules use the same control code, but the movement code is different.

The Yamor robot is mainly controlled by the controller function of each module. We set the parameters of each module and the module robot movement. Figure 1 is Yamor robot planning simulation process.

Figure 2 shows Yamor modular robot motion planning simulation, it chose worm and loop two exercise combining movement planning for validation test and

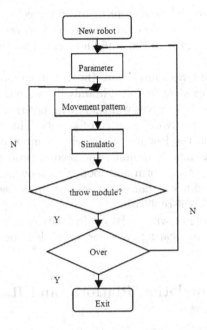

Fig. 1. Yamor Modular robot movement control process

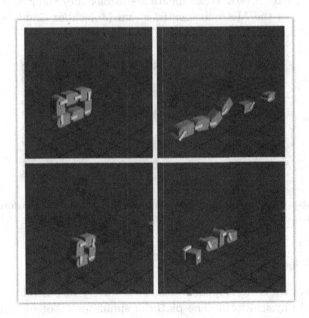

Fig. 2. Yamor Modular robot motion planning simulation screenshots

the impact of the lost module and implementation mechanism of the difference. Yamor modular robot can complete the function of self-reconfigurable with worm and loop two movements in both ring and linear configurations, and on this basis it is able to throw module.

Through experiments we found that the robot cannot complete the self-replicating behavior of existing features. Mainly by two reasons, the first problem is the receiver, lost module receives will change its position because of the emergence of up and down swing which caused of gravity, That will be different with the mother's status and the lost module receives cannot be completed connection operations such as matrix module; the second problem is the movement, the robot command receiver worm and loop, this movement restrain from the robot veering. But we can see that some are not in the same direction, so the robot cannot be able to self-replication.

Summary, Yamor robots with existing features can be self-reconstruction, but cannot achieve self-replicating behavior, which is important reason that the subject give up the robot.

4   Molecubes Simulation Platform and Its Molecubes Modular Robot

Molecubes modular robot is developed by Victor Zykov and his research partners. In 2007, Victor Zykov et al. mentioned Molecubes robot system in their article. Figure 3 is the Molecubes entities and Molecubes simulation platform 3D realistic renderings.

Fig. 3. Molecubes module robot entity diagram and simulation platform for three-dimensional view

4.1   Molecubes

Molecubes platform is developed to achieve a variety of possible combinations of systems and physical form. The platform simulates robot movement by the AGEIA PhysX physics engine and the OGRE picture engine. Therefore, it can provide an effective intuitive way to test a variety of physical combination of the possible features and physical characteristics, and it also allows the user to design a variety of intelligent modules for the simulation without care about the connection between the physical modules. The users do not have to combination various physical modules too, so that will reduce a lot of time in the design of complex modular system.

Initial state of a module is just a simple cube which composed of two and a half module. The Molecubes simulation platform has battery module, basic module, gripper modules, actuator module, control module. This experiment study the modular robot self-replicating behavior, therefore, basic modules, and gripper modules do not use. Only actuator module in the platform has 6-DOF connection, and only a 6-DOF connection module can be driven function-driven, the actuator module must be used; control module is the module system in the physical communication with the computer module, so it should be added.

4.2 Robot Design

The module design is particularly important during the experiment, which is the basis for the experiment. The main subject of the experiment is modular robot self-replication, so we choose to use the actuator module and control module. To achieve self-replication experiment, the experiment is needed to be able to generate a set of modules. In order to simplify the complexity of the experiment, the experimental process generates only two modules of the robot system.

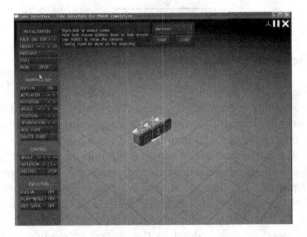

Fig. 4. Molecubes modular robot design

The modular robot is composed of three modules, two actuator modules and a control module shown in Figure 4. The experiment produced only two modules of the robot to complete a replication process. Modular robot called the "mother", a single module called "seed". The mother complete their own copy needs to find the seeds in the population, and move to the seed position to eat the seeds and grow the corresponding module. In this way the mother complete their own copy to self-replicating behavior.

4.3   Robot Self-replication Planning

Determine the desired robot configuration and properties of this topic; according to the experimental we have designed the self-replication behavior policy and processes. Figure 5 shows a flow diagram of Molecubes modular robot simulation. Its including a total of four main parts: the platform initialization, the robot configuration settings, Place seed module, and Simulation.

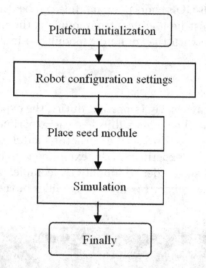

Fig. 5. Molecubes modular robot simulation flow diagram

Simulation, modular robot logic to judge the state of their own needs and the environment, the first step, if the simulation conditions is right then go to the next step, otherwise end the simulation. In the second step, according to the state and the relative position of the seed modules make sure the movement was ready. The third step, according to the second step of the preparation of adjusting their status to adjust the direction of movement to the seed module position, and by the moment, this modular robot is a continuous adjusts their status in order to achieve the best state in the fourth step. The fourth step is to receive the seed module, added to the matrix of modular robot, remove the module and the seed population. The fifth step, the end of the conditional, qualified to the end of the simulation, otherwise returns the logical judgment that the first step.

4.4   Motion Planning

We select Sliding movement on Molecubes modular robot. For the module program gives a continuous force and direction of the vector. With force and vector moment, the modular robot is gliding. Therefore, the movement planning of the experimental is to adjust their own state and movement in the simulation stage, including the two parts the modular robot state adjustments and modular robot. Which are mainly related to the modular robot vector, motion vectors (motion vectors of the mother movement to seed module) et al.

As mentioned above, the mother like straight line consists of three modules. From the mother head to tail constitute the vector of the modular robot, and mother to seed module constitutes a new direction vector. If mother want to complete the movement to eat module in accordance with the motion vectors, then mother move to the seed position. It is need for the adjustment of its own vector and motion vectors to select the mother to complete this action.

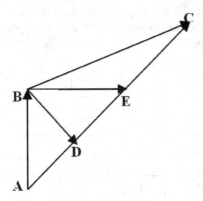

Fig. 6. Modular robot motion planning diagram, A is hear, B is tail. C represents the location of the target seed module, BC, BD, BE, respectively, three different sports planning

Simulation experiment, **AB** want to reach point C, there are three different movement strategies. The first one is the first rotation in accordance with the **BD** vector, then in accordance with the **DC** vector movement; The second is the first rotation in accordance with the BE vector, and then follow the vector movement; The third way is according to the **BC** way movement.

The first and second movement planning in principle is to the adjustment of the mother vector and exercise has better simulation results that compared to the third movement planning. But we tested by experiment in the first and second movement strategies, due to the particularity of the module slide by the force, resulting in adjust itself inaccurate, sports instability phenomenon, so that the mother take a long time in the adjustment planning. The another **DC** or **EC** stage run, due to the mother module design reasons also phenomena such as roll, and ultimately effect the same with the third planning. Considerations of integrated time-effective and simulation results, the experimental final choose the third movement planning to complete the modular robot self-replication behavior.

4.5   Molecubes Simulation Control

Already know from the previous introduction of the modular system design need to complete the design of modular robot, the issuance of the seed modules and modular robot motion simulation of these three functions. This platform enhancements, interface design, add the FEED RUN, STOP the three buttons; robot storage section, additional storage of the seed modules, modular robot system storage, the mother configuration storage; control section, add a search, maternal motion control, the receiver control and other control operations.

Modular robot movement simulation in Figure 7 shows, in accordance with the search, the mother movement control, receive control divided into five main parts: Logical judgment, Move preparation, Adjustment and exercise, Receiver module, End judgment. The following will introduce each part of the design ideas and their functions.

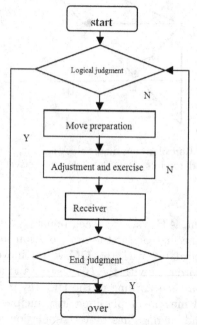

Fig. 7. Simulation flow chart

Logical judgment, modular robot simulation to do the logical judgment, first determine whether it has reached simulation requirements, followed to determine the system conditions, such as population and resources, whether the simulation requirements, if all meet the requirements to qualify, the next step, otherwise the end of simulation.

Movement preparation, modular robots first traversal itself to determine its own configuration, the module sequence, self-vector, the receiver module and other related information, followed through the population module to identify the modules to meet their own needs, and to develop sports planning, including the rotation vector their own motion vectors.

Adjustment and movement, according to various parameters of movement preparation phase, adjust their posture, movement and motion vectors. Module need to keep their own adjustments to achieve the best at this stage. This section also needs to determine the destination of the module movement to prepare for the next stage.

Receiver module, the module robot move to the destination location of the target seed, module simulation to receive the seed module and added to the mother, then removed the module and in the population to prepare for the next phase.

End of the judgment to determine the need to end the simulation, if it continues to simulation returns the first step, the next stage of the simulation logic judgment, otherwise end the simulation.

5   Results

According to the Molecubes modular robot design ideas, as well as self-replicating planning and movement planning, Program to achieve self-replicating. Figure 8 shows the modular robot simulation initialized from the simulation results of the replication experiment, that can be seen from Figure, the mother configuration initialization, scattering the seed modules; and then the mother robots through self-replicating planning, traverse the population, to find qualified seed vector and motion vectors, to behavior of self-replicating. After a successful replication, end the simulation experiments.

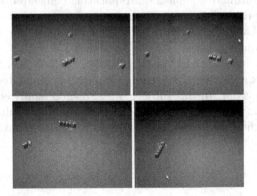

Fig. 8. Modular robot self-replication of simulation results

6   Conclusion

Modular robot based self-reconfigurable and self-replication has been a hot area of research of International Robot. Reconfigurable modular robot research is mainly carried out for the module, through a combination of a single module to replicate robot with the form and function or reconstruction the robot with different shapes, sizes and functions.

Through this experimental study, we have the following conclusions:

- Yamor robot under the existing functionality can be self-reconstruction, but cannot achieve self-replicating behavior. Molecubes simulation platform can provide an effective way to test a variety of physical combinations of possible features and a variety of physical characteristics, allowing the user to design a variety of intelligent modules for simulation. Molecubes modular robot has the ability to self-replication and self-reconstruction, and provides a variety of sports; it can be achieved through improvements to the platform, module self-replicating behavior.
- Sliding movement is selected for the features of Molecubes robot, the robot is composed by actuator module and control module. Revises the simulation platform based on demand. On this basis, with the control algorithm to control the robot motion of each module, so the modular robots can automatically search for the seed module and through the campaign planning to change their own status and state of motion, complete absorption of the function module, and the final copy to a new robot.

7 Outlook

In this study, we achieved self-replication behavior through molecube simulation platform, but like most of research situations at this stage just achieve self-replication as the ultimate goal, less self-replicating behavior optimization study. However, we find the robot configuration, movement, and strategies involving self-replicating behavior, explore the influencing factors will be great significance in the field of modular robots.

In addition, the modular robot self-replication behavior has a variety of replication. Victor Zykov in the paper using a replication simulated modular robot self-replication behavior. In this subject, we also successfully complete a modular robot self-replication. Research on Multiple individual replications, different replication, and efficiency comparison will also be a future research direction.

References

1. Fukuda, T., Nakagawa, S.: Dynamically reconfigurable robotics system. In: Proceedings of IEEE International Conference on Robotics and Automation, pp. 1581–1586 (1988)
2. Zykov, V., Mytilinaios, E., Adams, B., Lipson, H.: Self-reproducing machines. Nature 435(7038), 163–164
3. Mytilinaios, E., Marcus, D., Desnoyer, M., Lipson, H.: Designed and Evolved Blueprints For Physical Self-Replicating Machines. In: Ninth Int. Conference on Artificial Life (ALIFE IX), pp. 15–20 (2004)

4. Adams, B., Lipson, H.: A Universal Framework for Self-Replication. In: Banzhaf, W., Ziegler, J., Christaller, T., Dittrich, P., Kim, J.T. (eds.) ECAL 2003. LNCS (LNAI), vol. 2801, pp. 1–9. Springer, Heidelberg (2003)
5. LeClerc, V., Parkes, A., et al.: Senspectra: A Computationally Augmented Physical Modeling Toolkit for Sensing and Visualization of Structural Strain. In: CHI 2007. ACM Press, San Jose (2007)
6. Gro, R., Bonani, M., Mondada, F., Dorigo, M.: Autonomous Self-Assembly in Swarm-Bots. IEEE Transactions on Robotics 22(6), 1115–1128 (2006)
7. Tuci, E., Groß, R., Trianni, V., Bonani, M.: Cooperation through self-assembling in multi-robot systems. ACM Trans. Auton. Adapt. Syst. 1(2) (2006)
8. O'Grady, R., Groß, R., Mondada, F., Bonani, M., Dorigo, M.: Self-assembly on Demand in a Group of Physical Autonomous Mobile Robots Navigating Rough Terrain. In: Capcarrère, M.S., Freitas, A.A., Bentley, P.J., Johnson, C.G., Timmis, J. (eds.) ECAL 2005. LNCS (LNAI), vol. 3630, pp. 272–281. Springer, Heidelberg (2005)
9. Yim, M., Zhang, Y., Duff, D.: Modular robots. IEEE Spectrum 39(2), 30–34 (2002)
10. Whitesides, G.M., Grzybowski, B.: Self-assembly at all scales. Science 295(5564), 2418–2421 (2002)
11. Zykov, V., Mytilinaios, S., Desnoyer, M., Lipson, H.: Evolved and Designed Self-Reproducing Modular Robotics. IEEE Transactions on Robotics 23, 308–319 (2007)
12. Chen, A.: Modeling Molecubes with AGEIA PhysX (2007)
13. White, P.J., Kopanski, K., Lipson, H.: Stochastic Self-Reconfigurable Cellular Robotics. In: IEEE International Conference on Robotics and Automation (ICRA 2004), pp. 2888–2893 (2004)
14. Self-reconfiguring modular robot, http://en.wikipedia.org/wiki/Self-reconfiguring_modular_robot #Some_current_systems

Wavelet Application in Classification of Strata

Yueqin Dun[1], Yu Kong[2], and Wei Zhang[1]

[1] School of Electrical Engineering and Automation
Shandong, Jinan 250353, P.R. China
dunyq828@163.com,
13706418879@126.com
[2] Information Center of Shandong Medical College
Shandong, Jinan 250002, P.R. China
kongy@sdmc.net.cn

Abstract. Geophysical survey needs logging data to speculate stratum situation. The classification of strata based on logging curve is the first thing for the inversion problem. Electric logging curve is a stochastic time series. Because of the complex strata structure, the time series consist of real signal and white Gaussian noise. In order to get the real logging curves, the noise should be removed. In this paper, wavelet is used to filter the electric logging curves and the filtered signal properly reserves the curve edge and effectively wipes off the mutations of noise. The classification of strata based on the wavelet pre-treatment of electric logging curve becomes more ideal than the original curve.

Keywords: Classification of Strata, Wavelet, Gaussian White Noise, Threshold.

1 Introduction

To judge which layer of the earth is of petroleum or gas, a measurement sonde is put into a borehole and moved down to measure the physical characteristic of the soil, which is called logging. Electrical logging is an important logging method, which research the rock properties and distinguish them according to the difference of rock conductive ability. The resistivity curve with well depth can be obtained by the electrical logging, which is used to inverse calculate the real value of strata resistivity, and then the different lithology can be distinguished. The classification of strata according to logging curves is the first and basic step for the logging inversion. Usually this work is completed artificially, that is so-called artificial layered method. This method is time-consuming, and the empirical knowledge and proficiency of analyzer also influence the classification results. Different people often have different classification.

Establishing reasonable mathematic model and realizing the classification of strata by artificial intelligence processing is automatic stratification. Relative to the artificial stratification, the automatic stratification can avoid factitious optionally classification and improve the efficiency greatly. Due to the complexity

Z. Li et al. (Eds.): ISICA 2012, CCIS 316, pp. 490–497, 2012.

of the layers, the logging curves will be affected by a lot of interference, and sometime the obvious errors will appear in the automatic stratification. To get good results, the interference should be removed. The wavelet transform can be used to decrease disturbance. Some scholars used soft threshold to limit all detail coefficients to remove disturbance[1,2,3].

2   Introduction of the Wavelet Transform

2.1   Continuous Wavelet Transform

$$W_\psi f(a,b) = \frac{1}{\sqrt{|a|}} \int_{-\infty}^{+\infty} f(t)\psi_{a,b}^*(\frac{t-b}{a})dt \tag{1}$$

Where ψ denotes wavelet function, ψ^* is complex conjugate of ψ, $a \in R$ and $a \neq 0$ means scale factor, $b \in R$ denotes time shift of wavelet.

2.2   Dyadic Wavelet Transform

In discrete wavelet transform, a is dispersed according to the power series. Namely, let $a_j = 2^j$. b is dispersed evenly in scale. When j add 1, the scale doubled and sampling frequency reduced by half. In case of $a_j = 2^j$ and $b = 2^j kT$. Dyadic wavelet transform of continuous signal is defined as:

$$WT_f(j,k) = \frac{1}{\sqrt{2^j}} \int_{-\infty}^{+\infty} f(t)\psi^*(\frac{t}{2^j} - k \cdot T)dt \tag{2}$$

Where i is frequency scale, k is time scale, and T is constant which depends on mother wavelet. Dyadic wavelet transform with discrete time(of discrete signal) is defined as:

$$y_j[n] = \sum_{k=-\infty}^{\infty} f[k]h_j[2^j n - k] \tag{3}$$

Where h_j is impulse characteristic of filter which is identical to ψ_j^* for given j.

2.3   Fast Wavelet Transform

Fast wavelet transform (FWT) turns a signal in the time domain into a sequence of coefficients[4].

At the decomposition,

$$\begin{cases} a_{j+1}[k] = \sum_{n=-\infty}^{\infty} h[n - 2k]a_j[n] \\ d_{j+1}[k] = \sum_{n=-\infty}^{\infty} g[n - 2k]a_j[n] \end{cases} \tag{4}$$

At the reconstruction,

$$a_j[k] = \sum_{n=-\infty}^{\infty} h[k - 2n]a_{j+1}[n] + \sum_{n=-\infty}^{\infty} g[k - 2n]d_{j+1}[n] \tag{5}$$

Where $a_j[k]$ is j level approximation coefficients, $d_j[k]$ is j level detail coefficients. The filter h removes the higher frequencies and g collects the remaining highest frequencies.

3 Wavelet Filtering for Electric Logging Curves

The stratification of electric logging curves is according to the apparent difference between the different layers and the difference is small in the same layer. Even within a same layer, the noise often increases the diversity of the logging curves, which results in the increase of the number of automatic stratification. Therefore, it is necessary to pre-treat the logging curves before automatic stratification.

The electric logging curve is a stochastic time series

$$T = \{x(t) = r(t) + \sigma\varepsilon(t)|1 \le t \le n\} \tag{6}$$

The characteristics of wavelet transform for Gaussian white noise are as follows: 1) The expected value of wavelet transforms for noise is inversely proportional to the scale. 2) The average density of wavelet transform modulus maxima is inversely proportional to the scale[5]. So, the Gaussian white noise will gradually disappear with the scale increasing.

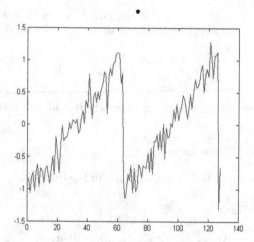

Fig. 1. Saw-tooth wave with 10db Gaussian white noise

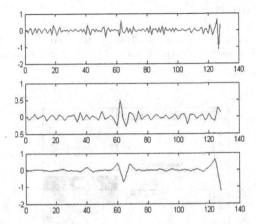

Fig. 2. Detail coefficients computed over 3 levels

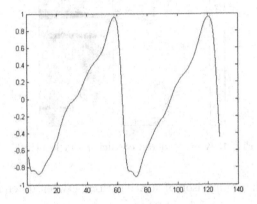

Fig. 3. The reconstruction signal

Fig. 1 was obtained by adding a Gaussian white noise to a saw-tooth wave and the SNR is 10db. Fig. 2 shows its FWT computed over 3 levels. The high-frequencies mainly focus on level1, 2. The reconstruction signal is in Fig. 3, in which the high-frequencies within level 1, 2 are taken out and the high-frequencies within level 3 are disposed by soft threshold.

According to the feature of logging curves, the process of pre-treatment is as follows:

1. Decomposing logging curves into 3 levels, the noise is usually included in detail coefficients.
2. Getting rid of the detail coefficients at level 1 and level 2 and limiting the detail coefficients at level 3 by soft threshold.
3. Reconstructing signal by approximation coefficients at levels 3 and detail ones produced in step 2.

4   Stratification Results

In this paper, the fisher's clustering method is used to the stratification for the logging curves with wavelet pre-treatment. The stratification results are shown from Fig. 4 to Fig. 7.

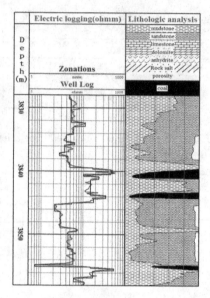

Fig. 4. Keeping detail coefficients at level 1,2,3

Table 1 gives the comparison of stratification between different pre-treatment signals.

Table 1. Comparison of stratification on different pre-treatment signal[a]

	Handling detail coefficients			Stratification Results
	level1	level 2	level 3	
Fig. 4[b]	Unchanged	Unchanged	Unchanged	Mutations layer appears at 3840
Fig. 5	Soft threshold	Soft threshold	Soft threshold	Mutations layer appears at 3841.
Fig. 6[c]	Removed	Removed	Removed	A zonation is omitted at 3850.
Fig. 7	Removed	Removed	Soft threshold	Stratification results and the lithologic analysis coincide well.

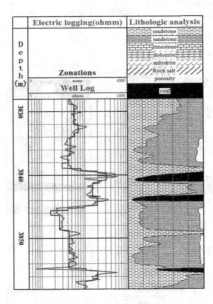

Fig. 5. Detail coefficients at level 1,2,3 disposed by soft threshold

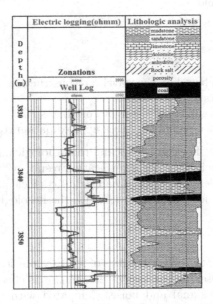

Fig. 6. Removing detail coefficients at level 1,2,3

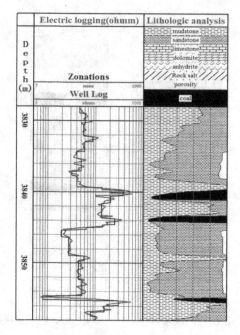

Fig. 7. Removing detail coefficients at level 1,2 and detail coefficients at level 1,2,3 disposed by soft threshold

Notes:

- a. Stratification is based on the wavelet reconstruction, in which the approximation coefficients stay the same.
- b. The reconstructed signal is same to original signal.
- c. The reconstructed signal only uses approximation coefficients at level 3.

By comparing the stratification results and the lithologic analysis from Fig. 4 to Fig. 7, we get the conclusion: The stratification result, reconstructing signal by approximation coefficients at levels 3 and detail coefficients at level 3 which is disposed by soft threshold, coincides well with the lithologic analysis.

5   Conclusion

Wavelet transform has the characteristics of multi-resolution analysis. Noise and its location can be observed from spectrum, thus it is effectively to distinguish the mutation part of signal and noise. Compared with the traditional filter method, wavelet filter is simple and effective. The method of disposing detail coefficients should change with different levels. It will get better reconstruct signal for the subsequent classification. After wavelet pretreatment of the electric logging signal, the classification results by using fisher's clustering method are more reasonable.

Acknowledgements. The work was supported by the Program for the Natural Science Foundation of Shandong Province (ZR2010EQ012) and Shandong Province Higher Educational Science and Technology program (J11LG30).

References

1. Zou, C., Yang, X., Pan, L., Zhu, J., Li, Y.: A new technique for denoising log curve on the basis of wavelet transform. J. Geophysical and Geochemical Exploration 23, 462–466 (1999)
2. Zhao, J.-J., Hu, W.-W., Gu, X.-G., Yang, P.: The Application in the Fault of High Voltage Electric Power Measurement System Based on Wavelet Analysis with the Improved Threshold Algorithm. In: 2011 3rd International Workshop on Intelligent Systems and Applications (ISA), pp. 342–345. IEEE Conference Publications, Kaifeng (2011)
3. Yang, H., Zhang, D., Huang, W., Gao, Z., Yang, X., Li, C., Wang, J.: Application and Evaluation of Wavelet-Based Denoising Method in Hyperspectral Imagery Data. In: Li, D., Chen, Y. (eds.) CCTA 2011, Part II. IFIP AICT, vol. 369, pp. 461–469. Springer, Heidelberg (2012)
4. Mallat, S.: A Wavelet Tour of Signal Processing, 3rd edn. Academic Press, Burlington (2008)
5. Mallat, S., Hwang, W.: Singularity detection and processing with wavelets. J. IEEE Trans. Inform. Theory 38, 617–643 (1992)

Duality Results of Nonlinear Symmetric Cone Programming*

Xiaoqin Jiang

Department of Public Basic
Wuhan Yangtze Business University,
Wuhan. 430065, China
xqjiang123@126.com

Abstract. There recently has been much interest in studying some optimization problems over symmetric cones. In this paper, we discuss the Lagrange dual theory of tonlinear symmetric cone programming, including the weak duality theorem, the strong duality theorem, and the saddle point theorem.

Keywords: Nonlinear symmetric cone programming, weak duality theorem, strong duality, saddle point theorem.

1 Introduction

Let X, Y, and Z be three finite dimensional vector spaces with inner products $\langle,\rangle_X$, $\langle,\rangle_Y$, and $\langle,\rangle_Z$, respectively, and $C \subset X$ and $D \subset Y$ be two symmetric cones. Recall that a cone C is called a symmetric cone if C is self-dual and its automorphism group acts transitively on its interior, i.e., for each $y, z \in \text{int}(C)$, the interior of C, there is a linear transformation $\mathcal{A}$ such that $\mathcal{A}(y) = z$ and $\mathcal{A}(C) = C$ (see an excellent monograph [1]). In this paper, we consider the following Nonlinear Symmetric Cone Programming (denoted by (NLSCP)):

$$\min f(x) \quad \text{s.t.} \quad x \succeq_C \mathbf{0}_X, \ g(x) \succeq_D \mathbf{0}_Y, \ h(x) = \mathbf{0}_Z, \tag{1}$$

where $f : X \to \Re$, $g : X \to Y$, $h : X \to Z$; $\mathbf{0}_X$, $\mathbf{0}_Y$, and $\mathbf{0}_Z$ are the zero elements of X, Y, and Z, respectively; and the order $\succeq_C$ is induced by cone C, i.e., for any $y, z \in X$, $y \succeq_C z$ means $y - z \in C$, and similarly, for any $y, z \in Y$, $y \succeq_D z$ means $y - z \in D$.

It is well known that the symmetric cones include, as special cases, the non-negative cone in $\Re^n$ ($\Re^n_+$), the second-order cone ($\mathcal{L}^n$), and the cone of positive semi-definite symmetric matrices ($\mathcal{S}^n_+$). So the NLSCP is a broad class of optimization problems, which contains the nonlinear programming, the nonlinear second-order cone programming, and the nonlinear semidefinite programming as special cases.

* This work was partially supported by the National Natural Science Foundation of China (Grant No. 10871144).

Z. Li et al. (Eds.): ISICA 2012, CCIS 316, pp. 498–502, 2012.
© Springer-Verlag Berlin Heidelberg 2012

Recently, there is much interest in studying some optimization problems over symmetric cones (see, for examples, [2–12]). Among them, one of the main issues is to investigate how to solve this kind of problems; and another main issue is to investigate some properties involving numerical methods for the problems. It is well-known that duality theory plays an important role in the development of the theory as well as solution methods for optimization problems. In this short note, we are interested in discussing some duality results of the NLSCP, including the weak duality theorem, the strong duality theorem, and the saddle point theorem.

2   Main Results

Let the Lagrange function L be defined by

$$L(x, u, v, w) = f(x) - \langle u, x \rangle_X - \langle v, g(x) \rangle_Y - \langle w, h(x) \rangle_Z, \quad \forall u \in X, \forall v \in Y, \forall w \in Z.$$

Denote

$$\mathcal{F}_p = \{x \in X : x \succeq_C \mathbf{0}_X, \; g(x) \succeq_D \mathbf{0}_Y, \; h(x) = \mathbf{0}_Z\};$$

$$\mathcal{F}_d = \left\{ (u, v, w) \in X \times Y \times Z : \begin{array}{l} \min_{x \in X} L(x, u, v, w) \text{ is solvable for} \\ u \succeq_C \mathbf{0}_X, \; v \succeq_D \mathbf{0}_Y, \text{ and } w \in Z \end{array} \right\}.$$

Then, we call the following programming as a dual programming problem of the NLSCP (1) (denoted by DNLSCP):

$$\max_{(u, v, w) \in \mathcal{F}_d} \mathcal{L}(u, v, w) \tag{2}$$

where the function $\mathcal{L}$ is defined by $\mathcal{L}(u, v, w) = \min_{x \in X} L(x, u, v, w)$ for any $(u, v, w) \in \mathcal{F}_d$.

Theorem 1. *(Weak Duality Theorem) Suppose that $\mathcal{F}_p \neq \emptyset$ and $\mathcal{F}_d \neq \emptyset$. Then, for any $x \in \mathcal{F}_p$ and any $(u, v, w) \in \mathcal{F}_d$, we have $\mathcal{L}(u, v, w) \leq f(x)$.*

Proof. For any $x \in \mathcal{F}_p$ and any $(u, v, w) \in \mathcal{F}_d$, by the definitions of $\mathcal{F}_p$ and $\mathcal{F}_d$ we have

$$x \succeq_C \mathbf{0}_X; \; g(x) \succeq_D \mathbf{0}_Y; \; h(x) = \mathbf{0}_Z; \; u \succeq_C \mathbf{0}_X; \; v \succeq_D \mathbf{0}_Y.$$

Since C and D are self-dual cones, the above inequalities imply that

$$\langle u, x \rangle_X \geq 0, \quad \langle v, g(x) \rangle_Y \geq 0, \quad \text{and} \quad \langle w, h(x) \rangle_X = 0, \quad \forall x \in \mathcal{F}_p, \; \forall (u, v, w) \in \mathcal{F}_d.$$

These, together with the definition of the function L, imply that

$$L(x, u, v, w) = f(x) - \langle u, x \rangle_X - \langle v, g(x) \rangle_Y - \langle w, h(x) \rangle_X$$
$$\leq f(x), \quad \forall x \in \mathcal{F}_p, \; \forall (u, v, w) \in \mathcal{F}_d.$$

Furthermore, by the definition of the function $\mathcal{L}$ we obtain that

$$\mathcal{L}(u, v, w) = \min_{x \in X} L(x, u, v, w) \leq L(x, u, v, w) \leq f(x), \quad \forall x \in \mathcal{F}_p, \; \forall (u, v, w) \in \mathcal{F}_d.$$

The proof is complete.   □

Theorem 2. *(Strong Duality Theorem) Suppose that $x^* \in \mathcal{F}_p$, $(u^*, v^*, w^*) \in \mathcal{F}_d$, and $f(x^*) = \mathcal{L}(u^*, v^*, w^*)$. Then, x^* and (u^*, v^*, w^*) are optimal solutions of (1) and (2), respectively.*

Proof. We first show that x^* is an optimal solution of the NLSCP (1). From the condition that $f(x^*) = \mathcal{L}(u^*, v^*, w^*)$ and the definition of the function $\mathcal{L}$ it follows that

$$
\begin{aligned}
f(x^*) &= \mathcal{L}(u^*, v^*, w^*) \\
&= \min_{x \in X}\{f(x) - \langle u^*, x \rangle_X - \langle v^*, g(x) \rangle_Y - \langle w^*, h(x) \rangle_Z\} \\
&\leq f(x) - \langle u^*, x \rangle_X - \langle v^*, g(x) \rangle_Y - \langle w^*, h(x) \rangle_Z, \qquad \forall x \in X. \qquad (3)
\end{aligned}
$$

For any $x \in \mathcal{F}_p$, by combining $x^* \in \mathcal{F}_p$ and $(u^*, v^*, w^*) \in \mathcal{F}_d$ with the fact that C and D are symmetric cones, it is easy to obtain that

$$
\langle u^*, x \rangle_X \geq 0, \quad \langle v^*, g(x) \rangle_Y \geq 0, \quad \text{and} \quad \langle w^*, h(x) \rangle_Z = 0.
$$

These and (3) imply that $f(x^*) \leq f(x)$, $\forall x \in \mathcal{F}_p$, which demonstrates that x^* is an optimal solution of the NLSCP (1).

Now, we show that (u^*, v^*, w^*) is an optimal solution of the DNLSCP (2). By Theorem 1 it follows that

$$
\mathcal{L}(u, v, w) \leq f(x), \qquad \forall x \in \mathcal{F}_p, \; \forall (u, v, w) \in \mathcal{F}_d,
$$

and hence $\mathcal{L}(u, v, w) \leq f(x^*)$ holds for any $(u, v, w) \in \mathcal{F}_d$ since $x^* \in \mathcal{F}_p$. This, together with the condition that $f(x^*) = \mathcal{L}(u^*, v^*, w^*)$, implies that

$$
\mathcal{L}(u, v, w) \leq \mathcal{L}(u^*, v^*, w^*), \qquad \forall (u, v, w) \in \mathcal{F}_d,
$$

i.e., (u^*, v^*, w^*) is an optimal solution of the DNLSCP (2).

The proof is complete. □

Suppose that $x^* \in \mathcal{F}_p$ and $(u^*, v^*, w^*) \in \mathcal{F}_d$. Then, (x^*, u^*, v^*, w^*) is called a saddle point of the function L if

$$
L(x^*, u, v, w) \leq L(x^*, u^*, v^*, w^*) \leq L(x, u^*, v^*, w^*) \qquad (4)
$$

holds for any (x, u, v, w) with $x \in X$ and $(u, v, w) \in \mathcal{F}_d$.

Theorem 3. *(Saddle Point Theorem) (x^*, u^*, v^*, w^*) with $x^* \in \mathcal{F}_p$ and $(u^*, v^*, w^*) \in \mathcal{F}_d$ is a saddle point of the function L if and only if*
(i) x^* is an optimal solution to the NLSCP (1);
(ii) (u^*, v^*, w^*) is an optimal solution to the DNLSCP (2); and
(iii) $f(x^*) = \mathcal{L}(u^*, v^*, w^*)$.

Proof. We first show that if the rsults (i)–(iii) hold, then (x^*, u^*, v^*, w^*) is a saddle point of the function L. From the results (i) and (ii), it follows that

$x^* \in \mathcal{F}_p$ and $(u^*, v^*, w^*) \in \mathcal{F}_d$. Thus, by (4) and the definition of the function L we only need to show that

$$f(x^*) - \langle u, x^* \rangle_X - \langle v, g(x^*) \rangle_Y - \langle w, h(x^*) \rangle_Z$$
$$\leq f(x^*) - \langle u^*, x^* \rangle_X - \langle v^*, g(x^*) \rangle_Y - \langle w^*, h(x^*) \rangle_Z \qquad (5)$$

is satisfied for any $(u, v, w) \in \mathcal{F}_d$; and that

$$f(x^*) - \langle u^*, x^* \rangle_X - \langle v^*, g(x^*) \rangle_Y - \langle w^*, h(x^*) \rangle_Z$$
$$\leq f(x) - \langle u^*, x \rangle_X - \langle v^*, g(x) \rangle_Y - \langle w^*, h(x) \rangle_Z \qquad (6)$$

is satisfied for any $x \in X$. Noticing that

$$f(x^*) = \mathcal{L}(u^*, v^*, w^*) = \min_{x \in X} L(x, u^*, v^*, w^*)$$
$$= \min_{x \in X} \{f(x) - \langle u^*, x \rangle_X - \langle v^*, g(x) \rangle_Y - \langle w^*, h(x) \rangle_Z\}$$
$$\leq f(x) - \langle u^*, x \rangle_X - \langle v^*, g(x) \rangle_Y - \langle w^*, h(x) \rangle_Z \qquad (7)$$

for all $x \in X$, where the first equality follows from the condition; and the second equality and the third equality from the definition of the function L.

By taking $x = x^*$, it follows from (7) that

$$0 \leq -\langle u^*, x^* \rangle_X - \langle v^*, g(x^*) \rangle_Y - \langle w^*, h(x^*) \rangle_Z.$$

However, it is easy to show that $\langle u^*, x^* \rangle_X \geq 0$, $\langle v^*, g(x^*) \rangle_Y \geq 0$, and $\langle w^*, h(x^*) \rangle_Z = 0$ since $x^* \in \mathcal{F}_p$, $(u^*, v^*, w^*) \in \mathcal{F}_d$, and C, D are symmetric cones. Thus,

$$\langle u^*, x^* \rangle_X = 0, \quad \langle v^*, g(x^*) \rangle_Y = 0, \quad \text{and} \quad \langle w^*, h(x^*) \rangle_Z = 0. \qquad (8)$$

By combining (7) with (8), we obtain that (6) holds for any $x \in X$. In addition, it is easy to show that $\langle u, x^* \rangle_X \geq 0$, $\langle v, g(x^*) \rangle_Y \geq 0$, and $\langle w, h(x^*) \rangle_Z = 0$ hold for any $(u, v, w) \in \mathcal{F}_d$. This and (8) imply that (5) holds for any $(u, v, w) \in \mathcal{F}_d$. Therefore, (x^*, u^*, v^*, w^*) is a saddle point of the function L, where $x^* \in \mathcal{F}_p$ and $(u^*, v^*, w^*) \in \mathcal{F}_d$.

Now, we show that if (x^*, u^*, v^*, w^*) is a saddle point of the function L where $x^* \in \mathcal{F}_p$ and $(u^*, v^*, w^*) \in \mathcal{F}_d$, then the results (i)–(iii) hold. In this case, (5) hold for any $(u, v, w) \in \mathcal{F}_d$ and (6) hold for any $x \in X$. Since $\langle w, h(x^*) \rangle_Z = 0$ for any $w \in Z$, by taking $u = u^*$ it is easy to obtain from (5) that $\langle v, g(x^*) \rangle_Y \geq \langle v^*, g(x^*) \rangle_Y$ for any $(u^*, v, w) \in \mathcal{F}_p$, and hence $\langle v^*, g(x^*) \rangle_Y \leq 0$ by taking $v = \frac{1}{2}v^*$. However, it is easy to show that $\langle v^*, g(x^*) \rangle_Y \geq 0$. Thus, $\langle v^*, g(x^*) \rangle_Y = 0$. Similarly, we can obtain that $\langle u^*, x^* \rangle_X = 0$. That is, we obtain that

$$\langle u^*, x^* \rangle_X = 0, \quad \langle v^*, g(x^*) \rangle_Y = 0, \quad \text{and} \quad \langle w^*, h(x^*) \rangle_Z = 0. \qquad (9)$$

In the following, we show that the results (i)–(iii) hold. By (6) we know that x^* is an optimal solution of the problem $\min_{x \in X} L(x, u^*, v^*)$, and hence $\mathcal{L}(u^*, v^*, w^*) = L(x^*, u^*, v^*, w^*)$. Thus, by the definition of the function $\mathcal{L}$ and (9) we obtain that

$$\mathcal{L}(u^*, v^*, w^*) = L(x^*, u^*, v^*, w^*)$$
$$= f(x^*) - \langle u^*, x^* \rangle_X - \langle v^*, g(x^*) \rangle_Y - \langle w^*, h(x^*) \rangle_Z$$
$$= f(x^*),$$

which demonstrates that the result (iii) holds. By combining this result with $x^* \in \mathcal{F}_p$, $(u^*, v^*, w^*) \in \mathcal{F}_d$ and Theorem 2, we obtain that the results (i) and (ii) hold.

The proof is complete. □

3   Conclusions

In this paper, we have considered the nonlinear symmetric cone programming and its Lagrange-type dual problem, and established the weak duality theorem, the strong duality theorem, and the saddle point theorem of the problems concerned. Since the nonlinear symmetric cone programming is a broad class of optimization problems, which contains the nonlinear programming, the nonlinear second-order cone programming, and the nonlinear semidefinite programming as special cases, it deserves further study the duality results for other-type dual problems.

References

1. Faraut, J., Korányi, A.: Analysis on Symmetric Cones. Oxford Mathematical Monographs. Oxford University Press, New York (1994)
2. Faybusovich, L.: Euclidean Jordan algebras and interior-point algorithms. Positivity 1, 331–357 (1997)
3. Gowda, M.S., Sznajder, R., Tao, J.: Some P-properties for linear transformations on Euclidean Jordan algebras. Linear Alg. Appl. 393, 203–232 (2004)
4. Huang, Z.H., Hu, S.L., Han, J.: Global convergence of a smoothing algorithm for symmetric cone complementarity problems with a nonmonotone line search. Sci. China, Ser. A 52, 833–848 (2009)
5. Huang, Z.H., Liu, X.H.: Extension of smoothing Newton algorithms to solve linear programming over symmetric cones. J. Syst. Sci. Complex 24, 195–206 (2011)
6. Huang, Z.H., Lu, N.: Global and global linear convergence of a smoothing algorithm for the Cartesian $P_*(\kappa)$-SCLCP. J. Ind. Manag. Optim. 8, 67–86 (2012)
7. Huang, Z.H., Ni, T.: Smoothing algorithms for complementarity problems over symmetric cones. Comput. Optim. Appl. 45, 557–579 (2010)
8. Kong, L.C., Sun, J., Xiu, N.H.: A regularized smoothing Newton method for symmetric cone complementarity problems. SIAM J. Optim. 19, 1028–1047 (2008)
9. Liu, X.H., Huang, Z.H.: A smoothing Newton algorithm based on a one-parametric class of smoothing functions for linear programming over symmetric cones. Math. Meth. Oper. Res. 70, 385–404 (2009)
10. Schmieta, S.H., Alizadeh, F.: Extension of primal-dual interior-point algorithms to symmetric cones. Math. Program. 96, 409–438 (2003)
11. Sun, D., Sun, J.: Löwner's operator and spectral functions in Euclidean Joedan algebras. Math. Oper. Res. 33, 421–445 (2008)
12. Yoshise, A.: Interior point trajectories and a homogeneous model for nonlinear complementarity problem over symmetric cones. SIAM J. Optim. 17, 1129–1153 (2006)

The Linear Convergence of a Merit Function Method for Nonlinear Complementarity Problems

Xiaoqin Jiang[1] and Liyong Lu[2,*]

[1] Department of Public Basic
Wuhan Yangtze Business University,
Wuhan. 430065, China
xqjiang123@126.com
[2] Department of Mathematics
Tianjin University of Technology,
Tianjin. 300191, China
mathluliyong@yahoo.cn

Abstract. Based on a family of generalized merit functions, a merit function method for solving nonlinear complementarity problems was proposed by Lu, Huang and Hu [*Properties of a family of merit functions and a merit function method for the NCP, Appl. Math.- J. Chinese Univ., 2010, 25: 379–390*], where, the global convergence of the method was proved. However, no the result on the convergence rate of the method was reported. In this short paper, we show that the method proposed in the above paper is globally linearly convergent under suitable assumptions.

Keywords: Complementarity problems, merit function, derivative-free methods, linear convergence.

1 Introduction

The nonlinear complementarity problem (NCP for short) is to find a vector $(x, s) \in \Re^{2n}$ such that

$$(x, s) \geq 0, \quad s = F(x) \geq 0, \quad \text{and} \quad \langle x, s \rangle = 0,$$

where $F : \Re^n \to \Re^n$ is a given vector function and $\langle \cdot, \cdot \rangle$ denotes the usual inner product in $\Re^n$. The NCP has been studied extensively due to its various applications in many fields, such as mathematical programming, economics, engineering and mechanics (see, for example, [4, 5, 9, 10]). We refer the interested readers to see the excellent monographes by Facchinei and Pang [4] and by Han, Xiu and Qi [9]. Various methods for solving the NCP have been proposed in the literature (see, for example, [1–4, 6–11, 15, 16]). Among them, one of the most

---

* Corresponding author.

Z. Li et al. (Eds.): ISICA 2012, CCIS 316, pp. 503–511, 2012.

popular and powerful approaches is to reformulate the NCP as an unconstrained minimization problem (see, for example, [1–3, 6–8, 11, 15, 16]).

Recall that a function $\phi : \Re^2 \to \Re$ is called an NCP-function, if for any $a, b \in \Re$, $\phi(a, b) = 0$ if and only if $a \geq 0$, $b \geq 0$, and $ab = 0$. If $\phi(a, b) \geq 0$ then the NCP-function ϕ is called a nonnegative NCP-function. In addition, if a function $\Psi : \Re^n \to \Re$ is nonnegative and $\Psi(x) = 0$ if and only if x solves the NCP, then Ψ is called a merit function for the NCP. If the NCP-function ϕ is nonnegative on $\Re^2$, then it is easy to see that the function $\Psi : \Re^n \to \Re$ defined by $\Psi(x) = \sum_{i=1}^{n} \phi(x_i, F_i(x))$ is a merit function for the NCP. Thus, finding a solution of the NCP is equivalent to finding a global minimum of the unconstrained minimization $\min_{x \in \Re^n} \Psi(x)$ with the optimal objective value zero. Thus, one may find a solution of the NCP by solving the corresponding unconstrained minimization problem. This kind of methods is called the merit function method.

Recently, Lu, Huang, and Hu [15] proposed the following merit function:

$$\Psi_{\alpha\beta}(x) = \sum_{i=1}^{n} \phi_{\alpha\beta}(x_i, F_i(x)), \quad \forall x \in \Re^n, \tag{1}$$

where $\phi_{\alpha\beta} : \Re^2 \to \Re$ is defined by $\forall (a, b)^T \in \Re^2$

$$\phi_{\alpha\beta}(a, b) = \frac{\alpha}{2}(ab)_+^2 + \frac{1}{2}\left(a + b - \sqrt{(1 - \beta)(a^2 + b^2) + \beta(a - b)^2}\right)^2, \tag{2}$$

with $\alpha \geq 0$ and $1 \geq \beta \geq 0$. Thus, finding a solution of the NCP is equivalent to finding a global minimum of the unconstrained minimization $\min_{x \in \Re^n} \Psi_{\alpha\beta}(x)$ with the optimal objective value zero. It is easy to see that if $\alpha = 0$ and $\beta = 1$, then the function $\Psi_{\alpha\beta}(\cdot)$ defined by (1) reduces to the natural residual merit function $\Psi_{NR} : \Re^n \to \Re$ given by

$$\Psi_{NR}(x) = \sum_{i=1}^{n} \phi_{NR}(x_i, F_i(x)) \tag{3}$$

where $\phi_{NR} : \Re^2 \to \Re$ is a nonnegative NCP-function given by $\phi_{NR}(a, b) = 2(\min\{a, b\})^2$.

Based on the function defined by (1), a merit function method for the NCP was proposed in [15]. The investigated method is a derivative-free descent method. The derivative-free method has attracted much attention in the literature since it does not require computation of derivatives of F (see, for example, [1–3, 6–8, 11, 15–17]). This kind of methods is particularly suitable for problems where the derivatives of F are not available or are extremely expensive to compute. Based on the analysis on properties of the function defined by (1), the authors in [15] showed that the proposed method is globally convergent under suitable assumptions. Some promising numerical results were also reported. However, no the result on the convergence rate of the method was reported in [15]. In this short paper, we further investigate the method proposed in [15]. We will show that the method proposed in [15] is globally linearly convergent under suitable assumptions.

The rest of this paper is organized as follows. In Section 2, we recall some basic concepts, the investigated algorithm and the related results. In Section 3, we establish the linear rate of convergence for the method proposed in [15]. Some final remarks are given in Section 4.

Throughout this paper, $\Re^n$ denotes the space of n-dimensional real column vectors and A^T denotes the transpose of the real-valued matrix A. For any differentiable function $F : \Re^n \to \Re$, $\nabla F(x)$ denotes the gradient of F at x. For any differentiable mapping $F = (F_1, \ldots, F_m)^T : \Re^n \to \Re^m$, $\nabla F(x) = [\nabla F_1(x), \cdots, \nabla F_m(x)]^T$ denotes the Jacobian of F at x. The norm $\| \cdot \|$ denotes the Euclidean norm. $\lceil z \rceil$ denotes the smallest integer no less than z, where $z \in \Re$. The level set of a function $\Psi : \Re^n \to \Re$ is denoted by $\mathcal{L}(\Psi, \gamma) = \{x \in \Re^n \mid \Psi(x) \le \gamma\}$, where γ is a given real number.

2 Preliminaries

We first recall some basic concepts [4], which will be used in the sequent analysis.

Definition 1. *Let the function $F : \Re^n \to \Re^n$, we say that F is monotone if $\langle x-y, f(x)-F(y)\rangle \ge 0$ for all $x, y \in \Re^n$; F is strongly monotone with a modulus $\lambda > 0$ if $\langle x - y, F(x) - F(y)\rangle \ge \lambda\|x - y\|^2$ for all $x, y \in \Re^n$; or equivalently, $\langle \nabla F(x)y, y\rangle \ge \lambda\|y\|^2$ for all $y \in \Re^n$ if F is continuously differentiable; F is a uniform P-function with modulus $\kappa > 0$ if $\max_{\substack{1 \le i \le n \\ x_i \ne y_i}} (x_i - y_i)(F_i(x) - F_i(y)) \ge \kappa\|x - y\|^2$ for all $x, y \in \Re^n$; F is Lipschitz continuous with a Lipschitz constant $L > 0$ if $\|F(x) - F(y)\| \le L\|x - y\|$ for all $x, y \in \Re^n$.*

It is well-known that every strongly monotone function is a uniform P-function. The following merit function method was proposed in [15].

Algorithm 1. *Give $\alpha, \varepsilon \in [0, +\infty), \beta, \gamma, \eta, \sigma \in (0, 1), x^0 \in \Re^n$. Set $k := 0$.*

Step 1. *If $\Psi_{\alpha\beta}(x^k) \le \varepsilon$, then stop.*

Step 2. *Let $d^k(\eta^{t_k}) := -\frac{\partial \phi_{\alpha\beta}}{\partial b}(x^k, F(x^k)) - \eta^{t_k}\frac{\partial \phi_{\alpha\beta}}{\partial a}(x^k, F(x^k))$.*

Step 3. *Set $x^{k+1} := x^k + \gamma^{t_k}d^k(\eta^{t_k})$, where t_k is the smallest nonnegative integer t satisfying*

$$\Psi_{\alpha\beta}(x^k + \gamma^t d^k(\eta^t)) - \Psi_{\alpha\beta}(x^k) \le -\sigma\gamma^{2t}\left\|\frac{\partial \phi_{\alpha\beta}}{\partial b}(x^k, F(x^k)) + \frac{\partial \phi_{\alpha\beta}}{\partial a}(x^k, F(x^k))\right\|^2.$$

Step 4. *Set $k := k + 1$ and go to Step 1.*

After discussed several important properties related the function defined by (1), the authors in [15] obtained the following global convergence result of Algorithm 1.

Theorem 1. *(See [15, Theorem 2.3]) If one of the following conditions holds:*

(i) $\alpha \in (0, +\infty), \beta \in [0, 1]$, F is monotone and the NCP is strictly feasible (i.e., there exists an $x^0 \in \Re^n$ such that $x^0 > 0$ and $y = F(x^0) > 0$).

(ii) $\alpha \in [0, +\infty), \beta \in [0, 1)$, F is strongly monotone.

Then Algorithm 1 is well-defined. Furthermore, the sequence $\{x^k\}$ generated by Algorithm 1 has at least one accumulation point and every accumulation point of $\{x^k\}$ is a solution of the NCP.

3 Main Result

The following concepts will be used in our analysis later.

Definition 2. *Suppose that the sequence $\{x^k\}$ converges to x^*. $\{x^k\}$ is said to be globally Q-linearly convergent, if there is a constant $\hat{\beta} \in (0,1)$ which is independent of the iterative index k, such that $\|x^{k+1} - x^*\| \leq \hat{\beta}\|x^k - x^*\|$ holds for all k. $\{x^k\}$ is said to be globally R-linearly convergent, if there is a sequence of nonnegative sequence of scalars, $\{q^k\}$, such that $\|x^k - x^*\| \leq q^k$ for all k, and $\{q^k\}$ converges to zero globally Q-linearly.*

In this section, we show that Algorithm 1 is globally linearly convergent under suitable assumptions. For this purpose, we need the following two lemmas.

Lemma 1. *Let $\phi_{\alpha\beta}$, $\Psi_{\alpha\beta}$, and Ψ_{NR} be defined by (2), (1), and (3), respectively; and $\alpha \geq 0$. Then, the following results hold.*

(i) *If $1 > \beta \geq 0$, then $\frac{\partial \phi_{\alpha\beta}}{\partial a}(a,b)\frac{\partial \phi_{\alpha\beta}}{\partial b}(a,b) \geq 0$ for all $(a,b)^T \in \Re^2$.*

(ii) *If $1 \geq \beta \geq 0$ and F is a uniform P-function, then there exists a $\mu > 0$ such that $\|x - x^*\| \leq \mu \Psi_{\alpha\beta}^{\frac{1}{4}}(x)$ holds for all $x \in \Re^n$, where x^* is the unique solution of the NCP.*

(iii) *If $1 \geq \beta \geq 0$ and F is strongly monotone, then $\mathcal{L}(\Psi_{\alpha\beta}, \gamma)$ is compact for any $\gamma \in \Re$.*

(iv) *If $1 \geq \beta \geq 0$, then for any bounded set $\mathcal{L}(\Psi, \Psi(x_0))$,*

$$c_1 \Psi_{NR}(x) \leq \Psi_{\alpha\beta}(x) \leq c_2 \Psi_{NR}(x),$$

where $c_1 = \frac{(2-\sqrt{2})^2}{2} < 1$, $c_2 = \frac{\alpha}{2}\sup_{x \in S}\{\max_{1 \leq i \leq n}\{\max|x_i|, |F_i(x)|\}\} + \frac{1}{2}(2 + \sqrt{2})^2 \geq 4$, and $x \in \mathcal{L}(\Psi, \Psi(x_0))$.

(v) *If $1 > \beta \geq 0$, then*

$$\left\|\frac{\partial \phi_{\alpha\beta}}{\partial a}(a,b) + \frac{\partial \phi_{\alpha\beta}}{\partial b}(a,b)\right\|^2 \geq \frac{1}{2}((2 - \sqrt{2})(2 - \sqrt{2(1-\beta)}))^2 \phi_{NR}(a,b)$$

for all $(a,b) \in \Re^2$.

Proof. The results (i) and (ii) of Lemma 1 were taken from [15, Lemma 2.2] and [15, Theorem 2.2]; and results (iii), (iv) and (v) of Lemma 1 can be easily proved from [15, Theorem 2.3], [15, Lemma 2.3] and [15, Lemma 3.1], respectively. We omit them here. □

Lemma 2. *Let $\alpha \geq 0, 1 > \beta \geq 0$ and sequences $\{x^k\}$ and $\{d^k\}$ be generated by Algorithm 1. Suppose that F is continuously differentiable and strongly monotone with modulus $\lambda > 0$. Then*

$$\|d^k\|^2 \leq \left\|\frac{\partial \phi_{\alpha\beta}}{\partial b}(x^k, F(x^k)) + \frac{\partial \phi_{\alpha\beta}}{\partial a}(x^k, F(x^k))\right\|^2; \tag{4}$$

and there exists a constant $\varrho > 0$ such that

$$\langle \nabla \Psi_{\alpha\beta}(x^k), d^k \rangle \leq -\varrho \left\|\frac{\partial \phi_{\alpha\beta}}{\partial b}(x^k, F(x^k)) + \frac{\partial \phi_{\alpha\beta}}{\partial a}(x^k, F(x^k))\right\|^2. \tag{5}$$

Proof. Since $\eta^{t_k} \in (0,1)$, and $\left\langle \frac{\partial \phi_{\alpha\beta}}{\partial a}(x^k, F(x^k)), \frac{\partial \phi_{\alpha\beta}}{\partial b}(x^k, F(x^k)) \right\rangle \geq 0$ by Algorithm 1 and Lemma 1(i), we have

$$
\begin{aligned}
\|d^k\|^2 &= \left\| -\frac{\partial \phi_{\alpha\beta}}{\partial b}(x^k, F(x^k)) - \eta^{t_k} \frac{\partial \phi_{\alpha\beta}}{\partial a}(x^k, F(x^k)) \right\|^2 \\
&= \left\| \frac{\partial \phi_{\alpha\beta}}{\partial b}(x^k, F(x^k)) \right\|^2 + 2\eta^{t_k} \left\langle \frac{\partial \phi_{\alpha\beta}}{\partial a}(x^k, F(x^k)), \frac{\partial \phi_{\alpha\beta}}{\partial b}(x^k, F(x^k)) \right\rangle \\
&\quad + (\eta^{t_k})^2 \left\| \frac{\partial \phi_{\alpha\beta}}{\partial a}(x^k, F(x^k)) \right\|^2 \\
&\leq \left\| \frac{\partial \phi_{\alpha\beta}}{\partial b}(x^k, F(x^k)) \right\|^2 + 2 \left\langle \frac{\partial \phi_{\alpha\beta}}{\partial a}(x^k, F(x^k)), \frac{\partial \phi_{\alpha\beta}}{\partial b}(x^k, F(x^k)) \right\rangle \\
&\quad + \left\| \frac{\partial \phi_{\alpha\beta}}{\partial a}(x^k, F(x^k)) \right\|^2 \\
&= \left\| \frac{\partial \phi_{\alpha\beta}}{\partial b}(x^k, F(x^k)) + \frac{\partial \phi_{\alpha\beta}}{\partial a}(x^k, F(x^k)) \right\|^2,
\end{aligned}
$$

i.e., (4) holds. In the following, we show that (5) holds. It follows that

$$
\begin{aligned}
\langle \nabla \Psi_{\alpha\beta}(x^k), d^k \rangle &= \left\langle \frac{\partial \phi_{\alpha\beta}}{\partial a}(x^k, F(x^k)) + \nabla F(x^k) \frac{\partial \phi_{\alpha\beta}}{\partial b}(x^k, F(x^k)), \right. \\
&\qquad \left. -\frac{\partial \phi_{\alpha\beta}}{\partial b}(x^k, F(x^k)) - \eta^{t_k} \frac{\partial \phi_{\alpha\beta}}{\partial a}(x^k, F(x^k)) \right\rangle \\
&= -\left\langle \frac{\partial \phi_{\alpha\beta}}{\partial a}(x^k, F(x^k)), \frac{\partial \phi_{\alpha\beta}}{\partial b}(x^k, F(x^k)) \right\rangle \\
&\quad -\eta^{t_k} \left\langle \frac{\partial \phi_{\alpha\beta}}{\partial a}(x^k, F(x^k)), \frac{\partial \phi_{\alpha\beta}}{\partial a}(x^k, F(x^k)) \right\rangle \\
&\quad -\left\langle \nabla F(x^k) \frac{\partial \phi_{\alpha\beta}}{\partial b}(x^k, F(x^k)), \frac{\partial \phi_{\alpha\beta}}{\partial b}(x^k, F(x^k)) \right\rangle \\
&\quad -\eta^{t_k} \left\langle \nabla F(x^k) \frac{\partial \phi_{\alpha\beta}}{\partial b}(x^k, F(x^k)), \frac{\partial \phi_{\alpha\beta}}{\partial a}(x^k, F(x^k)) \right\rangle \\
&\leq -\eta^{t_k} \left\langle \frac{\partial \phi_{\alpha\beta}}{\partial a}(x^k, F(x^k)), \frac{\partial \phi_{\alpha\beta}}{\partial a}(x^k, F(x^k)) \right\rangle \\
&\quad -\left\langle \nabla F(x^k) \frac{\partial \phi_{\alpha\beta}}{\partial b}(x^k, F(x^k)), \frac{\partial \phi_{\alpha\beta}}{\partial b}(x^k, F(x^k)) \right\rangle \\
&\quad -\eta^{t_k} \left\langle \nabla F(x^k) \frac{\partial \phi_{\alpha\beta}}{\partial b}(x^k, F(x^k)), \frac{\partial \phi_{\alpha\beta}}{\partial a}(x^k, F(x^k)) \right\rangle,
\end{aligned}
$$

where the inequality follows from $\left\langle \frac{\partial \phi_{\alpha\beta}}{\partial a}(x^k, F(x^k)), \frac{\partial \phi_{\alpha\beta}}{\partial b}(x^k, F(x^k)) \right\rangle \geq 0$ by Lemma 1(i). Since the level set $\mathcal{L}(\Psi, \Psi(x^0))$ is bounded and closed by Lemma 1(iii) and the function F is continuously differentiable, it follows that there exists a constant $\rho > 0$ such that $\|\nabla F(x)\| \leq \rho$ for any $x \in \mathcal{L}(\Psi, \Psi(x^0))$. So, by the strong monotonicity of F and the Cauchy-Schwarz inequality, we further obtain that

$$\langle \nabla \Psi_{\alpha\beta}(x^k), d^k \rangle \leq -\eta^{t_k} \left\| \frac{\partial \phi_{\alpha\beta}}{\partial a}(x^k, F(x^k)) \right\|^2 - \lambda \left\| \frac{\partial \phi_{\alpha\beta}}{\partial b}(x^k, F(x^k)) \right\|^2$$
$$+ \eta^{t_k} \rho \left\| \frac{\partial \phi_{\alpha\beta}}{\partial a}(x^k, F(x^k)) \right\| \left\| \frac{\partial \phi_{\alpha\beta}}{\partial b}(x^k, F(x^k)) \right\|. \tag{6}$$

In the following, we will show that for a suitable η^{t_k} by choice in Step 3 of Algorithm 1, there exists a constant $\varrho > 0$ such that the following inequality holds:

$$-\eta^{t_k} \left\| \frac{\partial \phi_{\alpha\beta}}{\partial a}(x^k, F(x^k)) \right\|^2 - \lambda \left\| \frac{\partial \phi_{\alpha\beta}}{\partial b}(x^k, F(x^k)) \right\|^2$$
$$+ \eta^{t_k} \rho \left\| \frac{\partial \phi_{\alpha\beta}}{\partial a}(x^k, F(x^k)) \right\| \left\| \frac{\partial \phi_{\alpha\beta}}{\partial b}(x^k, F(x^k)) \right\|$$
$$\leq -\varrho \left\| \frac{\partial \phi_{\alpha\beta}}{\partial b}(x^k, F(x^k)) + \frac{\partial \phi_{\alpha\beta}}{\partial a}(x^k, F(x^k)) \right\|^2. \tag{7}$$

Since

$$\left\| \frac{\partial \phi_{\alpha\beta}}{\partial b}(x^k, F(x^k)) + \frac{\partial \phi_{\alpha\beta}}{\partial a}(x^k, F(x^k)) \right\|^2$$
$$\leq \left\| \frac{\partial \phi_{\alpha\beta}}{\partial a}(x^k, F(x^k)) \right\|^2 + \left\| \frac{\partial \phi_{\alpha\beta}}{\partial b}(x^k, F(x^k)) \right\|^2$$
$$+ 2 \left\| \frac{\partial \phi_{\alpha\beta}}{\partial a}(x^k, F(x^k)) \right\| \left\| \frac{\partial \phi_{\alpha\beta}}{\partial b}(x^k, F(x^k)) \right\|,$$

it is sufficient to show that

$$(\eta^{t_k} - \varrho) \left\| \frac{\partial \phi_{\alpha\beta}}{\partial a}(x^k, F(x^k)) \right\|^2 + (\lambda - \varrho) \left\| \frac{\partial \phi_{\alpha\beta}}{\partial b}(x^k, F(x^k)) \right\|^2$$
$$- (\eta^{t_k} \rho + 2\varrho) \left\| \frac{\partial \phi_{\alpha\beta}}{\partial a}(x^k, F(x^k)) \right\| \left\| \frac{\partial \phi_{\alpha\beta}}{\partial b}(x^k, F(x^k)) \right\| \geq 0. \tag{8}$$

It is easy to see that (8) holds if and only if $\eta^{t_k} - \varrho \geq 0$, $\lambda - \varrho \geq 0$, and $\triangle = (\eta^{t_k} \rho + 2\varrho)^2 - 4(\eta^{t_k} - \varrho)(\lambda - \varrho) \leq 0$. Through a direct calculation, it is easy to show that (8) holds if and only if $\varrho \leq \eta^{t_k}$, $\varrho \leq \lambda$, and $\varrho \leq \frac{\eta^{t_k}\lambda - (\eta^{t_k}\rho/2)^2}{\eta^{t_k}(\rho+1)+\lambda}$. Hence, if we choice

$$\varrho \leq \min \left\{ \eta^{t_k}, \lambda, \frac{\eta^{t_k}\lambda - (\eta^{t_k}\rho/2)^2}{\eta^{t_k}(\rho+1)+\lambda} \right\},$$

then the inequality (7) holds. Combining (6) and (7), we obtain that (5) holds. $\square$

Now, we show the main result in this paper.

Theorem 2. *Suppose that all conditions of Lemma 2 are met. Let F and ∇F be Lipschitz continuous on $\Re^n$. Then the sequence $\{\Psi_{\alpha\beta}(x^k)\}$ converges to zero globally Q-linearly, and the sequence $\{x^k\}$ converges to a solution of the NCP globally R-linearly.*

Proof. Since F is strongly monotone, it follows from Theorem 1 that the sequence $\{x^k\}$ converges to the unique solution of the NCP, denoted by x^*. In the following, we show the global linear convergence of Algorithm 1. We use k to denote an arbitrary iterative index in the following analysis.

It is easy to see that $x^k, x^k + \gamma^{t_k} d^k(\eta^{t_k}) \in \mathcal{L}(\Psi, \Psi(x^0))$, where x^0 is the starting point. Since F and ∇F are Lipschitz continuous on $\Re^n$, it follows that $\nabla \Psi_{\alpha\beta}(x)$ is Lipschitz continuous on $\Re^n$ with some Lipschitz constant $\mathcal{C} > 0$, i.e., $\|\nabla \Psi_{\alpha\beta}(x) - \nabla \Psi_{\alpha\beta}(y)\| \le \mathcal{C}\|x - y\|$ holds for any $x, y \in \Re^n$. Thus, we obtain that

$$
\begin{aligned}
\Psi_{\alpha\beta}(x^k) - \Psi_{\alpha\beta}(x^k + \gamma^{t_k} d^k) &= -\int_0^{\gamma^{t_k}} \langle \nabla\Psi_{\alpha\beta}(x^k + \mu d^k), d^k \rangle \, d\mu \\
&= -\int_0^{\gamma^{t_k}} \langle \nabla\Psi_{\alpha\beta}(x^k + \mu d^k) - \nabla\Psi_{\alpha\beta}(x^k), d^k \rangle \, d\mu \\
&\quad -\gamma^{t_k} \langle \nabla\Psi_{\alpha\beta}(x^k), d^k \rangle \\
&\ge -\gamma^{t_k} \langle \nabla\Psi_{\alpha\beta}(x^k), d^k \rangle - \int_0^{\gamma^{t_k}} \langle \mathcal{C}\mu d^k, d^k \rangle \, d\mu \\
&= -\gamma^{t_k} \langle \nabla\Psi_{\alpha\beta}(x^k), d^k \rangle - \mathcal{C} \int_0^{\gamma^{t_k}} \mu \|d^k\|^2 d\mu \\
&= -\gamma^{t_k} \langle \nabla\Psi_{\alpha\beta}(x^k), d^k \rangle - \frac{\mathcal{C}\gamma^{2t_k}}{2} \|d^k\|^2.
\end{aligned}
$$

This, together with (5) and (4), implies that

$$
\begin{aligned}
\Psi_{\alpha\beta}(x^k) &- \Psi_{\alpha\beta}(x^k + \gamma^{t_k} d^k) \\
&\ge -\gamma^{t_k} \langle \nabla\Psi_{\alpha\beta}(x^k), d^k \rangle - \frac{\mathcal{C}\gamma^{2t_k}}{2}\|d^k\|^2 \\
&\ge \gamma^{t_k}\varrho \left\| \frac{\partial\phi_{\alpha\beta}}{\partial b}(x^k, F(x^k)) + \frac{\partial\phi_{\alpha\beta}}{\partial a}(x^k, F(x^k)) \right\|^2 \\
&\quad - \frac{\mathcal{C}\gamma^{2t_k}}{2} \left\| \frac{\partial\phi_{\alpha\beta}}{\partial b}(x^k, F(x^k)) + \frac{\partial\phi_{\alpha\beta}}{\partial a}(x^k, F(x^k)) \right\|^2 \\
&\quad \gamma^{t_k}(\varrho - \frac{\mathcal{C}\gamma^{t_k}}{2}) \left\| \frac{\partial\phi_{\alpha\beta}}{\partial b}(x^k, F(x^k)) + \frac{\partial\phi_{\alpha\beta}}{\partial a}(x^k, F(x^k)) \right\|^2,
\end{aligned}
$$

which imply that the inequality in Step 3 of Algorithm 1 is satisfied whenever $\gamma^{t_k}(\varrho - \frac{\mathcal{C}\gamma^{t_k}}{2}) \ge \sigma\gamma^{2t_k}$, i.e., $\gamma^{t_k} \le \frac{2\varrho}{2\sigma+\mathcal{C}}$. So, it is obvious that the inequality in Step 3 of Algorithm 1 holds if $t_k \ge \lceil \log_r \frac{2\varrho}{2\sigma+\mathcal{C}} \rceil$.

Now, from Step 3 of Algorithm 1 and Lemma 1(v), we obtain that

$$
\Psi_{\alpha\beta}(x^k) - \Psi_{\alpha\beta}(x^k + \gamma^{t_k} d^k) \ge \sigma\gamma^{2t_k} \| \frac{\partial\phi_{\alpha\beta}}{\partial b}(x^k, F(x^k)) + \frac{\partial\phi_{\alpha\beta}}{\partial a}(x^k, F(x^k)) \|^2
$$

$$\geq \sigma\gamma^{2t_k} \frac{\left((2-\sqrt{2})(2-\sqrt{2(1-\beta)})\right)^2}{2} \Psi_{NR}(x^k).$$

By Lemma 1(iv), we further obtain that

$$\Psi_{\alpha\beta}(x^k) - \Psi_{\alpha\beta}(x^k + \gamma^{t_k}d^k) \geq \sigma\gamma^{2t_k} \frac{\left((2-\sqrt{2})(2-\sqrt{2(1-\beta)})\right)^2}{2c_2} \Psi_{\alpha\beta}(x^k)$$

$$\geq \sigma\gamma^{2t_k} \left(1 - \frac{\sqrt{2}}{2}\right)^3 \Psi_{\alpha\beta}(x^k),$$

where the second inequality from the $\beta \in (0,1)$ and $c_2 \geq 4$. Thus,

$$0 \leq \Psi_{\alpha\beta}(x^k + \gamma^{t_k}d^k) \leq \left(1 - \sigma\gamma^{2t_k}(1 - \frac{\sqrt{2}}{2})^3\right) \Psi_{\alpha\beta}(x^k).$$

This demonstrates that $\{\Psi_{\alpha\beta}(x^k)\}$ converges to $\Psi_{\alpha\beta}(x^*) = 0$ globally Q-linearly.

Furthermore, by Lemma 1(ii) we obtain that $\|x^k - x^*\| \leq \chi\Psi_{\alpha\beta}^{\frac{1}{4}}(x^k)$, where χ is a positive constant. Since the sequence $\{\Psi_{\alpha\beta}(x^k)\}$ converges to zero globally Q-linearly, the sequence $\{x^k\}$ converges to the solution x^* of the NCP globally R-linearly. □

4   Some Final Remarks

Under suitable assumptions, we showed in this paper that the iterative point sequence $\{x^k\}$ generated by Algorithm 1 is globally R-linearly convergent and the corresponding merit function sequence $\{\Psi_{\alpha\beta}(x^k)\}$ is globally Q-linearly convergent.

Recently, two classes of generalized complementarity functions for the NCP and the second-order cone complementarity problem (SOCCP) were proposed in [11] and [12], respectively. It is possible that some more general complementarity functions could be proposed by combining those given in [11, 12, 15] and the convergence results of the corresponding merit function methods could be obtained for the NCP; or more generally, for the SOCCP [12]. In addition, as a generalization of the NCP and the SOCCP, the symmetric cone complementarity problem has been studied extensively, and various algorithms for solving this class problems have been proposed (such as smoothing Newton algorithms [13, 14]). It is possible that the analysis given in [11, 12, 15] and in the paper can be extended to investigate this class of problems.

References

1. Chen, J.S.: On Some NCP-Function Based On The Generalized Fischer-Burmeister Function. Asia-Pac. J. Oper. Res. 24, 401–420 (2007)
2. Chen, J.S., Pan, S.: A family of NCP-functions and a descent method for the nonlinear complementarity problem. Comput. Optim. Appl. 40, 389–404 (2008)

3. Du, S.Q., Gao, Y.: Merit functions for nonsmooth complementarity problems and related descent algorithm. Appl. Math. – J. Chinese Univ. 25, 78–84 (2010)
4. Facchinei, F., Pang, J.S.: Finite-dimensional variational inequalities and complementarity problems. Springer, New York (2003)
5. Ferris, M.C., Pang, J.S.: Engineering and economic applications of complementarity problems. SIAM Review 39, 669–713 (1997)
6. Fischer, A.: Solution of monotone complementarity problems with Lipschitzian functions. Math. Program. 76, 513–532 (1997)
7. Fischer, A., Jeyakumar, V., Luc, D.: Solution point characterizations and convergence analysis of a descent algorithm for nonsmooth continuous complementarity problems. J. Optim. Theory Appl. 110, 493–513 (2001)
8. Geiger, C., Kanzow, C.: On the resolution of monotone complementarity problems. Comput. Optim. Appl. 5, 155–173 (1996)
9. Han, J.Y., Xiu, N.H., Qi, H.D.: Theory and Methods for Nonlinear Complementarity Problems. Shanghai Science and Technology Press (2006) (in Chinese)
10. Harker, P.T., Pang, J.S.: Finite dimensional variational inequality and nonlinear complementarity problem: A survey of theory, algorithms and applications. Math. Program. 48, 161–220 (1990)
11. Hu, S.L., Huang, Z.H., Chen, J.S.: Properties of a family of generalized NCP-functions and a derivative free algorithm for complementarity problems. J. Comput. Appl. Math. 230, 69–82 (2009)
12. Hu, S.L., Huang, Z.H., Lu, N.: Smoothness of a class of generalized merit functions for the second-order cone complementarity problem. Pacific J. Optim. 6, 551–571 (2010)
13. Huang, Z.H., Ni, T.: Smoothing algorithms for complementarity problems over symmetric cones. Comput. Optim. Appl. 45, 557–579 (2010)
14. Liu, Y.J., Zhang, L.W., Liu, M.J.: Extension of smoothing functions to symmetric cone complementarity problems. Appl. Math. – J. Chinese Univ. 22, 245–252 (2007)
15. Lu, L.Y., Huang, Z.H., Hu, S.L.: Properties of a family of merit functions and a merit function method for the NCP. Appl. Math. – J. Chinese Univ. 25, 379–390 (2010)
16. Mangasarian, O.L., Solodov, M.V.: A linearly convergent derivative-free descent method for strongly monotone complementarity problems. Comput. Optim. Appl. 14, 5–16 (1999)
17. Peng, Y.H., Liu, Z.H.: A derivative-free algorithm for unconstrained optimization. Appl. Math. – J. Chinese Univ. 20, 491–498 (2005)

A Knowledge Representation in Possible World

Yangxin Ou[1,2,*], Ping Zou[1], and Chunyan Shuai[1]

[1] Kunming University of Science and Technology
Kunming City, P.R. China
[2] Yunnan Key Lab of Computer Technology Application
Kunming City, P.R. China
kmoyx@hotmail.com

Abstract. In order to represent the knowledge model of the combination of time and actions in a possible world,this article proposes a knowledge representation(KR) method which is based on first order logic and its semantic can be interpreted by production rules. First,we discuss the basic theory of knowledge representation(KR), and next,we construct a meta language through which semantic can be explained by the rules of semantic interpretation and predicates, and then,by the meta language and its semantic we describe the concepts in the possible world. At last,we use experiments in the possible world to demonstrate the effectiveness of the presented method.

Keywords: Semantic, Knowledge Representation, Model Theory, Expert System.

1 Introduction

With the development of computer technologies , people not only want to get higher computing performance,but also wish to get better semantic capacities. Formal semantics[1] [2] [3] [4] is a logical method to represent the true meaning of natural language semantic.

But there are some problems in intelligent systems which are not to be solved by the field of semantic representation, For example how to present the actions and roles ,which are very important parts of the knowledge of the scene,and how to represent the formal semantics in it.

We construct a possible world ,in which the problems about representing semantic is show potentially. To get there,we consider that there are not only the static objects existing in the possible world ,but also dynamic elements remained in it. See Fig.1. The pen in the world stands for the static object and the roles are John and Mike,Who are the subjects of the actions in the world and the actions is expressed as predicates. The clock implies the time going by. Our purpose is not only limited to representing the states of the world,but we also hope that when asking some questions about the the states of the world,we should get the

---

* Yangxin Ou is with Kunming University of Science and Technology, Kunming 650000, Yunnan, P.R. China

Z. Li et al. (Eds.): ISICA 2012, CCIS 316, pp. 512–521, 2012.

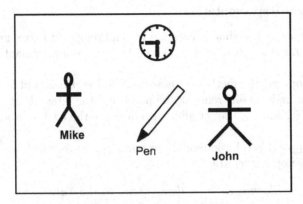

Fig. 1. A possible world of Mike, John, Pen and time

proper answers in one way or another. Focusing on the problem ,we will choose a proper way to represent the world.

We consider that during the period of watching the world ,the owner of the pen will be changed between Mike and John. What will we do for describing the world and the processes of the change of the ownership through time? Suppose that, there is an observer who is always watching the possible world outside the possible world,and he records everything in this world and can answer any question with formal language about the world. If the answer from the observer can reflect the states of the world correctly, we say that the observer can understand the possible world,in other word,the knowledge representation of the possible world is comprehensible knowledge with semantic.

For the purpose of describing the world in formal language ,we will divide the task of understanding the possible world into three parts: knowledge representation of the world, semantic of the formal language,implementing of the method.

2   Related Solving Methods

Now,we will discuss several representations that perhaps can be used in the possible world as follows,and we will choose a better method from them.

2.1   Describing in Classical First-Order Logic

First-order logic (FOL) [5] is a formal deductive system, uses mathematical structures to explain the form of unambiguous language.

The premise of using the classic FOL predicate is that environment of predicates are not relation to time and space,which is static, stable. But, the world we want to describe is dynamic, changeable. Therefore, it is impossible for us to use classical logic to achieve our purpose of describing the evolution world proposed in Fig 1.

2.2 Temporal Representation

In the possible world,the time expression (or call temporal representation) is as a very important role when we handle the knowledge and semantic related to time.

Allen [6] proposed the method of how temporal expression of knowledge can be represented ,and the difference of the point and the intervals of time in logic. A.Artale et al[7]. and A. Cau et al[8]. also describe the content about temporal logic.

Pnueli [9] applied two basic modal temporal operators which are now widely used .These two operators are:

$$\Box\varphi \ means \ \varphi \ is \ always \ true \ in \ the \ future \qquad (1)$$

$$\Diamond\varphi \ means \ \varphi \ is \ true \ at \ some \ time \ in \ the \ future \qquad (2)$$

The $\Box$ has a significant reference while we consider the predicates that is never changed as the time goes by in the possible world.

2.3 Describing in Ontology Language

Semantic Web [10] is a knowledge representation based on object-oriented method, the ontology description language of semantic Web is OWL [11] ,which could be used to describe the knowledge of based on classes and attributes.

The core of Semantic Web is based on Description logics (DLs)[12], which are subset of the first order predication logics. Description logic provides semantic Web formalized basis and automatic reasoning ability, but it has restricted logical reasoning ability.

2.4 Summary

As mentioned above, The methods of the knowledge representation that we discussed could not describe completely the knowledge in the possible world in Fig. 1. To solve the problem of semantic description in the possible world,based on semantics of Kripe,this paper proposes a method to construct and reason the semantic in a possible world evolving dynamically. We first present the theory reference to [13] as following.

3 Semantic Representation

3.1 Formal Language

Definition 1: A Formal Language L is given by the following data:

1. a set of logical operators $\{ \land, \lor, \neg, \rightarrow \}$;
2. a set of range: {**true, false**}, normally, we use $1, 0$ representing them respectively;

3. a set of predicates P ;
4. a set of functions F ;
5. a set of constants C ;
6. a set of variables V ;
7. **item** and **formula**(also call **sentence**): for each $c \in C$, c is an item; for each $p \in P$, p is an **item** too; **formula** is a combination of items and logical operators.

Definition 2: Meta functions and Meta predicates.

Meta functions are predefined functions in language L, which are original semantic routines, can be used to interpret each new concept. A set of meta functions denotes as F_{meta}, $F_{meta} \subseteq F$, **Meta predicates** are predefined predicates in language L, can be used to present formulas(statement). A set of **Meta predicates** donates as P_{meta}, $P_{meta} \subseteq P$.

Definition 3: for any formula φ, $W \models_p \varphi$, if and only if there exists an assignment or a predicate in possible world W, let $\varphi = true$.

3.2   Formal Semantic

Definition 4:   A possible world is defined as:

$$W = \{CE, AP, SP, F, D, I_T\}, \text{where:}$$

CE is a set of concepts that appear in the world; AP is a set of action predicates that can be taken by the roles in the world; SP is a set of state predicates that show the the states in the wolrd and $P = AP \cup SP$; F is a set of functions that can be used in AP and SP; D is a nonempty set ,called domain; a one-to-one map: $W \times (P \cup F \cup C) \rightarrow D^n$.

Setting, the sets of AP, SP, F in W are denoted separately by $W.AP$, $W.SP$, $W.F$, and so on.

Setting $|A.AP|$ is the cardinality of AP in W.

In Model Theory[14], the semantic can be defined as follows:

For a formal language $L = \langle \lambda, \mu, K \rangle$, a **L-structured** M is a family: $M = \langle A, \langle R_i^M \rangle i \in I, \langle f_j^M \rangle, j \in J, \langle c_k^M \rangle, k \in K \rangle$ where: $A = |M|$ is a non-empty set, the universe of M; for every $i \in I, R_i^M \subseteq A^{\lambda(i)}$; for every $j \in J, f_j^M : A^{\mu(j)} \rightarrow A$; for every $k \in K, c_k^M \in A.R_i^M, f_j^M, c_k^M$, are called the interpretation of R_i, f_j, c_k in M.

Generally speaking, the formal semantic of a word or predicate can be expressed as one of the following cases:

1. an assignment of meanings to its predicates and truth-conditions to its sentences,that is,there exists an interpretation satisfying the sentences in the possible world;
2. the process of taking the actions attached to the predicate;
3. the synonyms of a word, and the word has been interpreted correctly;
4. the antonym of another word,which is antonym too.

Definition 5: Suppose that w is an instance of W,

1. for each $f \in w.F$, there exists a function in D which can interpret f: $I_T(f) \in D$
2. for each $p \in (w.AP \cup w.SP)$ there exists a predicate D which can interpret p: $I_T(p) \in (0, 1)$
3. for each $c \in w.C$, we have $I_T(c) \in D$

Definition 6: $W = \{w_1, w_2, ..., w_m\}$ is a sorted set of possible worlds.

Definition 7: $\langle t_k, w_k, Acts \rangle \overset{o}{\to} w_{k+1}$ is defined as an evolution of the possible world, where t_k is a time pint in the possible world, w_k and w_{k+1} both are possible worlds, $Acts \in w_k.AP$.

This means the world w_k will change to w_{k+1} under the affect of $Acts$.

$\langle t_0, w_0, Acts_1 \rangle \overset{o}{\to} w_1; \langle t_1, w_1, Acts_2 \rangle \overset{o}{\to} w_2; ...; \langle t_n, w_n, Acts_{n+1} \rangle \overset{o}{\to} w_{n+1};$ is a chain of possible worlds evolving, denote $\langle t_0, w_0, Acts_1 \rangle \overset{o}{\to} \langle t_1, w_1, Acts_2 \rangle \overset{o}{\to} ... \overset{o}{\to} \langle t_n, w_n, Acts_n \rangle$.

If the possible world is changed only by time without any actions, such as: $\langle t_k, w_k, Null \rangle \overset{o}{\to} w_{k+1}$ it is called **conservative evolution**, denoted: $\langle t_k, w_k \rangle \overset{o}{\to}_r w_{k+1}$.

Obviously, in a **conservative evolution**, the time-independent predicates remain unchanged, For some $s \in w_i.SP$, implies $s \in w_{i+1}.SP$ too, if s is a time-independent predicate.

If a possible world is time-independent, and actions only exist in it, then the change omits time factor in the evolution and expresses as: $\langle w_k, Acts \rangle \overset{o}{\to} w_{k+1}$.

Theorem 1: It is not unique way that from a possible world evolves to a new possible world.

Proof. Collecting the actions $Acts_i, ..., Acts_k$ from the evolution of w_i to w_k, we assign the actions to a set of named $Acts_i, ..., Acts_k$ is a permutation of a division of $Acts$, the division of $Acts$ are not unique, paths from a possible world evolve to a new possible world is not unique. **Q.E.D**

Definition 8: Predicates set P automatically extend.

Because of $P = AP \cup SP$, the P expands with the expansion of AP and SP. As we know, the action taken will affect the state being changed in the world, therefore, the expansion of AP will result in the changes of SP in some case, but if any action in AP is not taken, the state in SP will not change. Envisaging a working mechanism, it can infer new knowledge by self-refection, which can implement the expansion of knowledge automatically.

3.3   The Algorithm of Semantic Modeling

Established Interpretation Set. To design a method of semantic interpretation, at first, a domain D is constructed, and the algorithm is as following:

1. for each $c \in C$, denotes as $C.c$, is added into domain D, and denotes as $D.c$. Then the interpretation is $I_T(C.c) = D.c$

2. for each $p \in P_{meta}$, is added into D, then the interpretation is $I_T(p) = ture$;
   $p \in P_{meta}$
3. for each $f \in F_{meta}$, is added into D, then the interpretation is $I_T(f) \in D$;
   $f \in F_{meta}$

The Algorithm of Semantic Interpretation. We can divide the interpretation into three different parts:

1. The first part is the interpretation of the interpretation of the atomic concepts. If atomic concepts existed, then interpret is success;
2. The second part is the interpretation of the combination concepts, Calling explain to make sure whether each part of the combination concepts exists and whether each concept in combination concepts exists, if so, explain is success;
3. The last one is the interpretation of the formula (call it as sentence) which is combination of concept and predicates, including:
   (a) A sentence is not **true**,while some predicates in the sentence are **false**;
   (b) A sentence increases reliability of the knowledge,and return value is **true**;
   (c) Sentences are a subset of knowledge base. In order to keep the consistency and integrity of the information,some information in the statement need to be satisfied with.

4 The Evolution of the Knowledge Expression in a Possible World

4.1 Definition of Meta Functions

1. $Value(w, x)$, is a function used to calculate the value of x in world w;
2. $Now(w)$ means the time when the function is being invoked in w;
3. $currentWorld()$ means now talking about (or we focus on) the possible world;
4. function $After(w, t)$, $\boldsymbol{After}(w, t) = \{x | time(w, x) \wedge x > t\}$;
5. function $Before(w, t)$, $\boldsymbol{Before}(w, t) = \{x | time(w, x) \wedge x < t\}$;
6. function $Between(w, t_1, t_2)$, $\boldsymbol{Between}(w, t_1, t_2) = \{x | time(w, x) \wedge x < t_2 \wedge x > t_1\}$.

4.2 Definition of Meta Predicates

1. The predicate $\boldsymbol{BelongTo}(w, x, y) \in \{0, 1\}$ means, in the world w, something named x belongs to y;
2. The predicate $\boldsymbol{Exist}(w, x) \in \{0, 1\}$ means, in the world w, exists an object named x;
3. The semantic of Own, $I_T(\boldsymbol{Own}(w, x, y)) = \{Exist(w, x) \wedge Exist(w, y) \wedge BelongTo(w, x, y)\}$;
4. Define the semantic of $\boldsymbol{Give}(w, x, y, sth)$ predicate: $I_T(\boldsymbol{Give}(w, x, y, sth)) = \{Exist(w, x) \wedge Exist(w, y) \wedge Exist(w, sth) \wedge BelongTo(w, sth, y) \wedge \neg Belong To(w, sth, x)\}$.

Agreed, if there is no world in any function or predicate, the world refer to current possible world.

4.3   Verifying with Experiment

This paper uses CLIPS[15] to demonstrate the evolution of the possible world in Fig. 1. The CLIPS is a method of knowledge representations based on production, which can be transformed into FOL language equivalently. Fundamentally, the experiment is based on FOL, therefore, the result of the experiment is universality.

The Expression of Facts, Functions and Predicates. The expression of facts, functions and predicates in CLIPS inherited from LISP, a well-know AI language, the format of command in CLIPS likes:

<predicate or function or fact-name> <argument-list>

Examples as following: (name Mike 'Mike')(Is Mike Person)(Is John Person)(Own Mike Pen)

Defining Give meta function as:(Give?Person1?Person2?Something), which means *'Person1 give Something to Person 2'*.

In the front of the word Persion1 and Person2 there are two symbols '?', which indicate that Persion1 and Persion2 are variables in CLIPS.

The Expression of the Time Semantic. It is more complex to represent the semantic of the time. Because the time is probably a range, or a point. An additional variable *Env* is taken into account, as a context descriptor in any possible world. *Env* is added to the facts that related to time, which are consist of a series of two-tuples structures just like $\langle StartTime, EndTime \rangle$, and denotes as $\langle Env.StartTime, Env.EndTime \rangle$.

Then the inference is as following:

1. If the time of sentence is before $t1$, then the values to $Env.StartTime$ and $Env.EndTime$ are given respectively
   $Env.StartTime = 0$, $Env.EndTime = t1$
2. If the time of sentence is between $t1$ and $t2$, the values are assigned:
   $Env.StartTime = t1$, $Env.EndTime = t2$
3. If the time of sentence is after $t2$, then the values are:
   $Env.StartTime = t2$, $Env.EndTime = 0$

The command *assert* in CLIPS is used to represent the facts in possible world. *(assert (Own Mike Pen [t1 0]))* means *'Mike has owned a Pen since the time t1'*.

After executing the action *(assert(Give Mike John Pen))*, suppose that the time is $t2$, the following facts will be add to the knowledge base automatically: *(assert(Own Mike Pen [t1, t2]))*.

If the facts is time-independent, then $Env.StartTime = 0$, $Env.EndTime = 0$ invariably.

Remark: the relation of the time in real world and possible world can be seen in Fig. 2.

In Fig. 2, real world is that we are living in, and the possible world is a virtual world only exists in computer, the clock in the real world is different from that

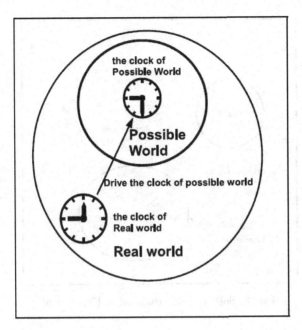

Fig. 2. The clocks of real world and possible world

in any possible world, but the clock in real world can drive the clock in possible world. The time in the real world is shown by the clock in CLIPS' running environment, which can be got by calling the routine or function in CLIPS. We use *standard unit of time (SUT)* as a benchmark unit of time in the possible world, and define every n seconds in real world is equal to 1 *SUT* in the possible world. For example, we can let n=5 in our experiment. *TichketEvent()* function is executed, it has gone one SUT in the possible world, some predicates and fucntions are added into *TicketEvent()*.

4.4   The Process of the Experiments

We construct a environment that consists of Java and CLIPS ,in which we use two threads which are concurrent and independent working in Java language to simulate the actions of the roles of John and Mike, who only live in possible world, and the Pen can be expressed with a class in Java language. The facts that Mike and John have known in possible world also can be represented in CLIPS. The CLIPS stores the rules which can be used to represent the semantic of predicates or facts that describe the states and time in the possible world.

In Fig. 3, we can show how the experiment can handle the information and the knowledge in a possible world. The roles in the possible world are Mike and John, who only exist as two threads running in Java, their knowledge (or memory in brains like human) can be stored in CLIPS dynamically,and can be retrieved at any time. We can ask questions,which should be expression(or formula) in formal language, and get the feedback through the console. If we want to known

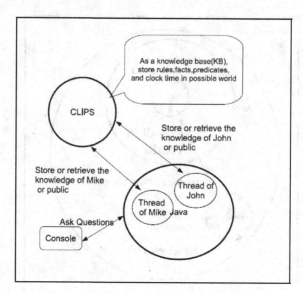

Fig. 3. Software Environment of Experiment

Table 1. Process of Environment

STU	W	Q/Action	A	KB
1	w1	Q:(Own ? Who Pen)	John	(Own Mike Pen [0 1])
2	w2	Q:(Own ? Who Pen)	John	(Own Mike Pen [0 2])
...				
50	w50	Q: (Give Mike John Pen)		(Own Mike Pen [0 50]) (Own John Pen [50 0])
51	w51	Q: (Own ? Who Pen)	Mike	(Own Mike Pen [0 50]) (Own John Pen [50 51])
...				
70	w70	Q: Before(50)∧ (Own ? Who Pen)	John	(Own Mike Pen [0 50]) (Own John Pen [50 70])
...				
90	w90	Q: Between(60, 70)∧ (Own ? Who Pen)	John	(Own Mike Pen [0 50]) (Own John Pen [50 90])

who has the pen,then we type a question: *(Own?Who Pen)* in the console of Fig. 3. The console related to the software in Java accepts the question and handle it internally and shows the the result. Initial facts(or states): Keeping the current state *(Own Mike Pen [t0 0])* in in period of time,and taking the action *(Give Mike John Pen)*, and then asking a question in a random time. We can get the dialogue records, and list them in Table 1. **Q/A**: Q means Question, A means Answer, W means worlds, KB means Knowledge base.

In Table 1 ,When asking the system in the time of 1,2,..70,90,representing the time point of standard unit of time in the possible world, because the time in possible world has changed,the possible world also changed, Using $w1, w2, ..., w70$, $w90$ to denote. At each time point,the same problems "Who owns the pen?" is asked,expression as (Own?Who Pen) in CLIPS language. Difference is additional condition of time. The answer is correct[15].

5   Conclusion

This article, by constructing a possible world,defines a meta language with first order logic for the possible world and describes the concepts , actions and time relations in the possible world. By describing the knowledge representation and semantic of concepts and actions and states in the possible world,we try to find a a proper method expressing actions and facts. The experiment in possible world proves that the theory is effective. Our next work is that researching how to make the roles in possible world understand the concepts and the actions in the world.

References

1. Montague, R.: On the nature of certain philosophical entities. The Monist 53(2), 159–194 (1969)
2. Montague, R.: The Proper Treatment of Quantification in Ordinary English. Formal Semantics, 17–34 (1973)
3. Kripke, S.A.: Naming and necessity. Wiley-Blackwell (1981)
4. Kripke, S.: Semantical considerations on modal logic. Acta Philosophica Fennica 16, 83–94 (1963)
5. Majkic, Z.: First-order Logic: Modality and Intensionality, arXiv: 1103-0680 (March 2011)
6. Allen, J.F.: Maintaining knowledge about temporal intervals. Communications of the ACM 26(11), 832–843 (1983)
7. Artale, A., Franconi, E.: A temporal description logic for reasoning about actions and plans. Arxiv preprint arXiv: 1105-5446 (2011)
8. Cau, A., Moszkowski, B., Zedan, H.: Interval temporal logic (2010), http://www.cms.dmu.ac.uk/cau/itlhomepage/itlhomepage.html
9. Pnueli, A.: The temporal logic of programs. In: 18th Annual Symposium on Foundations of Computer Science, pp. 46–57 (1997)
10. Lee, T.B., Hendler, J., Lassila, O., et al.: The semantic web. Scientific American 284(5), 34–43 (2001)
11. Bechhofer, S., Van Harmelen, F., Hendler, J., Horrocks, I., McGuinness, D.L., Patel-Schneider, P.F., Stein, L.A., et al.: OWL web ontology language reference. W3C Recommendation 10, 2006-01 (2004)
12. Baader, F., Horrocks, I., Sattler, U.: Description Logics as Ontology Languages for the Semantic Web. Mechanizing Mathematical Reasoning: Essays in Honor of Jorg H. Siekmann on the Occasion of His 60th Birthday (2005)
13. Srivastava, S.M.: A course on mathematical logic. Springer (2008)
14. Prestel, A., Delzell, C.N.: Mathematical Logic and Model Theory. Springer (2011)
15. Giarratano, J.C., et al.: CLIPS User's guide, Artificial Intelligence Section, Lyndon B. Johnson Space Center (1989)

Grid Resource Scheduling Method
Based on BP Neural Network

Min Li[1,2] and Zhenhua Li[2]

[1] Jingchu University of Technology
Jingmen, Hubei 448000 China
Ziye9670_cn@sina.com
[2] China University of Geosciences
WuHan 430074, China
zhli@cug.edu.cn

Abstract. The grid mesh is a high performance calculation of the main direction, the influence of the grid function and performance of the main factors for the efficiency of the grid resources scheduling, because of the complexity of the grid, the resource management compared with the traditional distributed network more complicated, so efficient grid resources scheduling algorithm is grid research hot spot and the difficulty. This paper puts forward a layered resource scheduling model and a simple structure function complete resources scheduling method, and put forward feedback of the BP neural network algorithm applied to the grid resources scheduling of better solve the grid resources scheduling problem.

Keywords: Grid, Resource Scheduling, BP.

1 Introduction

Grid can provide than current machine more powerful computation ability, and therefore, can satisfy the science and the business field of many requirements. A large number of application and cluster configuration in network environment played an important role. Resource scheduling is the core issue of grid computing, it includes resources organization, positioning, discovery, dispatching, distribution, validation, process to create and preparation of the resources required to other activities, in the grid so that a highly dynamic environment, a resource malfunctions or the possibility of failure is higher, the resources of the system will continue to expand, the application will be increasing, the system's whole structure and overall performance will be constantly changing, at any time and unpredictable system behavior occurs, and the intermediate system information service function provided may be already outdated system information, so based on the information of a resource scheduling strategy will produce the bad influence, this requires resources scheduling should provide a wide area resources to adapt to the environment to support adaptive resource scheduling strategy; Must dynamic monitoring and scheduling grid resources, from the available resources select the best resource service, as far as possible to reduce this due

Z. Li et al. (Eds.): ISICA 2012, CCIS 316, pp. 522–530, 2012.

to the fault or failure, the overall structure and overall performance changes or unpredictable behavior of a system to grid problems such as the influence of the overall performance.

The neural network has strong adaptability and learning ability, robustness and fault tolerant ability, thus can replace complex time-consuming traditional algorithm, so that the signal processing process more close to human thinking activity. By using the neural networks' highly parallel computation ability, can real-time realize it is difficult to use digital calculation and technology to realize the optimal signal processing algorithm.

In recent years, no matter in grid computing or neural network aspects of experts and scholars at home and abroad have done a great deal of research, 2008 Zhong ShaoBo is proposed based on dynamic load balancing strategy of the grid task scheduling optimization model and algorithm[1] in 2009, such as Hu ZhiGang is proposed based on resource performance evaluation of the grid task scheduling algorithm[2], and in the same year, such as Huang Wen Ming is proposed based on improved ant colony algorithm of the grid resources scheduling[3]. Based on the study of the grid structure and BP neural network is put forward on the basis of using the BP neural network algorithm to solve grid resources scheduling optimization method.

2   Grid Structure and Resource Analysis

2.1   Grid System Structure

Grid architecture is about how to build network technology. It gives the basic composition and function of the grid, stipulated the grid various part of relations and integrated ways or methods, depicting the support grid operative mechanism.

Grid system structure is mainly have abstract hierarchical structure, block structure, concept space structure and mixed structure and OGSA (Open Grid Services Architecture) structure and form. So far, more important and influential network system structure is mainly has two kinds: five layer sandglass structure and open grid services OGSA.

2.2   Five Layer Sandglass Structure

Five layer sandglass structure is based on the United States national laboratory grid r&d project Globus carry out, is a kind of in the agreement as the center "agreement structure". It is a kind of influence is very wide range of structure, the structure emphasize agreement in grid resource sharing and interoperation of the position, through agreement to realize a kind of mechanism. It focuses on the qualitative description and not specific agreement definition, so easy to understand on the whole. As shown in Fig. 1 shows for five layer sandglass structure shape diagram.

The structure of the salient features are "hourglass" shape, because each part number agreement is different, the core part of the need to be able to realize

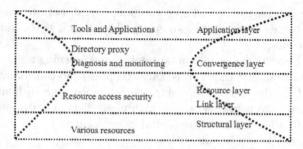

Fig. 1. Five layer sandglass structure

the mapping of the various agreements, and at the same time to realize the core agreement to lower other agreement mapping, core agreement in all support grid node can get support. Therefore, the number of core agreement should not be too much. In this way, the core protocol are formed in the protocol hierarchy a bottleneck, in the diagram above resources layer and link layer composed the core part of the bottleneck.

Five layer sandglass structure of the five layers (from the bottom up, respectively, constructions of the connection layer, resource layer, convergence layer, application layer) meaning is as follows:

1. Structural layer is physical and logical entity, basic function is to control the local resources, upward provides access to resources of the interface. At the same time provide control service quality resources scheduling ability and query mechanism, etc.
2. Connection layer basic function is to realize the mutual communication. It defines the core of the communication and the authentication protocol used to mesh network transaction processing. Constructions of all kinds of resources to provide the data exchange between in this layer under the control of the implementation, the resources of between authentication, safety control is realized in here.
3. Resources layer basic function is to achieve a single resource sharing. This layer defines in a separate resources provide sharing operation of the agreement, the agreement to realize the function of the structural layer to access and control the local resources.
4. Convergence layer basic function is to coordinate a variety of resources sharing. Explain different resources between set is how to interaction.
5. The application layer exists in virtual organization environment. According to the application in a layer of definition of service to structure to achieve certain function.

Open Grid Services ArchitectureOGSA. OGSA is based on service oriented architecture. Known as is the next generation of grid structure. OGSA is by the Global Grid Forum (GGF) Open Grid Services Infrastructure (OGSI) working group in June 2002 the formulated. Five layer sandglass structures based on agreement as the center "agreement structure", and OGSA is service as the center "service structure". This kind of structure is the key of network technology and Web service mechanism integrated up, facing the grid service and creates a distribution system framework.

OGSA architecture by four major layer form, as shown in Fig. 2 from the bottom up is as follows: the physical resource and the logical resources; Web services and definition of grid services OGSI extended; Based on OGSA architecture service; Network applications.

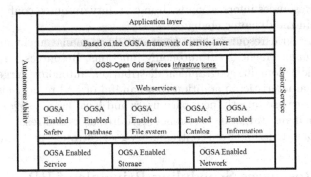

Fig. 2. The main frame of OGSA

1. Physical and logical resource layer this layer is the central part of grid computing, including physical resources are constitute the ability of grid resources, the logical resources located in the physical resource on, they through the virtualization and polymerization of physical resources to offer additional functionality.
2. The Web service layer OGSA important principle: will all grid resources (including logic and physical) modeling become service. This service is by the OGSI (Open Grid Services infrastructure, Open Grid Services infrastructure) completed. OGSI also further expanded the definition of Web service, provides a dynamic, state and can management Web service capacity.
3. Based on OGSA architecture of grid services layer Web service layer and its OGSI extension based on OGSA architecture of grid service provides infrastructure.
4. Grid application layer later will develop a group of rich framework based on grid services, the use of one or more based on grid architecture service of the new grid application also can appear.

2.3   Grid Resource Analysis

Definition of Resource. Grid resource is that the network can be Shared and use of any ability, is the grid discernible "state" entity. Including the traditional physical resources (such as equipment and instrument, computer software, computer hardware, etc.), including virtual services (such as database, data transmission software applications).

The Characteristics of Grid Resources. Grid the environment resources have the following features:

1. Isomerism: resources is of great variety, have different types and performance characteristics, direct access interface is endless also and same. The local management system is different, different sharing rules.
2. The dynamic: the grid resources may at any time to join or leave the grid system. Resources information with uncertainty, resource configuration and capacity in the dynamic change.
3. Autonomy: grid resources have their own local management institution or at local management, local autonomy functions. Resources across multiple management domain, geographical distribution, autonomy, large scale.
4. Binary characteristic: in addition to a handful of grid resource is to provide network user use, mostly net space users and local users use jointly.
5. The participants target inconsistency: resource users and resources providers are not consistent, and even contradictory goals, strategy and demand model.

3   Grid Resources Scheduling Principle of BP

3.1   BP Neural Network Structure

Network structure design is a comprehensive problem now, there are no fixed follow steps, and there are many parameters by experience to choose. The BP neural Network that is Error Back Propagation Network (Error Back Propagation Network), is a typical multiple Networks, divided into input layer, hidden layer and output layer, between layer and layer adopts full interconnection way. Use the neural network based on BP algorithm to solve practical problems, is the network structure optimization decision network characteristic enhancement, become fitting is the key to success. BP nerve network basic structure as shown in Fig. 3 shows.

If three layer BP network, the number of input layer and output layer number is sure, only the number of hidden layer nodes cannot generally be determined, and the number of hidden layer nodes, approximation accuracy, so usually by trial and error method and empirical method to determine the number of hidden layer nodes. So determine its network structure, is in fact the input and output layer node selection; The determination of hidden layer; The number of hidden layer nodes to determine.

1. Input and output layer node selection;
   BP network's input and output layer node number completely according to user's request to design. If used BP network classifier and the number of categories for m, then input generally take m a neurons, the training

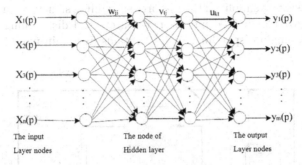

$X_1(p)$ W_{ji} V_{tj} U_{kt} $y_1(p)$

$X_2(p)$ $y_2(p)$

$X_3(p)$ $y_3(p)$

$X_n(p)$ $y_m(p)$

The input Layer nodes   The node of Hidden layer   The output Layer nodes

Fig. 3. BP Neural network structure

sample concentration of $XP1$ belong to the first j class, ask its output is $Y = (0, ..., 1j, 0, ..., 0m)T$, so for a n d output $X \in Rn$ classification mapping $Y \in Rn$ meet $Y1 = 1(XP1$ belong to 1 type) or $Y1 = 0(XP1$ does not belong to 1 class). Input node can also according to solving problems and data representation way and decide.

2. The number of hidden layer determined;
   BP neural network is the most major characteristic is a nonlinear mapping function. To mathematical two Weierstrass approximation theorem as the basis, for any closed interval in a continuous function, you can use a implicit layers of BP network to approximation. Thus a three layer BP network can complete any n d to m dimensional mapping. The practical application, the general take hidden for a layer, constitutes a three layer BP network.

3. The number of hidden layer nodes to determine;
   The neural network based on BP algorithm in each layer node number choice to the network performance is affected, so, layer number of nodes in the need for appropriate choices. Usually, a multi-layer network need how many implicit layers, each layer of how many hidden units, it shall be determined by the network use. Generally speaking, the hidden unit number is too little, network training may not come out; But hidden unit number is too much, and make learning time is too long, error are not the best, so, in fact, exist to determine the best number of hidden units.

3.2 BP Neural Network Scheduling Model

The traditional parallel scheduling problem is an application son task scheduling to some parallel computer in order to reduce the running time, the scheduling problem is a from many different users to application flow scheduling to a computing resources to maximize utilization of collection system.

We choose to hierarchical model to realize scheduling has centralized scheduling, assignments, be submitted to the centralized scheduling program, and every machine resources use a separate scheduling program for local scheduling.

This model system simplified the resource scheduling structure but its function is still complete, it is divided into user, resources and dispatching center three main parts, the main dispatching center contains five main modules. As shown in Fig. 4.

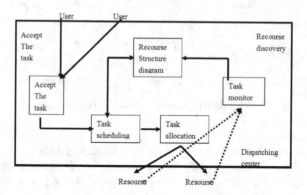

Fig. 4. Neural network scheduling model

Each function module function is as follows:

1. The task receiver: user receiver to task put forward application requirement description, including the calculation of application Load, communication load, and time limit, etc.
2. Resources structure: describe the current resources situation of the state.
3. The task scheduler: the task scheduler according to the resource conditions and task scheduling strategy according to certain demand of selecting a kind of optimal allocation strategy, and the corresponding task to task distributor.
4. The task distributor: the task distributor maintain a allocating task queue, record assigned task name, task requirements, was assigned resource name, etc.
5. Resources monitor: resources monitor testing the latest resource status and accordingly maintenance resource state diagram.

3.3   The Analysis of Experimental Results

We use SimJava distributions of the software package formation on behalf of users and resources of the pseudo random sequence, to carry out the experimental training BP learning of the grid resources scheduling model.

1. The number of resource for eight, resource calculation force in (350, 500) based on selected, the resources the longest running time in (8, 12) interval selection, the shortest run time in (2, 4.5) interval selection. As shown in table 1 shows:

Table 1. Resources random sequence

Resource name	Resource total computing power	The longest operation time	The shortest operation time
R1	373	9.98	3.64
R2	427	10.68	3.22
R3	496	11.50	3.65
R4	356	9.67	2.05
R5	485	11.29	2.00
R6	429	11.50	3.25
R7	391	8.30	3.44
R8	413	8.10	2.03

2. The user and user operation users for sixteen, the user operation length in (2000500) interval with uniform distribution formation, homework time (30, 60) interval with uniform distribution formation, assignments, budget in (15000, 50000) interval with uniform distribution formation. As shown in table 2 shows:

Table 2. Users' random sequence

User	Working length	Finish time	Operating budget
User1	3750.000000	39.000000	43676.230000
User2	4680.000000	57.000000	26612.570000
User3	2653.000000	46.000000	52629.350000
User4	4538.000000	53.000000	34796.470000
User5	2199.000000	44.000000	18764.680000
User6	3547.000000	45.000000	38734.440000
User7	2657.000000	33.000000	46593.620000
User8	3978.000000	50.000000	20912.350000
User9	2593.000000	36.000000	24657.430000
User10	4077.000000	42.000000	35997.630000
User11	3267.000000	38.000000	25436.310000
User12	5143.000000	43.000000	56447.670000
User13	2478.000000	51.000000	32561.450000
User14	3750.000000	39.000000	43676.230000
User15	4680.000000	57.000000	26612.570000
User16	2653.000000	46.000000	52629.350000

4   Conclusion

Grid is a highly dynamic environment, the overall structure of the system and the overall performance will be constantly changing, at any time and unpredictable system behavior occurs, must dynamic monitoring and scheduling grid resources, selects the best resource service, as far as possible to reduce this due to the fault or failure. In this paper, based on the BP neural network grid resources scheduling optimization method is solved this problem well, experiments show that: based on the BP neural network scheduling method task response time quickly, assigned tasks more reasonable, can effective use of grid computing resources.

References

1. Hu, Z., Hu, Z., Li, L.: Grid Task Scheduling Algorithm Using Resource Performance Evaluation. Journal of System Simulation, 3542–3548 (November 12, 2009)
2. Zhong, S.: Optimization models and algorithms based on dynamic load balancing strategy for grid task scheduling. Computer Applications, 28(11) (2008)
3. Liu, H., Hao, W., Gao, Q.: Distributed deployment of grid services scheduling algorithm and QoS Performance Analysis. Journal of Computer Science (November 6, 2011)
4. Li, X., Sun, Z.: Selective disposition of grid resources. Journal of Computer Science (November 4, 2010)
5. Gao, Z.: Grid task scheduling, quality of service issues related to research. Beijing Jiaotong University (2010)
6. Niu, D., Shi, H., Li, J., Xu, C.: Research on power load forecasting based on combined model of Markov and BP neural network. In: 2010 8th World Congress on Intelligent Control and Automation (WCICA 2010), pp. 4372–4375 (2010)
7. Minh, T., Le, N., Cao, J.: Flexible and Semantics-Based Support for Web Services Transaction Protocols. In: Wu, S., Yang, L.T., Xu, T.L. (eds.) GPC 2008. LNCS, vol. 5036, pp. 492–503. Springer, Heidelberg (2008)
8. Nan, X., Yu, J., You, Z., Li, Q.: Wind speed forecasting based on combination forecasting model. In: 2010 International Conference of Information Science and Management Engineering (ISME 2010), vol. 2, pp. 185–189 (2010)
9. Zhang, M.-G., Li, L.-R.: Short-term load combined forecasting method based on BPNN and LS-SVM. In: 2011 IEEE Power Engineering and Automation Conference, vol. 1, pp. 319–322 (2011)
10. Lv, G.: Grid resource scheduling based on BP algorithm. Harbin Polytechnic University (March 2007)
11. Li, Z., Xie, L.: Grid architecture and its development. Computer Engineering (July 2005)

A Comparison of Artificial Neural Networks and Support Vector Machines on Land Cover Classification

Yan Guo[1], Kenneth De Jong[2], Fujiang Liu[3], Xiaopan Wang[3], and Chan Li[3]

[1] School of Computer Science
China University of Geosciences
Wuhan 430074, P.R. China
guoyanwuhan@yahoo.com.cn
[2] The Krasnow Institute for Advanced Study
George Mason University
Fairfax, VA 22030, USA
kdejong@gmu.edu
[3] Faculty of Information Engineering
China University of Geosciences
Wuhan 430074, P.R. China
{felixwuhan,wxpwuhan}@163.com,
921906301@qq.com

Abstract. Artificial Neural Networks (ANNs) as well as Support Vector Machines (SVMs) are very powerful tools which can be utilized for remote sensing classification. This paper exemplifies the applicability of ANNs and SVMs in land cover classification. A brief introduction to ANNs and SVMs were given. The ANN and SVM methods for land cover classification using satellite remote sensing data sets were developed. Both methods were tested and their results of land cover classification from a Landsat Enhanced Thematic Mapper Plus image of Wuhan city in China were presented and compared. The overall accuracy values of ANN classifiers and SVM classifiers were over than 97%. SVM classifiers had slightly higher accuracy than ANN classifiers. With demonstrated capability to produce reliable cover results, the ANN and SVM methods should be especially useful for land cover classification.

Keywords: Artificial Neural Networks, Support Vector Machines, Land Cover, Landsat, Remote Sensing Classification, Wuhan.

1 Introduction

Land cover has been identified as one of the crucial data components for global studies and environmental applications[1]. The derivation of land cover information increasingly relies on remote sensing technology due to its ability to acquire measurements of land surfaces at various spatial scales. One of the major approaches to deriving such information from remotely sensed images is classification. Numerous classification algorithms have been developed. Among the most

Z. Li et al. (Eds.): ISICA 2012, CCIS 316, pp. 531–539, 2012.

popular are artificial neural networks and support vector machines. Artificial neural networks (ANNs) are commonly used for classification of remote sensing data. ANNs could be used to improve the classification accuracy of remote sensing image. Their abilities to approximate nonlinear functions and capture complex relationships in data sets are instrumental abilities which could support the remote sensing process domain[2–5]. The feed-forward back-propagation multilayer perceptrons is the type of ANNs most commonly encountered in remote sensing image classifier[6].

Support vector machines (SVMs) are newly developed machine learning method based on strict theoretical fundamentals. It has good adaptive capacity for solving non-linear problems with high dimensions and small samples. SVMs use kernel functions to map non-linear decision boundaries in the original data space into linear ones in a high-dimensional space and employ optimization algorithms to locate the optimal boundaries between classes. Many types of SVMs have been developed. The most widely used in the classification of remotely sensed images is radial basis function SVMs[7].

The purpose of this paper is to demonstrate the applicability of ANNs and SVMs of deriving land cover information from Landsat Enhanced Thematic Mapper Plus image and to evaluate their performances in comparison to each other. A brief introduction to ANNs and SVMs is given in the following section. The dataset and experimental design are presented in section 4. Experimental results are discussed in the following sections. The paper concludes with the results of this study and plans for future work.

2   Artificial Neural Networks

An ANN has nodes (neurons) and interconnections (directed edges) with weights between neuron outputs and neuron inputs. ANNs can learn complex nonlinear input-output relationships. ANN learning/training involves using samples in updating and optimization of the architecture and weights to efficiently perform good classification later. For a specific selected architecture, training involves optimization of the weight parameters.

The most widely used in the classification of remotely sensed images is multilayer perceptrons (MLPs).A typical MLP has three layers of units. Each layer is nourished with the previous layers. MLPs can have any number of weighted connections, but networks with only two weighted connections are very much capable of approximating just about any functional mapping[8]. The MLP is mathematically represented by:

$$y_k = f_{outer} \left[\sum_{j=1}^{M} w_{kj}^{(2)} f_{inner} \left[\sum_{i=1}^{d} w_{ji}^{(1)} x_i + w_{j0}^{(1)} \right] + w_{k0}^{(2)} \right] \quad (1)$$

Where y_k represents the k-th output, f_{outer} represents the output layer transfer function, f_{inner} represents the input layer transfer function, w represents the weights and biases, (i) represent the i-th layer.

3 Support Vector Machines

A SVM is a machine learning algorithm proposed by V.N. Vapnik[9], based on statistical learning theory and structural risk minimization principle and integrated the optimal separating hyperplane with the kernel method. The original idea of SVMs is to use a linear separating hyper-plane which maximizes the margin distance between two classes to create a classifier. For a binary SVM, the training data consist of n pairs $(x_i, y_i), ..., (x_n, y_n)$, where $x_i \in \Re^m$ and $y_i \in \{-1, 1\}, i = 1, ..., n$. Computing the hyper-plan is equivalent to solving the following optimization problem:

Optimization Problem 1 (Soft-Margin SVM (primal))

$$Minimize_{w,\xi} \frac{1}{2}(w \cdot w) + C \sum_{i=1}^{n} \xi_i \tag{2}$$

$$Subject \ to : y_i(w \cdot \varphi(x_i) + b) - 1 + \xi_i \geq 0, \ \xi_i \geq 0, \ i = 1, 2, ..., n$$

Where $w \in \Re^m$ is a vector of weights of training instances, C is a parameter that allows trading-off training error *vs.* model complexity, ξ is a penalty parameter vector (slack variable), φ is a transformation function maps input vector x into a feature space, and b is a constant. The SVM of (2) is a nonlinear SVM when φ maps x into a higher-dimensional feature space. As a special example, if $\varphi(x) = x$, the SVM of (2) is a linear SVM finding a linear separating hyperplane with the maximal margin. For computational reasons it is useful to solve the Wolfe dual of optimization problem 1 instead of solving optimization problem 1 directly.

Optimization Problem 2 (Soft-Margin SVM (dual))

$$Minimize_{\alpha} \sum_{i=1}^{n} \alpha_i - \frac{1}{2} \sum_{i=1}^{n} \sum_{j=1}^{n} \alpha_i \alpha_j y_i y_j K(x_i, x_j) \tag{3}$$

$$Subject \ to : \sum_{i=1}^{n} \alpha_i y_i = 0 \ and \ 0 \leq \alpha \leq C, i = 1, 2, ..., n$$

Where α_i are Lagrange multipliers, $K(x_i, y_j)$ is a kernel function.

$$K(x_i, y_j) = \phi(x_i) \cdot \phi(y_j) \tag{4}$$

From the solution α^* of optimization problem 2 the SVM decision function can be computed is:

$$f(x) = \sum_{i=1}^{n} \alpha_i^* y_i K(x_i, x) + b^* \tag{5}$$

$$w^* = \sum_{i=1}^{n} \alpha_i y_i x_i \ and \ b^* = y_i - w^* \cdot x_i$$

The vectors x_i with $\alpha_i > 0$ are called support vectors. The most widely used kernel function is radial basis function (RBF):

$$K(x_i; x_j) = exp\left(-\gamma ||x_i - x_j||^2\right) \gamma > 0 \tag{6}$$

Several SVM programs have been developed and made publicly available. In this study, we used the LIBSVM program developed by Chang and Lin[10].

4 Data Sets and Experimental Design

The ANNs and SVMs were applied to land cover classification tasks of Landsat Enhanced Thematic Mapper Plus images. The obtained optimal solutions are compared each other. All the programs were written in MATLAB Version 7.8.0.347 (R2009a), and executed on a PC with 2.53 GHz Intel(R) Core(TM) 2 Duo CPU P8700 processor with 1.89 GB of memory.

4.1 Study Area and Data Sets

In this study we used the multi-spectral remote sensing data acquired by the Landsat Enhanced Thematic Mapper Plus (ETM+) sensor. The study area included Wuhan city, the capital of Hubei province of China. It lies between 29○58′ and 31○22′ North Latitude and 113○41′ and 115○05′ East Longitude. It is situated in the intersection of the middle reaches of the Yangtze River and Hanshui River. The Yangtze River and Hanshui River divide Wuhan into three parts: Hankou, Hanyang and Wuchang, which are generally known as Wuhan's Three Towns.

A Landsat ETM+ image of 892 × 898 pixels covering the study area was selected. Six spectral bands including blue (Band.1), green (Band 2), red (Band 3), near-infrared (Band 4) and two mid-infrared (Band 5 and 7) were used for the study (spatial resolution 30m). The image was pre-processed in ERDAS Imagine 8.7 by using RGB color composition, geometric correction, radiometric corrections and principal component analysis. Three bands were used as input variables in this study.

According to the visual interpretation of the pre-processed image (as shown in Fig. 2), the classification scheme included five types: (1) vegetation, (2) City, (3) River, (4) Pond and (5) road & bare land.

In order to present the spectral variation of each land cover type to classifiers, sample sets for each class were selected from the image. A commonly used sampling method was employed by identifying and labeling small patches of homogeneous pixels in an image. Equal sample size was employed in which a fixed number of pixels are sampled from each class as training data. In this experiment there were 185 samples from each class and a total of 925 samples were selected. The input attributes used in this work were rescaled in the range [0, 1].

4.2 Experiments and Results

For ANNs, a three-layer feed-forward backpropagation network with a resilient backpropagation algorithm was used in MATLAB. The first layer had three

Table 1. Training samples :User's accuracies, producer's accuracies and overall accuracies of ANN classifiers and SVM classifiers after 10-fold cross validation on the training data sets

Land cover	User's accuracy(%)		Producer's accuracy(%)		Overall accuracy(%)		Kappa coefficient	
	ANN	SVM	ANN	SVM	ANN	SVM	ANN	SVM
Vegetation	99.94	100.00	100.00	100.00				
City	99.81	99.82	95.38	97.36				
River	96.52	99.58	100.00	100.00	98.05	99.35	0.9757	0.9919
Pond and bare land	98.34	100.00	99.64	100.00				
Road	95.77	97.41	95.26	99.40				

Table 2. Testing samples: Users accuracies, producers accuracies and overall accuracies of ANN classifiers and SVM classifiers after 10-fold cross validation on the testing data sets

Land cover	User's accuracy(%)		Producer's accuracy(%)		Overall accuracy(%)		Kappa coefficient	
	ANN	SVM	ANN	SVM	ANN	SVM	ANN	SVM
Vegetation	100.00	100.00	100.00	100.00				
City	99.44	100.00	95.14	96.22				
River	96.35	99.46	100.00	100.00	97.95	98.81	0.9743	0.9851
Pond and bare land	97.88	100.00	100.00	100.00				
Road	95.60	96.34	94.05	99.46				

tansig neurons that correspond to each of the features in the input data, the second layer had 14 logsig neuron, and the three layer had 5 purelin neuron that correspond to each type of land cover. The trainrp network training function was used. The target training performance was set to 0.001. Each network was trained using maximum 1000 epochs with 6 maximum validation failures and 50 maximum weight changes.

For SVMs, the RBF kernel was used in this study because it had fewer parameter values to predefine and yet had been found at least as robust as other kernel types[11]. We used the LIBSVM 3.12. The training data sets and testing data sets were scaled and parameters were selected by cross-validation and grid-search.

The performances of each algorithm were evaluated in terms of algorithm accuracy and stability. Two widely used accuracy measures-overall accuracy and the kappa coefficient-were used in this study. The overall accuracy has the advantage of being directly interpretable as the proportion of pixels being classified correctly, while the kappa coefficient allows for a statistical test of the significance of the divergence between two algorithms.

To evaluate the performance of the two classifiers, a ten-fold cross validation trial was performed; that is, the algorithms were executed ten times, each time using a random split on the data with 90% of the total dataset for training while the remaining 10% for testing. For each fold, an ANN and a SVM were trained using a training data set and evaluated with a testing data set.

Table 1 and Table 2 show the users accuracies, producers accuracies and over-all accuracies of the two algorithms after 10-fold cross validation on the training data sets and testing data sets separately. Table 3 gives the mean and standard deviation of the overall accuracies and Kappa coefficient of classifications devel-oped through 10-fold cross validation. The classified image of Fig. 2 by an ANN classifier and a SVM classifier are shown in Fig.3 and Fig. 4 respectively.

Table 3. Mean and standard deviation (σ) of the overall accuracies (%) and Kappa coefficients of ANN and SVM classifications using training samples and testing samples selected from the Wuhan data set through 10-fold cross validation

Procedure	Overall accuracy (%)				Kappa coefficient			
	ANN		SVM		ANN		SVM	
	Mean	σ	Mean	σ	Mean	σ	Mean	σ
Train samples	98.05	0.4209	99.35	0.1164	0.9757	0.0053	0.9919	0.0015
Test Samples	97.82	2.2472	99.13	0.9987	0.9728	0.0281	0.9891	0.0125

Several patterns can be observed from all tables and Fig. 1 as follows.

1. Both ANNs and SVMs had good performance. Table 1 and 2 presented the overall accuracy values of the ANNs and SVMs were over than 97%, the users accuracies and producers accuracies of the ANNs and SVMs were over 94%.
2. Generally SVM classifiers, though insignificantly, gave higher accurate than ANN classifiers. On average when three variables were used, the overall accu-racy of SVMs was 1-2% higher than that of ANNs. This is expected because, as discussed in section 3, SVMs is designed to locate an optimal separating hyperplane, while ANNs may not be able to locate this separating hyper-plane. Statistically the optimal separating hyperplane located by the SVM should be generalized to unseen samples with least errors among all sep-arating hyper planes. The training of ANNs uses empirical risk minimiza-tion which stops training once learning error is within a specified margin. This leads to non-optimal model and the solution is often plagued by local minimum problems. Also different training session lead to different weights parameters.
3. SVM classifiers did not give significantly higher accuracies than ANN clas-sifiers. The average overall accuracies of SVM classifiers were slightly higher than those of ANN classifiers shown as Table 3.

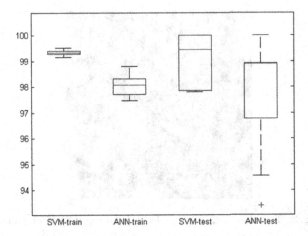

Fig. 1. Boxplots of the overall accuracies of the SVM and ANN classifiers developed using 10-fold cross validation on the training samples and testing samples. Red lines show the median.

Fig. 2. A Landsat-7 ETM+ image of Wuhan, Hubei, China

4. In terms of algorithm stability, the standard deviation of the overall accuracy of an algorithm estimated in cross validation is a quantitative measure of its relative stability (Table 3). Fig. 1 shows the variations of the accuracies of the two types of classifiers. Table 3 and Fig. 1 reveal that the stability of SVM classifiers is basically identical to the range of ANN classifiers. That is because the stabilities of the algorithms were affected by training data size and number of input variables[1]. In this study the training data size is small with only three variables. If the overall accuracies of the SVM classifiers were more stable than ANN classifiers, more training data and variables are needed.

Fig. 3. The classified image of Fig.2 by using a SVM classifier

Fig. 4. The classified image of Fig.2 by using an ANN classifier

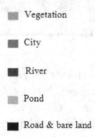

Vegetation

City

River

Pond

Road & bare land

5   Conclusion and Future Work

Artificial neural networks (ANNs) and support vector machines (SVMs) are the effective tools on dealing with various imprecise and incomplete data. In this study, an experiment was performed to evaluate the ANNs and SVMs performances in land cover classification. Of the two algorithms evaluated, SVMs had higher accuracies than the ANNs. The higher accuracies of SVMs should be attributed to its ability to locate an optimal separating hyperplane.

There are several avenues to extend this work. In this paper we used multi-layer perceptrons (MLP) and radial basis function (RBF) SVMs to derive land cover classification from satellite images. While the accuracy and stability of SVMs are affected by training data size, kernel function and parameters and class separate ability. In addition, the accuracy and stability of ANNs depended on network structure, momentum rate, learning rate and converging criteria. A natural extension of this work would be to evaluate the impacts of SVM kernels and selection of training data and input variables on its performance.

References

1. Huang, C., Davis, L.S., Townshend, J.R.G.: An assessment of support vector machines for land cover classification. International Journal of Remote Sensing 23(4) (2002)
2. Anderson, J.A.: An Introduction to Neural Networks. MIT Press (1995)
3. Liu, Z., Liu, A., Wang, C., Niu, Z.: Evolving neural network using real coded genetic algorithm (GA) for multispectral image classification. Future Generation Computer System 20(7), 1119–1129 (2004)
4. Li, Z.-Y.: Supervised classification of multispectral remote sensing image using BP neural network. Journal of Infrared and Millimeter Waves 17(2), 153–156 (1998)
5. Bayaer, A., Shen, Y.-J., Zhu, L., Tateishi, R., Wang, Y.-M.: SPOT/Vegetation NDVI images large scale neural networks classification supported by GIS. Journal of Infrared and Millimeter Wave 24(6), 427–431 (2005)
6. Foody, G.M., Cutler, M.E., McMorrow, J., Pelz, D., Tangki, H., et al.: Mapping the Biomass of Bornean Tropical Rain Forest from Remotely Sensed Data. Global Ecology and Biogeography 10, 379–387 (2001)
7. Paola, J.D., Schowengerdt, R.A.: A review and analysis of backpropagation neural networks for classification of remotely sensed multi-spectral imagery. International Journal of Remote Sensing, 16, 3033-3058
8. Bishop, C.M.: Neural Networks for Pattern Recognition, 1st edn. Oxford University Press (1995)
9. Vapnik, V.N.: The Nature of Statistical Learning Theory. Springer, Berlin (1995)
10. Chang, C.-C., Lin, C.-J.: LIBSVM: a library for support vector machines. ACM Transactions on Intelligent Systems and Technology 2, 27:1–27:27 (2011)
11. Joachims, T.: Text categorization with support vector machines-learning with many relevant features. In: European Conference on Machine Learning, Chemnitz, Germany, pp. 137–142

A Neural Network for Episodic Memory
with Pattern Interrelation

Min Xia*, Liguo Weng, Xiaoling Ye, and An Wang

College of Information and Control Science
Nanjing University of Information Science and Technology
Nanjing, P.R. China
xiamin_wh@hotmail.com

Abstract. Episodic memory refers to the ability to encode and represent the temporal order of discrete elements occurring in a sequence. At the neural level, reproducible sequential neural activity has been shown to be crucial in a variety of cases. In this paper, the ordered pattern's interrelation has been proposed to realized the episodic memory. The ordered pattern's interrelation can help the network oscillate between the stored pattern, leading to the episodic memory.

Keywords: Neural Activity, Neural Network, Episodic Memory, Interrelation.

1   Introduction

Episodic information processing, for instance the episodic memory, plays an important role on many functions of brain[1]. At the neural level, reproducible sequential neural activity has been shown to be crucial in a variety of cases, such as processing of sensory information[2], animal communication[3], and motor control and coordination[4]. All these intriguing experimental observations pose the problem about how to generate robust sequences of neural activity. Many of the behaviors that we produce are sequential in nature.

Neural networks are often utilized to model the episodic memory. Unlike conventional auto-associative neural networks which evolve to settle at a stable steady state[5–8, 18, 19], a neural network model for episodic association switches orderly among the patterns that stored in the network. This requires the neural network model to have an ability to get out of a stable equilibrium[9]. Several neural network models[10-17][20, 21] have been proposed to model the sequential learning and memory. For instance, H. Sompolinsky and I. Kanter[10] used the asymmetric synapses to achieve the temporal association; Philip's model[11] was implemented in a two-layer neural network, with one layer acting as sensory neurons and another layer acting as principal neurons; M. Rehn[12] proposed a temporal association with dynamic depression synapses; O.A.S. Carpinteiro gave a hierarchical self-organizing map model for episodic recognition[21].

---

* Corresponding author.

Z. Li et al. (Eds.): ISICA 2012, CCIS 316, pp. 540–548, 2012.
© Springer-Verlag Berlin Heidelberg 2012

In the real brain function, the stored patterns are always correlative with each other, but in many past neural network models of memory, including that of Winder's model, the interrelation between the stored patterns is ignored. In the most traditional memory model, the stored memories would typically tend to be fixed point attractor states[6, 19, 26, 27] which can no realize the episodic memory. The interrelation between the stored patterns is very important in the memory process, it is always regarded as a mechanism that have an ability to get out of a stable equilibrium[24, 25]. In this paper, the ordered pattern's interrelation can help the network oscillate between the stored pattern, realizing the episodic memory memory.

The remaining paper is organized as follows. In Sect. 2, the episodic memory model based on ordered pattern interrelation is described. And the simulation results of the model are discussed in Sect. 3. Finally, Sect. 4 concludes this paper.

2 Methods

2.1 Method Description

The neuron in the network is characterized by binary state $\{S_i\}_{i=1}^N = \{1, -1\}$, where N is the number of the nodes in the network. A set of patterns $\{X^u = (x_1^u, x_2^u, ..., x_N^u)\}$ are stored in the network, where $u = 1, 2, ..., p$ denotes the label of the patterns. Each element $\{X^u\}$ of the uth memory pattern is generated independently with $prob[x_i^u = 1] = 1 - prob[x_i^u = -1] = k$, where k is coding level of the network. In this work, k is set as 0.5. For the task of the recognition of p stored patterns, the synaptic coupling strength w_{ij} is usually given by the Hebbian learning rule. In such rule, connection strength for network are kept in a $N \times N$ weight matrix W, where each weight w_{ij} is a real-valued number. The weight are incremented proportional to the activity of the framing nodes:

$$\Delta w_{ij} \propto (x_i - \langle x_i \rangle)(x_j - \langle x_j \rangle) \tag{1}$$

Where $\langle . \rangle$ indicates mean node activity. In In the following, we simplify the notation by assuming rate representations with zero mean and learning procedures which clamp the network state to the training patterns. And the w_{ij} defined as:

$$\begin{cases} w_{ij}^0 = \frac{1}{N} \sum_{u=1}^p x_i^u x_j^u \ for \ i \neq j \\ w_{ii}^0 \end{cases} \tag{2}$$

In this work, weights on self-connections are not required, so $w_{ii} = 0$. The state of each neuron in the network is updated as follows:

$$s_i(t+1) = \begin{cases} 1 & with \ probability \ (1 + e^{-h_i(t/T)})^{-1} \\ -1 & with \ probability \ (1 + e^{h_i(t/T)})^{-1} \end{cases} \tag{3}$$

Where $s_i(t)$ is the output of neuron i at time t. T is a temperature parameter, it is assumed to be 0.1 in our simulations. Updating is done synchronously. $h_i(t)$ is the local field of neuron i:

$$h_i(t) = \sum_{j=1}^{N} w_{ij}(t)s_i(t) - \theta_i(t) - \beta K_i(t) \tag{4}$$

Where $\theta_i(t)$ is the threshold associated with local field of neuron i, $K_i(t)$ is a biasing factor that compensates for the unequal numbers of +1 and -1 values among the memory patterns, and β is the scale parameter. Based on the Horn and Usher's model[22], the factor $K_i(t)$ was described as:

$$K_i(t) = \sum_{u=1}^{p} A \cdot M \cdot x_i^u + 2(M - A) \tag{5}$$

where A is the average activity level of all nodes over all memory patterns that were used, and M is the average activity of the network at a given time step. Unlike traditional fixed-point attractor networks where after learning a nodes threshold is fixed, the threshold values here are changed with each time step t, describing as:

$$\theta_i(t) = k_r r_i(t) \tag{6}$$

$$r_i(t) = \frac{r_i(t-1)}{c} + s_i(t) \tag{7}$$

where $r_i(t) = 0$ and $c > 1$. At every time step, when the state of neuron i is +1, then the threshold of neuron i can be increased, making it more likely that the state of the neuron will become negative during the next time step. When the state of the neuron i is -1, the threshold can be decreased, making it more likely that the state of the neuron will become positive. Thus, the network state can oscillate and explore different stored patterns. In order to present the episodic memory during a running memory span task, when each presented memory state is transiently present, the connection strengths w_{ij} are all concurrently updated as the weight change rule:

$$w_{ij}(t) = (1 - \tau)w_{ij}(t-1) + \frac{1}{N}\tau s_i(t-1)s_j(t-1)(1 - \delta_{ij}) + \tau m_{ij} \tag{8}$$

Where $\tau(0 \leq \tau < 1)$ is a decay rate capturing how weights diminish over time, and δ_{ij} is Kronecker's delta (if $i = j$, that $\delta_{ij} = 1$; if $i \neq j$, that $\delta_{ij} = 0$). The part $-\tau w_{ij}(t-1)$ gradually reduces the influences of old memory patterns. In this work, the initial weight of $w_{ij}(t)$ is set as $w_{ij}(0) = w_{ij}^0$. The second term on the right side of this weight change rule is following the Winder's model[23], but in this letter, only $\frac{1}{N}\tau s_i(t-1)s_j(t-1)$ can be added to the new connection weight, m_{ij} of the third term on the right side of this weight change rule is the pattern's interrelation, which is described as:

$$m_{ij} = \frac{1}{N} \sum_{u=1}^{p} x_i^{u+1} x_j^u \qquad (9)$$

The cycle can be incorporated by definition of $x_i^{p+1} = x_i^1$. The pattern's interrelation lead to hetero-assocaitions between different patterns in a sequence, resulting in episodic memory. Both the part of $(1 - \tau)w_{ij}(t - 1)$ and the part of τm_{ij} are necessary to achieve the robust sequence memory, $(1 - \tau)w_{ij}(t - 1)$ trend to stay at a stable equilibrium, and the w_{ij} weights can drive the state of the network from one pattern to the next stored pattern.

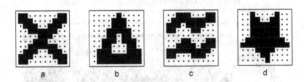

a   b   c   d

Fig. 1. Four stored patterns used for network memory states, where the networks 100 nodes are pictured as a 10×10 array of elements for illustrative purposes. Nodes with an activation level of +1 are indicated by a black block, and those with an activation level of -1 are in a dot

2.2 Measuring the Overlap between Stored Patterns and Network's States in Episodic Memory

In order to evaluate the performance of the network in episodic memory, the similarity measure for the stored patterns and neurons' states should be proposed. Firstly, the distance between the state of the network $s(t)$ and the stored pattern x^u at time t must be computed. In this paper, we use the Hamming distance, which is characterized as follow:

$$z^u(t) = \frac{1}{2} \sum_{i=1}^{N} |s_i(t) - x_i^u| \qquad (10)$$

The similarity of current state $s(t)$ to the stored pattern X^u is then computed as follow:

$$F^u(t) = 1 - \frac{z^u(t)}{N} \qquad (11)$$

which lies between 0.0 and 1.0. If the similarity $F^u(t)$ is closed to 1, the stored pattern x^u is assumed to be retrieved at time step t, while progressively lower values of $F^u(t)$ indicate progressively worse matches.

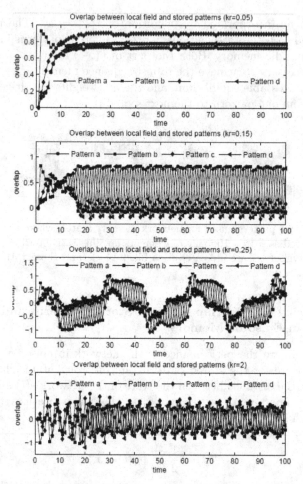

Fig. 2. Overlaps of local field $h(t)$ and network's state $s(t)$ during a run of the network over 100 time steps starting from initial state of pattern b. The values of k_r are 0.05, 0.15, 0.25, and 2 from top to bottom respectively.

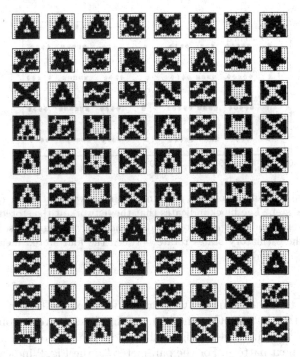

Fig. 3. The sequence of the output patterns of the network, starting from initial state of pattern b, from 1 time step to 80 time step

3   Results

In order to facilitate visual interpretation of the networks state, the nodes in the network are displayed as a 10×10 array, each represented so as to pictorially resemble the figure visually, as illustrated in Fig. 1. In this study, there are four patterns shown in Fig. 1 employed as the stored patterns or learning patterns. Each pattern comprises 10×10 binary pixels in the neural network with 100 neurons, that is, $N = 100$. A neuron will be represented by a black block, when its output, x_i, is equal to 1, which indicates that the neuron is "firing", while a neuron is denoted by a dot when its output is equal to -1, which indicates that the neuron is "resting". According to the synaptic weights learning rule given in Eq. (1), the four stored patterns shown in Fig. 1 are embedded in the neural network.

With the episodic memory model proposed in Sect. 2, firstly we must find appropriate values of parameters τ and k_r. With small decay τ, the interrelation between the stored memory patterns is too small, preventing the network getting out the stable equilibrium. In contrast, when the decay rate is very large, the sequence is not stable, and the network tends to attract to the next pattern in sequence, causing it to overlap a number of consecutive patterns and consequently lose the sequence. In this paper, the parameter τ should be set in $[0.15, 0.3]$.

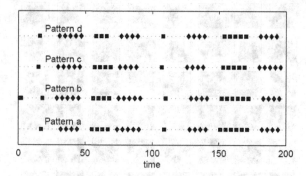

Fig. 4. The sequence of the output patterns of the network, starting from 101 time step to 180 time step. The block indicates that the network at the stored pattern, and the reverse pattern recalled is indicated by diamond. By viewing the oscillation peaks, it shows that the oscillatory states rotate between the four stored memory patterns and their reverse patterns.

In order to transfer from a pattern x^u to the next pattern x^{u+1}, $h(t)x^{u+1}$ must be more than $h(t)x^u$, that the $h(t)$ can drive the network from pattern x^u to pattern x^{u+1} in time $t+1$. Fig. 4 show how the network transfers from a pattern to the next pattern. For the parameter k_r, the Fig. 2 indicated that the value of k_r can not be too large or to small. Fig. 2 present the simulation result for overlap of the local field $h(t)$ with the stored patterns in a sequence with parameter value $k_r = 0.05, k_r = 0.15, k_r = 0.25$ and $k_r = 2$ respectively. The Fig. 2 shows that when $k_r = 0.05$, the network is in the stable equilibrium, that transitions between the patterns do not occur, the part of the hetero-association is not sucient for making the sequence evolve to the next pattern. As the value of $k_r = 0.15$, the episodic memory can be realized perfectly. When the value of k_r comes to 0.25, the Fig. 2 indicates that not only the stored patterns can be memorized, but the reverse patterns of stored patterns can be memorized. When the parameter k_r is 2, the network can not be stable in a stored pattern or a reverse pattern, it is random in space, the episodic memory can not be achieved. So, from the discussion above, the value of k_r should be more than 0.12 and less than 0.6, that the episodic memory can happened.

The Fig. 3 gives the sequence of the output patterns of the network starting from initial state of pattern b with the parameter $k_r = 0.25$, from 1 time step to 80 time step (In fact this simulation is run in 200 time steps). From Figure 3, we can see that the stored patterns can be remembered orderly, and the reverse patterns of the stored patterns can also be memorized. Fig. 4 shows when the oscillations in $F^u(t)$ values peaked at 0.98 or above during just the middle of the testing period of Fig. 3. Peaks in the oscillations associated with the different recalled patterns alternate with each other, unlike with fixed-point neural associative memories.

Fig. 5 presents the overlaps of stored patterns and network's state $s(t)$ during a run of the network over 100 time steps with $k_r = 0.15$, starting from initial

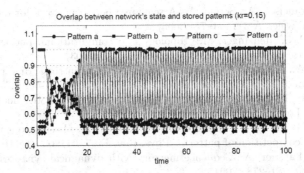

Fig. 5. Overlaps of stored patterns and network's state $s(t)$ during a run of the network over 100 time steps with $k_r = 0.15$

state of pattern b after being presented with the temporal sequence in the order. The pattern a, b, c and d are recalled in the process, their oscillations peak indicates that the network perfectly match the patterns.

4   Conclusion

Sequence memory refers to the ability to encode and represent the temporal order of discrete elements occurring in a sequence. This paper proposed an episodic model based on the ordered pattern interrelation. The proposed model can provide a new method to simulate the memory function of the real brain. Ascribing to the ordered pattern interrelation, the network have an ability to get out of a stable equilibrium, which drive the network transfers from a pattern to the next pattern, leading to the episodic memory.

Acknowledgment. This work is supported was partially supported by the National Natural Science Foundation of PR China (grant No. 61105115), and the national department public benet research foundation (grant No. GYHY200806017).

References

1. Anderson, J.A.: Learning and Memory. John Wiley & Sons, New York (1995)
2. Laurent, G., et al.: Odor encoding as an active, dynamical process: experiments, computation, and theory. Annu. Rev. Neurosci. 24, 263–297 (2001)
3. Hahnloser, R.H.R., et al.: An ultra-sparse code underlies the generation of neural sequences in a songbird. Nature 419, 65–70 (2002)
4. Selverston, A.: General principles of rhythmic motor pattern generation derived from invertebrate CPGs. Prog. Brain Res. 123, 247–257 (1999)
5. Bohland, J.W., Minai, A.A.: Efficient associative memory using small-world architecture. Neurocomputing 38, 489–496 (2001)
6. Hopfield, J.J.: neural networks and physical systems with emergent collective computation abilities. Proc. Nat. Acad. Sci., USA 79, 2445–2558 (1982)

7. Juan, I., Francisco, A., Sergio, A.: A scale-free neural network for modelling neurogenesis. Physica A 371, 71–75 (2006)
8. McGraw, P.N., Menzinger, M.: Topology and computational performance of attractor neural networks. Phys. Rev. E 68, 047102 (2003)
9. Sandberg, A., Lansner, A.: Synaptic depression as an intrinsic driver of reinstatement dynamics in an attractor network. Neurocomputing 44-46, 615–622 (2002)
10. Sompolinsky, H., Kanter, I.: Temporal association in asymmetric neural networks. Phys. Rev. Lett. 57, 2861–2864 (1986)
11. Seliger, P., Tsimring, L.S., Rabinnovich, M.I.: Dynamics-based sequential memory: Winnerless competition of patterns. Phys. Rev. E 67, 011905 (2003)
12. Rehn, M., Lansner, A.: Sequence memory with dynamical synapses. Neurocomputing 58-60, 271–278 (2004)
13. Tank, D.W., Hopfield, J.J.: Neural computation by concentrating information in time. Proc. Nat. Acad. Sci. 84, 1896–1900 (1987)
14. Kleinfeld, D.: Sequential state generation by model neural networks. Proc. Nat. Acad. Sci. 83, 9469–9473 (1986)
15. Gutfreund, H., Mezard, M.: Processing of temporal sequences in neural networks. Phys. Rev. Lett. 61, 235–238 (1988)
16. Lawrence, M., Trappenberg, T., Fine, A.: Rapid learning and robust recall of long sequences in modular associator networks. Neurocomputing 69, 634–641 (2006)
17. Huerta, R., Rabinovich, M.: Reproducible Sequence Generation In Random Neural Ensembles. Phys. Rev. Lett. 93, 238104 (2004)
18. Amit, D.J.: Attractor neural networks and biological reality: associative memory and learning. Futur. Gener. Comp. Syst. 6, 111–119 (1990)
19. Xia, M., Fang, J., Yang, T., Wang, Z.: Dynamic depression control of chaotic neural networks for associative memory. Neurocomputing 73, 776–783 (2010)
20. Kleinfeld, D., Sompolinsky, H.: Associative neural network model for the generation of temporal patterns. Theory and application to central pattern generators. Biophys. J. 54, 1039–1051 (1988)
21. Carpinteiro, O.A.S.: A Hierarchical Self-Organizing Map Model for Sequence Recognition. Neural Process. Lett. 9, 209–220 (1999)
22. Horn, D., Usher, M.: Parallel activation ofmemories in an oscillatory neural network. Neural Computation 3, 31–43 (1991)
23. Winder, R.K., Reggia, J.A., Weems, S.A., Bunting, M.F.: An Oscillatory Hebbian Network Model of Short-Term Memory. Neural Computation 21, 741–761 (2009)
24. Xia, M., Fang, J., Pan, F., Bai, E.: Robust sequence memory in sparsely-connected networks with controllable steady-state period. Neurocomputing 72, 3123–3130 (2009)
25. Xia, M., Tang, Y., Fang, J., Pan, F.: Efficient multi-sequence memory with controllable steady-state period and high sequence storage capacity. Neural Computing and Application 20, 17–24 (2011)
26. Wickramasinghe, L.K., Alahakoon, L.D., Smith-Miles, K.: A novel Episodic Associative Memory model for enhanced classiffication accuracy. Pattern Recognition Letters 28, 1193–1202 (2007)
27. Amari, S.: Characteristics of sparsely encoded associative memory. Neural Networks 2, 451–457 (1989)

The Application of Improved RHT
in High-voltage Transmission Lines Detection

Fang Li, Yishui Shui, Lan Liu, and Zhiqiang Guo

Key Laboratory of Fiber Optic Sensing Technology and Information Processing
University of Technology, Ministry of Education
Hubei, Wuhan 430070, P.R. China
{otmvictory,shuiyishui}@gmail.com,
whekon@163.com,
guozhiqiang@whut.edu.cn

Abstract. The detection of high-voltage transmission lines was one of the important part of Smart Grid.The image was acquired by the industrial cameras set up on a high-voltage transmission line towers.The algorithm was proposed based on improved Randomized Hough Transform (RHT).According to the particularity of the collected images, image enhancement,filtering and edge detection preprocessing was needed before detection.And,cause of the linear characteristics of the high voltage transmission lines and angle characteristics,we used improved RHT combined with the angle to complete transmission line detection. Compared with traditional Hough Transform, the algorithm could save testing time,improve the detection accuracy, laid the foundation for real-time detection.Through processing pictures, it was well improved that the algorithm can detect the high-voltage transmission lines effectively.

Keywords: High-voltage Transmission Line, Pre-processing, Edge Detection, Hough Transform, Angle Match.

1 Introduction

In remote mountainous areas or regions with arduous conditions, we usually use manpower to monitor the high-voltage transmission system in the traditional mode .Owing to the development of image processing technology in the Industrial field and the Smart Grid booms, remote monitoring has also been used in Smart Grid to save a lot of labor costs. We used image processing techniques here to detect the transmission line in order to reduce the labor intensity effectively. And it also laid the foundation of the research of Windage Yaw. Due to natural factors, photos of transmission lines which was collected by remote monitoring which have low contrast. The gray values of images was similar to the sky, and the high-voltage electric towers have linear structure too. These reasons leaded to the decline of image quality, and it would not only affect the detection of wires, but also make it difficult to detect the line. So we should use image processing to deal with them. From paper[1], we could see the author use edge detection in the

Z. Li et al. (Eds.): ISICA 2012, CCIS 316, pp. 549–556, 2012.

image, but he didn't identify the high-voltage transmission lines. There is only some researches in the wire which have the ice cover phenomenon in paper [2] and paper [3], as for general lines there wasn't more introductions, which is the first step to do research on Windage Yaw.Therefore, the paper proposed improved Random Hough Transform to detect the line which was useful to the value of practical application of high-voltage transmission lines detection. Transmission lines are generally linear like, and linear won't change whenever it translate or rotate. The paper used RHT to detect lines in the image,and angles to limit the result to improve, and make it more suitable in the transmission line detection.

2   Theory

Hough Transform[4] is the classic method of line detection. Owning to its insensitive to random noise and partially covered and some other advantages like this, it is widely used in areas such as machine vision and pattern recognition. The algorithm is for the overall relationship between every pixel, and it was divided into standard Hough Transform (SHT) and random Hough Transform(RHT)[5,6].

2.1   Standard Hough Transform

The traditional HT was proposed by P.V.C.Hough, and the algorithm based on slop and intercept used the Dotted line Dual to detect lines. A straight line can be expressed as the follow formula in the image space X-Y:

$$y = mx + c \tag{1}$$

In formula (1), m is the slope of the straight line, and c is the intercept, it can also be described like this:

$$c = -mx + y \tag{2}$$

According to formula (2), it can be considered as a straight line equation in parameter space, and x is the slope, y is the intercept.

Moreover, from Fig. 1(a) and Fig. 2(b) we could see that a point (x, y) in image space X-Y correspond to a straight line in the parameter space, and a point (m, c) in parameter space M-C correspond to a straight line in the image space.

We usually discrete the parameter space M-C into two-dimensional accumulate unit when calculating. The steps are as described below.

1. Transform pixels of image space into parametric linear of parameter space.
2. Press each parameters which was transformed from the image parameter into the accumulation unit.
3. Find the partial maximum in the accumulation unit, and its argument coordinate- es is the parameters of the straight lines in the image space. Then determine the values of m and c according to the location of the peak point.

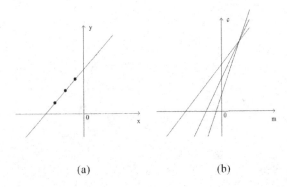

Fig. 1. The mutual transformation of the image space and parameter space(SHT)

2.2   Randomized Hough Transform

However, using SHT to detect lines would lead the value of m and c to be infinite when we faced to vertical lines. What's more, it needs large parameter space and requires huge amount of computation and some other defects like this. So in response to these problems, Lei XU had proposed Randomized Hough Transform, RHT used Multi-to-one mapping to avoid the huge amount of computation in the SHT.

In image space, the mapping relationship of the polar coordinates of a straight line could be expressed as follows:

$$\rho = x\cos\theta + y\sin\theta \tag{3}$$

In formula(3), (x, y) is a feature point in image space; ρ is the vertical distance between the origin and the line in the image space; θ is the angle between the normal of the straight line and the x-axis.

As shown in Figure 2(a) and Figure 2(b),we used formula(3) to do this corresponds to the relationship. In Figure 2(a),each point on the straight line in image space would be a sinusoid in the parameter space. And then, the sinusoid as was shown in Figure 2(b), the point of intersection on the sinusoid lines indicated these point are on the same line when in the image space.

Steps of Randomize Hough Transform were as follows.

1. Transform the image space into parameter space $\rho - \theta$, each straight line in the image space would be a point in the parameter space;
2. Disperse the parameter space $\rho - \theta$ and press it into accumulation unit;
3. Find the partial peak point in the accumulation unit, and the point express one straight line in the image space.

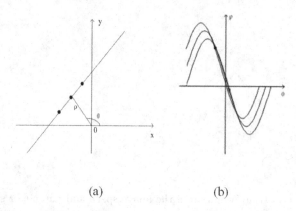

(a)   (b)

Fig. 2. The mutual transformation of the image space and parameter space(RHT)

3   Experiment

3.1   Pre-process

For the particularity of the images of high-voltage transmission line, Pre-processing was needed to achieve the best result. Pre-process used in this paper mainly were image enhancement, median filtering, edge detection of canny operator, and so on.

Image Enhancement. The images we collected were 24-bit true color pictures, and stored as JPG format. The calculate capacity will be huge if we use this format. In order to facilitate the calculation, we transformed the true color picture into 8-bit grayscale first.

The illumination had a strong effect on the high-voltage transmission lines image. As the images were collected under somber weather, which had a low contrast and that would have a significant impact on the detection. So we used Histogram Equalization to enhance the pictures first. Histogram Equalization is an algorithm that enhance the images indirectly, which is non-linear stretched. The method distributes the gray-scale which was concentrated into holistic gray-scale range. From experiment we made a conclusion: Histogram Equalization had a good effect on the images we collected.

Median Filtering. The algorithm is a typical non-linear method that can filter out high frequency component efficiently. Median Filtering needs to set the neighborhood of pixels, so that we can get a median of the former's neighborhood. What's more, it uses a sliding window containing an odd number of points, and the pixel in the window will be arranged to get a median, then, the former pixels in the window will be replaced by the median. In this paper, we used [3 × 3] template for filtering. The Median filter formula was as below[8]:

$$\widetilde{f}(x,y) = med_{S_M(i,j)}[\widetilde{f}(k,l)] \qquad (4)$$

From the formula (4) above we could see that: the widow of median filtering was S_M, and the neighborhood of pixel (i, j) is $S_M(i, j)$, and M is the number of pixels in the sliding window, $\tilde{f}(i, j)$ is the output of (i, j) after Median Filter.

Edge Detect. As for the high-voltage transmission lines images, we found the arithmetic operator -'Sobel' was most efficient for the pictures. The operator has excellent characteristics such as: faster calculate speed, simple to implement, and can detect the edge of the image accurately. If the binary image to be detected was $f(x, y)$, and the image after edge detecting was F:

$$F_x = \begin{bmatrix} -1 & -2 & -1 \\ 0 & 0 & 0 \\ 1 & 2 & 1 \end{bmatrix} * f(x, y) \tag{5}$$

$$F_y = \begin{bmatrix} -1 & 0 & 1 \\ -2 & 0 & 2 \\ -1 & 0 & 1 \end{bmatrix} * f(x, y) \tag{6}$$

$$F = \sqrt{F_x^2 + F_y^2} \tag{7}$$

Formula(5) expressed images after edge detect used 'Sobel' operator, which was in horizontal orientation. Formula(6) expressed the image after edge detect, which was in the portrait orientation. And in formula(7) F is the final image after edge detect.

3.2   Improved Hough Transform

Considered the special environment and the angle of the camera, we found that the transmission line in the image extended to the edge of the picture, and had a certain range of angle. As for the special environment, we took the improved Hough Transform based on matching the angle.

The formula transform image space to parameter space is as follows:

$$\theta = arctan(\frac{y}{x}) \tag{8}$$

$$\rho = \sqrt{x^2 + y^2} \tag{9}$$

From formula (8)(9) we could get the parameter space $\rho - \theta$,and we got the threshold (T_{min}, T_{max}) through experiments. And then, we could decide which line was useful by using the threshold. The algorithm could be described like follows:

1. Set a limit to the image while calculating the grads to get a binary image.
2. Transform image space into parameter space, and ensure the useful area through the threshold.
3. Disperse the parameter space $\rho - \theta$, and press it into accumulation unit, if there have already been the point, add 1 to the cell.
4. Chose the partial peak point when the accumulation unit exceeds the threshold, then the point (ρ, θ) corresponds to the right line.

4   Results and Analysis

Because of the linear characteristics of the high voltage power line, we could do some operation such as gray processing, grayscale, image enhancement, median filtering, and edge detection to the original image. After all these pre-processing, we could find the straight line through by using improved Hough transform.

From the figure we could see that, Figure 3(a) was the result after edge detecting which based on 'Sobel' operator, and from the result we could see that edge information could be well distilled based on 'Sobel' operator. Next, We used the same operator to do the same processing to the original image of Figure 4, then use improved RH to detect the lines. The results came as follows:

Fig. 3.(a) Result after edge detect

Fig. 3(b). Result after SHT

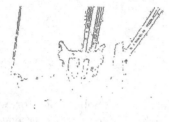

Fig. 3.(c) Result after RHT

Fig. 3(d). Result after improved RHT

Fig. 3.

However, we could detect the power line by using SHT, but from the figure we also found that the false positive rate was too high. What's more, the pre-processing has played an important role to the detection, and we found the 'Sobel' operator was adaptive to these projects.

Experiment data in Table 1 was the conclusion of figures appeared above. And from Figure 3,4 we could see that the original Hough transform keep in a low precision in detecting wire. But it could detect the useful transmission lines effectively by using the method Standard Hough transform. This also has shown that the algorithm was insensitive to random noise and partially covered From Figure 3(a) we can see, although the drooping abandoned wires were not

Fig. 4. The Second figure which shows the result after improved RHT

Table 1. Experiment data

Figure number	The actual number of lines	The number of lines detected	False detection rate	Detection rate
Fig.3(b)	7	13	46.1%	30.8%
Fig.3(c)	7	8	24.8%	62.5%
Fig.3(d)	7	6	0	85.7%
Fig.4	9	7	0	77.8%

detected, but the false positive rate was high. After improved Hough detection and angle filter, the accuracy of detection of lines has been greatly improved, and the false positive rate is almost 0. What's more, it eliminated the influence of transmission line towers which has achieved desiring results.But the detection rate still needs to improve.

5   Conclusion

With the development of image processing technology in the Smart Grid, Image Process also has been applied to this field. Remote monitoring of transmission lines can not only save labor costs but also can give alarm at real-time. Improved RHT applied to the detection of the transmission line provides an important foundation for the monitoring system, so that to judge Windage Yaw and ice cover by detecting the Power Transmission Line. The method of angle limited improved RHT proposed by the paper can effectively shorten the detection time and reduce the false detection rate, improve the detection accuracy and have higher application value. From experiment results we can see that this method can be applied in the monitoring system of high-voltage Transmission lines efficiently. But the accuracy still need to be improved, we are considering use Genetic Algorithm or Particle Swarm Optimization to optimize the threshold.

References

1. Tong, W.-G., Li, B.-S., Yuan, J.-S., Zhao, S.-T.: Transmission line extraction and recognition from natural complex background. In: Proceedings of the International Conference on Machine Learning and Cybernetics, pp. 473–477 (2009)
2. Zhu, Y.: Icing Detection of Transmission Lines Based on Image-Processing Technology. Southern Power System Technology, 106–110 (2010)
3. Lu, J., Luo, J., Zhang, H., Li, B., Li, F.: An image recognition algorithm based on thickness of ice cover of transmission line. In: Proceeding of 2011 International Conference on Image Analysis and Signal Processing, IASP 2011, pp. 210–213 (2011)
4. Hough, P.V.C.: A method and means for recognizing complex patterns. US Patent, vol. 3,069,654 (1962)
5. Xu, L., Oja, E.: Randomized Hough Transform (RHT): Basic Mechanisms, Algorithms, and Computational Complexities. Image Understanding, 131–153 (March 1993)
6. Xu, L., Oja, E., Kultanen, P.: A new curve detection method: Randomized Hough Transform (RHT). Pattern Recognition Letters, 31-338 (May 1990)
7. Lehmann, A., Leibe, B., Van Gool, L.: Fast PRISM: Branch and Bound Hough Transform for Object Class Detection, pp. 175–197 (2009)
8. Duan, D., Xie, M., Mo, Q., Han, Z., Wan, Y.: An improved Hough transform for line detection. In: Proceedings of the International Conference on Computer Application and System Modeling, ICCASM 2010, pp. 354–357 (2010)
9. Singh, M., Panigrahi, B.K., Maheshwari, R.P.: Transmission line of fault detection and classification. Emerging Trends in Electrical and Computer Technology (ICETECT), 15–20 (2011)
10. Han, J., Li, E., Tao, B., Lv, M.: Reading recognition method of analog measuring instruments based on improved hough transform. In: IEEE 201110th International Conference on Electronic Measurement and Instruments, ICEMI 2011, pp. 337–340 (2011)
11. Gonzalez, R.C., Woods, R.E.: Digital Image Processing, 3rd edn. (January 2010)

VQ Codebook Design Using Genetic Algorithms for Speech Line Spectral Frequencies

Fatiha Merazka

Electronic & Computer Engineering Faculty
University of Science & Technology Houari Boumediene
P.O. Box 32, El Alia, 16111 Algiers, Algeria
fmerazka@usthb.dz

Abstract. Vector quantization (VQ) is a data compression and coding technology widely used in speech coding, speech recognition and image data compression. One of the key problems occurring in VQ is the codebook design problem. This paper presents the design of codebooks for VQ of the line spectrum frequencies (LSF) using genetic algorithms (GA). The algorithms performance is experimentally evaluated by creating codebooks for speech LSF quantization. A comparison with LBG and GLA algorithms is made. Simulation results show that a hybrid algorithm based on the GA+GLA provides superior performance compared with GLA and LBG.

Keywords: LSF, Genetic Algorithm, GLA, LBG, MSE.

1 Introduction

Line spectral frequencies (LSF) coefficients vector quantization (VQ) is very important in various speech coding systems that are based on the linear prediction (LP), such as code-excited linear prediction (CELP), mixed-excitation linear prediction (MELP), especially in medium and low bit rates applications[1,2]. Since the LP model coefficients are very important parameters and are critical for the quality of speech synthesis, many quantization bits are used to code them. Compared with the other LP model coefficients, the LSF coefficients have better performance on anti-jamming and quantization capability[3]. The LSF are related to the poles of the LPC filter (or the zeros of the inverse filter) in the Z-plane. For a p-th order LPC analysis, the Z-transform of the LP inverse filter is denoted by :

$$A_p(z) = 1 + a_1 z^{-1} + ... + a_p z^{-p} \qquad (1)$$

The parameters $\{a_i\}, i = 1, 2, ..., p$, are commonly referred to as the LP coefficients [3], From (1) two new polynomials are defined:

$$\left.\begin{array}{c} P(z) \\ Q(z) \end{array}\right\} = A_p(z) \pm z^{-(p+1)} A_p(z^{-1}) \qquad (2)$$

The roots of these polynomials are usually called the Line Spectrum Pairs (LSP). These polynomials have the following properties:

Z. Li et al. (Eds.): ISICA 2012, CCIS 316, pp. 557–566, 2012.

1. All zeros of LSF polynomials are on the unit circle.
2. Zeros of $P(z)$ and $Q(z)$ are interlaced with each other on the unit circle.
3. The minimum phase property of $A_p(z)$ can be easily preserved if the first two properties are intact after quantization.

Some important properties are described in detail in[3].

The LSF codebook design is one of the key problems in VQ. The traditional codebook design method LBG Algorithm[4] is affected by the initial codebook and often generates a local optimal codebook. The generalized Lloyd algorithm (GLA)[5] had been widely used in the design of the codebook. It suffers the drawback that the user must provide the number of clusters in advance while the user in general has no idea about how many clusters there should be in the data set. The quality of codebook can be improved by using simulated annealing (SA)[6,7], which attempts to find a better codebook by shaking the codebook off the local valley in the hope that it will converge to another valley that gives less error. But it is difficult to set the relational parameters in SA to get more improvement. Codebook design algorithms based on evolutionary computation are new methods. Partition-based mechanism[8,9] is proposed whose coding string is the indices of the all training vectors; codebook-based mechanism[10] is proposed whose coding string is the all codebook vectors.

The genetic algorithm (GA) is effective for solving various kinds of heuristic and optimal problems. GA is a random optimization algorithm based on the process of biological evolution by natural selection and genetic variation[11]. GA has strong global search ability, but a weak local optimum capacity and slow convergence rate. As a new global optimization search method, GA has advantages of easy use, universality and wide range of application[12].

Based on the concept of GA, we investigate a method based on the combination of GA+GLA for generating codebooks for LSF VQ.

This paper is organized as follows. In section 2 we review some basics of GA. In section 3 we give the design of codebooks using genetic algorithms. In section 4 the results are presented and discussed. Section 5 concludes the paper.

2   Genetic Algorithms

In this section, we will briefly review the GA. The current genetic algorithms in science and engineering refer to a model introduced and investigated by Holland[13]. In genetic algorithms, a set of solutions to a problem is called chromosomes. A chromosome (string of solution) is composed of genes (features, characters or detectors). Usually, the individual of the whole population contains only one chromosome.

The performance of the solution is called fitness. The fitness of chromosomes is evaluated, ordered and then new chromosomes are produced by using the selected candidates as parents and applying mutation and crossover operations. The new set of chromosomes is then evaluated and ordered again. This cycle continues until a suitable solution is found. The conventional genetic algorithm is described as the following steps:

Step 1-Initialization: To apply GAs to solve a problem, the fitness function and some parameters for the problem have to be defined. Usually, the parameters that affect the fitness function are regarded as the genes. All the genes are combined together to form a chromosome. An individual may consist of more than one chromosome. In this step, the number of individuals and chromosomes, the selection rate, the crossover rate, the mutation rate, and the considered GA iteration are defined. Also, the initial genes are generated (within the considered range) randomly.

Step 2-Evaluation: The performance of each individual is measured according to the fitness function and its genes. Obviously, an individual with good genes usually has good performance. It therefore is more fit in the environment.

Step 3-Selection: Similar to the natural world that fitter individuals have a higher chance to survive and produce off springs, some bad-performance individuals are discarded according to a selection rate in this step.

Step 4-Crossover: Some new individuals, called children, are produced from the survived individuals in this step. First of all, two individuals, called parents, are selected from the survived individuals using the well-known roulette-wheel-selection scheme[14] or other type of scheme. Then, a pre-defined crossover rate is referred to decide whether the crossover operator is performed or not. That is, if a random-generated probability is higher than crossover rate, the crossover operator is then skipped and the selected parents are simply copied to the children without any change. Otherwise, a crossover point (or N-point crossover sometimes) between the first and the last chromosomes is determined randomly. The fractions of each parent are then exchanged after the point, and two child individuals are produced. This process is repeated until the needed numbers of child individuals are produced.

Step 5-Mutation: A mutation operator is considered in order to avoid GAs get trapped on a local optimum. This usually happens when most individuals in one certain.

3   Codebook Design Using Genetic Algorithms

It has been shown[5] that two conditions are necessary but not sufficient for the existence of an optimal minimum mean squared error (MSE) quantizer MSE given by:

$$MSE = \frac{1}{k} \sum_{i=0}^{k-1} \frac{1}{n} ||(x_i - y_i)||^2 \tag{3}$$

where x_i is the i-th input sample and y_i is the i-th output codeword, n is the dimension of the vectors and k is the size of training sequence.

1. the codewords should be the centroids of the partitions of the vector space.
2. the centroid is the nearest neighbor for the data vectors in the partitioned set.

These conditions have been applied to codebook design in GLA[5]. Since these conditions are necessary but not sufficient, there is no guarantee that the resulting codebook is optimal. The GLA is widely used in codebook generation for VQ. It is a descent algorithm in the sense that at each iteration the average distortion is reduced. For this reason, GLA tends to get trapped in local minima. The performance of the GLA dependent on the number of minima and on the choice of the initial conditions.

This section describes the combination of GA and GLA for codebook design to produce better optimum codebook vectors. It is referred to as GA1+GLA algorithm if the evaluation, selection and crossover are adopted in combination with GLA to produce an enhanced codebook design algorithm.

The fitness of GA can be represented by MSE. In the VQ operation, a chromosome is designated as the centroid of the cluster. The individuals of the population is the codebook. GA1+GLA algorithm consists of the following steps:

Step 1-Initialization: Compute the centroid C_0 from the training sequence of vectors $X_i(i = 1, ..., n)$ and Select L chromosomes $C_j(i = 1, ..., L)$ for every member of the population using random number generator. L corresponds to the codebook size, so that each codebook consists of L single-vector chromosomes. P sets of L chromosomes are generated in this step, P is the population size.

Step 2-Update: GLA is used to update L chromosomes for every individual of the population.

Step 3-Evaluation: MSE of every member of the population is evaluated in this step.

Step 4-Selection: The survivors of the current population are decided from the survival rate (here 0.5). A random number generator is used to generate random numbers between 0 and 1. If the random number is smaller than survival rate, this codebook survives; otherwise, it does not survive. The smallest MSE of the population always survives. Pairs of parents are selected from these survivors and endure a subsequent crossover operation to produce the child chromosomes that form a new population in the next generation.

Step 5-Crossover: The chromosomes of each survivor are sorted in decreasing order according to the squared error (SE) between the chromosome C_j of the current population and the central chromosome C_0. Without sorting here, it is difficult to jump out of the local minima. The 1-point or 2-point crossover technique (Goldberg, 1989) is used to produce the next generation from the selected parents.

Step 6-Termination: Step 2 to step 5 are repeated until the predefined number of generations have been reached. After termination, the optimal codebook is generated from L chromosomes in the best member of the current population.

Another combination of GA and GLA is called GA2+GLA algorithm. It is similar to the GA1+GLA algorithm except that a mutation is applied after crossover.

A random value is added to selected genes in the mutation step. This perturbation gives the GA2+GLA algorithm more opportunity to jump off the local optimum. The added value of the perturbation can be a normal distribution, uniform distribution or any other possible distributions.

Mutation The genes in the chromosomes of the population are mutated according to the mutation rate. Here, the total number of mutations is set to population size P^* number of chromosomes L^* mutation rate. When one chromosome is selected to be mutated from random generation number, the new genes are generated from the old genes by adding the random value α_n with $1 < n < k$ and k is the number of genes in one chromosome:

$$-0.5\sigma_n\eta^m \leq \alpha_n \leq 0.5\sigma_n\eta^m \tag{4}$$

$$\sigma_n(x) = \sqrt{\frac{\sum_{i=1}^{i=n}(x_i - \mu)^2}{n}} \tag{5}$$

$$\mu(x) = \frac{\sum_{i=1}^{k} x_i}{k} \tag{6}$$

σ_n is the standard deviation of the nth dimension of the vector, m is the number of generations processed at present and $\eta < 1$.

4 Experimental Results

LSF coefficients are used as the test features for the codebook generation experiments. The test materials for our experiments consist of speech obtained from TIMIT database[15].LSF vectors are extracted on speech frames of 10ms with Hamming window after high pass. First the LP coefficients are extracted by using of the Levinson-Durbin algorithm; then LP coefficients are converted to LSF coefficients for quantization[16]. Firstly experiments are carried out to compute the MSE versus different number of generation in order to fix this number. Secondly, we calculate the MSE versus different population size in order to fix the population size. Then, we compare different crossover methods. Finally, we investigate the influence of mutation on the combination of GA with GLA.

4.1 Number of Generations

The number of generation can be assimilated to the iteration. We can obtain this number according to the variation of MSE versus the number of generation. It corresponds to the value for which the MSE becomes constant.

Figure 1 shows the MSE of GA1+GLA with N-point crossover and 50 individuals for codebooks of size 64 and 1024 respectively.

From Figure 1, we notice that MSE is almost constant after the 80th generation, then the generation number is fixed to 80 in our experiments.

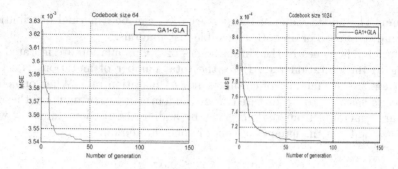

Fig. 1. MSE of codebooks of size 64 and 1024 for different number of generations

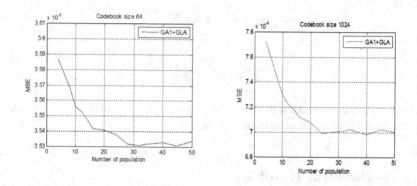

Fig. 2. MSE of codebooks of size 64 and 1024 for different population size

4.2 Population Size

The population size is obtained when the variation of MSE versus the population size became constant. Figure 2 shows the MSE of GA1+GLA with N-point crossover and 50 individuals for codebooks of size 64 and 1024 respectively.

From Figure 2 the population size is fixed to 30.

4.3 Different Type of Crossover

We compare different types of crossover in terms of MSE as follows:

a)point crossover, b)2-point crossover, c)Uniform crossover and d)N-point crossover

The parameters for number of generations, population size and the survival rate are 80, 30 and 0.5 respectively.

A comparison of different crossover is given in Tables 1 and 2 for codebooks of size 64 and 128 respectively.

From Tables 1 and 2, we can see that N-point crossover give the best results.

In Table 3, we give the average MSE (Avg. MSE) for codebooks of size 64 and 128 respectively. The results presented in Table 3 confirm that N-point crossover is the best.

Table 1. MSE of GA1+GLA in comparison with GLA and LBG algorithms for different crossover for a codebook of size 128

GLA	0.004231			
LBG	0.003628			
init population	GA_GLA1 1-point crossover	GA_GLA1 2-point crossover	GA_GLA1 Uniform crossover	GA_GLA1 N-points crossover
1	0.003534	0.003537	0.003533	0.003531
2	0.003543	0.003533	0.003536	0.003539
3	0.003546	0.003545	0.003537	0.003533
4	0.003536	0.003535	0.003532	0.003531
5	0.003537	0.003535	0.003530	0.003529
6	0.003546	0.003542	0.003538	0.003539
7	0.003540	0.003535	0.003536	0.003533
8	0.003541	0.00 3537	0.003540	0.003534
9	0.00 3531	0.003533	0.003529	0.003529
10	0.003539	0.003536	0.003536	0.003532

Table 2. MSE of GA1+GLA in comparison with GLA and LBG algorithms for different crossover for a codebook of size 128

GLA	0.002992			
LBG	0.002737			
init population	GA_GLA1 1-point crossover	GA_GLA1 2-point crossover	GA_GLA1 Uniform crossover	GA_GLA1 N-points crossover
1	0.002623	0.002619	0.002614	0.002607
2	0.002624	0.002623	0.002613	0.002611
3	0.002623	0.002624	0.002621	0.002614
4	0.002623	0.002614	0.002619	0.002619
5	0.002618	0.002617	0.002611	0.002611
6	0.002623	0.002624	0.002619	0.002618
7	0.002628	0.002622	0.002617	0.002616
8	0.002620	0.002619	0.002619	0.002614
9	0.002626	0.002614	0.002624	0.002615
10	0.002627	0.002624	0.002612	0.002617

Table 3. Avg. MSE of GA1+GLA in comparison with GLA and LBG algorithms for different crossover for codebook of size 64 and 128

		Codebook 64	Codebook 128
GLA		0.004231	0.002992
LBG		0.003628	0.002737
GA+GLA1	1- point crossover	0.003539	0.002623
	2- points crossover	0.003537	0.002619
	Uniform crossover	0.003535	0.002617
	N-points crossover	0.003533	0.002614

4.4 Mutation

In this section, we present a comparison between GA1+GLA and GA+GLA with mutation given by GA2+GLA in terms of MSE. Tables 4 and 5 present comparisons for codebooks of size 64 and 128 respectively.

From Tables 4 and 5, we can see that results obtained with crossover with mutation are better that those obtained for crossover without mutation for two codebook sizes 64 and 128 respectively.

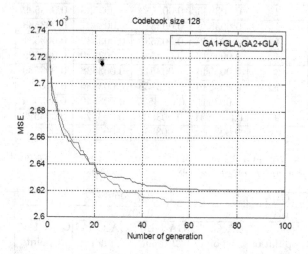

Fig. 3. MSE of codebooks of size 128 for different number of generations for GA1+GLA and GA2+GLA

Table 4. MSE for ten tests with GA1+GLA and GA2+GLA for codebooks size 64and 128 respectively

	Codebook64		Codebook128	
GLA	0.004231		0.002992	
LBG	0.003628		0.002737	
Init population	GA1+GLA without mutation Codebook 64	GA2+GLA with mutation Codebook 64	GA1+GLA without mutation Codebook 128	GA2+GLA with mutation Codebook 128
1	0.003538	0.003528	0.002624	0.002616
2	0.003539	0.003532	0.002617	0.002613
3	0.003543	0.003531	0.002618	0.002609
4	0.003540	0.003540	0.002615	0.002606
5	0.003538	0.003534	0.002616	0.002610
6	0.003539	0.003532	0.002618	0.002605
7	0.003542	0.003534	0.002610	0.002612
8	0.003538	0.003530	0.002614	0.002606
9	0.003541	0.003543	0.002627	0.002620
10	0.003537	0.003528	0.002616	0.002615

Figure 3 show the variation of MSE versus number of generation for codebook size 128. It is clear, from Figure 3, that GA2+GLA is the best. From this study we can see the importance of mutation after crossover which plays a key role in escaping a local minima and approach as much as possible a global minimum.

Table 5. Performance Comparison of GA1+GLA and GA2+GLA for a codebook size 64 and 128 respectively

Algorithm	GA1+GLA	GA2+GLA
Avg. MSE Codebook 64	0.003540	0.003533
Avg. MSE Codebook 128	0.002617	0.002611

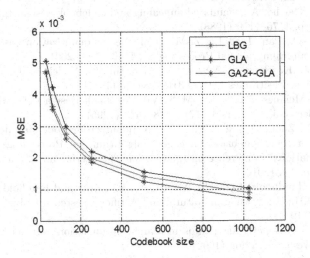

Fig. 4. Comparison between LBG, GLA and GA2+GLA for different codebook size

In Figure 4 we present results obtained for a comparison between LBG, GLA and GA2+GLA. It is clear, from Figure 4, that GA2+GLA is the best.

5   Conclusion

We have presented in this paper the design of codebooks by the use of genetic algorithms. We have shown that by a combination of GA without mutation with GLA, , namely GA1+GLA, we obtain better performance than GLA and LBG. We have added mutation to GA and combine it with GLA, namely GA2+GLA. We have shown that GA2+ GLA give better results compared to GA1+GA2+GLA. The obtained codebooks with GA2+GLA can be used for an efficient quantization of the LSF parameters.

References

1. Makhoul, J.: Linear prediction: A tutorial review speech. Proc. IEEE 63, 124–143 (1975)
2. Paliwal, K.K., Atal, B.S.: Efficient vector quantization of LPC parameters at 24 bits/frame. IEEE Trans. Speech Audio Process. 1(1), 3–14 (1993)
3. Itakura, F.: Line spectrum representation of linear predictor coefficients of speech signal's. J. Acoust. SOC. Amer. 57, S35(A) (1975)
4. Linde, Y., Buzo, A., Gray, R.M.: An Algorithm for Vector Quantizer Design. IEEE Transactions on Communications COM-28(1), 84–95 (1980)
5. Lloyd, S.P.: Least Squares Quantization in PCM. IEEE Transactions on Information Theory IT-28(2), 129–137 (1982)
6. Kirkpatrick, S., Gellatt Jr., C.D., Vecchi, M.P.: Optimization by simulated annealing. Science 220, 671–680 (1983)
7. Vaisey, J., Gersho, A.: Simulated annealing and codebook design. In: Proc. IEEE ICASSP, pp. 1176–1179 (1988)
8. Huang, H.-C., Pan, J.-S., Lu, Z.-M., et al.: Vector quantization based on genetic simulated annealing. Signal Processing 81, 1513–1523 (2001)
9. Delport, V., Koschorreck, N.: Genetic Algorithm for Codebook Design in Vector Quantisation. Electronics Letters 31(2), 84–85 (1995)
10. Pan, J.S., Mcinnes, F.R., Jack, M.A.: VQ Codebook Design Using Genetic Algorithms. Electronics Letters 31(17), 1418–1419 (1995)
11. Yuan, Y.J., Zhou, Q., Zhao, P.H.: Vector quantization codebook design method for speech recognition based on genetic algorithm. In: Proceedings of the 2010 2nd International Conference on Information Engineering and Computer Science, Wuhan, pp. 1–4 (2010)
12. Davis, L.: Handbook of Genetic Algorithms. Van Nostrand Reinhold (1991)
13. Holland, J.H.: Adaptation in Natural and Artificial Systems. University of Michigan Press (1975)
14. Goldberg, D.: Genetic algorithms in search, optimization and machine learning. Addison-Wesley, Reading (1992)
15. NIST, Timit Speech Corpus, NIST (1990)
16. ITU, ITU-T G.729: CS-ACELP Speech Coding at 8 kbit/s, ITU(1998)

Numerical Simulation of an Optimized Xiaermen Oilfield Adjustment Plan

Zhenliang Guan, Congjiao Xie, and Guoping Luo

Key Laboratory of Tectonics and Petroleum Resources
(China University of Geosciences)
Ministry of Education, Wuhan, Hubei, P.R. China
{guanzl,cxie2004,gpluo}@cug.edu.cn

Abstract. In the oil and gas industry, reservoir characterization, modeling, dynamic production systems simulation and optimization has achieved a tremendous success in reducing challenges faced by the industry. This work is concerned with setting up of a numerical model and optimizing the best plan associated with two development scenarios, plan2006B and plan 2005 using professional software Eclipse. The Eclipse software was used for application on Xiaermen oilfield. A model of complexity in between that of the simple decline curve analysis and the complex multidimensional, multi-flow simulator is used. The results obtained proved to be successful for plan2006B beyond plan2005 by an enhanced oil recovery of 3.8%.

Keywords: Numerical model, optimization, Eclipse.

1 Introduction

Numerical simulation has been performed to study the oilfield development plan especially for aged mature field adjustment. Reservoir simulation is a very good technique in which a computer-based mathematical representation of the reservoir is constructed and used to predict its dynamic behavior. One of the most important tasks of the reservoir engineer is the prediction of future production rates from a given reservoir.

The oil industry has developed several methods to accomplish this task. The methods range from simple decline curve analysis techniques to sophisticated multidimensional, multi-flow reservoir simulators. Whether a simple or complex method is used, the general approach taken to predict production rates is first to calculate producing rates for a period for which the engineer already has production information. If the calculated rates match the actual rates, the calculation is assumed to be correct and can then be used to make future predictions. If the calculated rates do not match the existing production data, some of the process parameters are modified and the calculation repeated[1]. This production prediction method was successfully applied in Xiaermen oilfield adjustment development.

Z. Li et al. (Eds.): ISICA 2012, CCIS 316, pp. 567–575, 2012.

The Xiaermen oil field lies in Biyang County, Henan Province. It is located in the middle of a large northeast fault edge, to the east of Biyang Sag, Nanxiang Basin (Fig. 1). The structure of this field is a brachy-anticline situated in an east-west direction complicated by four major and minor faults adjacent to the main oil-bearing areas. Among these faults, a series of small faults to the east were originated from the big fault (Fig. 1). The maximum height of the trap is 275 m.

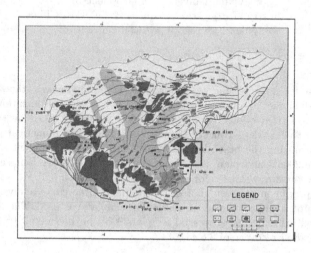

Fig. 1. The location of Xiaermen oil field(Source: JCPT2009, p.37)

The Xiaermen sand was deposited by deltaic fans[2]. Considerable variation in sand thickness occurring over short lateral distances is characteristic of this particular deposit and thick layers with high permeability are patchy and isolated. The sediment underlying the project area is finely grained and relatively cleaned quartz sand. The average porosity is about 24% whilst the mean geometric permeability is 2 μm^2. The values of the porosity and permeability are generally high, but fairly heterogeneous.

To increase the Xiaermen oilfield oil recovery, a study procedure was given by the authors proposed as Fig.2. The numerical simulation is the most important in this route. The process of modifying these parameters to match the calculated oil rates with the actual observed rates is referred to as history matching. Based on oil and water well history match, residual oil saturation was calculated and 2 development scenarios were presented by using professional software Eclipse for Xiaermen oilfield. Compared with the base plan, an optimized adjustment plan from the 2 scenarios was produced. It was found that the oil recovery was enhanced 3.8% after 1 year development.

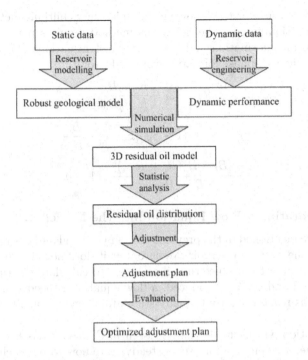

Fig. 2. The diagram of research technical route

2 Development of the Model

The material balance equations do not yield information on future production rates because the equations do not have a time dimension associated with them. These equations simply relate average reservoir pressure to cumulative production. To obtain rate information, a method is needed whereby time can be related to either the average reservoir pressure or cumulative production. A single-phase flow in porous media and equations were developed for several situations that relate flow rate to average reservoir pressure. In view of this development, it became necessary to combine the material balance equations with the flow equations in a model or simulator that could provide a relationship for flow rates as a function of time. The model will require accurate fluid and rock property data and past production data. It is worthy of mention that once a model has been tested for a particular well or reservoir system and found to reproduce actual past production data, it can be used to predict future production rates. The importance of the data used in the model cannot be over emphasized. If the data are correct, the prediction of production rates will be fairly accurate.

2.1 The Material Balance Part of the Model

The problem involves a volumetric, internal gas-drive reservoir. There are several methods for calculating the oil recovery as a function of reservoir pressure for this type of reservoir. For the following example, the Schilthuis method is used. It is worth mentioning that the Schilthuis method requires permeability ratio versus saturation information and the solution of Equations (1), (2), and (3) (written with the two-phase formation volume factor):

$$R = R_{so} + \frac{K_g \mu_o B_o}{K_o \mu_g B_g} \tag{1}$$

$$S_L = S_w + (1 - S_w)[1 - \frac{N_p}{N}])\frac{B_o}{B_{oi}} \tag{2}$$

$$\frac{\frac{N_p}{N}[B_t + B_g(R_p - R_{soi})]}{B_t - B_{ti}} - 1 = 0 \tag{3}$$

2.2   Incorporating a Flow Equation into the Model

The procedure mentioned in the previous section yields oil and gas production as a function of the average reservoir pressure, but it does not give any indication of the time required to produce the oil and gas. To calculate the time and rate at which the oil and gas are produced, a flow equation is needed. It was found that most wells reach the pseudosteady-state after flowing for a few hours to a few days.

An assumption will be made that the well in this work has been producing for a long time, long enough for pseudosteady-state flow to be reached. For this case, Equation (4) can therefore be used to describe the oil flow rate into the wellbore:

$$q_0 = \frac{0.00708 K_o h}{\mu_o B_o}[\frac{p - p_{wf}}{ln(\frac{r_e}{r_w}) - 0.75}] \tag{4}$$

This equation assumes pseudosteady-state, radial geometry for an incompressible fluid. The subscript, o, refers to oil, and the average reservoir pressure, P, is the pressure used to determine the production, Np, in the Schilthuis material balance equation. The increase in time required to produce an increment of oil for a given pressure drop is found by simply dividing the incremental oil recovery by the rate computed from Equation (4) at the corresponding average pressure:

$$\Delta t = \frac{\Delta N}{q_0} \tag{5}$$

The total time that corresponds to a particular average reservoir pressure can be determined by summing the incremental times for each of the incremental pressure drops until the average reservoir pressure of interest is reached. Since Equation (5) requires Np and the Schilthuis equation determines Np/N, N, the initial oil in place, must be estimated according to the volumetric approach by the use of Equation (6).

$$N = \frac{7758 A h \Phi (1 - S_{wi})}{B_{oi}} \tag{6}$$

Combining these equations with the solution of the Schilthuis material balance equation yields the necessary production rates of both oil and gas.

2.3   Procedure for a History Match

Before a history match was performed, a fine 3D reservoir geological model was set up. The initial data of a reservoir generally needs to be adjusted, or tuned, for the simulation model to predict reservoir performance adequately. These data adjustments are performed during the history-matching phase of the simulation study. Fig.3 shows the 3D net pay model of Xiaermen oilfield and others like porosity, permeability models were built as well.

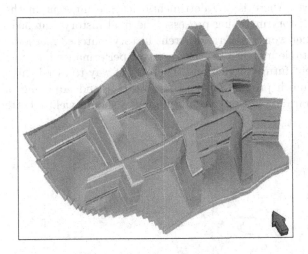

Fig. 3. The 3D net pay model of Xiaermen oilfield

The following eight stages were followed to attain a successful history matching:

1. Set the objectives of the history -matching process;
2. Determine the method to use in the history match. This is dictated by the objectives of the history match, company resources available for the history match, the deadlines for the history match, and data availability;
3. Determine the historical production data to be matched and the criteria to be used to describe a successful match. This should be dictated by the availability and quality of the production data and by the objectives of the simulation study;
4. Determine the reservoir data that can be adjusted during the history match and the confidence range for these data. The data chosen should be those that are the least accurately known in the yield but that have the most significant impact on reservoir performance. This step should be performed in conjunction with the reservoir engineers, geologists, and field operations staff working on the field under study;
5. Run the simulation model with the best available input data. During the pressure-match stage of the history match, the reservoir-voidage rates are

specified. During the saturation stage of the history match, oil rates at standard conditions are specified;

6. Compare the result of the history match run with the historical production data chosen in Step 3;
7. Change the reservoir data selected in Step 4 within the range of confidence;
8. Continue with Step 5 through 7 until the criteria established in Step 3 are met.

These procedures should lead to an improved prediction from the simulation model. However, there is one detrimental feature inherent in the procedures, as with any history-matching process. The final history-matched model is not unique. In other words, several different history-matched methods may provide equally acceptable matches to past reservoir performance but may yield significantly different future predictions. There is no way to avoid this problem, but matching as much production data as available and adjusting only the least known reservoir data within acceptable ranges should yield a better match.

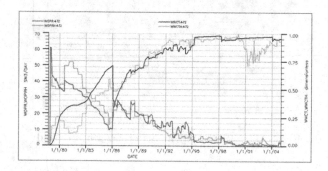

Fig. 4. History match curves of No.4-72 well of Xiaermen oilfield (WOPR– simulated oil rate, SM3/Day; WOPRH– actual oil rate, SM3/Day; WWCT– simulated water cut, f; WWCTH– actual water cut, f.)

Although there are no rules for conducting a history match, several features are common to all successful history-matching exercises. The engineers, geologists, and operations staff of the subject field should be involved intimately in the history-matching phases of the simulation study. The role of the field staff should be to define the confidence intervals for the production data being matched (step 3), to help select the reservoir data to be adjusted, and to determine the acceptable range for adjustment of the reservoir data (step 4). In addition, the field staff should provide knowledge of the field that the simulation engineers may not have.

Although relative permeability can be a powerful history-matching parameter, it should be used only as a last resort. The best approximation for relative permeability should be incorporated during the model-construction phases of the study and, if possible, need to be modified unless technically justified (Fig. 4).

Fig.4 showed No.4-72 well of Xiaermen oilfield simulated oil rate, water cut fit actual performances very well. Especially the water cut curve matched well in different stages.

3   Numerical Simulations

After all wells were matched well, the numerical model was set up and the Eclipse software was used to simulate the oil recovery of the Xiaermen oil reservoir. A series of models were designed to study the effect of the geological characterization and development performances. Fig. 5 is a series of curves of development index of Xiaermen oilfield.

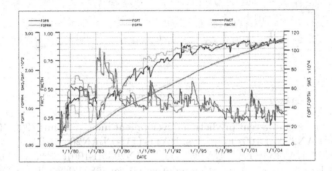

Fig. 5. The curves of development index of Xiaermen oilfield (FOPR– simulated oil rate,SM3/DAY; FOPRH–actual oil rate,SM3/DAY; FOPT– simulated cumulative oil rate, SM3; FOPTH– actual cumulative oil rate, SM3; FWCT– simulated water cut, f; FWCTH– actual water cut, f.)

The amount of detailed input and the type of simulation model depends upon the issues to be investigated, and the amount of data available. It is in view of this development that two adjustment plans were designed and the development index was predicted to optimize the Xiaermen oilfield development.

4   Results and Suggestions

Two main scenarios were considered. Adjustment plan2005 and adjustment plan2006B were designed to optimize the Xiaermen oilfield development.

– Adjustment plan2005: At current oil rate and injection volume produced;
– Adjustment plan2006B: Infilling 13 oil wells and 5 water injections produced.

Running the software Eclipse with the numerical model above, two production parameters were calculated until 1/1/2020 as shown in Fig.6. It indicated that the adjustment plan2006B was much better than the adjustment plan2005 in three parameters like oil rate, cumulative oil rate and water cut.

The Xiaermen Oil Field was developed in 1978. It went through many stages, such as development with natural energy and a wide-spaced wells phase; subdivision series of strata phase and well pattern thickening and adjusting phases. The adjustment plan2006B was used to enhance oil recovery in the H2VII formation since the water cut reached 74.9% in 2006. The total oil production was 224 104 t, with a recovery factor of 22.3%. With 3 years of water injection, the field increased its recover factor from 16.8% to 26.1%, enhanced oil recovery is 3.8%. The obtained water cut is 80.8% in 2020, lower than when the trial started.

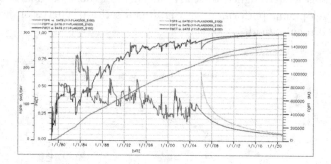

Fig. 6. The curves of different prediction development index vs. time (2020)(FOPR vs. DATE (111-PLAN2005_E100) FOPR vs. DATE (111-PLAN2006B_E100) - plan1plan2 of oil rate, SM3/DAYFOPT vs. DATE (111-PLAN2005_E100)FOPT vs. DATE (111-PLAN2006B_E100) - plan1plan2 of cumulative oil rate, SM3FWCT vs. DATE (111-PLAN2005_E100)FWCT vs. DATE (111-PLAN2006B_E100) - plan1plan2 of water cut, f.)

As expected, the simulated results rely on skills and experience of the engineer. One will need to be careful with prediction results if they rely on parameters that have not been validated in the history match. Different reservoir descriptions can produce the same history match but provide different forecasts.

Acknowledgement. We are grateful to the Exploration and Development Research Institute of SINOPEC Henan Oil Field for providing us with the Xiaermen data used in this work and for financial support. These contributions are gratefully acknowledged. We would also like to thank our colleagues from China University of Geosciences for their guidance and suggestions for this paper.

References

1. Zhu, F., Xie, C.: Petroleum Engineering Special English, pp. 54–56. China University of Geosciences Press, Wuhan (2005)
2. Xie, C.-J., Cuan, Z.-L., Blunt, M., Zhou, H.: Numerical simulation for oil Reservoir after cross-linked Polymer Flooding. Journal of Canadian Petroleum Technology 48(4), 37–41 (2009)

3. Xie, C., Guan, Z., Luo, G.: Decline Analysis of Oil Well Stimulation Rule Using Matlab. In: The 2010 International Conference on Web Information Systems and Mining, WISM 2010-AICI 2010, Sanya, China, pp. 23–24 (2010)
4. Hu, Y.T., et al.: The Distribution of Remaining Oil after Cross-Linked Polymer Flooding. Geological Technical Information 25(4), 51–56 (2006)
5. Vargo, J., Turner, J., Vergnani, B., Pitts, M.J., Wyatt, H., Patterson, D.: Alkaline-Surfactant-Polymer Flooding of the Cambridge Minnelusa Field. SPE Reservoir Evaluation & Engineering 3(6), 552–558 (2000)
6. Giese, S.W., Powers, S.E.: Using Polymer Solutions to Enhance Recovery of Mobile Coal Tar and Creosote DNAPLs. Journal of Contaminant Hydrology 58(1-2), 147–167 (2002)
7. Smith, J.E., Liu, H., Guo, Z.-D.: Laboratory Studies of In-Depth Colloidal Dispersion Gel Technology for Daqing Oil Field. Paper SPE 62610 Presented at the SPE/AAPG Western Regional Meeting, Long Beach, CA, pp. 19–22 (2000)
8. Shen, P.P., Song, J., Zhu, B.: The Achievements and Challenges of EOR Technology for Onshore Oil Fields in China. In: Proceedings of the Fifteenth World Petroleum Congress; Exploration, Production and Downstream (Refining and Petrochemicals), World Petroleum Congress, vol. 15(2), pp. 363–372 (1998)
9. Wang, D.M., Zhang, Z.-H., Cheng, J.-C., Yang, J.-C., Gao, S.-T., Li, L.: Pilot Tests of Alkaline/ Surfactant/ Polymer Flooding in Daqing Oil Field. SPE Reservoir Engineering 12(4), 229–233 (1997)
10. Lu, X.G., Song, K.P., Zhao, R.B.: Numerical Simulation of Effect of Reservoir Heterogeneity on the Residual Oil Distribution after Polymer Flooding. Scientia Geologica Sinica 5(2), 245–253 (1996)
11. Zhong, L.: Experiments of deoxidizing NO with re-burning of pulverized coal and char and research on the chemical kinetic mechanism (PhD thesis). North China Electric Power University, Baoding (2003)
12. Connolly, T., Begg, C.: Database systems. A practical approach to design, implementation, and management. Addison Wesley, Reading (2002)
13. Effect of the Spreading Coefficient on Three-Phase Flow in Porous Media. Journal of Colloid and Interface Science 187(1), 45–56 (1997)

Research of Grid Map Services Implementation for Spatial Information Grid

Jing Zhu[1], Zheng Liu[2], and Junqing Fan[1]

[1] School of Computer Science
China University of Geosciences
Hubei, Wuhan 430070, P.R. China
jingzhu723@gmail.com,
fanjunqing@sina.com
[2] Division of Technology & Science
China University of Geosciences
Hubei, Wuhan 430070, P.R. China
liuzheng_1028@163.com

Abstract. The key issue in implementation of Grid Map Service based Spatial Information Grid (SIG) is combing geospatial information services with grid framework WSRF. The spatial information service specifications developed by OGC based on Web Service such as WMS, WFS, WCS are stateless services, so they cant record the content of the concrete operation of services, service address, service time and other runtime information about service. WSRF framework is characterized by combing stateless services with stateful resources. This article research on how to combine grid technology with geospatial specifications and technologies, add stateful property resources through wrapping OGC services for grid to solve runtime problem of OGC service, and how to implement spatial information services in WSRF grid framework. The major work of this article includes follows:1. Researches on how to capture and analysis OGC Web Services use XML and java technology, get the service metadata. 2. Researches on how to build spatial information service model under the grid framework WSRF.3. Researches on the combination method of grid and OGC map service, then on this basis, put forward a resources packaging method of spatial information service according to WSRF specification and discuss implementation of grid map service.

Keywords: SIG, WSRF, Grid Service, OGC, Globus.

1 Introduction

Currently, the implements of spatial information service are mostly based on Web Service technology. According to Spatial information services specifications, such as Web Map Service, Web Feature Service, and Web Coverage service, which established by OGC, the spatial information services support sharing and inter operability for spatial information resource. But these specifications encountered many bottleneck difficult to overcome in processing massive distributed spatial

Z. Li et al. (Eds.): ISICA 2012, CCIS 316, pp. 576–586, 2012.

information and providing continuous transparent services for users, and there are also many shortages on computing capability and inter operability. The main reason of problem is the OGC services are stateless, it cant record the concrete operation content, address and time of services, and so it cant build a service chain or implement inter operation easily. The spatial information service based on WSRF framework added stateful property resources for services can solve effectively the problem on implement of transient services.

Releasing and running spatial information services based on WSRF in Spatial Information Grid can make the services being stateful services in line with WSRF specifications and achieve purpose of real-time management and scheduling by monitoring and managing these spatial information services in grid platform.

2   Grid Architecture Implementation Mechanism Based WSRF

WSRF solve the time problem of services by creating a state concept and the methods of processing status. WSRF add state resource in stateless environment of Web services and manage it by Web services. The state resources been named WSResource, it maybe is anything-database or mouse. In fact, we can process anything whose properties can be changed based WSRF. The WSRF with service stateful properties is illustrated in Fig. 1.

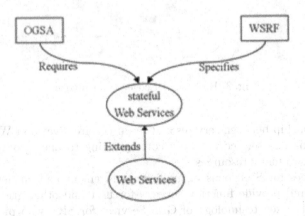

Fig. 1. WSRF with stateful service properties

The stateful resource is the core technology feature of WSRF grid framework, and also is most important feature different from normal Web services. Under this framework, the resources of SIG are discovered and managed by Resource Home.

The interface of Resource Home is illustrated by Fig. 2.

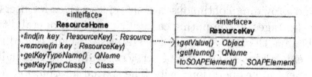

Fig. 2. Resource Home Interface Map

The services and stateful resources are apart from each other in WSRF, so services must access Registry Center of resources firstly and get Resource Home, then it can get resources through management and assignment of Resource Home.

The steps are illustrated by Fig. 3.

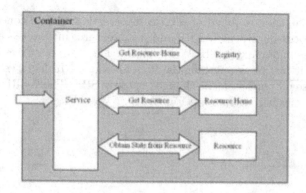

Fig. 3. Resource Discovery and Invoke

The relationship between services and resources are diverse in WSRF framework, it means one service not only corresponding to one resource, but also corresponding to many resources.

Grid Services for SIG focus on unified management and efficient accessing of Grid GIS and provide function services for user and other grids. Metadata Catalogue is the key technology of Grid Services for SIG, who provides available information accessing interfaces for virtual view design, resources selection, resources location and other operations. Grid Service expands Web Service and the Web Service is base of Grid Service. Therefore, the implementation of Grid Service should accord to the four level architecture refer to the Web Service to enroll, discover and search services, in which the underlying transmission is realized by HTTP protocol, the services are triggered by SOAP and described by WSDL language, the metadata catalog of spatial services are constructed by UDDI.

3   Grid Services Implementation

The implement of Grid Services on SIG means transform services relate with spatial data to grid services which can be managed and used in grid environment. The Grid Services include two classifications: one is OWS services of OGC, the other is Web Services based Web Services technology.

The technology routing to transform services is illustrated by Fig. 4.

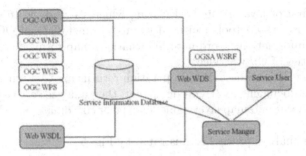

Fig. 4. Grid Spatial Services Roadmap

As Fig. 4, arrange OWS services and Web Service through key technology of searching services build services information base, then analysis these services and transform it to Grid Services supporting WRSF, at last Release the Grid services and manage and monitor it.

Globus Toolkit improved OGC services by according to WSRF, it added state property for spatial information services to establish long-lasting mechanism of map operations, and implement operations in OGC services interface as Grid services and construct Spatial Information Grid infrastructure based services.

3.1   Getting and Managing OGC Services Information

OGC and ISO/T211C launched jointly some spatial data inter operability specifications based on Web Services such as WMS, WFS, WCS and GML using which to transmit and transform spatial data. OGC standard operations can be executed directly in browser through inputting URL. For example to WMS service requests, OGC defines three standard operations for WMS: GetCapabilities returns metadata about service, which describes content and acceptable parameters of service information GetMap returns a map, whose parameters about geospatial and size being defined clearly, GetFeatureInfo (optional) returns information about special factors information displayed on the map.

The information returned for GetCapabilities request of WMS is metadata in XML format.CCapabilities XML. The first part of Capabilities XML is element ¡Service¿ which provides basic metadata of all services, it should include a Name, a Title, an online Resource URL and several optional elements such as AbstractKeyword ListContact InformationFees and Access Constraints. The most pivotal part of Capabilities XML is Layers and Styles. Any kind of accessible map is announced in Capabilities XML through an element ¡Layer¿A parent layer includes many different child layers. Some properties defined in parent layer can be inherited by its child layers. Child layers can redefine these properties inherited from parent, or add some new properties.

3.2   Services Analysis

JDOM is a kind of java-oriented API using which to read, write and operate XML document. JDOM tools can used to analyze metadata of OGC services, extract layer information, coordinate information, map format, version, as resource properties of Grid Service.

Through the URL request, WMS operation returns an XML documenttransform the XML document to JDOM document, then analyze service information in element ¡service¿ and map layer name, map layer coordinate, map layer format and pattern.

The service analysis process is illustrated by Fig. 5.

Fig. 5. Analysis of Services' Information

3.3   Wrapping and Deploying Grid Services

WSDL describes public interfaces of Web Service. It defines how to communicate with Web Service based XML, and describes binding protocol and information format used in interaction between public interfaces and Web Services listed in catalog.

Constructing Services Description WSDL Document and Resource Properties document Comply with WSRF.

WSDL document mainly define the service's input and output parameters, methods of operation. The most important part in WSDL document is WSRP-Web Service Resource Properties document. WSRP is a XML document including resource properties document represents logical combination of resource properties elements, its a mapping of resource state associated with the portType of WSDL. WSRP defines a series of message exchange format to operate state of resources: access, insert and modify attributes.

According to the standards of WSRF, its a convenient way to design and implement Grid Map Service based unified interaction model between services and resources of WSRF by designing and reforming the interfaces of WMS, detaching operating states and service results from stateless services as resources for the use of other services.

There are no resource types in WSDL, this paper redefines interfaces of WMS based extended WSDL of Globus Toolkit and describes the design and implementation for OGC services based Singleton model. In network service design based Singleton model, the network service is composed of a stateless Web Service and a stateful resource class expressing the property of corresponding instance.

As follows we define a WSRP document in WSDL of Grid Service.

```
<xsd:elementname=WMSResourceProperties>
<xsd:complexType >
<xsd:sequenee>
<xsd:elementref=tns:WidthminOccurs=1maxOccurs=1/>
<xsd:elementref=tns:HeightminOccurs= 1maxOccurs=1/>
<xsd:elementref=tns:LayersminOccurs= 1maxOccurs=1/>
<xsd:elementref=tns:StylesminOccurs=1maxOccurs=1/ >
<xsd:elementref=tns:Format1minOccurs=1maxOccurs=1/>
<xsd:elementref=tns:SrsminOccurs=1maxOccurs=1/ >
<xsd:elementref=tns:HostminOccurs=1maxOccurs=1/>
<xsd:elementref=tns:Port1minOccurs=1maxOccurs=1/>
</xsd:sequence>
</xsd:complexTyPe>
</xsd:element>
```

This XML document defines resources for Grid Map Service, such as Layer, Width and Height of map. Users can operate these resources by services.

Constructing Services Implementation Entity Classes.

The major custom classes include: QNames.javaprovides a interface to access resource properties, Resources.java, provides methods for initializing and operating resource properties, ResourcesHome.java is a class for managing resources and Service. java is a major function implement class, return data that user requested.

1. Constructing Service Deployment File of WSDD and JNDI

   WSDD deployment file references WSDL file and defines the mapping between service namespace and java classes by mapping the file path in services into service namespace. This file declares there are what services in Spatial Information Grid and by what classes to provide these services. To construct a WSDD file needs modify service name, class name and WSDL file position in sample service.

   JNDI is Web service registration file to register services. It's a "switch" in J2EEthe J2EE components search other components, resources or services indirectly in runtime and reference external resources out of system through definition and reference in JNDI file.

2. Archiving Service Files

   This step will use the native compilation command file -service.py in Globus, and archive WSDL file, java entity classes, and deployment files into a .gar file. The process is shown in Fig. 6.

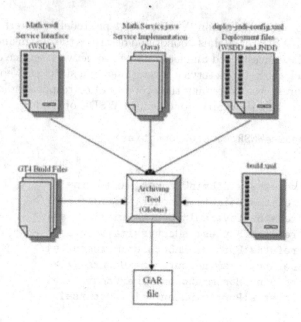

Fig. 6. Generating Grid Services Archive File

3. Deploying Services to Service Container
   This step runs command of Globus to deploy services into Globus Container
   directly.
4. Invoking Services
   The stub file complied by NetBeans find Endpoint of service, then call methods of services from portType to get results. The client create a instance of EndpointReferenceType to reference an endpoint of service, then create portType according to service address and generate service instance, users invoke a service by service instance. The process is described by follows:

```
EndpointReferenceType endpoint=new EndpointReferenceType();
endpoint.SetAddress(new Address(serviceURI);
WMSServicePortType
WMS=locator.getWMSServicePortTypePort(endpoint);
WMS=locator.getWMSServicePortTypePort(endpoint);
String mapURL=WMS.GetMap();
```

4   Grid Map Service Implementation Based WMS

The template will number citations consecutively within brackets [1]. The sentence punctuation follows the bracket [2]. Refer simply to the reference number, as in [3]do not use "Ref. [3]" or "reference [3]" except at the beginning of a sentence: "Reference[3] was the first ..."

The steps to implement WMS in SIG consistent with the Grid service development steps in Globus. It includes five major steps as follows:

- Create WSDL file, in which defines service interfaces
- Create service implement classes for WSDL
- Create service deployment file-JNDI, WSDD
- Compile and generate.gar file
- Deploy services to grid container

The key step is how to define interfaces of spatial service. In this paper, through wrap corresponding function of method GetMap and GetCapability in WMS to construct Grid Map Service in line with WSRF. In WSDL file, regard parameters for map request as ResourcesProperty, its described as follow:

```
<xsd:element name=Width type=xsd:int/>
<xsd:element name=Height type=xsd:int/>
    <xsd:element name=Layers type=xsd:string/>

    <xsd:element name=GridMapserviceResourceProperties>
    <xsd:complexTyPe>
    <xsd:sequence>
    <xsd:element ref=tns:Width minOccus=1maxOccurs=1/>
    <xsd:element ref=tns:Height minOccus=1maxOccurs=1/>
    <xsd:element ref=tns:Layers minOccus=1maxOccurs=1/>

</xsd:sequenee>
</xsd:complexType>
</xsd:element>
```

User can manipulate the resource attribute to query or change service status. The method Getmap responsible for discovering the map service address, it means Get map only return a network server address, the concrete map layer information should resolved from map metadata get by method GetCapability. The client use the information (layer, coordinate system, map range and so on) resolved from metadata to invoke services.

5   Map Service Implementation Based WFS

WFS response GML data to user's request through get/post model, then user can display map features by independent map client. The process is described by Fig. 7.

To establish WFS in grid, means reforming WFS interfaces in WSDL to in line with Grid Service specification. In this paper, the method is implement operation GetCapabilities and GetFeature in WSDL. Users invoke WFS by service endpoint and regard operation parameters as grid resources properties. The service call process is described by Fig. 8.

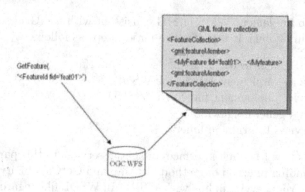

Fig. 7. Xml-based Service Request of WFS

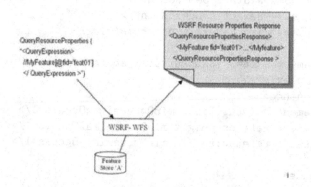

Fig. 8. Service Request of WFS under WSRF Framework

There is an important note in grid service implementation based WFS: bind WFS operation interface like getCapabilities and getGMLObject from SOAP to WSDL at first to make sure WFS being a service which can be accessed by normal network service. The binding process is described as follow:

```
<wsdl:binding name==wfs-SOAPtype=wfs-req:wfs>
    <wsdl:documentation>WFS interface bound to SOAP over
        HTTP/1.1.</wsdl:documentation>
<soap:binding style=document transPort=
        http://schemas.xmlsoap.org/soap/httP/>
<wsdl:operation name=wfs.getCapabilities>

<soap:operation/>
<wsdl:input>
<soap:body use=literal/>
</wsdl:inPut>
<wsdl:output>
```

```
<wsdl:fault name=ServiceExceptionReport>
<soap:fault name=ServiceExceptionReport use=literal/>
</wsdl:fault>
</wsdl:operation>

</wsdl:binding>
```

The stateful resource of Grid WFS-GridWfsResourceProperties includes four resource properties: Capability, Typename, Feature and FeatureTypeDescription. They are defined in WSRP file like follow:

```
<element name=Capability type=xsd:string/>
    <element name=Typename type=xsd:string/>
    <element name=Feature type=xsd:string/>
    <element name=FeatureTypedeseription type=xsd:string/>
    <element name=GridWFSResourceProperties>
    <complexType>
    <sequence>
    <element ref=tns: Capability/>
    <element ref=tns: Typename/>
    <element ref=tns: Feature/>
    <element ref=tns: FeatureTypedescription/>
    </sequence>
  </complexType>
</element>
```

Then user can request and get this file.

6   Conclusion

This article researches on the principle, methods and implementation of spatial information grid services based WSRF framework, enhances the geospatial service based OGC into grid platform, and put forward the grid packaging technology route using Globus Toolkit based WSRF framework. On this basis, the paper discusses the method of wrapping WMS and WFS to grid services.

On the purpose of defining resource type and status of grid service, it needs resolve OGC service metadata to get map layer information, coordinate information, map format and other service property information and defines these information as resource attributes of grid service. Then build WSDL and WSRP file according to WSRF specification, defines grid services consist of a stateless Web Service and a stataful resource class corresponding to service. Redefine interfaces of WMS and WFS, create service implementation class for WSDL and the service deployment file, deploy grid services into grid environment, provide client grid services by normal way to access network services.

The article doesnt discuss the implementation of WCS in SIG, and the implementation of geospatial information grid service also is involved in interoperability of grid services. It need further study based this article.

References

1. Xie, Z., Song, M.M., Luo, X.G.: Resource and Information Sharing Mechanism Based on Spatial Information Grid. In: The Seventh International Conference on Grid and Cooperative Computing (GCC 2008), October 24-26, pp. 221–223 (2008)
2. Czajkowski, K., Ferguson, D.F.: The WS-Resource Framework. Globus Alliance, IBM, HP (March 2004)
3. Liu, S.L., Liu, Y.X., Zhang, F., Tang, G.F., Jing, N.: A Dynamic Web Services Selection Algorithm with QoS Global Optimal in Web Services Composition. Journal of Software 18(3), 646–656 (2007)
4. Aydin, G., et al.: Building and Applying Geographical Information System Grids Concurrency and Computation: Practice and Experience. In: Special Issue on Geographical Information Systems and Grids based on GGF15 Workshop (2006)
5. Open Geospatial Consortium Inc., OpenGIS Catalogue Services Specification 2.0.2-ISO Metadata Application Profile, pp. 68, 07-045 (July 19, 2007)
6. Kunz, C., Groper, R.: Geospatial Workflows on the D-Grid Infrastructure. In: OGF25 Meeting (March 4, 2009)
7. Open Geospatial Consortium Inc., OpenGIS Web Map Service Specification. OGC 07- 006r1 (February 23, 2007)
8. Open Geospatial Consortium Inc., Web Feature Service Specification. OGC 07-110r2 (February 29, 2008)
9. Open Geospatial Consortium Inc., OpenGIS Catalogue Services Specification 2.0.2-ISO Metadata Application Profile, pp. 31, 07-045 (July 19, 2007)
10. Xie, Z., Song, M.M., Luo, X.G.: Resource and Information Sharing Mechanism Based on Spatial Information Grid. In: The Seventh International Conference on Grid and Cooperative Computing (GCC 2008), October 24-26, pp. 221–223 (2008)

Application and Research of Shortest Time Limit-Resource Leveling Optimization Problem Based on a New Modified Evolutionary Programming

Yuanfei Luo, Jiehui Tang, Si Xu, Li Zhu, and Xiang Li

Department of Computer Science
China University of Geosciences
Wuhan 430074, P.R. China
csluoyf@163.com

Abstract. Aiming at the optimization problem of shortest time limit - resource leveling, the paper first introduces Evolutionary Programming (EP) to solve it, and a new modified method based on evolutionary programming is proposed: the mutation operator of EP is improved by using the theory of Simulated An-nealing (SA), and without using repair operator. Then use Genetic Algorithm (GA), EP and the modified EP to solve this problem, the experimental results indicate that EP can optimize this problem effectively, and EP has better opti-mization performance than GA. The average evolution generation decreased significantly in the modified EP to approach the optimal solution, the variance after optimization decreases 42.64%.

Keywords: Shortest Time Limit-resource Leveling Optimization, Evolutionary Programming, Simulated Annealing, Modified, Optimization.

1 Introduction

Shortest time limit-resource leveling optimization problem is in the shortest time and a certain period of time, a variety of items which the total amount of multi-resource they required is limited, to find a reasonable scheduling, so that the resource consumption of each unit period is balanced as much as possible to achieve the best utilization of resource. The definition of schedule is a sequence composed of starting time of each process so that all projects can complete on time. This is a very difficult combinatorial optimization problem, which belongs to NP-hard problem[1].

Evolutionary programming is a parallel stochastic search technique that maps natural evolution to an effective multi-agent search strategy. This search technique was first described by Fogel et al.[2]. It is not based on gradient methods and can overcome local minima. Evolutionary programming has been successfully applied to difficult combinatorial problems such as the traveling salesman problem[3], training and designing neural networks and system identification[4].

Z. Li et al. (Eds.): ISICA 2012, CCIS 316, pp. 587–594, 2012.

There are three widely researched paradigms in simulated evolution: genetic algorithms, evolution strategies and evolutionary programming. Recently, evolutionary programming has received more attention[5]. In 1991, Fogel et al. proposed the meta-evolutionary programming that can evolve 'high-level' parameters while concurrently searching for the optimum solution[6]. In 1993, Yib and Pao[7] proposed the incorporation of the process of simulated annealing into the selection process of evolutionary programming which is called Guided Evolutionary Simulated Annealing (GESA) and gave a regional guidance to the stochastic search process of evolutionary programming. Based on[7], the Cauchy distribution is used in this paper to replace the Gaussian distribution for creating offspring from their parents, which dramatically accelerates the search procedure.

Solutions proposed at home and abroad in general is the use of genetic algorithms or artificially add virtual work to convert multi-project to a single project problem, and have not utilized evolutionary programming to resource leveling optimization, this paper applies evolutionary programming to the multi-project resource leveling problems, and achieves satisfactory result, whats more, integrate the idea of simulated annealing into the mutation operator of evolutionary programming algorithm, the experimental result show that not only the result is better, but also significantly speed up the rate of convergence.

2   The Mathematical Model

Modify the model described in literature[8], and then establish the model of shortest period of multi-project and multi-resource balance problem: Objective function:

$$minF = min \sum_{l=1}^{r} \partial_l * (\frac{1}{TC}) \sum_{t=1}^{TC} (R_l(t) - \overline{R}_l)^2 \tag{1}$$

s.t:

$$R_t = \sum_{k=1}^{m} \sum_{(i,j)} R_l(k)(t)(i,j) \quad (i,j) = 1, 2, ..., mk \tag{2}$$

$$R_l(k)(t)(i,j) = \begin{cases} R_l(k)(i,j) & TS(k)(i,j) \leq t \leq TS(k)(i,j) + T(k)(i,j) \\ 0 & t < TS(k)(i,j) \ or \ t > TS(k)(i,j) + T(k)(i,j) \end{cases} \tag{3}$$

$$TF(k)(i,j) = LS(k)(i,j) - ES(k)(i,j) \tag{4}$$

$$max\{ES(k)(h,i) + T(k)(h,i)\} \leq ES(k)(i,j) \leq LS(k)(i,j)$$
$$\forall(h,i) \in P(k)(i,j) \tag{5}$$

$$ES(k)(i,j) \leq TS(k)(i,j) \leq ES(k)(i,j) + TF(k)(i,j) \tag{6}$$

$$\overline{R}_l = \frac{1}{TC} \sum_{t=1}^{TC} R_l(t) \quad l = 1, 2, ..., r \tag{7}$$

$$TC = max\{T_k\} \quad k = 1, 2, ..., P \tag{8}$$

$$\partial_1 + \partial_2 + ... + \partial_r = 1 \quad 0 < \partial_i < 1 \tag{9}$$

r is a total number of resources, T_k represents the duration of the k-th item, TC represents the maximum value duration of the number of m projects, m_k said the total number of k-th item, $R_l(k)(i,j)$ on behalf of the k-th item's consumption of l kinds of resources per unit time to work (i,j), $R_l(t)$ is the all projects'consumption of l-th kind of resource in the k-th item in t-th unit time of work (i,j). $ES(k)(i,j), LS(k)(i,j), TS(k)(i,j), TF(k)(i,j)$ denotes the work (i,j) in the k-th project the earliest start time, latest start time, actual start time, duration, relaxation time respectively, $P(k)(i,j)$ is tight first set of the work (i,j) in the k-th item, $\overline{R}_l$ is the average consumption of the m items of work for the l-th kind of resource, ∂_i is the weight coefficients.

3   The Design of Improved Evolutionary Programming

3.1   Individual Chromosome Structure and Coding Design

A reasonable scheduling is designed to meet the time bound constraint i.e. the specific topology of relationships. For shortest duration of project, resource leveling with multi-project and multi-resource, each project's work can be considered separately, then put all projects together on the basis of specific topology of relationships to assess the solution with formulated evaluation function. As the analysis above, a gene in the chromosome is represented by the actual start time for each work in its project, then sort each work in accordance with the specific topology order to form a sequence of chromosome, at last merge all the sequence of chromosome to a single one, which is shown in the following table of chromosome encoding design program. In the Table 1, P_{ij} represents the j-th work in i-th project while T_{ij} means the actual start time of $P_{ij}(i = 1, 2, ..., m; j = 1, 2, ..., n)$. In this process must follow specific topological relations to arrange the actual start time for each project of work.

Table 1. Chromosomes encoding design program

Work	11	12	...	1n	...	m1	...	mn
Work Order	P_{11}	P_{12}	...	P_{1n}	...	P_{m1}	...	P_{mn}
Start Time	T_{11}	T_{12}	...	T_{1n}	...	T_{m1}	...	T_{mn}

There are various methods in chromosome encode, such as binary encoding, self-adaptive encoding, symbol encoding, decimal real encoding. Due to the evolutionary programming using traditional decimal expression of real issues, and real number has efficient performance time in the encoding and decoding at the same time, thus decimal real coding is proposed.

3.2   Mutation Operator

One solution for shortest time limit - resource leveling optimization in multi-project is to adjust the actual start time of each work to make resources to the minimum average level when specific topology relationships are satisfied. So can start with the critical path method to find out the earliest start and last happen time of each work in the project, thus determine each works slack time. As conventions, let be $ES_{ij}, LS_{ij}, TF_{ij}, TS_{ij}, T_{ij}, TB_{ij}, UB_{ij}$ are used to represent earliest start time, last happen time, slack time, actual start time, duration, the last finish time when all formal work completed and the earliest time when all latter work are completed of P_{ij}. In the literature[9] as follows, a new individual is expressed like this:

$$T_{ij} = LB_{ij} + (LS_{ij} - LB_{ij}) * random(0,1) \qquad (10)$$

This method is imprecise in two aspects: one is that there is no compare demonstration of LB_{ij} and ES_{ij} and another doesnt account for illegal solution when specific topology relationships are not satisfied while $T_{ij} > UB_{ij}$. So repair operator is needed to fix irrational individual. Through analysis above, the process of a method that does not need repair operator is listed as follows:

1. Calculate LB_{ij}(the last finish time when all formal work completed) and UB_{ij}(the earliest time when all latter work completed) of P_{ij} to get the formula:$LB_{ij} = MAX(TS_{ki} + T_{ki}), UB_{ij} = Min(T_{ki})$.
2. Take the greater one of LB_{ij} and ES_{ij}, smaller one of UB_{ij} with LS_{ij}, if $LB_{ij} > UB_{ij}$, the program throws exception.
3. $T_{ij} = random(LB_{ij}, UB_{ij})$.

This can result in a new self-employed for a particular topological relationships, avoiding the illegal to produce, the procedures has greater efficiency.

For every individual in the parent population, in accordance with the above procedure to acquire a new individual, when judging the individual accept or not, call simulated annealing operator.

3.3   Simulated Annealing Operator

In the algorithm of simulated annealing, the state producing function generates a new state, and then the state received function determines whether to accept the new state or not. Therefore, the design of simulated annealing operator is as follows:

1. Calculate current temperature t_k, get the parent individual Xi and anneal rate α;
2. Generate individual X_i using mutation operator, let be C_i, C_j represent the fitness of X_i, X_j respectively, receive the new individual of C_i is less than C_j, otherwise, check if $exp[-(C_j - C_i)/t_k]$ greater than a random number between $[0,1]$ or not. If it does, receive current individual. If not, repeat the operation until receive one new individual.
3. Update the temperature with the formula: $t_{k+1} = \alpha * t_k$, and assignment $k + 1$ to k.

3.4   Initiate all Individuals Chromosome and Parameters

Call mutation operator in the process of initiating all individuals chromosome, and ignore the calling of simulated annealing operator. Decode the every individual in the initiate population $P(0)$, let be $\Delta f = f_{max} - f_{min}$ in which f_{max}, f_{min} represent the maximum fitness in the population and minimum fitness in the population respectively. The initial temperature can be calculated by this formula easily: $t_0 = -\Delta f/Inx_0$, in which x_0 is the initial inferior solution acceptance probability.

4   Selection Strategy

The purpose of selection is to moderate elimination of inferior individuals, and gradually increase the proportion of the best individual groups, in order to achieve convergence. The typical selection strategy in evolutionary programming is the method named random q competitive selection. The specific process is:

1. For each individual in the collection composed of parent and offspring donated by $X \in P(t) \cup P'(t)$, select q individuals in total to form a set named Q and q satisfies no less than one. Then compare the fitness of X with each individual in Q, record how many times of the fitness of X is greater than the individuals in Q as the score of X, let s represent the score.
2. Arrange the set which is composed of the whole population in descending order according to the score of every individual. Choose the first μ individuals as the next generation of population.

4.1   Evaluation Function

The evaluation function can be deduced from the mathematical model of shortest time limit and resource balance, which is shown as follows:

$$minF = min \sum_{l=1}^{r} \partial_l * (\frac{1}{TC}) \sum_{t=1}^{TC}(R_l(t) - \overline{R}_l)^2 \qquad (11)$$

As the subject itself is the minimum optimization problem, therefore it can be transformed to maximize the fitness function. Stretch the objective function while the process of transforming, the conversion method is as follows:

$$g(v_k) = \frac{f_{max} - f(v_k) + \varepsilon}{f_{max} - f_{min} + \varepsilon} \qquad (12)$$

In the formula above, v_k represents the k-th chromosome in the current population, $g(v_k)$ represents the evalution function, $f(v_k)$ represents the objective function(resource unbalanced coefficient), f_{max}, f_{min} are the maximum target value and the minimum target value respectively. ε is a real number between zero and one, which can avoid the denominator is zero.

5   The Experimental Results

Here is an example of a resource leveling optimization problem composed of nine real works, each work requires three different kinds of resource. Network structure is shown in Fig. 1. The relationship between the various works, hours of each work needed and resources required are shown in Table 2:

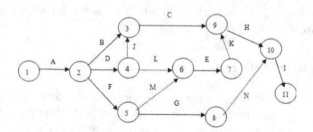

Fig. 1. Network structure

J, K, L, M, N are virtual works added artificially which represents the constraint relationship of all works.

Table 2. The time and resource every work required

Work ID	A	B	C	D	E	F	G	H	I	J	K	L	M	N
Tight first work		A	BJ	A	LM	A	F	CK	HN	D	E	D	F	G
Duration	5	7	13	10	14	19	14	8	9	0	0	0	0	0
First resource	5	4	3	10	11	13	3	6	5	0	0	0	0	0
Second resource	4	6	6	12	15	11	12	3	5	0	0	0	0	0
Third resource	5	2	3	14	12	3	5	2	5	0	0	0	0	0

Let be the weight of three resources all equal to 1/3. According to the critical path method (CPM) before optimization, arrange every work's real start time in accordance with the earliest start time of each job, we can obtain the shortest duration of 55, the greatest resource demand per unit time is 75, the variance is 76.0842.

In order to compare the evolutionary programming and modified evolutionary programming algorithm performance with the genetic algorithm, the experimental results are included in the following Table 3:

In the experiment, the settings of genetic algorithm parameter are: the population size is 100, the maximum evolution generation is 100, crossover probability is 0.85, and mutation probability is 0.01. The settings of EP parameters are: the population size is 100, the maximum evolution generation is 100, the beginning of the annealing rate is 0.95, and the initial acceptance probability of inferior solution is 0.75. What Table 2 shows is that: using the evolutionary programming algorithm without any improvements convergency in the 68th generation, the

Table 3. The results of different algorithm

Algorithm	Average evolution generation	Maximum resource demand per unit time	Variance
CPM	–	75	76.0842
GA	82	75	49.9943
EP	68	63	44.0798
Modified EP	24	63	43.6435

optimal chromosome structure is (0,16,19,5,24,5,32,38,46,18,38,21,24,46), after optimal the variance reduces by 42.06%. Evolutionary programming algorithm using the improvement convergency in 24th generation, the optimal chromosome structure is (0,15,17,5,24,5,32,38,46,17,38,18,24 46), the optimal variance reduces by 42.64%. Conclusion by experiment are as follows: EP has better performance in solving the shortest period-resource leveling problem, the improved EP can get better optimization results, which greatly reduces average evolution generation.

6   Conclusion

The evolutionary programming algorithm is applied to solving the shortest period-resources equilibrium problem, and using the theory of simulated annealing algorithm to the mutation operator of evolutionary programming to improve and speed up the convergence rate. Experiment through typical examples shows that the improved evolutionary programming algorithm can solve the shortest time limit - resource leveling optimization problem efficiently, whats more, the improved algorithm does not need repair operator, which has a obvious performance advantage in dealing with the large-scale network planning optimization problem.

References

1. Mitsuo, M., Cheng, R.: Genetic Algorithms and Engineering Optimization. Tsinghua University Press, Beijing (2004)
2. Fogel, L.J., Owens, A.J., Walsh, M.J.: Artificial intelligence through simulated evolution. Wiley, New York (1966)
3. Ambati, B.K., Ambati, J., Mokhtar, M.M.: Heuristic combinatorial optimization by simulated Darwinian evolution: a polynomial time algorithm for the traveling salesman problem. Biological Cybemetics (1991)
4. Fogel, D.B.: System identification through simulated evolution: a machine learning approach to modeling. Ginn, Neeham Heights (1991)
5. Fogel, D.B., Fogel, L.J., Atmar, J.W.: Meta-evolutionary programming. In: Proceedings of 25th Asilomar Conference on Signals, Systems and Computers (1991)
6. Michalewicz, Z.: Genetic Algorithms + Data Structures = Evolution Programs, 3rd edn. Springer, Heidelberg (1996)

7. Yib, P., Pao, Y.-H.: Combinatorial optimization with use of guided evolutionary simu-lated annealing. IEEE Trans. on Neural Networks (1995)
8. Li, X., Tan, W., Kang, L.S.: Research on balanced resource optimization based on genetic algorithm. Computer Engineering and Design, Beijing (2008)
9. Zhou, K., Jun, T.X., Xu, J.: A mathematical model and genetic algorithm for re-source optimization. J. Huazhong Univ. of Sci. & Tech. (Nature Science Edition), Wuhan (2005)

Design and Validation of a Parallel Parameter Inversion for Program Based on Genetic Algorithm

Yuan Cao[1,2], Wenke Wang[1], Tieliang Wang[2], and Feng Liu[2]

[1] Chang'an University
Xi'an, P.R. China
guoyuan2003@163.com,
wenkew@gmail.com
[2] Northwest Institute of Nuclear Technology
Xi'an, P.R. China
wangtiel@263.net,
Liufeng-laser@163.com

Abstract. Aiming at the demand for model parameter calibration of numerical simulation, a parallel parameter inversion program based on genetic algorithm is designed, and the inversion effect of this program is validated by two examples. The results show that this program has strong versatility, good input and output interfaces, high inversion precision and high calculated efficiency, so automatic parameter inversion can be realized by coupling of this program with various forward programs.

Keywords: Genetic Algorithm, Parallel, Parameter Inversion, Program Design.

1 Introduction

Parameter calibration is an important step of numerical simulation[1–3]. Manual parameter adjustment and intelligent parameter inversion are common methods of parameter calibration[4, 5]. Because of low efficiency and inherent defect in dealing with multi-parameters and nonlinear problem, the application of manual adjustment method is very little. There are some universal parameter inversion program developed abroad, like UCODE and PEST[6–8], which can be combined with various forward programs to realize automatic parameter inversion. Because a lot of forward simulations need to be done for inversion program, calculated efficiency has become an important problem to parameter inversion program.

Genetic algorithm is based on biological evolution idea and able to achieve global optimization. It adopts simple coding technology to describe kinds of complex structure, guides search direction through genetic operation of selection, crossover and variation, has advantages of self-organization, self-adaption and self-learning, searches solutions by way of population mode, and can avoid local convergence in search course, so it can be used in multi-parameters and

Z. Li et al. (Eds.): ISICA 2012, CCIS 316, pp. 595–602, 2012.
© Springer-Verlag Berlin Heidelberg 2012

nonlinear inversion problems[9]. Further more, its essential parallelism provides good condition for parallel program design[10].

A universal and parallel parameter inversion program (PPIP v1.0) is designed based on genetic algorithm, and the inversion effect of this program is validated by two examples in this paper.

2   Program Design

2.1   Design Objective

A universal, parallel and intelligent parameter inversion program needs to be designed based on Compaq Visual Fortran and genetic algorithm. Automatic multi-parameters calibration must be realized on a mainframe computer.

2.2   Function Design

Automatic Inversion Function. Automatic parameter adjustment and inversion are the main functions of the program. Intelligent and automatic calculation can be done, recurring to the heuristic character of genetic algorithm. Control parameters and modes of genetic algorithm can be changed according to actual need.

Parallel Calculation Function. A lot of forward simulations need to be done for parameter inversion. As all calculations are independent in an evolution generation, parallel function can be added to the inversion program to improve calculated efficiency. Then the parallel calculation advantage can be utilized to save calculation time. The process number can be adjusted according to available resource of the mainframe computer.

Universal Inversion Function. The inversion program must have good versatility and can be combined with various forward programs to do parameter inversion for different physical problems. Objective function and parameter number can be changed according to actual problem.

2.3   Module Division and Achievement

The program can be divided to input module, heuristic parameter adjustment module and output module according to the design objective and function of the inversion program, shown in Fig. 1.

In input module, users forward program is linked with the inversion program through objective function firstly, which adopts least square guideline usually. Secondly, parameter number and value range are set. Thirdly parallel calculated parameters like available processes and estimated calculation time are set. Lastly control parameters of genetic algorithm like population size, total generations, gene length, selection mode, crossover probability, variation mode, initial variation rate, least variation rate, most variation rate and breeding program are set.

In heuristic parameter adjustment module, selection modes include elite selection and non-elite selection mode. Coding and decoding mode is decimal coded mode. Crossover modes include one point crossover, two point crossover, multi-point crossover, uniformity crossover and arithmetic crossover. Variation modes include uniform variation, non-uniform variation, fixed rate variation and variable rate variation. The fitness of different parameter combinations are evaluated by fitness calculation. Population updating modes include all replacement, random replacement and worst replacement. New population is generated through evolution iteration.

In output module, screen and file are the two output position. No output, simple output and complex output are the three output modes. All sub-modules are realized by subroutines.

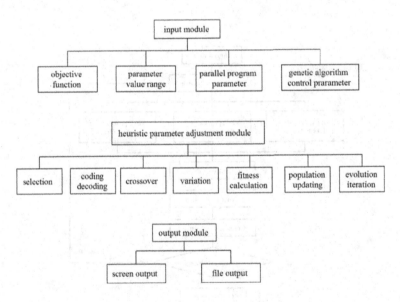

Fig. 1. Module division of the program

2.4 Calculation Flow of Kernel Parameter Adjustment Subroutine

The calculation flow of kernel parameter adjustment subroutine based on genetic algorithm is shown in Fig. 2. After parallel function was appended, the calculation flow is shown in Fig. 3.

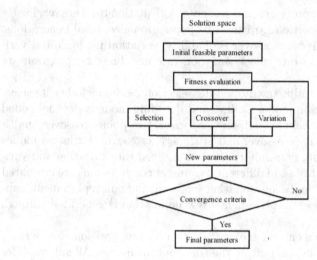

Fig. 2. Basic flow of serial parameter inversion subroutine

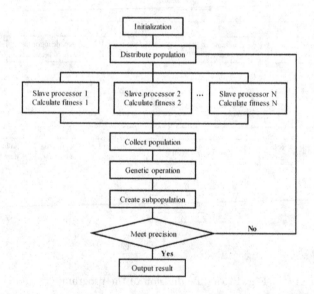

Fig. 3. Basic flow of parallel parameter inversion subroutine

2.5   Interface Design

The pre-processing interface of the program is designed as Fig. 4.

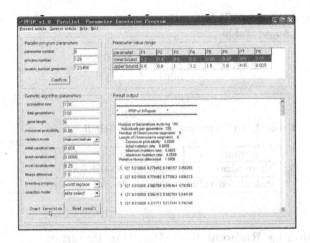

Fig. 4. Pre-processing interface of the program

3   Program Validations

3.1   Validation by Standard Test Function

Schaffer function is used to test inversion effect of the program firstly. Although the function is simple, it is difficult to identify its parameters for a completely nonlinear inversion program, because there are numberless suboptimal solutions

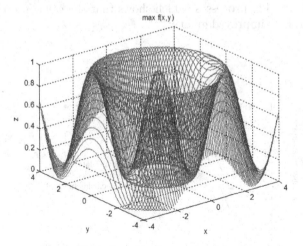

Fig. 5. Figure of Schaffer function

near the optimal solution. Global optimal solution is obtained by the program, as shown in Fig. 5 and Table 1.

$$maxf(x,y) = 0.5 - \frac{sin^2(x^2 + y^2)^{0.5} - 0.5}{(1 + 0.001(x^2 + y^2))^2}, \quad -4 < x, y < 4 \quad (1)$$

Table 1. PPIP inversion result to Schaffer function

parameters	x	y	f
theoretical value	0.00000	0.00000	1.00000
calculated value	0.00006	0.00040	1.00000

3.2 Validation by Radionuclide Migration Program

A true example in numerical simulation of radionuclide migration is used to test inversion effect of the program then. The calibrated parameters by manual adjustment method and PPIP are shown in Table 2(K_1 and K_2 are hydraulic conductivity, D_l is longitudinal dispersity, R_i) is initial distribution radius of radionuclide, N_e is effective porosity, and f is objective function). The objective function is optimized obviously, which shows the inversion precision is improved significantly. Fig. 6 and Fig. 7 show the comparison between observation data and simulation result with calibrated parameters in different observation holes. It also shows that the simulation result by PPIP is more close to observation data than that by manual adjustment method.

At the same time, absolute speedup of the program is tested. It is more than 100 on condition of 128 processes, which shows that the calculation precision of inversion program is improved greatly.

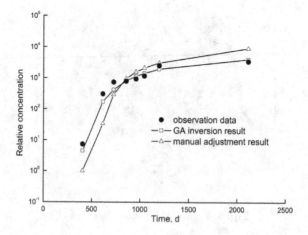

Fig. 6. Observation data and simulation result for hole 1#

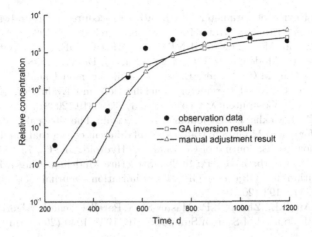

Fig. 7. Observation data and simulation result for hole 2#

Table 2. Calibrated parameters

Calibrated parameters	$K_1/m \cdot d^{-1}$	$K_2/m \cdot d^{-1}$	D_l/m	R_i/m	N_e	f
By manual adjustment	0.300	0.600	0.050	370	0.010	0.320
By PPIP	0.380	0.205	3.000	381	0.014	0.009

4   Conclusions

A parallel parameter inversion program is designed based on genetic algorithm, and the inversion effect of this program is validated by two examples. The results show that this program has strong versatility, good input and output interfaces, high inversion precision and high calculated efficiency, so automatic parameter inversion can be realized by coupling of this program with various forward programs.

References

1. Bonton, A., Bouchard, C., Rouleau, A., et al.: Calibration and validation of an integrated nitrate transport model within a well capture zone. Journal of Contaminant Hydrology 128, 1–18 (2012)
2. Dai, Z., Samper, J., Wolfsberg, A., et al.: Identification of relative conductivity models for water flow and solute transport in unsaturated bentonite. Physics and Chemistry of the Earth 33, S177–S185 (2008)
3. Geza, M., Murray, K.E., et al.: Watershed-Scale Impacts of Nitrogen from On-Site Wastewater Systems Parameter Sensitivity and Model Calibration. Journal of Environmental Engineering, 926–938 (2010)
4. Aster, R.C., Borchers, B., Thurber, C.H.: Parameter Estimation and Inverse Problems. Elsevier Academic Press, Burlington (2005)

5. Yeh, W.W.: Review of Parameter Identification Procedures in Groundwater Hydrology: The Inverse Problem. Water Resources Research 22, 95–108 (1986)
6. Poeter, E.P., Hill, M.C.: DOCUMENTATION OF UCODE, A Computer Code for Universal Inverse Modeling. U.S. Geological Survey, Denver (1998)
7. Banta, E.R., Hill, M.C., Poeter, E., et al.: Building model analysis applications with the Joint Universal Parameter IdenTification and Evaluation of Reliability (JUPITER) API. Computers & Geosciences 34, 310–319 (2008)
8. Zyvoloski, G., Kwicklis, E., Eddebbarh, A.A., et al.: The site-scale saturated zone flow model for Yucca Mountain: Calibration of different conceptual models and their impact on flow paths. Journal of Contaminant Hydrology 62-63, 731–750 (2003)
9. Metcalfe, T.S., Charbonneau, P.: Stellar Structure Modeling using a Parallel Genetic Algorithm for Objective Global Optimization. Journal of Computational Physics 185, 176–193 (2003)
10. Liu, X.P., An, Z.L., Zheng, L.P.: Master-Slave Parallel Genetic Algorithm Framework on MPI. Journal of System Simulation 16, 1938–1940 (2004) (in Chinese)

Charge Stations Deployment Strategy for Maximizing the Charge Oppurnity of Electric Vehicles (EVs)

Hong Yao, Zheng Zhao, Huawei Huang, and Lei Cong

School of Computer Science
China University of Geosciences
Wuhan 430074, P.R. China
yaohong@cug.edu.cn

Abstract. As explosive growth of private cars population, traditional cars have become a very important part of environment pollution. That is why we need electric vehicles (EVs) to replace the traditional cars. But limited to the development of Smart Grid and the technology of EVs battery pack, EVs need to be charged frequently that reduce the drivers comfort and the efficiency of the travel performance. This paper proposed a scheme which uses the Newton law of cooling to optimize the deployment of the charge stations on the map, the basic demand is to achieve the best quality of service with the least cost. And the result shows that this scheme can achieve great performance.

Keywords: Smart Grid, Electric Vehicles, Charge Stations Deployment Scheme.

1 Introduction

As environment preservation becomes a prominent issue around the world, EVs are poised to gain acceptance from the general public. According to the statistic data in U.S., it is forecasted that about 1.5 million EVs will be in the U.S. by 2015, and over 10 million by 2020[1]. Because the gene of the EVs, its use can provide many benefits such as low pollution gas emit, high energy utilization efficiency, potential to can be supported by renewable energy(Such as Wind and solar)and so on[2].

Although EVs have so many benefits, the battery technology limits the mass use of EVs. For example, the max distance some different EVs can achieve if they only been charged one time is like these: Wheeo Whip (160KM), Coda Automotive (193KM), Hyundai BlueOn(140KM), and Renault Fluence Z.E(160KM). It means that in our real life, we should charge our EVs frequently when we use the EVs, this is not we want to see. However, charge stations can ensure that EVs been driven on the road freedom. But deploying charge stations is too costly; it is not possible to deploy charge stations everywhere in a large scale. So how and where to deploy the charge stations on a real map becomes an issue that

Z. Li et al. (Eds.): ISICA 2012, CCIS 316, pp. 603–611, 2012.

we need to solve. This paper propose a new way to suggest where to deploy the charge stations can maximize the opportunity which can satisfy the EVs energy demand with the least cost.

How to deploy the charge stations is apartment of Smart Grid. Within these years, Smart Grid is developing rapidly in many countries. Most of the worlds existing electricity grids were decades-old, although they have significantly changed human life, their monitoring and control facilities gradually become out-of-date which may cause low energy efficiency/reliability in a world with ever-growing electricity demands. On the other hand, the information and communication technology has been significantly developed over the last few decades. By marrying the existing power grids with the state-of-art information and communication technology, Smart Grid can achieve electricity delivery more efficient, reliable, and secure. The Smart Grid can also facilitate the integration of renewable energy sources which is critical to secure the global economy towards gradually depleting fossil fuels, increasing energy costs, and worsening environmental conditions. When Smart Grid be used to manage the EVs, we can use the bi-direction communication character of Smart Grid to complete the delay-tolerant-network (DTN) in the future.

When EVs were driven on the road, they are the one of components of Vehicles-to-Grid (V2G). V2G is a type of the DTN. The EVs and other infrastructures consists the heterogeneous networks. Different from the traditional linked networks, it is a sparse mobile network in which the connections between nodes are intermittent and topology of the network often changed over time[3]. The deployment of charge stations is the network topology research of V2G. The conclusion can be used to design the protocol, increase the performance, and optimize the V2G.e.t. Our research proposed one way about how to deploy the charge stations in different scenarios.

The basic scenario is described as follows: EVs are driving on a real 2 dimension (2D) road, such as the map of city, or the ways of the villages. Every road should have the crossroads in different distance, we can deploy the charge stations at the crossroads because deployed at the crossroad can cover the bigger area then the other parts of one road. But in a real 2D map, it is not real and profitable to deploy charge stations at every crossroads. Under this scenario, our objective motivation is to design an effective location chose scheme that how to and where to deploy the charge stations can achieve the best performance with the least cost.

The remainder of this paper is organized as follows. In section 2, we will introduce the related work. In section 3, we describe the basic notions of the research model. Then, we present the proposed scheme in section 4. And in section 5, we will specify the simulation and the evaluation results. Finally, the conclusions and future work will be presented in section 6.

2   Related Work

In recent years, Smart Grid and V2G both are developed very quickly. The electricity sector has become the focus of heightened policy interest in China, as elsewhere, in the context of escalating concerns over emissions, security, and energy demand growth[4]. In this elevated policy context, the Smart Grid has been much discussed as panacea to solve the existing problem of traditional grids.

Smart Grid can use its bi-direction communication character to manage the EVs or other power facilities such as charge stations. According to the present research situation, Smart Grid is used to inject the new technology to the grid including the advanced communication technology, computer technology and information technology, automatic control technology and power engineering technology[5], thus making power grid has strong capacity and become a fully automated power supply network. But based on the conditions and the energy resource of our country, the research of Smart Grid is still in the initial stage. Although the policy is advocating the EVs or Hybrid Vehicles, the input of basic supporting infrastructures is still lacking.

So how to make every charge station has the max efficiency is the hot issue to research. A theoretical study has been conducted to formulate the waiting time minimized charge scheduling problem and drive a performance to the upper bound. One study of the profile of the load imposed on a Smart Grid by gird-charge of the onboard battery pack of electric and plug-in hybrid vehicles. The load profile is presented as an hourly probability of charge for each vehicle type[6]. One interesting research shows that V2G is expected to be one of the key technologies in Smart Grid strategies. It proposed an autonomous distributed V2G control scheme with scheduled charge providing a distributed spinning reserve for the unexpected intermittency of the renewable energy sources[7]. In the research[8], it developed a modeling and control paradigm for the aggregate charge dynamics of EVs. The central goal of the paper is to drive a control policy that can adapt the aggregate charge powers of EVs to highly intermittent renewable power. And in[9], it present a new metric called Contact Opportunity to max the contact opportunity for vehicular internet access. The purpose of the paper is to deploy the access point (AP) to the internet.

Our paper is to increase the charge opportunity of EVs because the lack of basic infrastructures such as the charge stations. In order to complete our research, we first proposed the deployment model use the Newton's law of cooling. The model is used to decide which crossroads can be deployed the charge stations, after that, we use the shortest path first algorithm to judge an EV whether can arrive the closet the charge stations, because the electricity one EV can store is limited. If the deployed charge stations can't provide the high-performance service, we will deploy more stations on that area.

3   Problem Formulation

The key point to improve the performance is improving the charge opportunity. Deploying more charge stations can achieve that certainly, but that is not to our profit. Once a EV driven on the road, it may not arrive at its destination with one charge because the limit of the cars technology. This disappointment experience limited the EVs apply seriously. So we thought of a improve solution to change this situation.

Deploying charge stations at the crossroads can cover bigger area with no doubt. But in a real map, deploying charge stations is not real for every country. We first abstract the real situation to our mode, and then use the model to simply the problem.

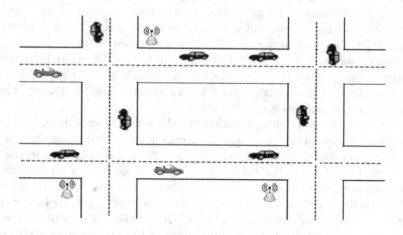

Fig. 1. The basic model of our research

As shows in Fig. 1, in our situation that EVs only can get electricity when they at the charge stations, we will specify the basic notions which include charge stations deployment model and DTNs network model as well.

The basic mode of EVs under Smart Grid is like this, EVs and charge stations are major nodes of the model, the source of electricity EVs can get all from the charge stations. And the EVs use the shortest path first to advise drivers go to the closet charge stations before electricity used up. As important parts of our research model, they have their individual characteristics as follows:

1. EV node: besides the mobility, every EV is resource constrained, for example, its storage electricity capacity is finite. Moreover, every vehicle has its current electricity storage and the unit power consumption every kilometer.
2. Charge Station: different from EV node, every charge station is a in a fixed location. Every charge station has outlets can be used to charge the EVs. Its duty is to try its best to satisfy EVs energy demand. To simplify the model, we assume that every charge station can charge all the EVs on the map. Because it is not real in a real world that every EV goes to a same charge station at the same time, so we can assume that.

The purpose of this paper is to come up with a novel scheme to solve the low charge opportunity under Smart Grid. The conclusion is to compose the changing performance with the number of charge station increasing by our scheme. If some areas need to deploy the charge stations, it can consult our result.

4   Problem Solution Scheme

According to the real situation when EVs are driven on the road, we divided our research model to a two stage model. First step, we need to give every crossroads a "temperature" T_n, n is the number of the crossroads. In order to give the T_n, we calculate the number of crossroads and give every crossroad a series number, and then assume some rulers about the charge stations model as follows:

1. In every moment, every crossroad has a current temperature T_{ci}, i is the series number of the crossroads. The crossroad has the highest T_{ci} ranks at the first number.
2. If a unit number of EVs passed the crossroad, the T_{ci} rose up one degree. The unit is designed based on the real travel condition on the area. If the travel condition is busy in the area, the unit should be bigger.
3. With the time goes by, the T_{ci} every crossroad has should be decrease. After the deadline we setting, the rank of the crossroads is the standard to deploy the charge stations. The crossroad ranks at the first will be chosen to deploy the first charge station.

Based on the rulers, we can use the Newtons law of cooling to design our model. We can use the existing formulations to build the function connection between "temperatures" with "time". In the real environment, the travel condition is changed repeat with the time. For example, in a real area, the travel often is busy at the time when people go to work or school or when they go to home after work or after school So we can use this connection to build an exponential decay progress in one day time travel condition.

In our model, the math describe of crossroads ranking is simply, we can use one word to represent:

$$T'(t) = \alpha(T(t) - H) \tag{1}$$

$T(t)$ denotes the temperature function of time t. The knowledge of calculus told us that the temperature change speed is $T'(t)$.

H represents the area init temperature in our model. Before beginning of the statistics, we assume that every crossroad has the same H. α represents the ratio relationship between current temperature and the change speed of temperature. In different areas may have the different α.

For calculate the current T_{ci}, we need to get the function expression of $T(t)$.

First step, change our equation:

$$\int \frac{T'(t)}{T(t) - H} dt = \int (\alpha) dt \tag{2}$$

Second step, get the result of the equation (C is the constant):

$$T(t) = H + Ce^{\alpha t} \tag{3}$$

Third step, assume the crossroad's temperature is $T(t_0)$ at time t_0. Drag in the up equation:

$$C = (T_0 + H)e^{\alpha t_0} \tag{4}$$

Forth step, use C to calculate the second step's equation, so, the equation can be simplified:

$$T = H + (T_0 + H)e^{\alpha(t-t_0)} \tag{5}$$

T_0 is the initial temperature of the crossroads and the H is the temperature of the area. We can use the last equation to calculate every crossroad's temperature of a day $T_i(1 \leq i \leq n)$, and rank the crossroads. We will deploy the charge stations based on the ranking of the crossroads, the highest temperature crossroad will be the first location charge station deployed at.

The second stage of our model is to judge an EV whether arrive at the closet charge station or not with the remainder electricity. The EV nodes and charge stations consists the graph. So we use the shortest path first algorithm to achieve this result.

When a EV is driven on the road, the current capacity of electricity it has is E_c, we know that different EVs' power consumption is different when they drive one kilometers. Assume the power consumption per kilometers is E_u, we can calculate the max distance it can be driven right now is:

$$D_{c-max} = \frac{E_c}{E_u} \tag{6}$$

We use the shortest path first algorithm calculate the distance between the EV and the closet charge station D_{e-c}, if the $D_{e-c} < D_{c-max}$, the EV can get the chance to charge, or the driver can't charge his EV in time. Use this way, we can judge the deployment of charge stations whether is efficiency or not, and also can be used to advise the drive goes to the closet charge station in time.

5 Simulation and Result

In this paper, we research the different performance of our scheme with different number of charge stations. This section evaluates the performance of the remedy scheme proposed in previous section.

To test our proposed scheme in practice, we coded the simulator based on the grid-road. It means that roads arrange as the grid graph. In future, we will test and improve our scheme based the road condition of the real environment.

Simulation settings are as follows, the parameters of EVs we use the characters of the EV: Wheego Whip. The max electricity it can storage is 30KWH, and the power consumption per kilometers we set is 0.2. The length of every road is 200 kilometers, The position of EVs is random created on the map, the current electricity is also random in the time begin the simulation.

The effective of different number of charge stations is shows in Fig. 2.

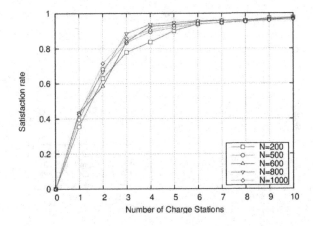

Fig. 2. The effective of different number of charge stations

As shows in Fig. 2, the X-axis represents the number of charge stations. The Y-axis represent the probability the EVs can get to the closet stations before their electricity used up, it means the probability of different developments can satisfy the travel's need. denotes the number of EVs on the map. From the figure we can know that the satisfaction rate is increase with the number of charge stations increasing. But the rate increasing approach 100 percentage and become smooth when the number of charge stations beyond 6. So we can get the conclusion that the best number of charge station deployed on the map is 6, it can achieve 95 percentage succeed ratio to satisfy EVs' energy demand.

The Fig. 3 shows the satisfaction rate of different time in one day, and in this situation the number of charge stations is chosen three different numbers.

Shows in Fig. 3, the Y-axis is also the Satisfaction rate. The X-axis represent the different time of one day. We only calculate the number of cars from 6 am to 21 pm. In other time of the day, the number of cars is few, so we didn't count it. We know that the mount of EVs is the biggest because people need go to work place or go to school at that clock, and the mount is smaller at 21 o'clock because it is late in the dark. So we set different number of EVs and different number of charge stations deployed to test our scheme. The different lines represent the different number of charge stations. As shows in Fig. 3, we can get the conclusion that when the number of charge stations is 6, the satisfaction can achieve almost 95 percentage, 3 charge stations can achieve almost beyond 80 percentage, if we only deployed 1 charge stations the satisfaction only can achieve 40 percentage. The ratio is too low to guarantee the travel's performance the EVs' energy demand.

Based on these conclusions, constructers can deploy the best number of charge stations when they complete the Smart Grid. In future work, we will consider the real trace of the travel and the real environment condition to test our scheme. The proposed scheme need be improved and compared with other proposed scheme. All of these will be our future work, and will be researched next time.

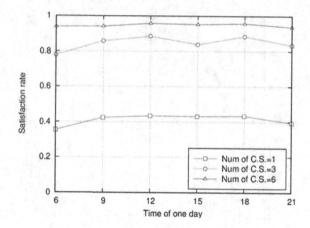

Fig. 3. The Satisfaction rate of different time

6   Conclusion

In this paper, we have observed that the proposed charge stations deployment scheme of EVs under Smart Grid. It actually helps solve the energy demand problem when EV driven on the road although it is a simple scheme and still has many aspects to complete. It can be useful to advise the constructer to build the Smart Grid, and proposed a thought to other researcher to consult. Therefore in the future work, we will continue to study the model of the real road situation, and we will attempt to improve the scheme to achieve better result.

Acknowledgements. The work was partially supported by Natural Science Foundation of Hubei Province of China. (No.2011CDB346, NO.2011CDB334); and the Fundamental Research Funds for the Central Universities, China University of Geosciences (Wuhan). (No.CUGL100232).

References

1. Top 10 Electric Car Makers for 2010 and 2011,
   http://www.cleanfleetreport.com/clean-fleet-aarticles/
   top-electric-cars-2010
2. Qin, H., Zhang, W.: Charge Scheduling with Minimal Waiting in A Net Work of Electric Vehicles and Charge Stations. In: VANET 2011 (September 2011)
3. Burleigh, H., Torgerson, L.: Delay-tolerant networking: an approach to Internet. In: VANET 2003 (June 2003)
4. Stefano Galli, Z.W., Scaglione, A.: For the Grid and Through the Grid: The Role of Power Line Communication in the Smart Grid. IEEE/ACM Transactions on Networking (2011)

5. Brools, A.N.: Vehicle-to-Grid Demonstration Project: Grid Regulation Ancillarg Service with a Battery Electric Vehicle, in 2002, http://smartgrid.com
6. Soheil Shahidinejad, S.F., Bibeau, E.: Profile of Charge Load on the Grid Due to Plug-in Vehicles. IEEE Transactions on Smart Grid
7. Yutaka Ota, T.N., Taniguchi, H.: Automous Distributed V2G(Vehicle-to-Grid) Satisfying Scheduled Charge. IEEE Transactions on Smart Grid
8. Bashsah, S., Fathy, H.K.: Transport-Based Load Modeling and Sliding Mode Control of Plug-In Electric Vehicles for Robust Renewable Power Tracking. IEEE Transactions on Smart Grid
9. Zheng, Z., Lu, Z., Sinha, P., Kumar, S.: Maximizing the Contact Opportunity for Vehicular Inter Access. In: INFOCOM (2010)

Efficient Arctangent Computation for Real-Time Histograms of Oriented Gradients Descriptor Extraction

Seung Eun Lee

Dept. of Electronic & Information Engineering
Seoul National University of Science and Technology
Seoul, Korea
seung.lee@seoultech.ac.kr

Abstract. In this paper, we propose the hardware architecture for the implementation of arctangent operation in Histograms of Oriented Gradients (HOG) descriptor extraction, targeting real-time pedestrian recognition with high accuracy on an embedded system with feasible hardware complexity.

Keywords: Histograms of Oriented Gradients (HOG), Pedestrian Recognition, Arctangent Operation, Hardware Accelerator, System-on-Chip (SoC).

1 Introduction

Pedestrian and vehicle recognition has become of significant interest for the usage model such as prevention of traffic accidents by using vehicle cameras and crime deterrence by using security cameras. In order to achieve this usage model, a recognition application is required to (a) acquire the image or video stream, (b) compute descriptor vectors from the image, (c) recognize object. Among these steps (b) and (c) are the most computationally challenging tasks.

In this paper, we propose architecture for the hardware implementation of arctangent operation in Histograms of Oriented Gradients (HOG)[1] descriptor extraction, targeting real-time pedestrian recognition with high accuracy on an embedded system with feasible hardware complexity. We first describe required computations in HOG descriptor extraction in Section 2. Section 3 proposes our hardware architecture for arctangent operation and demonstrates the implementation results. Section 4 concludes this paper by outlining the direction for future works on this topic.

2 Histograms of Oriented Gradients (HOG)

Pedestrian recognition uses feature descriptors as probability indicator of pedestrians. There are several feature descriptors that have been proposed to detect

Z. Li et al. (Eds.): ISICA 2012, CCIS 316, pp. 612–616, 2012.
© Springer-Verlag Berlin Heidelberg 2012

pedestrian recognition such as Harr wavelets[2], with Harr-like features[3], PCA-SHFT[4] and with Gabor filters[5]. In this paper, we chose HOG[1] descriptor for our pedestrian recognition because it is known to be an efficient features extraction scheme, is in combination with various classification algorithms, and has sufficient accuracy for our usage model of interest. HOG descriptor is a grey-level image feature formed by a set of normalized gradient histograms.

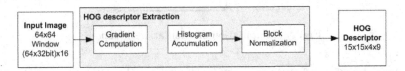

Fig. 1. Overview of HOG descriptor extraction

In the HOG pedestrian recognition scenario, we start with an input image that the recognition system takes with a camera. The intent is to extract HOG descriptors from the image. In order to do so, there are three major steps. Below we provide a brief explanation of the HOG descriptor extraction, although we refer the reader to[1] for a more detailed description. Fig. 1 illustrates the overview of the HOG descriptor extraction. Our system operates on the grayscale images and evaluates block of 8×8 pixels.

2.1 Gradient Computation

A 1-D spatial derivatives, $f_x(x, y)$ and $f_y(x, y)$, in x and y-direction are computed by applying the gradient filters M_x, M_y to all the pixels in the image $I(x, y)$,

$$f_x(x, y) = M_x \cdot I, \quad M_x = [-1, 0, 1]$$
$$f_y(x, y) = M_y \cdot I, \quad M_y = M_x^T \tag{1}$$

Fig. 2 shows the computation of the orientation direction. The gradient magnitude and orientation direction for each pixel are computed by

$$M(x, y) = \sqrt{f_x(x, y)^2 + f_y(x, y)^2}$$
$$\theta(x, y) = tan^{-1} \left(\frac{f_x(x, y)}{f_y(x, y)} \right) \tag{2}$$

2.2 Histogram Accumulation

The next step is the histogram generation. Each pixel within the cell casts a weighted vote given by equation (3) for an edge orientation histogram channel based on the orientation of the gradient element centered on it, and the votes are accumulated into orientation bins over local spatial regions (cells). In order

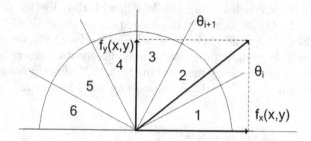

Fig. 2. Computation of orientation direction

to reduce the effect of aliasing, votes are interpolated tri-linearly among neighboring bin in both orientation direction and position. As a result, a histogram is generated by combining all generated histograms belonging to a block that consists of cells.

$$w_x = \left(\frac{x + 0.5}{width_{cell}} - 0.5 \right) - \left\lfloor \frac{x + 0.5}{width_{cell}} - 0.5 \right\rfloor$$

$$w_y = \left(\frac{y + 0.5}{width_{cell}} - 0.5 \right) - \left\lfloor \frac{y + 0.5}{width_{cell}} - 0.5 \right\rfloor$$

$$w_\theta = \left(\frac{\theta(x,y) + 0.5}{size_{bin}} - 0.5 \right) - \left\lfloor \frac{\theta(x,y) + 0.5}{size_{bin}} - 0.5 \right\rfloor \qquad (3)$$

$$lower \quad cell : \alpha = 1 - w$$

$$upper \quad cell : \alpha = w$$

$$w_{total} = \alpha_x \cdot \alpha_y \cdot \alpha_\theta$$

2.3   Block Normalization

Once all of the histograms within a block have been obtained, they are normalized due to the variance of gradient strengths over a wide range. When V_k is the vector corresponding to a combined histogram for the block region and ε is a very small closed to zero, the normalized HOG descriptor vector V is given by equation (4).

$$V = \frac{V_k}{\sqrt{\|V_k\|^2 + \varepsilon}} \qquad (4)$$

3   Implementation of Arctangent Operation

The HOG descriptor extractor operates on the grayscale images and evaluates block of 8 × 8 pixels, voting edge oriented gradient into 9 orientation bins in 0-180 degree. This requires the calculation of gradient magnitude and orientation direction for each pixel (See Eq. (2)). Our hardware accelerator calculates the gradient magnitude and orientation direction for eight pixels in parallel. The

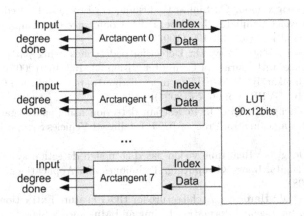

Fig. 3. Block diagram of arctangent calculation unit

square root calculation for gradient magnitude is implemented with iterative operation which requires 8 cycles.

Fig. 3 illustrates the block diagram of the arctangent calculation unit which consists of a LUT and eight calculation block (Arctangent 0-7). The LUT stores 90 entries, where each entry has 12bits to represent the tangent value. The Arctangent block takes inputs (value and start signal), and search the corresponding degree by searching the LUT. We adopted binary search algorithm to complete the arctangent operation in 7 cycles, which is less than the gradient magnitude calculation latency. Therefore, the multi-cycle operation of arctangent calculation unit does not slow down the overall performance of HOG descriptor extraction. The arctangent calculation unit was implemented using $Verilog^{TM}$ HDL and verified its functionality and feasibility through simulation.

4   Conclusions

In this paper, we proposed the hardware architecture for arctangent calculation unit, which is a sub-block of HOG descriptor extractor. Our design adopted LUT and binary search algorithm to find out the orientation direction, requiring multi-cycle operation. However, the latency for calculating arctangent is hidden by the computation time of magnitude, without slowing down the overall performance. Moreover, the proposed hardware calculates the orientation angle to be 1-degree precision, enabling the tri-linear interpolation which results superior classification performance. We plan to implement overall system for the pedestrian recognition on FPGA.

References

1. Dalal, N., Triggs, B.: Histograms of oriented gradients for human detection. In: Proc. IEEE Conference on Computer Vision and Patern Recognition, pp. 886–893 (2005)

2. Oren, M., Papageorgiou, C., Sinha, P., Osuna, E., Proggio, T.: Pedestrian detection using wavelet templates. In: Proc. IEEE Conference on Computer Vision and Pattern Recognition, pp. 193–199 (1997)
3. Viola, P., Jones, M.J., Snow, D.: Detecting pedestrians using patterns of motion and appearance. Intl Journal of Computer Vision 63, 153–161 (2005)
4. Ke, Y., Sukthankar, R.: PCA-SIFT: a more distinctive representation for local image descriptors. In: CVPR, pp. 66–75 (2004)
5. Cheng, H., Zheng, N., Qin, J.: Pedestrian detection using sparse gabor filter and support vector machine. In: Proc. of IEEE Intelligent Vehicles Symposium, pp. 583–587 (2005)
6. Cao, T.P., Deng, G.: Real-time vision-based stop sign detection system on FPGA. In: Proc. of Digital Image Computing: Techniques and Applications (DICTA), pp. 465–471 (2008)
7. Kadota, R., et al.: Hardware Architecture for HOG Feature Extraction. In: Proc. of Intl Conf. on Intelligent Information Hiding and Multimedia Signal Processing, pp. 1330–1333 (2009)

Single Photon Counting X-Ray Imaging System

Seung Eun Lee and Sang Don Kim

Dept. of Electronic & Information Engineering
Seoul National University of Science and Technology
Seoul, Korea

Abstract. In this paper, a prototype system for single photon counting X-ray imaging system with multiple energy discrimination is presented which consists of an image acquisition and pre-processing unit and GPGPU.

Keywords: X-ray Imaging, Single Photon Counting, Image Reconstruction, GPU.

1 Introduction

Modern medical X-ray imaging system allows investigation of organs and tissues in the human diagnostic field, visualizing the anatomical structures of a body. For the diagnosis purpose, does to the patient and high position resolution are important factors in medical X-ray imaging, requiring the minimum number of photons to obtain an image with sufficiently small statistical fluctuations. However, the dose given to the patient is normally significantly higher due to additional sources of noise, inefficiency, variations in the response of the detector to the wide energy spread of the incoming photons, scattered radiation that smears the image, etc.[1]. A type of X-lay imaging based on photon counting instead of charge integrating has been introduced offering improved image quality compared to the former devices[2]. This hybrid approach is capable of discerning and processing each single X-ray photon in addition to just counting, allowing a significant dose reduction and providing better image quality.

In response to the need and the opportunity, we are developing multi-energy imaging system for single X-ray photon detector, providing a fast image reconstruction through the material decomposition. In this paper, we introduce our ongoing work and propose hardware architecture for a prototype X-ray imaging system consisting of an image acquisition and pre-processing unit and GPU (Graphics Processing Unit) using NVIDIA CUDA (Compute Unified Device Architecture)[3] framework.

The rest of this paper is organized as follows. We first briefly introduce the single photon counting for X-ray imaging in Section 2 and present the hardware architecture for our prototype in Section 3. Section 4 concludes this paper by outlining the direction for future works on this topic.

Z. Li et al. (Eds.): ISICA 2012, CCIS 316, pp. 617–620, 2012.

2 Single Photon Counting (SPC)

X-rays of given energy are differently attenuated from different tissues, providing information on anatomical structures. The charge integration scheme sums the charge accumulated in a pixel corresponding to the total X-ray energy absorbed in the pixel. Therefore, the charge integrating detector underestimates the contribution of the low energy photons to the signal, resulting in underestimated beam hardening artefacts because changing the low energy part of the photon spectrum is primarily responsible for the appearance of the beam hardening artefacts[4]. However, SPC detector provides an accurate representation of the beam hardening effect by assigning the same weight to the all the detected photons, leading to a higher but more correct expression of the beam hardening effect[5]. Measuring low energy photons individually, by counting the number of phonons with various thresholds, increases the image contrast. Medical X-ray imaging based on SPC is required to keep the following things in key focus:

1. need to lower the patient dose;
2. provide a high count rate capability;
3. support a fast data transfer and image construction.

However, high image quality for a satisfactory visualization of soft tissues is the fundamental aspect to take into account.

In SPC, the digital pixel sensor is required to (a) acquire the charge of each individual event sensed by its corresponding pixel, (b) classify the total energy of each incoming charge packet into one of selectable energy bins, and (c) count every charge by energy. Then, the counter values for each energy bin are forwarded to the processing unit for image reconstruction. Finally, reconstructed images are overlaid the multiple slices by using visualization software.

3 A Prototype X-Ray Imaging System

3.1 Overview

To reconstruct X-ray image using SPC, a prototype X-ray imaging system (See Fig. 1) which consists of a multi-energy image acquisition, followed by pre-processing units and parallel processing unit (GPU) is under developing. The image acquisition and pre-processing unit provides the interface between the read-out module and the host computer, enabling control of the read-out module and multi-energy image acquisition. The read-out module allows simultaneous measurement and efficient readout from the sensor. The GPU completes the image reconstruction according to the users demands and supports some common image processing features including smoothing, sharpening and special filters. Our prototype acquires raw data from the image sensor and forwards them to the host PC through Ethernet.

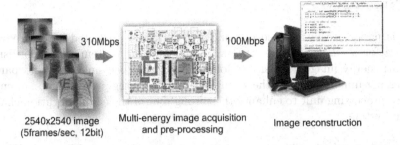

Fig. 1. System flow for multi-energy X-ray image reconstruction

3.2 Sensor Data Acquisition and Pre-processing Unit

The multi-energy image acquisition and pre-processing unit establishes communication channel between readout module and GPGPU, supporting the pre-processing of the raw data. It also provides the way to control the readout module from the host PC. Fig. 2 shows the block diagram and photograph of our prototype board. The data-path for pre-processing is implemented on FGPA providing concurrent processing of the raw data. An embedded ARM processor enables high level control of the acquisition and pre-processing unit and manages the data-path on the FPGA.

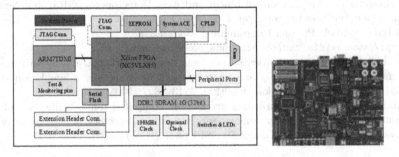

Fig. 2. Block diagram and photograph of our prototype board

3.3 Image Reconstruction Unit

GPGPU is many-core architecture, supporting thousands of threads in parallel and non-uniform access to memory. Image reconstruction is completed with GPU on the host PC by using NVIDIA CUDA framework which is an SDK, software stack and compiler that allows for parallel programming on a GPU. Using improved processing speed, scientists and researchers and software developers are using the CUDA for image and video processing, medical engineering, computational biology and chemistry, fluid dynamics simulation, ray tracing and much more[3]. In our implementation, we are using a GeForce GTX 460SE which has 288 CUDA cores, on Intel i7 machine. The CUDA devices accelerate the reconstruction of the image by harvesting a large amount of data parallelism.

4 Summary

In this paper, we introduced our ongoing work and presented hardware architecture for a single photon counting X-ray imaging system which consists of a multi-energy image acquisition, followed by pre-processing units and parallel processing unit (GPU). In the future, we plan to integrate compression engine on pre-processing unit to enhance the throughput of the communication channel in our system.

Acknowledgment. This work was supported by the Korean Ministry of Knowledge Economy and KEIT under contract No. 10038695.

References

1. Francke, T., Eklund, M., Ericsson, L., Kristoffersson, T., Peskov, V., Rantanen, J., Sokolov, S., Soderman, J., Ullberg, C.: Dose reduction in medical X-ray imaging using noise free photon counting. Nuclear Instruments and Methods in Physics Research Section A: Accelerators, Spectrometers, Detectors and Associated Equipment 471(1-2), 85–87 (2001)
2. Giersch, J.: Medical quantum X-ray imaging with 2D detectors. Nuclear Instruments and Methods in Physics Research Section A: Accelerators, Spectrometers, Detectors and Associated Equipment 551(1); Proceedings of the E-MRS Fall Meeting 2004, Symposium D - Applications of Linear and Area Detectors for X-Ray and Neutron Diffraction and Spectroscopy, pp. 125–138 (October 2005)
3. NVIDIA, CUDA: Parallel Programming Made Easy,
   http://www.nvidia.com/object/cuda_home_new.html
4. Shikhaliev, P.M.: Beam hardening artefacts in computed tomography with photon counting, charge integrating and energy weighting detectors: a simulation study. Physics in Medicine and Biology 50(24), 5813–5827 (2005)
5. Shikhaliev, P.M.: Beam hardening artefacts in computed tomography with photon counting, charge integrating and energy weighting detectors: a simulation study. Physics in Medicine and Biology 50, 5813–5827 (2005)

Evaluation of CUDA for X-Ray Imaging System

Seung Eun Lee and Dae-Young Park

Dept. of Electronic & Information Engineering
Seoul National University of Science and Technology
Seoul, Korea

Abstract. CUDA is being used in many areas of science and research. Pararell processing of CUDA is faster than CPU. But the CUDA could not be taking advantage depending on kind of program. Dvelopers using CUDA should be aware of the characteristic of CUDA and the difference of foundational architecture compared to CPU. In this paper, we investigate the performance of the CUDA for parallel processing in order to evaluate the potential use of it for x-ray image reconstruction.

Keywords: CUDA, GPGPU, X-ray Imaging, Image Processing, Analysis, Optimization.

1 Introduction

In recent years, developers have been finding effective method improving processing speed in the image processing area. NVIDIAs CUDA (Compute Unified Device Architecture) provides a massively parallel processing. Thanks to the improved processing performance, scientists and researchers are using the CUDA for image and video processing, medical engineering, computational biology and chemistry, fluid dynamics simulation, ray tracing and much more[1]. The CUDA shows exceptional performance in many areas.

Usually, most engineers using CUDA expect the higher performance of CUDA. In this paper, we evaluate CUDA for potential use of x-ray image reconstruction and present our experimental results.

2 Backgrounds

In recent years, GPUs (Graphics Processing Units) are increasingly developed using parallel architecture. The GPU has moved from being used solely for graphical tasks to the whole process of computation. GPGPU (General-Purpose computing on Graphics Processing Units) is extensively used for signal processing, using the GPU instead of CPU for floating point operation[2].

Fig. 1 shows the CUDA execution model where a host operates serial processing. In case of repeatedly operating algorithm, the data to be processed transfer from host(CPU) using serial processing to device (GPU, CUDA) using the parallel processing by calling kernel function to start parallel processing on the device

Z. Li et al. (Eds.): ISICA 2012, CCIS 316, pp. 621–625, 2012.

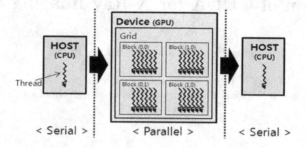

Fig. 1. Sequential Execution with CUDA

(CUDA). The kernel function makes the grid which is composed of many blocks of threads on the device. A thread is a sequence of instructions that can be executed on different data units in parallel. The data completed by the process on device are returned to host for serial operation. Such a sequence is repeated by the process until the program is completed. The CUDA is expected to offer improving performance by the parallel process rather than the serial process.

3   Performance Evaluation

3.1   Experimental Introduction and Expectation

In this paper, we implement that the three vector matrix are added in each of the CPU and CUDA for comparing computation speed. We expect CUDA operated by the parallel is to show the good processing performance rather than CPU operated by the serial.

To measure processing speed on CPU and CUDA, experiments are implemented in various environments.

1. We make three vectors (composing of per each 64×64, 128×128, 640×480, 800×600, 1024×768, 1280×1280 to measure processing speed according to the size of the matrix;
2. It is separated by thread 512 per block and thread 1024 per block to measure processing speed according to the number of threads;
3. We measure the processing speed of the parallel programs and sequential program by using a non-weight adding equation in order to (1)

$$S_{ij} = a_{ij} + b_{ij} + c_{ij} \tag{1}$$

3.2   Experimental Setup

The experiments were run onto a CPU *Intel i7 2600K 3.4GHz* Quad-Core, L1 Cache $4 \times 32KB$ / $2 \times 32KB$, 8GB DDR3 1.333GHz of main memory, and PCI-E 2.0. The GPU is NVIDIA GeForece GTX 460 SE with global memory 993 Mbytes and GPU clock speed 1.46GHz, L2 cache size 524288bytes, maximum number of threads per block are 1024. It has 6 multiprocessors, each multiprocessor has 48 CUDA cores, resulting in total of 288 CUDA cores. Windows 7 professional is used as Operation system. Developer tool is Microsoft Visual C++ 2008. Table 1 summarizes the experimental setup.

Table 1. Experimental setup

	CPU(host)	GPU_CUDA(device)
Model name	Intel i7 2600K	NVIDIA GeForece GTX 460 SE
Clock speed	3.4 GHz	1.46GHz
Memory	8GB DDR3(RAM)	993 Mbytes(global memory)
Operation System	Windows 7 Professional K	
Developer tool	Microsoft Visual C++ 2008	

3.3 Experimental Result and Discussion

Averages processing time for three times test for the equation (1) are described in Table 2, 3 and figured in the Fig. 2. The CUDA processing time is longer than serial processing of CPU which is unexpected result. However, CUDA shows almost constant performance while the problem size increases. On the other hand, CPU execution time is steadily increased according to the size of matrix.

Table 2. Computation time of CUDA and CPU(ms)[Thread 512]

	64 × 64	128 × 128	640 × 480	800 × 600	1024 × 768	1280 × 1280
First GPU Exp.	40.1935	54.0977	58.4948	41.3261	54.1398	42.2111
First CPU Exp.	0.0057	0.0203	0.3103	0.4670	0.8808	1.7420
Second GPU Exp.	41.0369	47.3134	43.5064	51.8256	42.5637	4.8153
Second CPU Exp.	0.0060	0.0193	0.3106	0.0494	0.8382	1.8147
Third GPU Exp.	40.0603	39.7697	41.6668	40.2304	56.8708	41.5335
Third CPU Exp.	0.0060	0.0199	0.3097	0.6396	0.9180	1.7494
GPU Average	40.4302	47.0603	47.8893	44.4607	50.1914	43.5200
CPU Average	0.0059	0.0198	0.3102	0.3853	0.8790	1.7687

Table 3. Computation time of CUDA and CPU(ms)[Thread 1024]

	64 × 64	128 × 128	640 × 480	800 × 600	1024 × 768	1280 × 1280
First GPU Exp.	41.2735	43.1759	39.7787	42.3518	49.4005	44.6546
First CPU Exp.	0.0054	0.0193	0.3064	0.5104	0.8618	2.0311
Second GPU Exp.	59.4656	51.0728	58.8399	40.5412	49.2363	39.7205
Second CPU Exp.	0.0051	0.0211	0.3194	0.4908	0.8546	2.0750
Third GPU Exp.	60.2622	53.8993	47.0942	51.8065	42.5643	42.7744
Third CPU Exp.	0.0054	0.0238	0.3133	0.4389	0.8639	1.9035
GPU Average	53.6671	49.3826	48.5709	44.8998	47.0670	42.3832
CPU Average	0.0053	0.0214	0.3130	0.4801	0.8601	2.0032

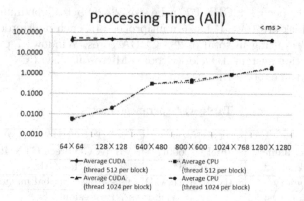

Fig. 2. Computation Time of CUDA and CPU

In order to see the performance of CPU and CUDA only, we changed the experimental setup to remove the data transfer time from the processing time. Fig 3. shows CPU and CUDA processing time without data transfer time between host and device. When vector matrix size is 64 × 64, CUDA is slower than CPU, and when vector matrix size is 128 × 128, the data operated in CPU and

Table 4. Computation time of CUDA and CPU excluding data transfer time(ms)[Thread 512]

	64 × 64	128 × 128	640 × 480	800 × 600	1024 × 768	1280 × 1280
First GPU Exp.	0.0163	0.0187	0.0187	0.0187	0.0253	0.0241
First CPU Exp.	0.0039	0.0175	0.3354	0.4721	0.0253	0.0241
Second GPU Exp.	0.0160	0.0169	0.0172	0.0163	0.0193	0.0214
Second CPU Exp.	0.0042	0.0196	0.3133	0.4923	0.9137	2.0146
Third GPU Exp.	0.0160	0.0166	0.0178	0.0178	0.0241	0.0223
Third CPU Exp.	0.0042	0.0154	0.3076	0.4392	0.8947	1.8748
GPU Average	0.0161	0.0174	0.0179	0.0176	0.0229	0.0226
CPU Average	0.0041	0.0175	0.3188	0.4679	0.9285	1.9321

Table 5. Computation time of CUDA and CPU excluding data transfer time(ms)[Thread 1024]

	64 × 64	128 × 128	640 × 480	800 × 600	1024 × 768	1280 × 1280
First GPU Exp.	0.0178	0.0166	0.0202	0.0184	0.0423	0.0217
First CPU Exp.	0.0039	0.0151	0.3139	0.4790	0.9635	1.9102
Second GPU Exp.	0.0157	0.0166	0.0172	0.0172	0.0211	0.0377
Second CPU Exp.	0.0039	0.0154	0.3139	0.4679	0.9430	1.8431
Third GPU Exp.	0.0154	0.0157	0.0175	0.0229	0.0190	0.0260
Third CPU Exp.	0.0042	0.0154	0.3127	0.5014	0.8575	1.8899
GPU Average	0.0163	0.0163	0.0183	0.0195	0.0275	0.0285
CPU Average	0.0040	0.0153	0.3135	0.4828	0.9213	1.8811

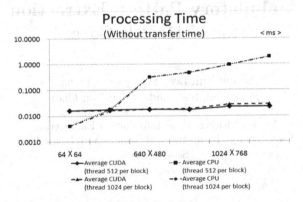

Fig. 3. Computation time of CUDA and CPU excluding data transfer time

CUDA are processed as similar processing time. After 640 × 480 of vector matrix size, the CUDA processing time is almost constant but The CPU processing time noticeably increase. We know that CUDA is little affected by matrix size. In contrast, the CPU is more affected by matrix size operates in this environment case. And thread numbers per block did not affect to the processing time. Although thread numbers per block are increased, we did not show the result that is noticeably the difference of processing time. That is because the operated function numbers are smaller than the core number of CUDA.

4   Conclusion

In this paper, we investigated the performance of the CUDA for parallel processing in order to evaluate the potential use of it for your x-ray image reconstruction. In the case of only processing time of the CUDA without the data transfer time, CUDA outperforms CPU for the large problems. Moreover, the CUDA processing time is nearly constant. However, the CPU processing time is steadily increased according to the matrix size. The CUDA for simple calculations like addition has the delay time which includes dynamic memory allocation & release time and transferring time between device and host. Our experimental results imply that CUDA is one of solutions for a large sized x-ray imaging and CPU is one of the solutions for a small sized x-ray imaging system such for dental clinic. We are going to evaluate the performance benefit of CUDA using more complex operations.

Acknowledgement. This work was supported by the Korean Ministry of Knowledge Economy and KEIT under contract No. 10038695.

References

1. NVIDIA Developer homepage, http://developer.nvidia.com/what-cuda
2. GPGPU homepage, http://gpgpu.org

Ambulatory Pattern Extraction
for U-Health Care

Seung Eun Lee, Yeong-seob Jeong,
Seung-jun Son, and Hyeon-Min Choi

Dept. of Electronic & Information Engineering
Seoul National University of Science and Technology
Seoul, Korea

Abstract. The incorrect ambulatory habit affects our body temporarily
as well as permanently. In this paper we propose Ambulatory Remedial
Clinic Helper (ARCH), shoes detecting wrong ambulatory habit. It is
easy to let patients know strange signals promptly after getting & analyzing the data of them in everyday life. ARCH system uses pressure
sensors and 3-axis acceleration sensor to measure degree of wrong ambulatory habit and report the acquired data to host PC. By implementing
this system, we can check the ambulatory pattern in real time, ward
off plenty of foot diseases, and prevent deterioration of the existing foot
problems.

Keywords: Pervasive Computing, Bio-signal Processing, Embedded
System.

1   Introduction

Most people usually spend time walking nearly in half a day, but we do not
care about importance of ambulation. Wrong ambulation affects not only our
body, but also height and balance. Korea Foot Health Association (KFHA) have
surveyed foot-printing of 323 women who got from 20 to 50 years old over the one
month. As a result, 84.5% of women suffered serious arch malformation state,
and only 15.2% of women with normal feet were surveyed[1]. Thus, inspection of
foot-printing and foot-pressure is widely used to predict disease symptoms and
to recommend good ambulatory habit which is known to be very relevant with
persons health. The gait has correlation with the meridian points, which are
the pathway of energy and are regarded as very sensitive and important spots
at sole. It is well known that good gait habits stimulate the Meridian points
and then make our body healthier. For instance, wrong gait habit can lead
to various complications. A fairly typical case, your feet would be numb by the
aftereffects which bad habit of walking cause. At this time, if you continue to walk
inaccurately or use only certain part of sole, it paved the way for foot paralysis.
And another case of Diabetes, when you did not distribute foot pressure on the
sole correctly, we find that the pressured spots from foot usually are moved to

Z. Li et al. (Eds.): ISICA 2012, CCIS 316, pp. 626–630, 2012.

unnatural spots on the sole and it occur blisters, calluses or ulceration. However, it is easy to keep the balance of the body if people know that the correct gait has an important role.

Recently, some equipments measuring foot pressure were developed and applied to different applications. Yonekawa et al. checked the fatigue of walking people using pressure sensors[2]. Velazquez et al. proposed a wearable interface for the foot which enables users to obtain information through the sense of touch of their feet[3]. Fujimoto et al. adopted accelerometers mounted on both shoes to recognize steps[4]. In this paper, we present Ambulatory Remedial Clinic Helper (ARCH) which has pressure sensors and 3-axis acceleration sensor to measure degree of wrong ambulatory habit and report the acquired data to host PC (it can be replaced with smart phone).

The rest of this paper is organized as follows. We first present the ARCH system including the detail description of each component in Section 2. Section 3 explains the experimental result with ARCH system and we conclude in Section 4 by outlining the direction for the future works on this topic.

2 System Overview

Fig. 1 shows the block diagram of our ARCH system. It takes signals from five pressure sensors which are attached under the sole of shoes and a 3-axis acceleration sensor at the front of shoes. The amplified analog signals are steered to A/D converter (ADC). The controller in the FPGA selects the input signals for an ADC and acquires value of the sensor to enter the register. Finally, the sampled data are shown through host PC using serial communication.

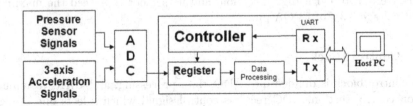

Fig. 1. Block diagram of our ARCH system

2.1 Sensors

ARCH uses eight sensors, the five FSR pressure sensors and 3-axis accelerometer, to measure the gait habit. FSR pressure sensors are placed according to position which receives a lot of the weight (See Fig. 2). We found proper spots for pressure sensors located in sole by testing a footprint at a paper in person. If people walk with these shoes, it is possible to make changes on signals because the resistance

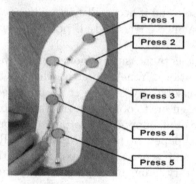

Fig. 2. ARCH Sensor placement

of weighted sensor is reduced. The pressure intensities from each part of the bottom of our shoes and 3-axis acceleration data are used to recognize the gait pattern.

2.2   A/D Converter

A/D converter which has 8 channels is embedded in our ARCH system. It converts sensor signals to digital. The A/D Converter is controlled by the control logic. The converted data are stored to the registers in FPGA sequentially in each sampling period. We set the (+) reference voltage to 3.3V and (−) reference voltage to 1.5V, in order to match the converted 8bits data inputs to rated 3.3V input voltage of FPGA port. Therefore, analog data between 1.5V∼3.3V are converted to 0∼FF digital section, and its set not to exceed the maximum rated voltage 3.3V of FPGA pin.

2.3   Control Block

The control block is divided into ADC, data processing, and UART communication control section. ADC receives control signal which selects one of eight analog input signals per 300microsecond and converts it to digital data. The data values converted into digital are stored in the internal register file. The data stored in the register can pass through a data processing before it is sent with the UART communication. It can be prepared to send data depending on user's choice. For example, when it receives the keyboard input, then it exports the hex data stored in registers sequentially or exports the desired register only.

2.4   Communication

RS232 serial communication transfers eight signals to the host PC at a speed of 9600bps. First, it starts to send the start signal. Next, the pressure and

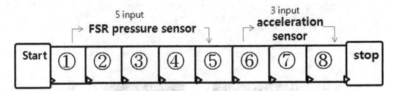

Fig. 3. Our packet format

acceleration sensor signals are transmitted sequentially. Finally, it sends the stop signal. Each of them is 1byte, so the 10byte signals are transmitted altogether, resulting 80 set of data forwarding to PC in a second.

2.5 Host PC

The data received through serial communication can be monitored by the Lab view. Each signal can be verified with real-time graph and it is possible to save decimal data as a text or Excel file. Therefore, the data stored in Excel can be checked by changing it into graphs in the future. The stored graphs and data transfer to the PCs which the hospital's doctors and specialists have, and the status of the patient can be examined at the desired time and place. Thanks to this, though the patient lives at home, they can receive remote examination by a doctor. And then, they can also be prescribed remotely.

3   Experiment Result

Fig. 4 shows our ARCH and experimental setup. We get data through experiment when person has different weight (55kg, 65kg, and 75kg) and walking speed (5km/h, 10km/h). The five pressure and acceleration values are displayed on the host PC in real time as shown in Fig. 5.

Fig. 4. Photograph of ARCH

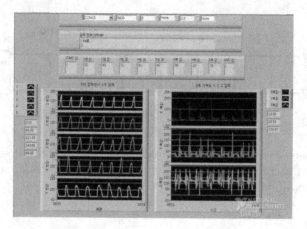

Fig. 5. The Output signals of pressure and Acceleration sensor on host PC

4   Conclusion

We have developed a personal remedial clinic shoe, a pair of shoes which captures incorrect ambulatory and displays users pressure waveform of each part of sole. User can easily straighten ambulatory habits as comparing normal and unnatural users pressure waveform. Besides, its suitable for everyday life because it embeds pressure sensors in a pair of shoes which allows users to wear them without any interference. We have a plan to analyze the acquired signals and integrate signal processing unit which makes decision whether a subject is under normal working or not. Also, we have a plan to forward the sensing data through Bluetooth to a smart phone.

References

1. No, M.-J., Lee, M.-H.: Korea foot health association
2. Yonekawa, K., Yonezawa, T., Nakazawa, J., Tokuda, H.: FASH: Detecting Tiredness of Walking People Using Pressure Sensors. In: 6th Annual International Mobile and Ubiquitous Systems: Networking & Services, MobiQuitous 2009, July 13-16, pp. 1–6 (2009)
3. Vel'azquez, R., Baz'an, O., Magana, M.: A Shoe-Integrated Tactile Display for Directional Navigation Intelligent Robots and Systems. In: IEEE/RSJ International Conference on IROS 2009, October 10-15, pp. 1235–1240 (2009)
4. Fujimoto, M., Fujita, N., Takegawa, Y., Terada, T., Tsukamoto, M.: A Motion Recognition Method for a wearable Dancing Musical Instrument. In: International Symposium on Wearable Computers, ISWC 2009, September 4-7, pp. 11–18 (2009)

Study on Signals Sources of Earth's Natural Pulse Electromagnetic Fields

Guocheng Hao[1] and Hongliang Wang[2]

[1] Faculty of Mechanical and Electronic Information
China University of Geosciences
WuHan 430074, China
apollo77@vip.qq.com
[2] Institute of Geophysics and Geomatics
China University of Geosciences
WuHan 430074, China
27735367@qq.com

Abstract. Earth's natural pulse electromagnetic fields (or ENPEMF) are geomagnetic theory or detection means for the Earth internal and external studies. Based on the study on this theory at domestic and overseas and combining with the data we have got from the special equipment, this paper analyzes the causes of the signals generation field, such as, Earth's induced magnetic field, exogenous disturbance field, abnormal magnetic field of the earth's crust and the Earth's rotation block interaction. The signals frequency of ENPEMF for the best observation is 14.5 KHz in Wuhan, and the pulse patterns are plotted combined with currently theory and actual measurement data. The theory has a verifiable control the standard waveform that has been found from reference. The article gives the measurement principle of the ENPEMF signals and provides a reference for subsequent research.

Keywords: ENPEMF, Earth's Induced Magnetic Field, Pulses Envelope Patterns, Observation Frequency.

1 Introduction

"Earth's natural pulse electromagnetic field" (ENPEMF) was introduced by A.A Vorobyov, who expressed a hypothesis that pulses can arise not only in the atmosphere but within the Earths crust due to processes of tectonic-to-electric energy conversion (Vorobyov, 1970; Vorobyov, 1979). According to it, an intensity of pulse flux was expected to be increased in the eve of and at the moment of large earthquakes[1]. A.Boryssenko have Carried out some experimental studies to detect and locate subsurface hidden objects by near-surface determination of statistical variations of the natural pulsed electromagnetic background field in the 0.1-50 kHz frequency range by using of necessary electronic equipment and statistical signal processing technique[2]. However, its signal source is not yet specifically identified, and there hasn't discoursed about such issues in the

Z. Li et al. (Eds.): ISICA 2012, CCIS 316, pp. 631–638, 2012.

domestic. The book ¡Physics of Electromagnetic Phenomena of the Earth¿ gives a very detailed analysis on the Earth's magnetic fields[3]. Our work on this paper is summarized from the present materials to find the various causes of the ENPEMF and draw the curve on experimental data to reflect the signal characteristics.

The ENPEMF signals shown in Fig. 1 can be understood as instantaneous changes that composed of the moment disturbance changes of the Earth's natural magnetic field. The amplitudes and frequencies of these disturbances are different. We choose the special frequency pulses as the object to analysis and setting the appropriate threshold, then record their number of daily 24 hours each time by electronic equipment, drawing out the envelope of the pulse numbers diagrams. If suddenly appeared inconsistent comparing with the standard curve, such as the number of pulses suddenly increased, there may be a precursor of the earthquake. This method can also be applied to a variety of other fields.

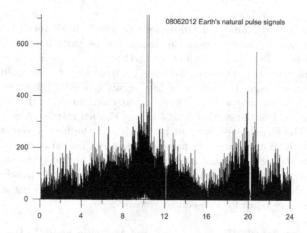

Fig. 1. Signals of ENPEMF in Wuhan

2   Analysis on the Field Sources

There are two views on the field sources of the natural impulse of the geomagnetic field at present from the materials available, one is the atmospheric effects (atmosphere electromagnetic field), and another is the crust movement. The latter are divided into two factions, some believe that the causes are the centrifugal rotation of the Earth's crust and the Earth's core, while another consider the gravity of the Earth's internal tidal waves be in the role.

In discussing the causes of the ENPEMF, we need to make a detailed analysis on the Earth's magnetic field. The magnetic field can be generated by the magnetic materials and currents as the present theory shows. So, Earth's magnetic field is superposition generated by the magnetic rocks of the Earth's interior and the Earth's internal and external currents magnetic field component. The origin of the magnetic field is different, so that a variety of magnetic field components of the temporal variations and spatial distribution are also different. According to modern theory, the hypothesis of Earth's electromagnetic field causes can be divided into two parts of the field of endogenous and exogenous in accordance with the field source location. Endogenous field originated in the magnetic material and the current below the surface, further divided into the Earth's core field, crustal field and induction field.

The Earth's core field is considered to be the main magnetic field of the geomagnetic field. The generally accepted argument is that the Earth's core field is produced by nuclear magnetic fluid generator process. The crustal field is also called lithospheric magnetic field or a local magnetic field anomaly is generated by the magnetic rocks of the crust and upper mantle. The main magnetic field and the local magnetic field anomaly change very slowly, which can be described as stable magnetic field. The induction field is generated by the induced current comes from Earth's interior that actually caused by the external variable magnetic field. And the induction field changes following its source field changes with a faster time characteristics.

2.1   The Induction Field Is One Component of the ENPEMF

Schuster spherical harmonic analysis concluded that the diurnal variation of the geomagnetic field affected by inside and outside the field. Exogenous field is main part originated by the earth's outer space current system (Steward, 1982) and which strength is about 2 times the endogenous field. The endogenous part is the inner earth's magnetic field comes from induced currents because of exogenous field changes. The Earth's induced magnetic field is less than 0.5% of the total Earth's magnetic field. But the induced magnetic field is very complex like Earth's main magnetic field. First, exogenous variation magnetic field and its corresponding induced magnetic field are numerous and complex. Secondly, Earth electrical is uneven global distribution, so, the same source field can produce different induced magnetic field in different regions.

The fundamental origin of the Earth's induced magnetic field is magnetosphere-ionosphere system's current system, also called primary inducing field, but the direct field source is the induced currents within the crust and mantle, also known as the secondary induction field. Therefore, the intensity and distribution of the induced magnetic field depends both on Earth electrical properties and the exogenous field strength and frequency. Induced magnetic field associated with the Earth's external electromagnetic environment and the Earth's interior electromagnetic nature. The actual measurements data we detect at the Earth's surface are the vector sum of the endogenous and the exogenous field.

The induced magnetic field is divided into global, regional and local magnetic field according to the different spatial scales. And these three magnetic fields may be having their contributions to ENPEMF.

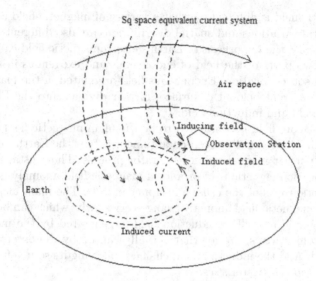

Fig. 2. Induction diagram of the field

2.2   Exogenous Perturbation Field Is the Main Component of ENPEMF

The second part of the geomagnetic field is external field which origins from the magnetosphere-ionosphere systems current system, and mainly exists in the ionosphere and magnetosphere. The ionospheric plasmas have complex movements in the Earth's magnetic field under the action of tidal forces, so these can generate the current and magnetic field, and resulting in the calm current system. At the same time, there are a variety of perturbations current systems in the ionosphere and corresponding changes in geomagnetic disturbance. There are some currents on the magnetosphere boundary surface and within the magnetosphere: the magnetotail neutral sheet current, magnetopause current, ring current and field-aligned current.

These exogenous magnetic fields are usually called variation magnetic field or transient magnetic field for its time-varying changes slightly faster. And variation magnetic field can be divided into the calm variation magnetic field and perturbed magnetic field According to the temporal variations of the magnetic field generated by the current system. From the global average, exogenous variation magnetic field and its induced magnetic field are only 1% of the total magnetic field and sourced from the ionosphere. The calm variation magnetic field is a global field, and is periodic field of 24 hours changing significantly during the daytime. This is very consistent with the existing measurement data. This also shows that the pulses may come from the ionosphere electromagnetic fields and which is considered to be one source of the ENPEMF.

At the same time, the pulses numbers are different at the same period every day if we observation of the measured data carefully, and that indicating the data

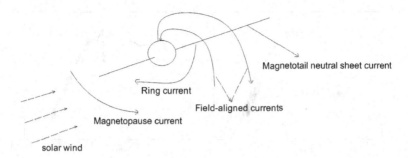

Fig. 3. Exogenous field current system

include disturbances changes of existence exogenous. These are mainly sourced from the magnetosphere and ionosphere, such as magnetic storms, sub-burst and geomagnetic pulsation. These fields occurs gradual change in space scale, reflecting the global field nature. Magnetic storms may be sustainable one day or a few days and the measurement data may also have great relevance. Sub-burst is concentrated in high latitudes, showing irregular changes for tens of minutes or several hours. Pulse disturbance is also a quasi-global field, quasi-periodic and second unit level. These parts may constitute one of the ingredients of the ENPEMF.

2.3   The Impact of Abnormal Crustal Magnetic Field to ENPEMF

Now there are some analysis about whether or not the crustal magnetic fields (also known as the lithospheric magnetic field) have some effects on the EN-PEMF. If the earth's core is seen as the main field normal magnetic field, the remaining parts are been called geomagnetic anomaly field for its deviation, variations and fluctuations. Geomagnetic anomaly field is very stable and its changing scale is to calculate with the geological time (millions of magnitude). Stable crust electromagnetic field cannot be superimposed on the ENPEMF signals.

However, some intense geological activity, such as earthquake gestation and occurrence or volcanic activitywill result in the rapid changes of the abnormal crustal magnetic field. Our study focuses on the impact of the earthquake.

The earthquake is the suddenly strong ground motion, and the earthquake shock waves spread out in all directions from the source. It is divided into transverse and longitudinal waves, also called the s-wave and p-wave. P-wave velocity is fast with small energy and s-wave velocity is slower with bigger energy.

Observational facts and theoretical studies have shown that there is a stress concentrated in the focal region before the earthquake, and especially prior to the impending earthquake seismogenic crust is in stress state of accelerated accumulation, the rocks of the seismogenic zone may occur micro-cracking or plastic phenomena. Therefore, the media physical properties will change in the seismogenic zone. In addition, the kinetic parameters changes in the small local

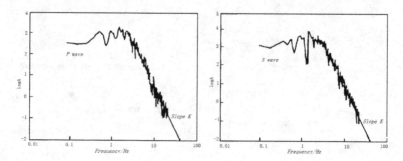

Fig. 4. Spectrum analysis of P wave (a) and S wave (b) of Yongdeng ML3.3 earthquake on Aug.8, 1995

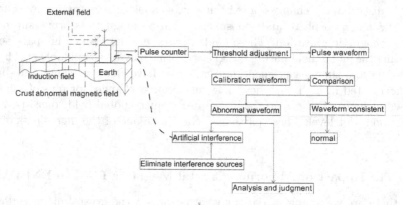

Fig. 5. Measuring working principle

earthquake source of the seismogenic zone may also cause some changes in the spectrum of seismic waves.

P-wave and s-wave spectrum analysis are shown as fig.4 that the earthquake has occurred at Yongdeng 1995 in 3.3[4]. It shows that the frequency spectrum within conventional analysis methods for seismic waves is not involved in our observed range. But we can found ENPEMF signals indeed before the earthquake. The geomagnetic intensity and magnitude of the pulse signal is proportional to the earthquake magnitude and inversely proportional to the distance of the source.

3   Signal Measurement Principle of ENPEMF

Table 1 shows the data of device in both directions form the specific measurement are not completely consistent. It can be seen from Fig.6, although the curves are not completely overlap, but they have the same trends.

Table 1. Observation data of the S-N and W-E Channel

No.	Time	N-S	W-E	No.	Time	N-S	W-E	No.	Time	N-S	W-E
1	0:2:47	19	3	16000	4:39:33	3	26	36000	10:17:28	2	5
2	0:2:48	0	25	16001	4:39:34	0	8	36001	10:17:29	10	8
3	0:2:49	0	3	16002	4:39:35	5	4	36002	10:17:30	7	35
4	0:2:50	7	7	16003	4:39:36	0	1	36003	10:17:31	19	10
...	...	...	...	...	...	...	...	...	...	...	...
7514	2:8:0	25	7	24076	6:54:9	110	138	78267	22:8:39	12	9
7515	2:8:1	20	52	24077	6:54:10	5	46	78268	22:8:40	10	11
7516	2:8:3	10	51	24078	6:54:11	19	37	78269	22:8:41	0	0
7517	2:8:4	0	0	24076	6:54:12	20	41	78270	22:8:42	0	0
...	...	...	...	...	...	...	...	...	...	...	...

Measuring principle like Fig. 5: drawing the envelope waveform of the pulse number of 24 hours every day, according to the instrumental record of the data and contrasting given normal reference envelope diagram. If the instrument received unusually large-scale changes in pulse, it may be most likely occurs earthquake precursor.

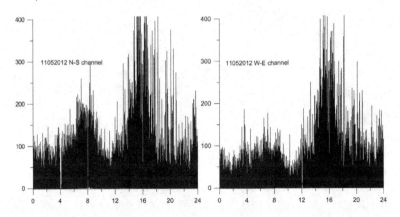

Fig. 6. May 11, 2012, pulse envelope diagram of 24 hours, N-S and W-E channel

ENPEMF have the global field nature so that the measure data diagram in Russia and Wuhan Jiufeng almost in an approximation envelope. Data frequencies in Wuhan have been tested in 14.5 KHz and 17 KHz. The two devices are placed in accordance with the S-N and W-E direction. As shown below, the total numbers of data are 65535 bits and we select a part of the observation.

4 Conclusion

ENPEMF signals may originate from the variation geomagnetic field, Earth's induced magnetic field, exogenous magnetic field and its induced magnetic field, geomagnetic pulsation disturbance and the abnormal crustal magnetic field pulse affected by the earthquake. Earth pulsed signal frequency is generally observed in the KHz range and Wuhan local signal is 14.5 KHz frequency. The measurement principle is feasible in this paper. But how to distinguish between noise and useful signals and relationship with solid tidal wave are the next further works.

References

1. Vorobyov, A.A., Zavadovskaya, E.K., Salnikov, V.N.: Report of the Academy of Science of the USSR, vol. 220(1), pp. 82–85 (1975) (in Russian)
2. Boryssenko, A., Polishchuk, V.I.: Earth near-surface passive probing by natural pulsed electromagnetic field. In: Proceedings of the Computational Electromagnetics and Its Applications, pp. 529–532 (1999)
3. Xu, W.: Physics of Electromagnetic Phenomena of the Earth, pp. 307–347. University of Science and Technology of China Press, Hefei (2009) (in Chinese)
4. Yin, Z., Xu, J., Li, P.: Study on the Seismic Spectral Variation Before and After YONGDENG Ms5. 8 Earthquake. Earthquake Research in Plateau 11, 56–60 (1996)

Author Index